KU-306-437

CONTENTS

ACKNOWLEDGEMENTS

The publication of this volume, the seventh in the series on the New Survey of Clare Island, has been made possible by the efforts of experts in field studies, supported by generous contributions of time and resources from their institutions and finance from our sponsors. The field workers made many ferry crossings to Clare Island, and at times worked under severe weather conditions to gather the data presented here. They were made most welcome by the islanders and were facilitated in very many ways, significantly enhancing the work of the Survey.

Thanks are due to the institutions that directly and indirectly supported the field workers, including the National Botanic Gardens, National University of Ireland Maynooth, University of Dublin Trinity College and University College Dublin. Their assistance and use of their facilities is greatly appreciated. We are very grateful for all the support provided by the staff of these institutions in assisting the field workers to undertake and complete their surveys.

The work of publishing the results of this survey of Plants and Fungi fell on the staff of the Publications Office, The Royal Irish Academy. In particular, Roisín Jones is to be congratulated on her excellent work in editing, correcting and presenting the complex data and analysis to be found in this volume.

Especial thanks are due to the landowners of Clare Island, whose enthusiastic co-operation allowed field workers unfettered access to their lands, enabling a complete survey of the island to be undertaken.

Martin Steer
Managing Editor, New Survey of Clare Island series

NEW SURVEY OF CLARE ISLAND

New Survey of Clare Island Executive Committee 2013

J. Breen
P. Coxon
L. Drury (President, 2011–)
J. Feehan
J.R. Graham
M. Jebb
T. Kelly
R.P. Kernan (Chairman, 1997–2004)
T.K. McCarthy
C. Manning (Deputy Secretary, 1996–2004; Secretary 2004–)
M.W. Steer (Managing Editor, 1994 -; Chairman 2004–)
D. Synnott

Former committee members: A. Clarke (President, 1990–3), D. Cabot (Chairman, 1989–97), G.J. Doyle (Secretary, 1991–2004), M.D.R. Guiry, J.S. Fairley, M. Herity (President, 1996–9), G.F. Imbusch, C. Mac Cárthaigh, W.I. Montgomery, A.A. Myers, M.E.F. Ryan (President, 2002–5), J.O. Scanlan (President, 1993–6), J.A. Slevin (President, 2005–2008), T.D. Spearman (President, 1999–2002), P.D. Sweetman, J. Waddell, K. Whelan.

Contributors to previous volumes in the New Survey of Clare Island series

Volume 1: History and Cultural Landscape: Timothy Collins, Críostóir Mac Cárthaigh (ed.), Nollaig Ó Muraíle, Kevin Whelan (ed.)

Volume 2: Geology: Peter Coxon, David Evans, John R. Graham (ed.), Kenneth T. Higgs, W.E.A. Phillips, C.J. Stillman, B.G.J. Upton, Michael Williams

Volume 3: Marine Intertidal Ecology: Michelle Cronin, Thomas Cross, Robert Cussen, Jane Delany Michael Guiry, Louise Harrington, Christine Maggs, David McGrath, Alan Myers (ed.), Julia Nunn, Sandy O'Driscoll, Ruth O'Riordan, Anne Marie Power, Fabio Rindi

Volume 4: The Abbey: Ann Buckley, Ian Cantwell, Fergus Gillespie, Paul Gosling (ed.), Conleth Manning (ed.), Karena Morton, Micheál Ó Comáin, Christoph Oldenbourg, Roger Stalley, John Waddell (ed.)

Volume 5: Archaeology: Markus Casey, Michelle Comber, Paul Gosling (ed.), Maureen McCorry, Conleth Manning (ed.), Karen Molloy, Sharon Nestor, Adrian Phillips, Mike Williams, Paula King, John Waddell (ed.)

Volume 6: The Freshwater and Terrestrial Algae: Jenny Bryant, Michael Guiry (ed.), David John (ed.), T.K. McCarthy (ed.), Fabio Rindi (ed.), Patricia Sims, Brian Whitton, David Williamson

Forthcoming volumes in the New Survey of Clare Island series

Volume 8: Soils and Soil Associations: J. Collins (ed.), G. Smillie (ed.) and W. Vullings (ed.)

Volume 9: Birds: T. Kelly (ed.)

Volume 10: Zoology: R. Anderson, T. Bolger, K. Bond, J. Breen (ed.), K. Creed, P. Cullen, E. de Eyto, S. Fahy, J. Fives, P. Giller, C. Lawton, J. Lusby, M. Cawley, T. Gallagher, T. Kelly, M. Kelly-Quinn, B. Keegan, R. Kennedy, K. McAney, T.K. McCarthy (ed.), S. McCormack, C. Müller-O'Toole, D. Murray, M. Nolan, J. O'Connor, A.M. Power, A. O'Grady, F. Tiernan, A. Trowanska, G. Walshe-Kemis, G. Woodward

Volume 11: Conclusions: J. Feehan (ed.)

NOTES ON CONTRIBUTORS

Ryan Corcoran completed his M.Sc. on paleoenvironmental investigations of Clare Island, Co. Mayo, in 2003. He now works as a teacher of geography.

Peter Coxon is a Professor in the Department of Geography TCD, a Fellow of Trinity College Dublin and a Member of the Royal Irish Academy. He was the Secretary-General of the International Union of Quaternary Research (INQUA) from 2003 to 2011. His interests in the Irish landscape were strongly influenced by working with the late Frank Mitchell. His research includes analysing Irish landscape evolution, Tertiary and Quaternary biostratigraphy, vegetational history and biogeography of Ireland, glacial and periglacial geomorphology and the analysis of flood events and mass movements in Ireland.

Paul Gibson is a senior lecturer in Physical Geography at the Department of Geography, National University of Ireland, Maynooth, where he formed the Environmental Geophysics Unit in 1997. His research interests are environmental geophysics, remote sensing and geomorphology. He is the Course Director of the M.Sc. in Geographical Information Systems and Remote Sensing, the longest running such course in Ireland, and is the author of four books.

Matthew Jebb is Director of the National Botanic Gardens in Glasnevin, Dublin. He is especially interested in the ant plant genera Squamellaria, Myrmecodia, Hydnophytum, Myrmephytum and Anthorrhiza, as well as the carnivorous plant genus Nepenthes.

Stephen McCarron is a Physical Geography lecturer in the Department of Geography at NUI Maynooth. His research is concerned with the record of past ice sheet activity in the Quaternary (glacial) geology of Ireland. He uses Geographical Information Systems and remotely sensed data to map glacial landforms and sediments both onshore and on the Irish Continental Shelf.

David Mitchel is an authority of fungi and an expert on waxap fungi in Ireland and Wales. He works in nature conservancy and is widely involved in informing and enthusing the public about fungi. He runs the Countryside Management System for exeGesIS, the environmental, ecological and heritage GIS consultancy. Prior to joining exeGesIS, David worked for the Environment & Heritage Service in Northern Ireland for 12 years. As Data Manager there he ran a GIS system and biological recording and Northern Ireland's input into the National Biodiversity Network.

D.H.S. Richardson is author of *The Vanishing Lichens* (1975), *The Biology of Mosses* (1981) and *Pollution Monitoring with Lichens* (1992), as well as many research papers and book chapters. He was professor of Botany at Trinity College Dublin from 1980 to 1992, and was subsequently Dean of Science at Saint Mary's University, Nova Scotia (1992–2007). As Dean Emeritus, he continues research on lichens and is a co-chair of COSEWIC (Committee on the Status of Endangered Wildlife in Canada). He was awarded the Ursula Duncan Award by the British Lichen Society and the George Dawson Medal by the Canadian Botanical Association. He is Editor-in-chief of the Journal Symbiosis and recently served as President of the Nova Scotian Institute of Science, founded in 1862.

Tim Ryle is a consultant ecologist who for nearly two decades has been involved in ecological surveying throughout Ireland for a number of environmental consultants and agencies. His chief interest is in vegetation descriptions and habitat mapping, although his work encompasses a wide range of environmental issues. In addition, he is a an occassional lecturer at University College Dublin and also leads fieldcourses for second level students in preparation for state exams.

Mark Seaward, Honorary Professor at Bradford and Lincoln Universities, has a strong interest in biomonitoring pollution, particularly heavy metals and radionuclides, for which he has been internationally honoured, being the recipient of the Acharius Medal and the Ursula Duncan Award and Doctor honoris causa of Wrocław University. He is the author and editor of numerous books, and has written more than 430 scientific papers. He has researched the Irish lichen flora for more then 40 years, being the author of three editions of the Census Catalogue of Irish Lichens.

Donal Synnott worked from 1963 as a botanist at the National Herbarium (DBN), first at the National Museum and then at the National Botanic Gardens, where he became Director in 1994. His interests include the history of Irish botany and the vascular and bryophyte floras of Ireland. Publications include popular booklets on ferns and flowering plants, accounts of the bryophytes of Meath and Westmeath and of the vascular plants of Mayo, the chapter 'Botany in Ireland' in *Nature in Ireland* (Foster 1997) and, with M.J.P. Scannell, *A Census Catalogue of the Flora of Ireland* (1987).

FOREWORD

The idea for the first survey of Clare Island may have occurred to Robert Lloyd Praeger when he spent a week recording the vascular plants of the island in 1903. Despite hinting that administrative duties prevented him from thoroughly studying the island's plants for the first survey, Praeger nevertheless wrote a very detailed account and analysis of the vascular plants and of the vegetation of the island, providing a good basis for further study (Praeger 1911, p. 2). Rather fittingly the proposal for a new survey would be made following a new listing of the island's vascular plants by Gerry Doyle and Peter Foss (1986).

In the present volume the plant lists, vascular plants, bryophytes, lichens and waxcap fungi are updated and further analysed and comparisons made with the listings from the first survey. In his review of the vascular plants, Matthew Jebb discusses the McArthur and Wilson (1967) theory that island species diversity is related to island size. Compared with the very thorough vascular plant surveys, the increased numbers of bryophytes and lichens recorded and the relatively small numbers of these cryptogamic species from the first survey that were not refound make difficult any analysis of change (the lichen list is more than doubled and the bryophyte list increased by a quarter). In the bryophyte chapter it is suggested that the increase in species numbers from the first survey is likely a function of the greater number of experienced bryologists to have visited the island in recent decades. In their chapter, Mark Seaward and David Richardson, however, emphasise that 'the sensitivity of lichens to natural and human disturbances makes them ideal environmental monitors' and recommend periodic re-evaluation of the island's lichen flora. The waxcap survey found that Clare Island is a rich and important location for those species of fungi. David Mitchel's chapter in the present volume places the island in the context of the rest of Ireland and further afield.

Tim Ryle's very extensive vegetation study presented in this volume gives detailed descriptions of the plant communities and will enable comparative studies to be made in the future. Praeger's study of the vegetation and the vegetation map published in the first survey (and presented as a fold-out map here) are sufficiently detailed to allow meaningful comparison with the present study. Changes in the vegetation related to changes in farming practice and other environmental factors are discussed at length.

Results from analysis of an eleven metre core at Lough Avullin have refined and expanded knowledge of the postglacial history of the vegetation presented by Pete Coxon in the *Geology* volume of the *New Survey of Clare Island* (2001). Further evidence for the influences of prehistoric peoples on the vegetation is presented and the correspondence with vegetational sequences in other parts of Ireland is confirmed in the chapter in this volume by Pete Coxon, Ryan Corcoran, Paul Gibson and Stephen McCarron.

I wish to thank the authors of chapters in the present volume for their forbearance in what has been a long process in arriving at publication. Some of the fieldwork was completed almost two decades ago but I feel confident that not much has been lost or rendered out of date by the delay.

I would like to pay tribute to Professor Gerry Doyle who began the organisation of this volume and who stimulated my interest in the project. Thanks are due to the many persons acknowledged in the separate papers in this volume. Additional thanks are due to Martin Steer for his cheerful and optimistic support and assistance throughout.

Thanks are in no small measure due to the people of Clare Island who have assisted us in our studies in ways too numerous to mention. The account of the visit of the British Bryological Society to the island in 1994 (Blockeel 1995) concluded 'After we boarded the boat for the crossing to the mainland, shower clouds descended over Clare Island. During the blustery crossing, with gannets above our heads, we were able to reflect on our good fortune on another favourable and productive day in such an excellent place'. There were many such good days.

Donal Synnott

REFERENCES

Blockeel, T.L. 1995 Summer field meeting, 1994, second week, Clifden. *Bulletin of the British Bryological Society* **65**, 12–18.

Coxon, P. 2001 The Quaternary history of Clare Island. In John R. Graham (ed.), *New Survey of Clare Island. Volume 2: geology*, 87–112. Dublin. Royal Irish Academy.

Doyle, G.J. and Foss, P.J. 1986 A resurvey of the Clare Island flora. *Irish Naturalists' Journal* **22**, 85–89.

MacArthur, R.H. and Wilson, E.O. 1967 *The theory of island biogeography*. New Jersey. Princeton University Press.

Praeger, R.L. 1911 Clare Island Survey. Part 10 Phanerogamia and Pteridophyta. *Proceedings of the Royal Irish Academy* **31**, 1–112.

A HOLOCENE POLLEN DIAGRAM FROM LOUGH AVULLIN, CLARE ISLAND, WESTERN IRELAND

Pete Coxon, Ryan Corcoran, Paul Gibson and Stephen McCarron

ABSTRACT

This paper describes the Holocene vegetational history of Clare Island elucidated from an eleven-metre sediment core from Lough Avullin. The core was radiocarbon dated and the forest succession and vegetational changes and disturbance of the forest cover were recorded. The palaeobotanical record shows a vegetational sequence with woodland conditions predominating in the earlier Holocene and with anthropogenic disturbance to the vegetation cover occurring after 5000 ^{14}C years BP. The paper highlights other palaeoecological work and the potential for further study.

Introduction

The diverse physical landscape of Clare Island and, in particular, the marvellous geomorphology, has been previously described as part of the New Survey of Clare Island (NSCI) by Graham (2001) and Coxon (2001). The undulating topography of much of the island lends itself to organic deposition but, despite the landscape having numerous hollows containing biogenic sediments and a widespread mantle of (blanket) peat deposits on the uplands, Clare Island has received scant attention from palaeoecologists.

The original Clare Island Survey, published as Volume 31 of the *Proceedings of the Royal Irish Academy*, contained sixty eight papers published between 1911 and 1915 (part 39 was in two sections) but one part, Part 8, Peat Deposits, was never published. Although the potential author (Lewis) visited the island there is no trace of his work or of his field notes. However, Forbes (1914) did map and comment upon the tree fossils found in various peats on the island and indeed mapped them and commented on woodland distribution and demise.

The absence of Lewis's contribution to the original survey led the late Professor Frank Mitchell to suggest that the Irish Quaternary Association (IQUA) should hold their Annual Field Meeting on the island in 1982, and Pete Coxon and Dr J.R. Creighton (of the Geological Survey of Ireland) visited the island in June of that year to carry out a reconnaissance of sites of Quaternary interest. Clare Island was then visited by the author, Catherine Coxon and Gina Hannon in July of 1982, when a number of cores for palaeobotanical work were collected. These two visits provided the basis for the 1982 IQUA *Fieldguide to Clare Island* (Coxon 1982).

This paper focuses on cores taken from Lough Avullin as part of the New Survey of Clare Island by Ryan Corcoran as part of his MSc research (Corcoran 2002).

Previous palaeobotanical research

The 1982 IQUA meeting was preceded by an intensive coring trip that involved sampling a diverse range of sites, including Lecarrow Townland (site 2, Fig. 1), peats near the summit of Knockmore (site 5, Fig. 1), Loughanaphuca (site 6, Fig. 1) and extensive peats in the col between Knocknaveen and Knockmore (site 4, Fig. 1 and Fig. 10 (see fold-out)). Not all of these cores were analysed in detail, although a preliminary pollen diagram from the Loughanaphuca core

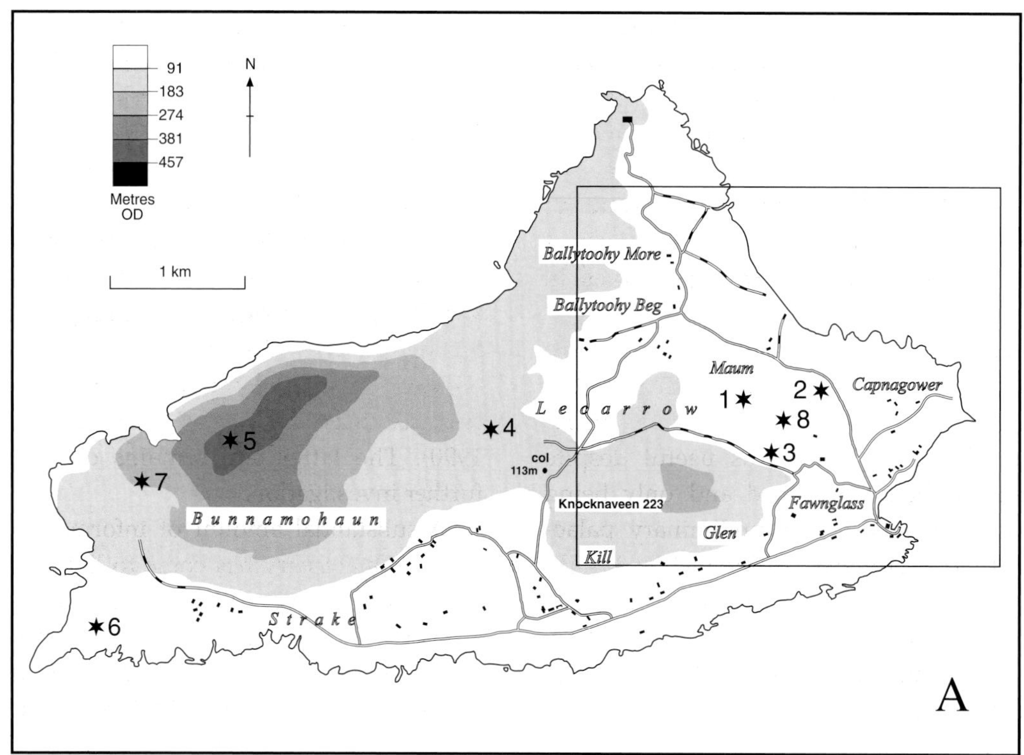

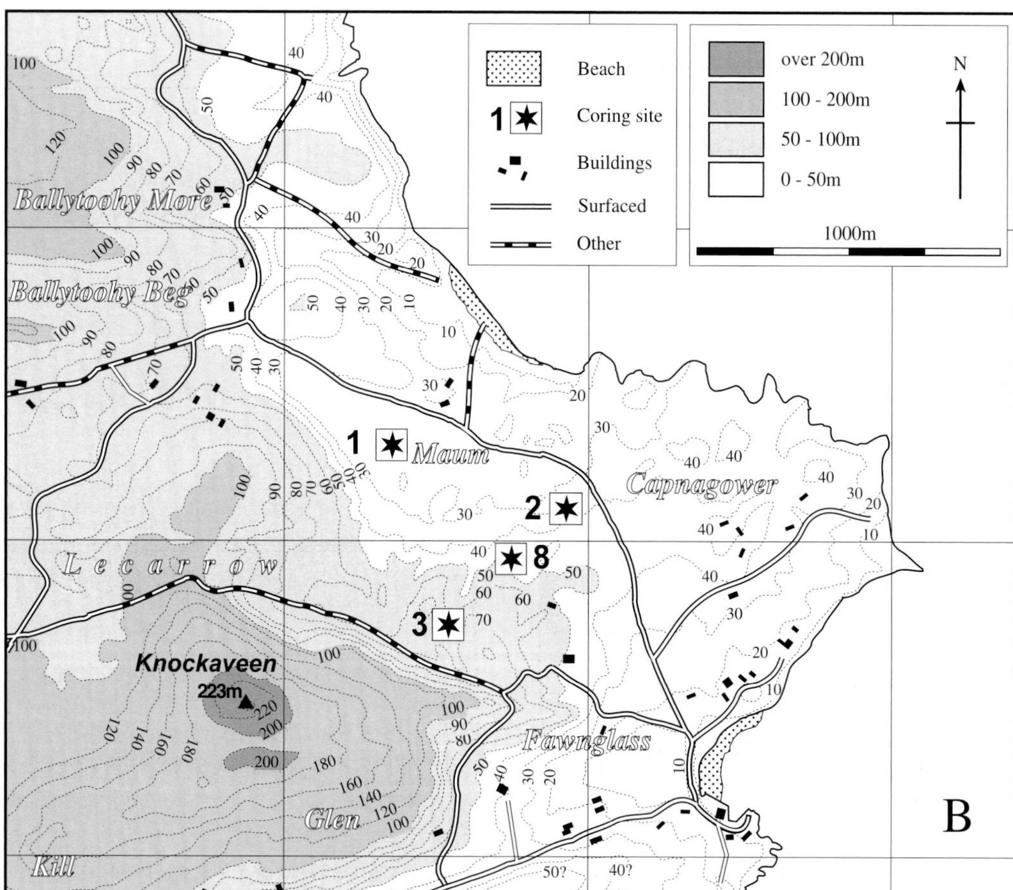

Fig. 1 A. Clare Island showing locations analysed for pollen that are referred to in the text;
B. Detail of the inset on Fig. 1A.

Site 1: Lough Avullin	Site 2: Lecarrow Townland
Site 3: Poirtín Fuinch (and megalithic tomb)	Site 4: Col between Knocknaveen and Knockmore
Site 5: Summit of Knockmore	Site 6: Loughanaphuca
Site 7: Bunnamohaun (Molloy 2007)	Site 8: Lecarrow fulacht fia (Molloy 2007)

suggested that the small arcuate end moraines present here may be Late-glacial in age (Coxon 1982; 2001). The transition between the Younger Dryas (blue/grey clays and silts) and the overlying organic muds is found lining the basin, but the absence of a full Younger Dryas sequence or any Late-glacial interstadial sediments suggests the moraines may be Younger Dryas in age (Fig. 2).

The Lecarrow Townland pollen diagram was analysed for the 1982 IQUA field meeting (in that publication it is incorrectly identified as being in 'Maum Townland') and was useful despite never being radiocarbon dated and only being a low-resolution study. This preliminary palaeoecological work provided an interesting insight to the vegetational history of the island, and the pollen diagram has appeared in a number of publications, including volume 2 of the *New Survey of Clare Island* (Coxon 1982; 1987; 2001; Coxon and O'Connell 1994).

A small lake adjacent to the megalithic court tomb at Poirtín Fuinch (site 3, Fig. 1, Pl. I and Fig. 10C on fold-out) was cored in 1990 and again in 1992. The lake basin contains just over 6m of sediment (mostly Holocene with a short Late-glacial sequence at the base). Disappointingly, given the lake's proximity to an important archaeological site (Gosling 2007a), the sediment had very variable pollen preservation throughout (Ulrike Huber and Chris Caseldine, pers. comm.) though it does contain remarkable macrofossil plant remains, including a layer of hazel nuts (Viney 1990). The latter site remains one worthy of further investigation.

A substantial amount of information regarding the Quaternary has come to light since these studies were made, and some of it was published in a second IQUA field guide in 1994 (Coxon and O'Connell 1994) and in volumes of the *New Survey of Clare Island*. More recently detailed palynological work has been carried out in association with

Fig. 2 A. An oblique aerial image generated using LIDAR data draped with orthophotographs. Illumination from the north. View looks eastwards towards the corrie on Knockmore. The moraines at Loughanaphuca are in the low ground to the right of the image (red dot); B. The most pronounced arcuate moraine at Loughanaphuca. The basin lies between the small stream to the left and the moraine ridge (centre foreground).

Pl. I A. Looking westwards over Poirtín Fuinch towards the megalithic tomb, with Knockmore in the background; B. The lake at Poirtín Fuinch and the view over the megalithic tomb (foreground) over the lake towards Croaghpatrick.

archaeological excavations (including radiocarbon dates) of a buried wall in Bunnamohaun and on a fulacht fia in Lecarrow (Molloy 2007).

Lough Avullin

Lough Avullin (derived from the Irish, Loch an Mhuilinn or 'lake of the mill') is located in Maum townland in the north-east of the island (site 1, Fig. 1 and Pl. II). Lough Avullin is almost completely infilled with sediment and there is a floating mat of *Phragmites* on the remaining surface water. Pl. II is an aerial photograph taken over the north of Clare Island and looking in a south-eastern direction towards the pier. The lake basin is principally fed by the River Dorree, which rises on the western flank of Knocknaveen. The River Dorree then exits Lough Avullin *via* a modified channel and enters the sea south-east of Portlea. The velocity of the Mill Stream was increased by the construction of a mill-race by Sir Samuel O'Malley at the start of the nineteenth century (Cullen and Gill 1992).

There is evidence in Pl. III of widespread deposits of hummocky moraines surrounding the site (Coxon 2001). The topography surrounding Lough Avullin contains a high density of cultivation ridges (Pl. II), which are scattered on flat low-lying fields. The mill is located next to the O'Toole residence, about 65m upstream from the mouth of the river.

Geomorphological setting and the basin at Lough Avullin

The basin of Lough Avullin is situated at the western margin of an extensive area of hummocky moraine that mantles and dominates the north-eastern area of the island (Coxon 2001). This hummocky moraine (Fig. 3 and Pl. III) is composed of a sandstone-rich diamicton (the Newport Till type) that drapes the land surface and the underlying till (the Roscahill Till type) to produce a prominent and remarkable 'kame and kettle' topography: this is particularly well brought out in Pl. IA. This landscape

Pl. II Oblique aerial photograph looking southeast over Lough Avullin (coring site was in the light coloured reeds centre background towards the harbour). The extensive cultivation ridges and later field walls of the island are clear in the photograph.

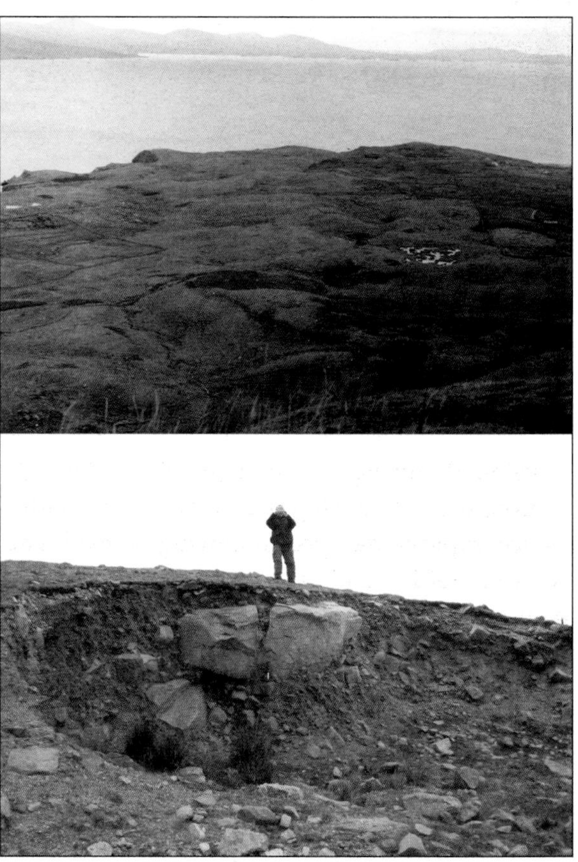

Pl. III A. The view northeastwards from the flank of Knocknaveen over the hummocky moraine. The lake at Poirtín Fuinch can be seen in the right midground; B. Section in the sandstone rich hummocky moraine (Newport Till type, see Coxon 2001).

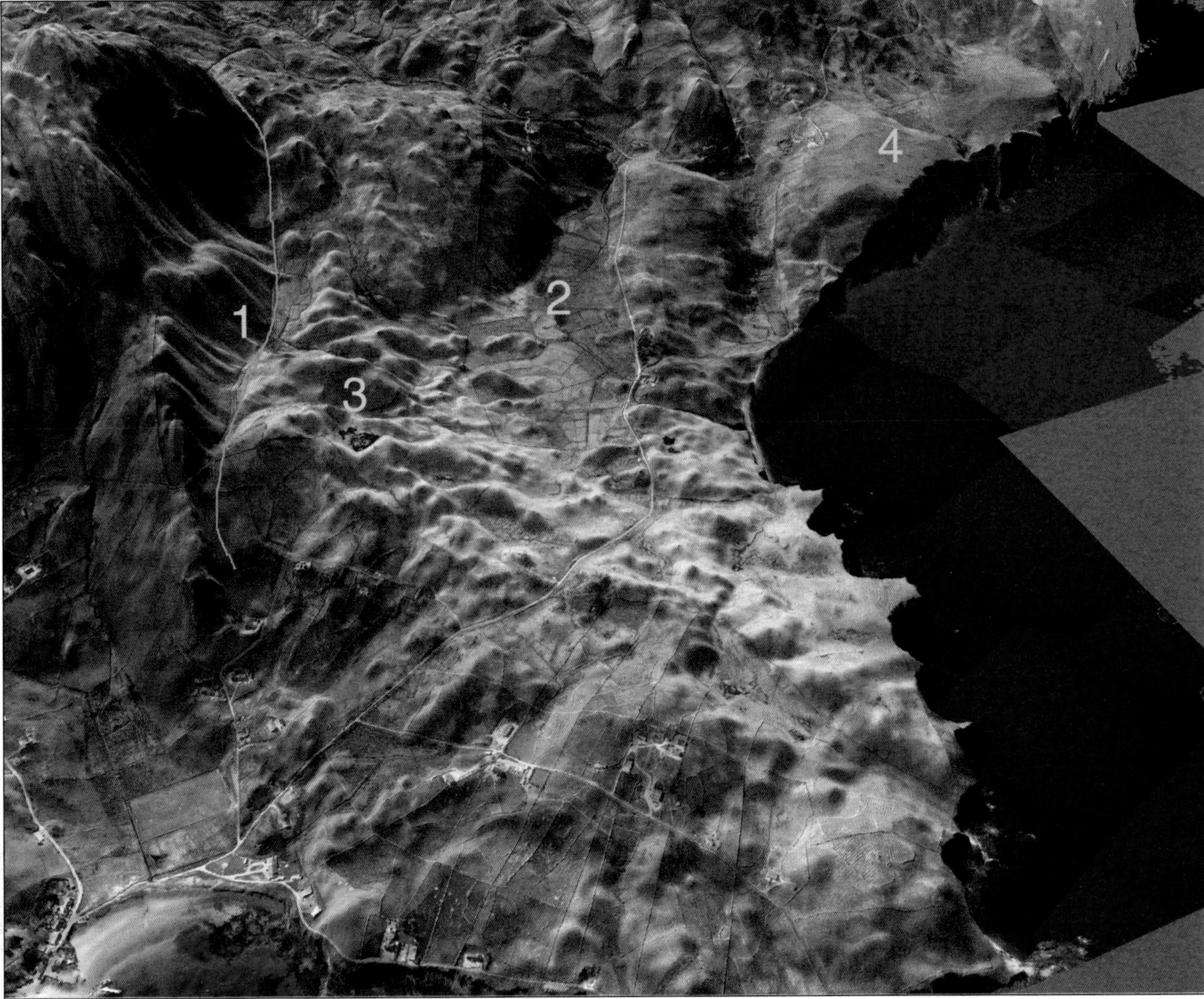

Fig. 3 Oblique aerial (LIDAR and draped orthophotographs) image looking north westwards from above the harbour into the extensive area of hummocky moraine in the north-eastern corner of the island. The flank of Knocknaveen is on the left hand side of the image. The numerous peat-filled hollows present many opportunities for further palaeoenvironmental work.
1 (green line): lateral moraine bordering the southern edge of the hummocky moraine.
2: Lough Avullin.
3: Poirtín Fuinch.
4: Flat-topped masses of lodgment till (Roscahill Till type, see Coxon 2001).

developed during a late stage of the last glaciation, possibly during the Killard Point Stadial, as ice rapidly decayed following a readvance (Coxon 2001), and subsequently mounds of melt-out till were deposited (Pl. IIIB). The resulting undulating topography contains numerous basins surrounded by hillocks with an amplitude (basin floors to summits) of 20–30m. The intermorainic basins to the east occupy areas within the morainic hills, while Lough Avullin is at the edge of the moraine and is less clearly delimited, as low ground occurs to the west and Lough Avullin stretches into this area.

From aerial photographs and a simple ground survey (including probing with coring rods) the basin of Lough Avullin was delimited and was seen to have a narrow neck lying to the west, widening to a second, larger and roughly circular, basin in the east (Fig. 4). Much of the basin's area is shallow, and the maximum extent of lake sediments >1m thick is approximately 250m × 115m. The exact limits of the lake basin are hard to determine as, at its greatest extent, the shallow margins of the lake occupied extensive low ground to the north and west. The deeper, circular basin to the east can be seen from the resistivity survey, and from coring, to be a compound structure that adjoins other, less deep, basins. The lines of the four resistivity surveys are marked on Fig. 4 and the results are shown in Fig. 5.

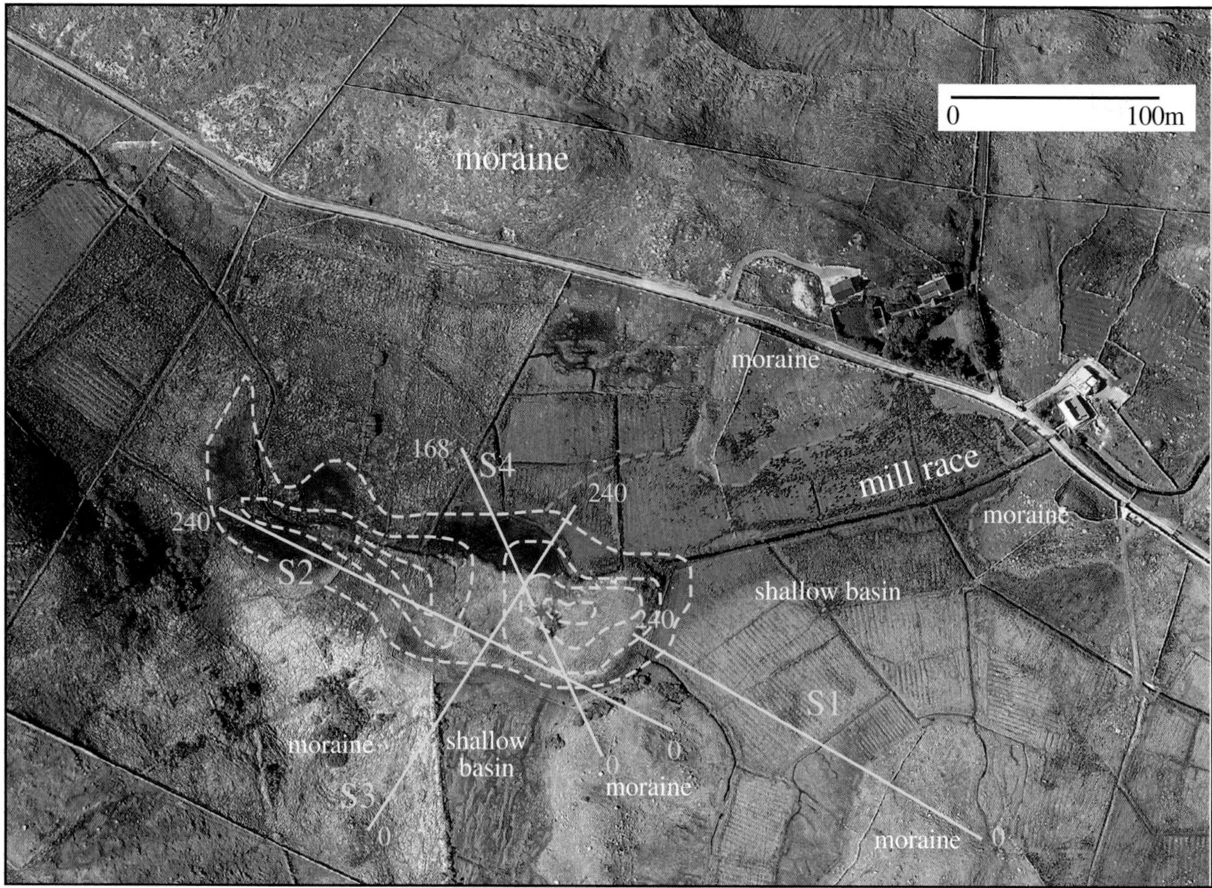

Fig. 4 An orthophotograph image of the area immediately around Lough Avullin. The lake basin is delimited by yellow dashed lines, the larger area of the shallow basins is delimited in orange and the hummocky moraines in the vicinity are labeled. The image is also overlain with the lines (and lengths in metres) of the resistivity surveys (S1–S4). The core location is marked by a red star.

Geophysics: resistivity imaging of Lough Avullin

Clare Island is characterised by an extensive cover of glacial and postglacial sediments, which greatly hinder a determination of the internal structure and variations in sedimentary facies. Conventional mapping generally relies on limited, scattered outcrops and some borehole data. Geophysics is a technique that allows the nature of the subsurface to be determined, and the application of resistivity surveying techniques is particularly applicable for deeper investigations (Gibson *et al.* 2004). Four transects (or lines) were taken on or in the vicinity of Lough Avullin (Fig. 4), and they are labelled S1–S4.

The model for profile S1 (Fig. 5) shows that resistivity values vary from a minimum of 17 ohm-m to a maximum of 1143 ohm-m (note that the colour scale is logarithmic). The highest values occur at depths greater than about 25m (shown in red) and probably represent the underlying rock surface. Near the surface the most prominent features are the two shallow (5m depth) basins (shown in blue and clearest on S1C, Fig. 5). These most likely represent postglacial basins infilled with relatively unconsolidated sediments. They are both characterised by very low resistivity values (20 ohm-m), showing that they are heavily saturated with water.

Three other resistivity survey lines were obtained on Clare Island. The resistivity range for Line S2 (Figs 4 and 5) is similar to S1 and also contains two shallow basins. One, located at 200m, is very compact, being approximately as wide as it is deep, whereas the other larger basin is more asymmetric, being approximately 40m wide and up to 10m deep. The highest resistivity values on this profile are not at depth but are located near the surface. This traverse commenced on morainic material containing sandstone blocks, hence the

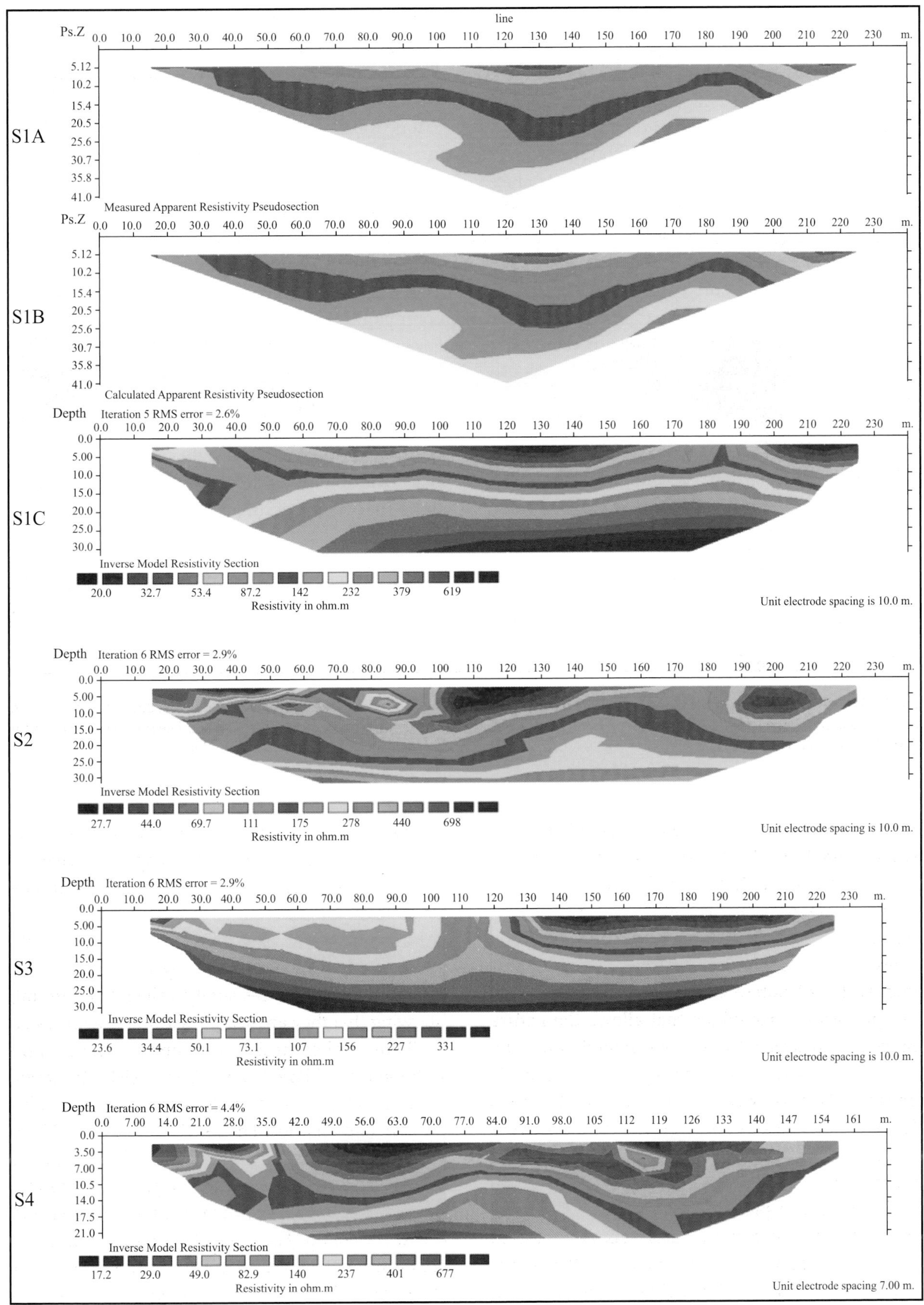

Fig. 5 Resistivity profiles S1–S4. S1A, S1B and S1C are the measured, calculated and modelled cross sections (see text).

high values. A buried glacial erratic is located at 80m and is indicated by the lozenge shape (such large blocks are common, see Pl. IIIB). Resistivity values remain low even to great depths, suggesting that the bedrock is at a greater depth here.

Line S3 is shown in Fig. 5, and although its lowest resistivity values are similar to other lines, its highest values are considerably lower, even to depths of 30m, suggesting different subsurface characteristics, possibly with greater porosity at depth. Again, the shallow basin near the surface is very evident.

Line S4 was taken with an electrode spacing of 7m (10m spacing was used for the earlier measurements). Although it gives a lower depth of penetration (21m), it provides more detail at shallow depths. The profile is dominated by a 100m long basin that is approximately 10m deep with an undulating floor. Very high values at the beginning of the profile are caused by a moraine containing sandstone boulders.

The sedimentological record at Lough Avullin

Methodology

The methodologies used in analysing the Lough Avullin cores are summarised here. A full account can be found in Corcoran (2002).

A. CORING

The site (see Pls IV and VB) was cored using a Livingstone piston sampler (Wright *et al.* 1965). One-metre cores were taken from the lake until the corer could no longer penetrate the sediment, and a second, overlapping, sequence of cores was also taken immediately adjacent to the first core. Each of the cores was wrapped in cling film to maintain moisture content and then covered in aluminium foil to avoid exposure to light. Subsampling took place within days of the coring to minimise alterations that might occur to the physical properties of the core.

Pl. IV A. The basin containing Lough Avullin. The deepest part of the basin is in the left midground; B. Pine stumps to the west of Lough Avullin. The westward dipping lateral moraine can be seen along the flank of Knocknaveen in the background. Forbes (1914, pl. II) included a photograph of these pine stumps in his paper in the first Clare Island Survey.

Pl. V A. Peat cuttings on the cliffs to the east of Knockmore; B. Coring Lough Avullin (October 1999); C. Probing 5m of lake sediment at Poirtín Fuinch (IGA field trip, September 2003. Photograph by kind permission of Dr Tom de Brit).

B. SEDIMENT DESCRIPTION, ORGANIC CONTENT AND GEOCHEMISTRY

The cored sediment was subsampled and a description made of the sediment using the method of Troels-Smith (1955). The colour was recorded using a Munsell chart. The sediment's characteristics are summarised in Fig. 6 and the detailed sedimentological description is in Appendix 2.

The core was subsequently subsampled at 5cm intervals for pollen analysis, loss on ignition, and geochemical analysis. Pollen samples of $1cm^3$ were placed into sample tubes and labelled. Larger samples with a mass of *c.* 10–20g were taken for both loss on ignition and geochemical tests and sealed in plastic sampling bags.

Samples were taken at 5cm intervals and subjected to loss on ignition tests using the procedure outlined by Davies (1974) allowing an estimate of the organic carbon content. The results of these analyses are shown in Fig. 6.

The charcoal content of each sample was calculated using the point count estimation procedure outlined by Clarke (1982). This method estimates the number of charred particles in a given sample of sediment but does not take the size of the particles into account (Tolonen 1986). The percentage charcoal is recorded in Fig. 6.

Geochemical analyses were carried out on each 5cm interval sample. The samples were dried and a disk measuring 3cm diameter was made using a Specac hydraulic press. A Link Systems 860 Electron Dispersant X-Ray Fluorescence Spectrometer (EDXRF) was used to measure the mineral component of each 3cm disk. The following elements were be measured: silicon (Si), aluminium (Al), titanium (Ti), potassium (K), calcium (Ca), iron (Fe) and manganese (Mn). The XRF analysis produces a number of counts for each element. From this number, the concentration of each element as a percentage of the total weight of the dry sample was calculated. The results are shown in Fig. 7.

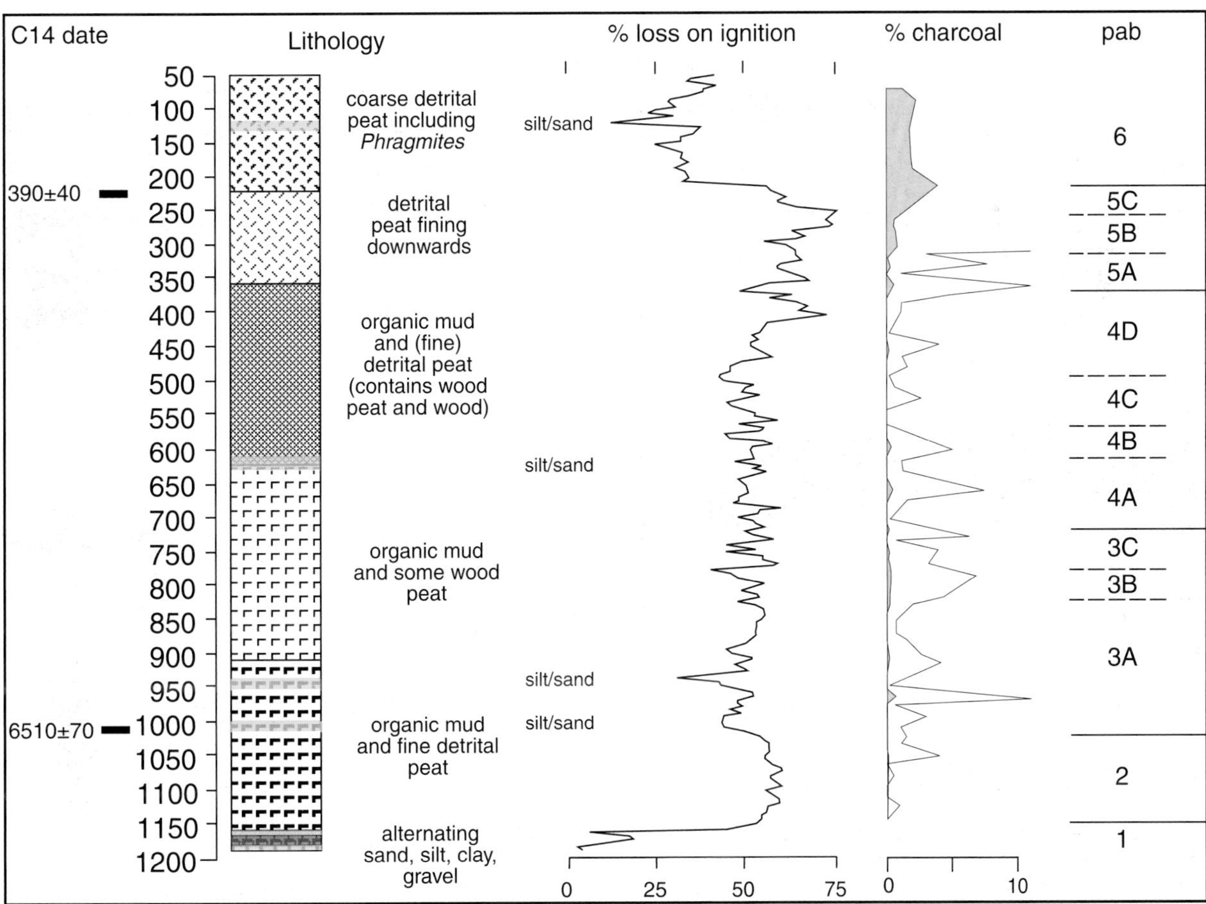

Fig. 6 Summary diagram of the lithology from the Lough Avullin core showing depth, ^{14}C ages, percentage loss on ignition, percentage charcoal (the unfilled, single, line is a × 20 exaggeration) and the pollen assemblage biozones (pab) (see Figs 9 and 10, on fold-out).

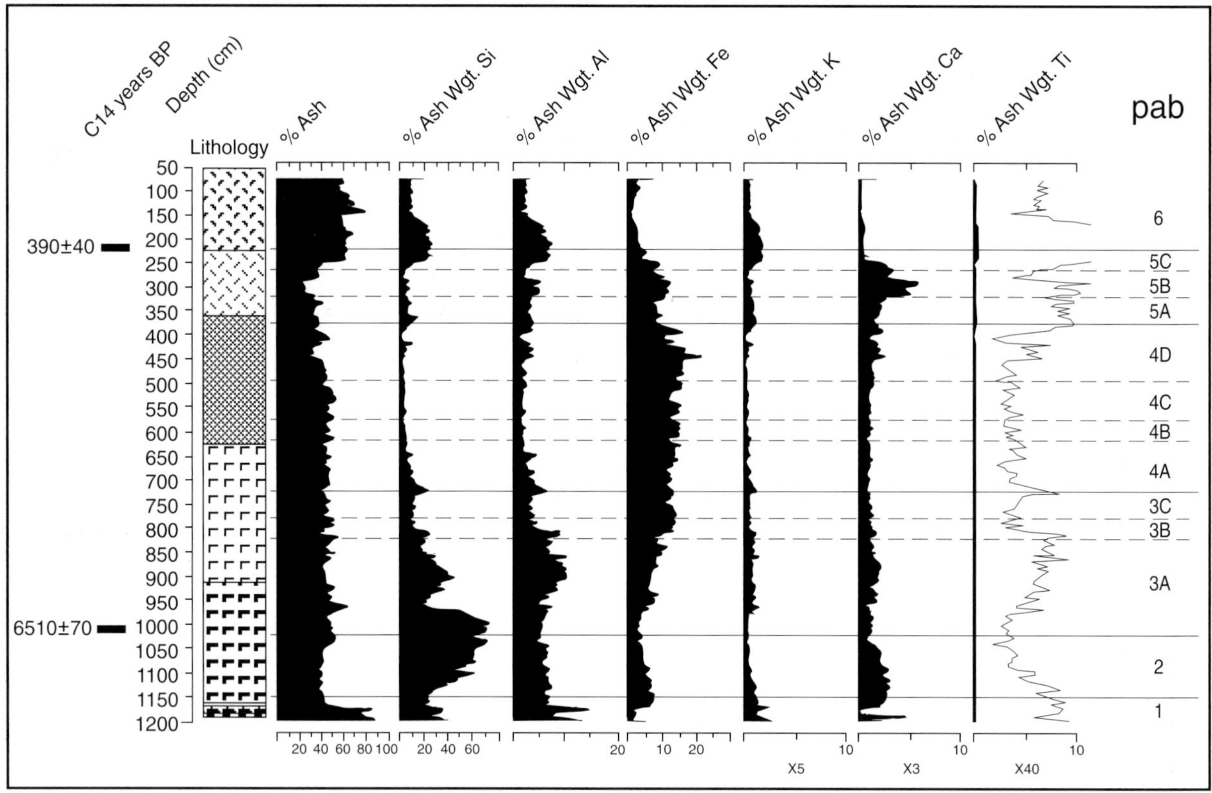

Fig. 7 Geochemical analyses.

C. RADIOCARBON DATING

Two bulk samples (5cm lengths of core) were submitted to Beta Analytic (4985 SW 74 Court, Miami, FL 33155, USA) for ^{14}C AMS analysis. The samples were taken at two significant points in the core where changes in the vegetation were identified in the pollen counts and where lithological change in the core was apparent.

The calibration of the ^{14}C ages (Calibration data set: IntCal09.14c—Reimer *et al.* 2009) should be taken into account when interpreting the dates. The ^{14}C dates have been used, along with characteristic lithostratigraphical and biostratigraphical events, to construct an age/depth curve for the sediment accumulation in the lake basin (Fig. 8). While useful in interpreting the ages of the pollen assemblage zone boundaries, the error inherent in using only two dates must be stressed.

D. PALYNOLOGY

Pollen and spores were isolated from the 1cm^3 subsamples using standard methods (Faegri and Iversen 1989; Moore *et al.* 1991). The prepared samples were unstained and residues were mixed with 2000cc silicone fluid and mounted on slides for light microscopy (LM). A smaller number of samples were also prepared for examination under scanning electron microscope (SEM) (Hitachi S-4300 Field Emission Scanning Electron Microscope, resolution: 1.5nm@15kv and 2.5nm@5kv), using the method outlined by Moore *et al.* (1991).

Taxa were identified using a standard pollen and spore key (Moore *et al.* 1991) and a using a modern reference collection. The plant nomenclature follows that of Stace (1997) and the pollen nomenclature that of Moore *et al.* (1991). Where possible pollen grains were identified to species level. Some photomicrographs of the taxa identified are shown in Pl. VI.

Each slide was counted by taking equidistant traverses across the microscope slide in order to count at least 500 pollen grains. Pollen counting was carried out using a Nikon Labophot microscope at 400 × magnification. The 1000 × magnification (using the oil immersion lens) was also used where deemed necessary. A calcareous tablet containing a spike taxa (in this case, *Lycopodium* spores) was added to each sample so that the pollen concentration could be calculated (Stockmarr 1971). The tablets were obtained from the University of Lund, Sweden (Batch number 201890) and each tablet consists of 11,300 ± 400 *Lycopodium* spores.

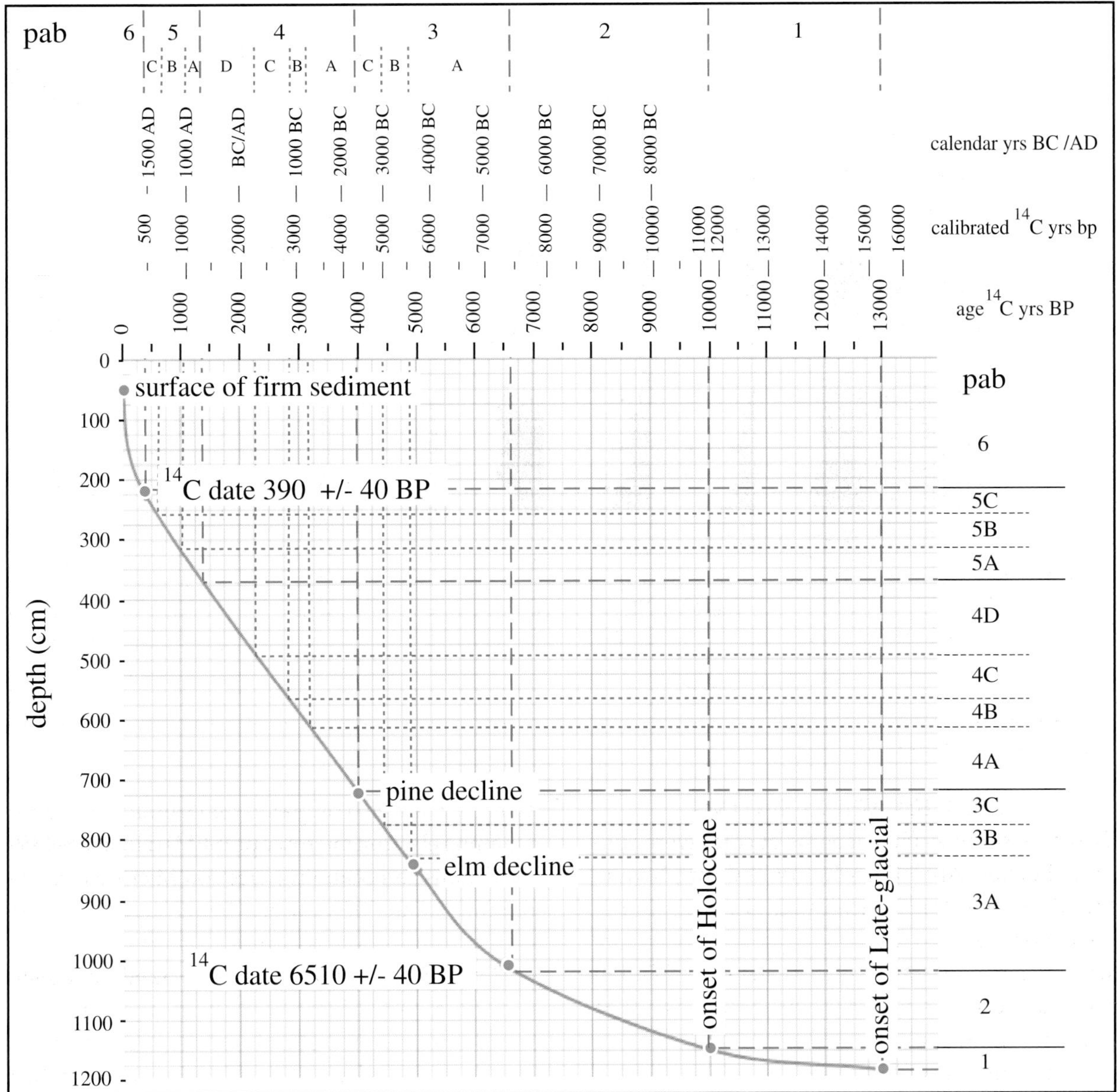

Fig. 8 Age/depth curve for Lough Avullin. Note that the age/depth comparison is made on only two radiocarbon ages and that these do not show the standard errors inherent once calibrated (see Table 1, p. 13). The curve was compiled using known lithostratigraphical and biostratigraphical markers and these are marked. The relationship between the suggested ^{14}C years BP and the calibrated (calendar) years BP and the AD/BC scale is not linear. The relationship between calendar years and radiocarbon age is complex and frequently misunderstood (for an explanation see Roberts 1998 and Walker 2005). The age/depth scale is useful in determining the timing of vegetational changes at the pollen assemblage biozone boundaries once the limitations above are taken into account.

Pollen diagrams were redrafted from the MS-DOS based programs Tilia Version 2.0 and Tilia Graph (TG) Version 2.0 (Grimm 1993) and modified to show additional data. The pollen sums used in the relative percentage diagrams were calculated as a percentage of all taxa (trees, shrubs and herbs) and the total sum is referred to here as 'P'. Aquatic taxa and spores from lower plants were calculated in their own sums as percentages of P + aquatics and P + lower plants. The pollen data are presented as a percentage pollen diagram (Fig. 9, see fold-out) and as a pollen concentration diagram (Fig. 10, see fold-out). The pollen concentration diagram allows observation of trends in each pollen taxon independently of other taxa as the counts are of each taxon's absolute frequency in each sample and not the relative percentage of taxa.

Results

The core taken from Lough Avullin came from the deepest part of the basin and was 11.84m in total length. The top 0.55m was not sampled.

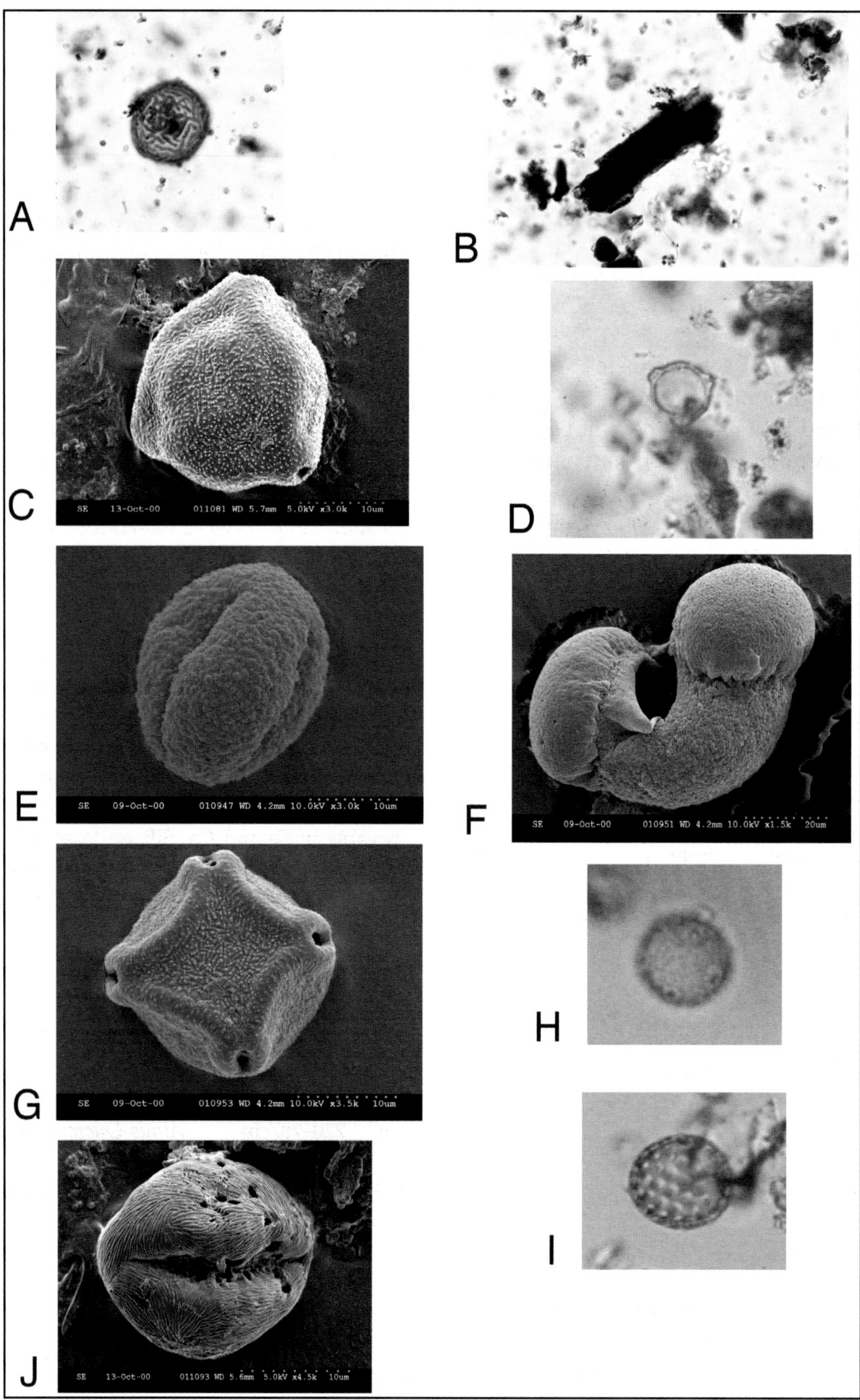

Pl. VI Photomicrographs from Clare Island cores. A. LM of a pollen grain of *Ulmus* (elm) from Lecarrow Townland, Clare Island (32µm diameter) (the pollen has been stained with 1% safranin); B. LM of organic fragments including wood and charcoal (darker material). Large fragment is 50µm long; C. SEM of *Corylus* sp. (hazel) pollen (27µm diameter); D. LM of *Betula* sp. (birch) pollen grain (30µm diameter); E. SEM of *Quercus* sp. (oak) pollen grain (25µm long); F. SEM of *Pinus* sp. Pollen grain (35µm long); G. SEM of a four-pored pollen grain of *Alnus* (alder) (this pollen usually has five pores); H. LM of *Plantago lanceolata* (ribwort plantain) pollen (20µm diameter); I. LM of Chenopodiaceae (chenopod family) pollen (22µm diameter); J. SEM of *Menyanthes trifoliata* (bog bean) pollen (25µm long). LM—light microscopy image; SEM—scanning electron microscopy image.

For pollen percentage data, see Fig. 9.

For pollen concentration data, see Fig. 10.

The palynological work on the core allowed the definition of distinct pollen assemblage biozones (pab) as follows and the content of these pab are commented on in turn. Subzones have been identified to aid description in the text (Table 1).

LA1 1153–1184cm *Empetrum-Poaceae-Juniperus* pab

Palynomorphs are very sparse at the base of this pab and the pollen concentration is very low. Some *Pinus, Salix* and *Juniperus* pollen grains are present and are followed by a marked peak in *Empetrum* reaching a maximum of 44% of the pollen sum (P). Poaceae and Cyperaceae are common at 33% and 4% respectively. *Rumex* accounts for 5%, and the top of the zone is marked by an increase in the percentage of *Polypodium* spores to a maximum of 11% (P + lower plants).

The geochemical content (Fig. 7) shows that amounts of Si (30%), Al (13%), K (2%) and Ti (trace) are relatively high in the early part of the zone and fall subsequently.

Table 1
Lough Avullin pollen assemblage biozones

Zone	Subzone	Depth (cm)	Pollen assemblage biozone (pab)
LA6		55–210	*Calluna–Corylus/Myrica*
	LA5C	210–265	Ericaceae–*Betula–Plantago*
LA5	LA5B	265–320	Poaceae–Cyperaceae–*Plantago*
	LA5A	320–380	*Betula–Corylus/Myrica*–Poaceae
	LA4D	380–500	*Betula–Alnus–Fraxinus*–Poaceae–*Salix*
LA4	LA4C	500–570	*Betula–Alnus–Fraxinus*–Poaceae
	LA4B	570–610	*Quercus–Betula–Alnus–Fraxinus*
	LA4A	610–720	*Betula–Quercus–Alnus–Fraxinus*–Poaceae
	LA3C	720–787	*Alnus–Quercus–Betula*–Poaceae
LA3	LA3B	787–833	*Alnus–Betula*–Poaceae
	LA3A	833–1025	*Pinus–Quercus–Alnus*
LA2		1025–1153	*Pinus–Quercus–Betula–Corylus*
LA1		1153–1184	*Empetrum*–Poaceae–*Juniperus*

LA2 1025–1153cm *Pinus-Quercus-Betula-Corylus* pab

The base of LA2 is characterised by the expansion of arboreal pollen (AP) species. *Betula* pollen increases accounting for 5%–15%. *Pinus* pollen rapidly reaches a maximum of 46% at 1025cm. *Ulmus* pollen also appears and accounts for nearly 10% late in the zone. *Quercus* represents a maximum of 20% at 1120cm. *Alnus* is the last of the tree species to appear in this zone and arrives at 1010cm, where the taxon accounts for 10.5%. Shrubs, particularly the *Corylus/Myrica* type, peak in this zone to a maximum of 67.4%. The pollen concentration data show high pollen accumulation in all of the arboreal taxa. Pl. VI shows standard SEM and LM images of the common pollen types during this woodland expansion.

There is a significant fall in most of the mineral matter components throughout the zone, namely K, Ti, and Al. Potassium (K) concentrations for example, decrease from a maximum ash weight of 2.75% in zone 1 to a minimum amount of 0.17% in this zone. Lesser falls in concentration are noticeable in Fe and Ca. Silicon levels fall at the beginning of this zone and then steadily rise again to a maximum concentration of 62% by ash weight at 980cm.

LA3 720–1025cm

PAB LA3 has been subdivided into three subzones LA3A, LA3B and LA3C.

LA3C	720–787cm	*Alnus-Quercus-Betula*-Poaceae
LA3B	787–833cm	*Alnus-Betula*-Poaceae
LA3A	833–1025cm	*Pinus-Quercus-Alnus*

The base of LA3A is marked by the appearance of *Alnus* and an increase in Poaceae, which peaks at 15% at 900cm. There is also a slight increase in the amount of total *Plantago*, which reaches 4% during this subzone (Pl. VI). The presence of charcoal also increases here relative to earlier levels and peaks early in the zone (Fig. 6). At the beginning of LA3A there is a decline in the overall percentage of tree pollen and an increase in overall herb pollen, although the concentration data show that trees remain an important component of the vegetation. Towards the end of the subzone, both Poaceae and *Plantago* pollen percentages decrease. The beginning of this zone also shows a significant peak in the concentrations of Al, K, and Ti, where the average percentage ash weights recorded for these elements are 8%, 1% and 0.15% respectively.

The base of LA3B is marked by a sharp decrease in *Ulmus* pollen. This decrease is noticeable in the relative pollen diagram for this zone (Fig. 9) and is even more apparent on the pollen concentration diagram as *Ulmus* concentrations fall from a count of 7484 grains cm^{-3} at 840cm to just 271 grains cm^{-3} at 825cm (Fig. 10). There is an overall drop in the pollen concentration of all tree and shrub taxa in this subzone and a rise in the levels of charcoal.

The base of LA3C is characterised by further increases in the pollen of Poaceae (23%), *Plantago lanceolata* (5%) and total *Plantago* (7%). There is also an increase in the diversity and percentage of herb pollen to 26.0% of P, while the concentration data show the deposition of arboreal pollen remains low. There is little change in the mineral matter concentrations except for the iron concentration, which has slightly increased.

LA4 380–720cm

PAB LA4 has been subdivided into four subzones:

LA4D	380–500cm	*Betula-Alnus-Fraxinus*-Poaceae-*Salix*
LA4C	500–570cm	*Betula-Alnus-Fraxinus*-Poaceae
LA4B	570–610cm	*Quercus-Betula-Alnus-Fraxinus*
LA4A	610–720cm	*Betula-Quercus-Alnus-Fraxinus*-Poaceae

The beginning of LA4A is marked by a final decline in *Pinus* from an average of 43% in the previous zone to an average of 2%. *Fraxinus* pollen grains appear and account for a maximum of 9% by LA4B. *Alnus* is still abundant but has decreased to 10% or less in LA4A, although the pollen becomes abundant again by the end of LA4D. *Quercus* is abundant during LA4B at 27%, and *Betula* remains important accounting for between 20%–30%. Although the *Corylus/Myrica* levels have fallen relative to earlier zones it is still a dominant taxon (10%–30%). Small amounts of *Salix* (6%), *Hedera* and *Ilex* pollen also occur. These changes are mirrored in the pollen concentration diagram.

Calluna pollen becomes more frequent and Poaceae remains high, accounting for 15% during LA4C. Some cereal-type grains were recorded at this point, and *Plantago lanceolata* (up to 3.7%) and total *Plantago* remain high. Lower plants also play an important role. *Osmunda* has increased to 11%, while other ferns, including Polypodiaceae and *Polypodium*, collectively represent 20% of P + lower plants.

The background charcoal content in this zone remains present, and Si, Al, K, Ca and Ti are relatively low at this stage. Average percentage ash weights for K, for example, have fallen from 0.46% in zone 3 to 0.24% in zone 4. The concentration of Fe has steadily increased to a maximum of 20.93% at 425cm and has an average percentage ash weight of 13.5% for this zone.

LA5 215–372cm

PAB LA5 has been subdivided into three subzones:

LA5C	210–265cm	Ericaceae-*Betula-Plantago*
LA5B	265–320cm	Poaceae-Cyperaceae-*Plantago*
LA5A	320–380cm	*Betula-Corylus/Myrica*-Poaceae

There is a drop in overall tree pollen at the beginning of LA5A, where it accounts for only 11.0% at 265cm. *Betula*, *Quercus* and *Alnus* all fall in representation in this zone, while virtually no *Pinus* pollen occurs. *Fraxinus* remains a contributor as does *Salix* (the latter expanding slightly). An increase in Poaceae is apparent and the family accounts for up to 50% in LA5B, where Cyperaceae also increases to 13%. The herbaceous taxa, cereal-type, *Plantago lanceolata*, total *Plantago* and *Rumex* and aquatic species such as *Nymphaea* also rise in LA5B. The pollen concentration diagram corroborates this further decline in tree/shrub pollen and a rise in the representation of herbaceous and aquatic taxa.

Charcoal presence rises throughout LA5 and peaks occur in geochemical components (e.g. Ca, where there is a 48% increase in average concentration values from LA zone 4 to zone 5). Increases are also apparent in the concentrations of K, Ti, Si, and Al, all of which have a similar trend.

LA6 055–215cm *Calluna-Corylus/Myrica* pab

The base of LA6 is marked by a substantial rise in *Calluna* from 4% at 255cm to 58% at 150cm (very notable from the concentration data). The pollen representing tree taxa decreases markedly, and very few species apart from *Betula*, *Alnus* and *Corylus*/ *Myrica* remain present. Poaceae decreases in abundance, while cereal-type pollen, Asteraceae (6%), Chenopodiaceae (1.4%) and *Plantago lanceolata* (8%) are all abundant. There is a further increase in lower plant species particularly Polypodiaceae (29.5%) and *Polypodium* (8%). There is also a significant increase in *Menyanthes* and *Sphagnum* (6.2%) (Pls XV, XVI). A further increase in charcoal is recorded, and a maximum value of 3.95% is reached. Geochemical components also increase, with significant peaks in Si (22.03%), Al (7.78%), K (1.90%) and Ti (0.40%). This is the most significant geochemical change in the Holocene record. The average value for Si % weight in LA6 is 13.71% compared to 7.06% in LA5. In contrast, there are decreases in the concentrations of Fe and Ca, with average concentrations of 2.90% and 0.36% in this zone.

Discussion

The oldest material obtained from the core was 30cm of a poorly defined tripartite sedimentary succession resembling that of the Midlandian Late-glacial period. These thin basal sediment units are coincident with PAB LA1 and represent parts of the Late-glacial (*c.* 13,000–10,000 ^{14}C years BP) deposition in the basin, albeit in very low resolution. Earlier trial boreholes from this site had obtained longer Late-glacial records (up to 100cm long) and other sites (including in the col between Knockmore and Knocknaveen—see site 4, Fig. 1) also record the last glacial–interglacial transition in more detail. The most likely explanation for the short sequence in this core is that the basin floor at Lough Avullin contains large clasts (including boulders, see above) and that it is a matter of chance whether the corer penetrates between these clasts to obtain the longest sedimentary record. The Late-glacial is well-known from elsewhere in Ireland (Gray and Coxon 1991; Coxon and McCarron 2009), and as it was not the main focus of interest for this study a second core to recover it was not taken.

The onset of the Holocene in the core is marked by the rapid increase in organic deposition (and in pollen sedimentation) recorded at 1158cm (Fig. 6) with the sediment becoming predominantly a fine mud containing plant fragments. The loss on ignition values rise sharply at this point to 55%–60%. The early Holocene vegetation of Clare Island appears to be very similar to many sites recorded around the country, with juniper being replaced by birch and hazel, pine, elm and oak immigrating subsequently (Mitchell 2006; 2009). The pollen record from Clare Island between 10,000 and 7000 ^{14}C years BP is dominated by pine, oak, hazel and birch, with elm also being important. The sedimentation rate in Lough Avullin at this time is very low (Fig. 8), and hence the vegetation record is very compressed in this early period, with almost 3500 years recorded in LA2, between 1158 and 1025cm. The slow sedimentation rate throughout the early Holocene is probably a consequence of stable soils, with forested conditions prevalent over most of the catchment of the lake. Similar early Holocene records are found throughout Ireland (both in terms of tree immigration and low sediment accumulation rate) and indeed have been reported from the island of Inishbofin some 25 km to the south-west (O'Connell and Ní Ghráinne 1994).

Alder appears at the onset of LA3A (at 1025cm—just preceding the radiocarbon date below) and the forest composition is predominantly pine, oak, birch, elm and alder, with hazel now less important but still probably forming a dense understorey. The presence of grasses and plantains as well as herbaceous taxa (including Asteraceae) suggest that there were open areas in the vicinity or that the woodland had an open aspect. Early in this zone there is a slight increase in charcoal present in the core, and this is associated with finely dispersed charcoal apparent in the sediment between 1009–1010cm. This layer of dark sediment was dated to 6510 ± 40 ^{14}C years BP (7322–7494 cal BP; see Table 2 and Fig. 8).

The dated charcoal horizon at 1010cm may represent the initial anthropogenic disturbance in the basin's catchment, as evidenced by charcoal, some washed-in clay and a slight drop in the organic content in the core (just above the base of LA3A), but this cannot be absolutely confirmed from the evidence available. It is equally possible that the rise in charcoal is due to natural forest fire, as suggested by Huang (2002) in Connemara.

Corcoran (2002) suggested that the rise in Al, K and Ti in the sediment above the radiocarbon

Table 2
Radiocarbon dates

Sample depth (cm)	Sample ID	Conventional radiocarbon age (^{14}C years BP)	2 sigma calibrated result (95.4% probability)	1 sigma calibrated result (68.3% probability)
225	Beta-155901	390 ± 40 BP	cal BP 316: cal BP 399	cal BP 333: cal BP 353
			cal BP 405: cal BP 405	cal BP 435: cal BP 505
			cal BP 422: cal BP 513	
			cal AD 1437: cal AD 1528	*cal AD 1445: cal AD 1515*
			cal AD 1545: cal AD 1545	*cal AD 1597: cal AD 1617*
			cal AD 1551: cal AD 1634	
1010	Beta-157717	6510 ± 40 BP	cal BP 7322: cal BP 7494	cal BP 7335: cal BP 7353
				cal BP 7416: cal BP 7475
			cal BC 5545: cal BC 5373	*cal BC 5526: cal BC 5467*
				cal BC 5404: cal BC 5386

date (noticeable at 950cm, Fig. 7), allied with the presence of certain open ground taxa (i.e. plantains), may establish human presence, represented by soil erosion and clearance. The change in geochemistry at *c.* 950cm as well as a distinct silt/sand horizon and a drop in the organic content of the core at this level (Figs 6, 7 and 8) all suggest a marked change in the catchment. The age–depth model used here (admittedly based on the assumed age of the elm decline) would place such a disturbance at *c.* 5750 ^{14}C years BP (6500 cal BP; see Fig. 8).

However, the evidence from the palaeoecology and sedimentology of this core alone is not strong enough to assign a precise date to the onset of anthropogenic disturbance on Clare Island. Proving human impact on the landscape from depositional records such as these is fraught with difficulty, and this will be better left to the discovery and dating of archaeological material in a relevant context.

The presence of open areas and washed-in charcoal suggest a disturbance of the island's vegetation during LA3, and this disruption becomes more pronounced at the onset of LA3B, when elm pollen shows a marked decline as does pine, birch, alder, ash and hazel pollen. The classical elm decline is dated from Irish sites, including a number of sites local to Clare Island, to *c.* 5000 ^{14}C years BP (5750 cal BP) (e.g. Browne 1986; Molloy and O'Connell 1987). Coincident with the elm decline is the appearance of consistent records of the pollen of *Plantago lanceolata*, increasing levels of charcoal and a rise in the pollen of grasses, sedges and *Calluna*. It is probable from this evidence that human interference in the vegetation of the island occurs during LA3, and although the exact timing is not known, the post-elm decline opening of the woodland cover during LA3B is most likely to reflect increasing human activity (i.e. post 5000 ^{14}C years BP). Gosling (2007a) suggests that '... Clare Island had a settled farming community of some duration during the early and/or middle Neolithic periods (*c.* 4000–3100 BC)'. However, this remains to be proven, despite an apparent opening of the woodland cover. The megalithic court tomb adjacent to Poirtín Fuinch in Lecarrow is probably of this age, and the period includes the latter part of LA3A (including the elm decline) and LA3B.

The base of LA3C (787–720cm, 4400–4000 ^{14}C years BP) records the decline in alder pollen and pine pollen remains at a lower level. The woodland still includes birch, oak, alder and pine, although the concentration data show they are all contributing less pollen (but high sedimentation rates could give this impression). However, ash maintains a significant presence. Hazel marginally increases in its abundance as do open ground taxa—especially plantains and grasses. There is little archaeology recorded during the time period covered by LA3C is (Gosling 2007a) but open ground taxa persist and most tree taxa decline in abundance (hazel is an exception). However, charcoal becomes less common in LA3C, and it is possible that the pollen assemblage and sediment record a decline in human impact, although disturbed ground is implied by the continued presence of *Plantago*.

LA4 covers the period from just before the virtual disappearance of pine (probably at *c*. 4000 ^{14}C years BP, 4500 cal BP, based on other Irish sites) until *c*. 1300 ^{14}C years BP (from the age–depth curve). Overall, the period is characterised by birch woodland, with oak, alder and ash maintaining a presence. Hazel and willow are also present, while open areas become more common, with grass, heathers and numerous herbs being important. It is worth noting that at the start of the zone, between 720cm and 610cm, grasses are important and plantains and Chenopodiaceae peak. These peaks in open ground taxa (dating to between 4000–3500 ^{14}C years BP) may represent local clearances (Pl. VII), although unfortunately the resolution here is not sufficient to analyse further or to be more precise. The archaeology from the second millennium BC (*c*. 3600 ^{14}C years BP) contains a '... veritable bounty of monuments and a relative abundance of artefacts.' (Gosling 2007a). The sedimentary record (peaks in charcoal, sandy horizons and falls in organic content) is mirrored by the pollen record of fluctuating tree cover and open ground taxa, suggesting phases of clearance and agriculture along with episodic burning (this can be seen within and between the zones of LA4 on the pollen and other diagrams). The archaeological survey recorded many burnt mounds or fulachtaí fia (Gosling 2007b, cat. no. 53), and there is a concentration of such features in Lecarrow, especially along streams feeding Lough Avullin (e.g. along Pollabrandy—see Gosling 2007c). The environmental impact of the use of such monuments on the landscape must have been severe, with large volumes of wood required for the heating of the stone clasts (see Pl. VII). Fulachtaí fia were dated during the archaeological survey (Gosling and Waddell 2007) with two in Lecarrow, cat. nos 57 and 37, giving ages of 3660 ± 33 and 3620 ± 32 ^{14}C years BP, respectively, i.e. during the middle of LA4A (and coincident with charcoal peaks in the core from Lough Avullin). The archaeological work on one fulacht fia (cat. no. 57) included associated palynological work (Molloy 2007). The pollen from this burnt mound site showed high percentages of arboreal pollen, including pine (in low percentages), oak, birch, hazel and alder, showing a woodland cover similar to that recorded in LA4A. The persistence of tree and shrub cover, both in the samples analysed by Molloy (2007) and in this study, suggests that the woodland disturbances were possibly localised in extent and that trees and shrubs maintained a presence throughout. The fluctuation in arboreal and non-arboreal pollen throughout LA4 confirms the sporadic nature of the clearances.

Pl. VII How did Clare Island come to look as it does now? A. A reconstruction of a fulacht fia by Christy Lawless and friends in Turlough, Co. Mayo (for a 1991 IQUA field excursion). Two legs of lamb wrapped in herbs and rushes were cooked by being lowered into the boiling trough produced with the heated clasts. The volume of wood burnt in the process of heating the stones is apparent in the foreground. Much of the washed-in charcoal in the Lough Avullin sediments probably originated in such fires and the number of fulachtaí fia, their distribution and age is discussed in detail in Vol. 5 of the *New Survey of Clare Island* (Gosling *et al*. 2007). B. An aerial photograph of Clare Island showing ancient cultivation ridges (exposed from below a peat cover) to the left, pre-Famine cultivation ridges cross cut by Congested Districts Board walls and a crop of potatoes on the re-used ridges. The incredible continuity of landscape use in this part of Ireland is beautifully illustrated in this image. (Fawnglass, Clare Island, Grid reference: L710857. Site identified for the authors by Anna Wettergren)

Two other fulachtaí fia that were dated are 1.5km south of Lough Avullin (and outside of the lake's catchment) in Glen. These monuments dated to 2922 ± 37 and 2957 ± 33 ^{14}C years BP (cat. nos 31 and 34 respectively), i.e. during LA4B.

Between 570–610cm in LA4B there is an increase in oak pollen, followed by an increase in birch pollen associated with a decline in open ground taxa, including grasses (this change is apparent in both the percentage and concentration data). This temporary and slight closing over of the woodland probably dates to between 3100 and 2700 ^{14}C years BP. (Possibly the use of fulachtaí fia on the southern side of the island (Glen) at this time suggests a move away from intense land use at Lough Avullin). The more closed vegetation cover is followed (in LA4C) by a return to more open conditions with grasses, *Rumex* and plantains present and with *Calluna* becoming more important, possibly indicating further clearance. LA4D again sees fluctuating open and closed plant cover but with trees and shrubs (especially birch and alder) maintaining a presence around the catchment of the lake. The detailed pollen analytical work carried out as part of the archaeological investigations on the island (Molloy 2007) provides more information regarding this time period (*c.* 2200 to 1300 ^{14}C years BP). Peat adjacent to the base of a pre-bog wall at Bunnamohaun (site 7, Fig. 1) was dated to 2468 ± 30 ^{14}C years BP and peat overlying the wall was dated to 1689 ± 30 ^{14}C years BP (Gosling *et al.* 2007). Molloy notes that the pollen and the radiocarbon dates suggest that the wall was built at the end of the Bronze Age (or in the Early Iron Age). The pollen shows that the very exposed landscape of the western end of the island carried an open vegetation dominated by Poaceae, *Plantago lanceolata* and *Calluna*, in contrast to the perseverance of shrub and tree cover recorded around the more-sheltered Lough Avullin.

The start of LA5 at 372cm (*c.* 1300 ^{14}C years BP) sees peaks in open ground taxa and notably in grasses, with heaths also becoming more important and alder, birch and hazel decreasing. This zone is also characterised by rising charcoal levels in the sediment, and probably represents increasing land-use (or a change in management practises) during the Early Christian period, continuing until the end of the zone. The peak in aquatic taxa (in both pollen diagrams) may suggest that open water has become more prevalent in the lake—perhaps due to management of the outlet of the lake.

At 390 ± 40 ^{14}C years BP, i.e. the start of LA6, oak, ash and shrub taxa decline markedly, alder and birch remain (probably around the wetter margins of the lake) as does *Corylus*/*Myrica* (here probably *Myrica*—bog myrtle) and the vegetation becomes dominated by *Calluna* and herbaceous taxa. These changes are especially clear in the pollen concentration diagram. The sediment contains a marked volume of charcoal, far higher than that recorded from older sediments, and a notable drop in the organic component of the core that continues from late in the last zone, suggesting greatly increased minerogenic inwash. The geochemical analyses also point to increased washing in of sediment.

This last zone shows a dramatic change in land use in the vicinity of Lough Avullin, with a clearance of most of the trees and shrubs (ash disappearing altogether) and *Calluna* and open ground vegetation becoming totally predominant. The disappearance of open water taxa and the increase in *Sphagnum* suggests that the lake itself had probably become mostly infilled and had changed in character to become a boggy area. The landscape has probably looked much as it does now for the last 400 years.

Conclusions

The pollen diagram from Lough Avullin has provided a more detailed record of Clare Island's vegetational history than that available from previous studies. The vegetation succession on the island was similar to that recorded on the nearby mainland of western Ireland and on other islands to the south. The Late-glacial vegetation was quickly succeeded at the onset of the Holocene (Pl. VIII) by woodland, and the colonisation by the vegetation follows the general pattern of the rest of the island of Ireland (Mitchell 2006; 2008).

The early woodland of the island is dominated by pine, oak, hazel, elm and birch, with alder appearing later. The evidence for anthropogenic disturbance of the vegetation on Clare Island is ambiguous prior to 6000 cal BP (5000 ^{14}C years BP), but it may have been disturbed by humans possibly as early as 6500 ^{14}C years BP ago but more likely by 5700 ^{14}C years BP. From 5000 ^{14}C years BP, after a pronounced elm decline and the marked reduction in other tree taxa presence, the woodland cover fluctuated (probably under human pressure, with sporadic regeneration of tree and shrub cover,

Pl. VIII A. Croaghpatrick from Clare Island (1991); B. The same view at the end of the Younger Dryas, some 11,700 (10,000 ^{14}C BP) years ago? The Irish landscape changed rapidly and dramatically from this frozen state at the end of the last glaciation. (This image shows Mount Searle on Horseshoe Island in Marguerite Bay, western side of the Antarctic Peninsula. Photograph with kind permission of Dr Alan Vaughan, British Antarctic Survey)

but also due to natural succession in and around previously disturbed areas). The impact of humans on the natural vegetation is corroborated by the archaeological evidence. There remain many isolated peat basins on the island that may contain more detailed and localised records.

Acknowledgements

Pete Coxon would like to acknowledge a number of people without whom his work on Clare Island would not have been possible. First and foremost he would like to thank his friend the late Frank Mitchell for his suggestion, if not insistence, that we visit the island in 1982 to complete some of the work left unfinished by the first Clare Island Survey. He would also like to thank Dr Ronnie Creighton of the GSI, Gina Hannon and Catherine Coxon, who all helped with the first visits in 1982 (the latter two even cored on the summit of Knockmore and at Loughanaphuca). John Graham (Geology, TCD) helped him take the first serious core at Lough Avullin in 1992. J.P. Corcoran assisted Ryan and Pete with the core described in this text in 1999. On the island, the help, accommodation, assistance and friendship of Ciara Cullen, Bernard McCabe and family, Chris O'Grady and family, Michael Bob O'Malley and Peter Gill, Anna Wettergren and family as well as many other landowners all made the visits and fieldwork far more enjoyable. Thank you.

REFERENCES

Browne, P. 1986 Vegetational history of the Nephin Beg mountains, County Mayo. Unpublished PhD thesis, University of Dublin, Trinity College.

Clark, R.L. 1982 Point count estimation of charcoal in pollen preparations and thin sections of sediments. *Pollen et Spores* **24**, 523–35.

Corcoran, R. 2002 Palaeoenvironmental investigations of Clare Island, Co. Mayo. Unpublished MSc. thesis, Geography Department, Trinity College Dublin.

Coxon, P. (ed.) 1982 *A fieldguide to Clare Island, Co. Mayo.* Dublin. Irish Association for Quaternary Studies.

Coxon, P. 1987 A post-glacial pollen diagram from Clare Island, Co. Mayo. *Irish Naturalists' Journal* **22**, 219–23.

Coxon, P. 2001 The Quaternary history of Clare Island. In John R. Graham (ed.), *New Survey of Clare Island. Volume 2: geology*, 87–112. Dublin. Royal Irish Academy.

Coxon, P. and McCarron, S.G. 2009 Cenozoic: Tertiary and Quaternary (until 11,700 years before 2000). In Charles H. Holland and Ian S. Sanders (eds), *The geology of Ireland*, 355–96. 2nd edn. Edinburgh. Dunedin Academic Press.

Coxon, P. and O'Connell, M. 1994 *Clare Island and Inishbofin.* Dublin. Irish Association for Quaternary Studies.

Cullen, C. and Gill, P. 1991 *Studying an island: Clare Island, Co. Mayo.* Clare Island Series 1. Clare Island. Centre for Island Studies.

Cullen, C. and Gill, P. 1992 *Recorded history: from Grace O'Malley to the present day on Clare Island, Co. Mayo.* Clare Island Series, No. 5. The Centre for Island Studies, Clare Island.

Davies, B.E. 1974 Loss on ignition as an estimate of soil organic matter. *Soil Science Society of America Proceedings* **38**, 150–51.

Edwards, K.J. 1981 The separation of *Corylus* and *Myrica* pollen in modern and fossil samples. *Pollen Spores* **23**, 205–18.

Edwards, L.S. 1977 A modified pseudosection for resistivity and IP. *Geophysics* **42** (5), 1020–36.

Faegri, K. and Iversen, J. 1989 *Textbook of pollen analysis.* 4th edn. Chichester. John Wiley and Sons.

Forbes, A.C. 1914 Clare Island Survey, Part 9. Tree growth. *Proceedings of the Royal Irish Academy*, 31, section 1: part **9**, 32pp.

Gibson, P.J. and George, D.M. 2004 *Environmental applications of geophysical surveying techniques*. Hauppauge, NY. Nova Science Publishers Inc.

Gibson, P.J., Lyle, P. and George, D.M. 2004 Application of resistivity and magnetometry geophysical techniques for near-surface investigations in karstic terranes in Ireland. *Journal of Cave and Karst Studies* **66** (2), 35–8.

Gosling, P. 2007 The human settlement history of Clare Island. In P. Gosling, C. Manning and J. Waddell (eds), *New Survey of Clare Island. Volume 5: archaeology*, 29–68. Dublin. Royal Irish Academy.

Gosling, P. 2007 Catalogue of archaeological and architectural monuments, sites and stray finds on Clare Island. In P. Gosling, C. Manning and J. Waddell (eds), *New Survey of Clare Island. Volume 5: archaeology*, 83–212. Dublin. Royal Irish Academy.

Gosling, P. 2007 A distributional and morphological analysis of fulachtaí fia on Clare Island. In P. Gosling, C. Manning and J. Waddell (eds), *New Survey of Clare Island. Volume 5: archaeology*, 69–80. Dublin. Royal Irish Academy.

Gosling, P. and Waddell, J. 2007 Appendix 3. Radiocarbon dates from archaeological sites on Clare Island. In P. Gosling, C. Manning and J. Waddell (eds), *New Survey of Clare Island. Volume 5: archaeology*, 331. Dublin. Royal Irish Academy.

Gosling, P., Manning, C. and Waddell, J. (eds) 2007 *New Survey of Clare Island. Volume 5: archaeology*. Dublin. Royal Irish Academy.

Graham, J.R. (ed.) 2001 *New Survey of Clare Island. Volume 2: geology*. Dublin. Royal Irish Academy.

Gray, J.M. and Coxon, P. 1991 The Loch Lomond Stadial Glaciation in Britain and Ireland. In J. Ehlers, P.L. Gibbard and J. Rose (eds), *Glacial deposits in Britain and Ireland*, 89–105. Rotterdam. Balkema.

Grimm, E.C. 1993 Tilia 2.0 and Tilia Graph 2. Illinois State Museum, Research Collections Centre, 1920 South 101/2 Street, IL 62703, USA.

Huang, C.C. 2002 Holocene landscape development and human impact in the Connemara uplands, western Ireland. *Journal of Biogeography* **29**, 153–65.

Loke, M. 2006 *RES2DINV version 3.55. Rapid 2-D Resistivity and IP inversion using the least-squares method. Instruction manual*. Geotomo Software.

Mitchell, F.J.G. 2006 Where did Ireland's trees come from? *Biology and Environment: Proceedings of the Royal Irish Academy* **106**B, 251–9.

Mitchell, F.J.G. 2008 Tree migration into Ireland. *Irish Naturalists' Journal* (special supplement) **2008**, 73–5.

Mitchell, F.J.G. 2009 The Holocene. In Charles H. Holland and Ian S. Sanders (eds), *The geology of Ireland*. 2nd edn. 397–404. Edinburgh. Dunedin Academic Press.

Molloy, K. 2007 Reconstruction of past vegetation history on Clare Island: palaeoenvironmental investigations undertaken in conjunction with the archaeological investigations. In P. Gosling, C. Manning and J. Waddell, (eds), *New survey of Clare Island. Volume 5: archaeology*, 297–310. Dublin. Royal Irish Academy.

Molloy, K. and O'Connell, M. 1987 The nature of the vegetational change at about 5000B.P. with particular reference to the Elm Decline: Fresh evidence from Connemara, western Ireland. *New Phytologist* **106**, 203–20.

Moore, P.D., Webb, J.A. and Collinson, M.E. 1991 *Pollen analysis*. 2nd edn. Oxford. Blackwell Science.

O'Connell, M. and Ni Ghráinne, E. 1994 Inishbofin palaeoecology. In P. Coxon and M. O'Connell (eds), *Clare Island and Inishbofin*, 61–101. Dublin. Irish Association for Quaternary Studies (IQUA).

Reimer, P.J., Baillie, M.G.L., Bard, E., Bayliss, A., Beck, J.W., Blackwell, P.G., Bronk Ramsey, C., Buck, C.E., Burr, G.S., Edwards, R.L., Friedrich, M., Grootes, P.M., Guilderson, T.P., Hajdas, I., Heaton, T.J., Hogg, A.G., Hughen, K.A., Kaiser, K.F., Kromer, B., McCormac, F.G., Manning, S.W., Reimer, R.W., Richards, D.A., Southon, J.R., Talamo, S., Turney, C.S.M., van der Plicht, J., Weyhenmeyer, C.E. 2009 IntCal09 and Marine09 Radiocarbon Age Calibration Curves, 0–50,000 Years cal BP. *Radiocarbon* **51** (4), 1111–50.

Roberts, N. 1998 *The Holocene—an environmental history*. Oxford. Blackwell Publishers.

Stace, C. 1997 *New flora of the British Isles*. 2nd edn. Cambridge. Cambridge University Press.

Stockmarr, J. 1971 Tablets with spores used in absolute pollen analysis. *Pollen et Spores* **13**, 614–21.

Tolonen, K. 1986 Charred particle analysis. In B.E. Berglund (ed.), *Handbook of Holocene palaeoecology and palaeohydrology*. Chichester. John Wiley and Sons.

Troels-Smith, J. 1955 Karakterising af løse jordater. *Danmarks Geologiske Undersøgelse* IV **3**, 1–73.

Viney, M. 1990 Landscape history in a nutshell. *Irish Times*, Saturday June 30, 1990.

Vullings, W., Collins, J.F. and Smillie, G. 2013 Soils and soil associations on Clare Island. New Survey of Clare Island. Dublin. Royal Irish Academy.

Walker, M.J.C. 2005 *Quaternary dating methods*. Chichester: J. Wiley.

Wright, H.E., Livingstone, D.A. and Cushing, E.J. 1965 Coring devices for lake sediments. In B. Kummel and D.M. Raup (eds), *Handbook of palaeontological techniques*, 494–520. San Francisco. Freeman.

APPENDIX 1

Resisitivity methodology

Resistivity is an intrinsic characteristic of the subsurface that can be measured by geophysical means. In the field, resistivity is obtained using four electrodes, a source and sink current electrode (C1 and C2) and two potential electrodes (P1 and P2).

It can be shown that the potential (V) at any point for the source and sink electrodes is given by

$$V = \rho\frac{I}{2\pi r_1} - \rho\frac{I}{2\pi r_2},$$

where r_1 is the distance from the source electrode, r_2 is the distance from the sink and ρ is the resistivity. The potential difference between P1 and P2 electrodes (ΔV) is measured and it is possible to show that the resistivity (ρ) is given by

$$\rho = \frac{2\pi\Delta V}{I}\left[\frac{1}{\frac{1}{r_1} - \frac{1}{r_2} - \frac{1}{r_3} + \frac{1}{r_4}}\right],$$

where r_1, r_2, r_3 and r_4 are the distances between various electrodes combinations (Gibson and George 2004). In this study, the Wenner electrode array was employed resulting in equal electrode spacings and thus the above equation reduces down to $\rho = 2\pi a[\Delta V / I]$, where a is the electrode spacing and $[\Delta V / I]$ is measured. Fig. 5 illustrates the changes in apparent resistivity measured along a traverse with an electrode spacing of 10m. Distinct changes in the subsurface can be determined from these data and they show an abrupt change in the nature of the subsurface. Apparent resistivity values average around 140 for 130m and drop over a very short distance to values of 42 for a 60m wide zone.

There are two main drawbacks with using the above approach. Firstly, collecting data along a traverse with a constant electrode spacing only allows a fixed depth of investigation, which for the Wenner array is about half the electrode spacing. Thus the median depth of investigation for Fig. 5 is 5m, and no information can be obtained for a greater depth. Secondly, the measured parameter is not true resistivity, but apparent resistivity because the acquired data are a function of both the variation in true resistivity and also the array configuration that was employed. These drawbacks can be eliminated by acquiring two-dimensional data, using a multicore electrode spread and modelling the resultant data. Two-dimensional electrical imaging (also termed tomography) allows the acquisition of apparent resistivity variations in the vertical and horizontal directions, effectively producing a 2D slice known as a pseudosection. Electrical imaging was undertaken on Clare Island using a Campus Geopulse system and a multicore cable with 25 fixed-interval take-off points to which the electrodes were connected. The system was computer controlled and a parameter file instructed in which sets of four electrodes were used for each reading. Data were initially automatically collected for a traverse with an a spacing of 10m (Level 1), then with an increased 'a' spacing for Level 2 (20m) and Level 3 (30m).

Level 1	Level 2	Level 3
C1, P1, P2, C2	C1, P1, P2, C2	C1, P1, P2, C2
1 2 3 4	1 3 5 7	1 4 7 10
2 3 4 5	2 4 6 8	2 5 8 11
....	....	....
21 23 24 25	19 21 23 25	16 19 22 25

This procedure was repeated for higher levels, which allowed the production of a pseudosection that displayed the apparent resistivity variations in two dimensions. The depth at which the data are plotted is generally the median depth of investigation for the particular electrode array for that specific level (Edwards 1977). In order to determine how the true resistivity varies with depth, the data must be modelled using an inversion program such as RES2DINV (Loke 2006). This program is based on the smoothness constrained least squares method of deGroot-Hedlin and Constable (1990) to solve the partial derivatives of the Jacobian matrix. This program is best explained by referring to Fig. 5. The topmost image in this figure shows the variation in measured apparent resistivity that was obtained on one of the Clare Island survey lines. Note that the depths given are only pseudodepths. The inversion program divides the subsurface into a number of boxes that are assigned true resistivity values, and an initial model of the surface is formed that shows the variation in true resistivity with real depth. The apparent resistivity that this model would yield for the electrode array adopted is then calculated and compared to the measured apparent resistivity. The model is progressively altered using a least-squares optimisation approach in order to reduce the root mean square (RMS) error between the measured and calculated apparent resistivity. Generally about five iterations reduce the RMS error to the order of 4%. Image S1B in Fig. 2 shows the calculated resistivity that the model (S1C) would produce. There is a very good match between the calculated and measured apparent resistivities, thus the model is a good representation of the subsurface variation in true resistivity.

APPENDIX 2

Sediment description

The sediment was described based on the Troels-Smith method (1955).

All depths are in terms of below ground level (bgl) and quoted in metres.

Troels-Smith analysis sediment component abbreviations:

As	Clay (<0.002mm)
Ag	Silt (0.002–0.06mm)
Ga	Fine sand (0.06–0.6mm)
Gs	Coarse sand (0.06–0.2mm)
Gg	Gravel (>2mm)
Ld	*Limus detrituosus* (mud <0.1mm)
Lc	*Limus calcareous* (calcareous mud <0.1mm)
Dl	*Detritus lignosus* (woody detritus >2mm)
Dh	*Detritus herbosus* (plant detritus >2mm)
Dg	*Detritus granosus* (fine organic detritus <2mm)
Tb	*Turfa bryophytica* (moss peat)
Tl	*Turfa lignosa* (wood peat)
Th	*Turfa herbacea* (herbaceous peat)

0.5–0.63m
Colour: Hue 5YR 3/1 (very dark grey)
Components: Dl 1; Dh 1; Tl 1; Th 1; Ld +; Tb +

0.63–0.71m
Colour: Hue 5YR 2.5/1 (black)
Components: Dh 1; Dg 1; Tb 1; Th 1; Lc +; Dl +

0.71–1.18m
Colour: Hue 10YR 3/2 (very dark greyish brown)
Components: Dg 1; Tb 1; Tl 1; Th 1; Ld++

1.18–1.25m
Colour: Hue 10YR 3/3 (dark brown)
Components: Ga 1; Gs 1; Dh 1; Tb 1; Ld +; Dg +

1.25–1.38m
Colour: Hue 10YR 3/2 (very dark greyish brown)
Components: Dh 1; Dg 1; Th 1; Tb ++; Dl +

1.5–1.74m
Colour: Hue 7.5YR 3/2 (dark brown)
Components: Tb 2; Dg 1; Th 1; Ga +; Ld +

1.74–2.2m
Colour: Hue 5YR 3/2 (dark reddish brown)
Components: Tb 2; Ld 1; Dg 1; Th +

2.2–2.49m
Colour: Hue 5YR 3/2 (dark reddish brown)
Components: Th 2; Dh 1; Tb 1; Tl ++; Dg ++; Ld +; Dl +

2.5–2.57m
Colour: Hue 10YR 2/2 (very dark brown)
Components: Dg 1; Tb 1; Tl 1; Th 1; Dh ++; Ld ++, Dl +

2.57–2.65m
Colour: Hue 7.5YR 3/2 (dark brown)
Components: Dl 1; Dh 1; Tb 1; Th 1; Dg ++

2.65–3.0m
Colour: Hue 5YR 3/2 (dark reddish brown)
Components: Dg++; Tb 1; Th 1; Dh++

3.0–3.01m
Colour: Hue 10YR 3/2 (very dark greyish brown)
Components: Dg 2; Tb 1; Th 1; Dh +;

3.01–3.48m
Colour: Hue 10YR 2/2 (very dark brown)
Components: Dh 1; Dg 1; Tb 1; Th 1

3.5–3.6m
Colour: Hue 10YR 2/2 (very dark brown)
Components: Dh 1; Dg 1; Tb 1; Th 1; Ld ++

3.6–3.81m
Colour: Hue 7.5YR 2/0 (black)
Components: Ld 1; Dh 1; Dg 1; Th 1; Dl +

3.81–3.93m
Colour: Hue 10YR 4/4 (dark yellowish brown)
Components: Ld 2; Tb 1; Th 1; Dg ++; Dh +

3.93–4.03m
Colour: Hue 7.5YR 2/0 (black)
Components: Ld 1; Dh 1; Dg 1; Th 1; Tb +

4.03–4.04m
Colour: Hue 10YR 4/3 (dark brown)
Components: Ld 1; Dh 1; Dg 1; Th 1; Dl +; Tb +

4.04–4.5m
Colour: Hue 7.5 YR 2/0 (black)
Components: Ld 3; Dg 1; Th +

4.5–4.71m
Colour: Hue 7.5 YR 2/0 (black)
Components: Dh 1; Dg 1; Tl 1; Th 1; Ld ++; Dl +; Tb +

4.71–4.72m
Colour: Hue 5YR 3/1 (very dark grey)
Components: Tb 2; Dl 1; Dh 1; Ga +

4.72–5.17m
Colour: Hue 7.5 YR 2/0 (black)
Components: Ld 2; Dg 2; Dl ++

5.17–5.19m
Colour: Hue 10YR 2/1 (black)
Components: Ld 2; Dg 2

5.19–5.4m
Colour: Hue 7.5YR N 2/0 (black)
Components: Ld2; Dh 1; Dg 1; Dl ++

5.5–5.7m
Colour: Hue 10YR 3/1 (very dark grey)
Components: Ld 2; Ga 1; Dg 1; Ag +; Th +

5.7–5.98m
Colour: Hue 7.5YR N 2/0 (black)
Components: Ld 2; Dh 1; Th 1; Dg +; Tl +

6.0–6.21m
Colour: Hue 7.5 YR N2/0 (black)
Components: Ld 3; Dg 1; Th +

6.21–6.22m
Colour: Hue 10YR 2/1 (black)
Components: Ag 2; Ld 2; Dg +; Th +

6.22–6.62m
Colour: Hue 7.5YR n 2/0 (black)
Components: Ld 3; Dg 1

6.62–6.64m
Colour: Hue 10YR 3/1 (very dark grey)
Components: As 1; Ag 1; Ld 2; Dg +; Th +

6.64–6.9m
Colour: Hue 7.5YR n 2/0 (black)
Components: Ld 3; Dg 1; Dh +; Tl +; Th +

7.0–7.45m
Colour: Hue 10YR 2/1 (black)
Components: Ld 3; Dg 1; Th ++

7.45–7.51m
Colour: Hue 10YR 3/2 (very dark greyish brown)
Components: Ld 2; Dg 1; Th 1

7.51–7.61m
Colour: Hue 10YR 2/1 (black)
Components: Ld 1; Dh 1; Dg 1; Th 1

7.61–7.7m
Colour: Hue 10YR 3/1 (black)
Components: Ag 1; Ld 2; Dg 1; Th +

7.7–7.74m
Colour: Hue 10YR 2/1 (black)
Components: Ld 3; Dh 1

7.74–7.76m
Colour: Hue 10YR 3/3 (dark brown)
Components: Ld 2; Ag 1; Dg 1; Dh +

7.76–8.0m
Colour: Hue 10YR 2/1 (black)
Components: Ld 3; Dg 1

8.0–8.35m
Colour: Hue 7.5YR 2/0 (black)
Components: Ld 3; Ag 1; DG +; Th +

8.35–8.55m
Colour: Hue 2.5 Yr 2.5/0 (black)
Components: Dg 2; Ld 1; Th 1; Ag +

8.55–8.75m
Colour: Hue 5YR 3/2 (dark reddish brown)
Components: Dg 2; Ag 1; Ld 1; Th +

8.75–9.0m
Colour: Hue 5YR 3/2 (dark reddish brown)
Components: Ld 2; Dg 2; Th +

9.0–9.1m
Colour: Hue 7.5 YR n 2/0 (black)
Components: Ld 4; Ag 1; Dh +; Dg +

9.1–9.38m
Colour: Hue 10YR 2/1 (black)
Components: Ld 1; Dh 1; Dg 1; Th 1; Ag +

9.38–9.43m
Colour: Hue 5YR 2.5/1 (black)
Components: Ld 1; Dh 1; Dg 1; Tb 1; Ag +; Th +

9.43–9.44m
Colour: Hue 5YR 4/2 (dark reddish grey)
Components: Dh 1; Dg 1; Dg 1; Th 1; As +; Ld +

9.44–9.52m
Colour: Hue 5YR 2.5/1 (black)
Components: Dg 2; Ld 1; Tb 1; Ag +; Gh +

9.52–9.99m
Colour: Hue 5YR 3/3 (dark reddish brown)
Components: Dg 2; Tb 2; Ld +; Dh +

10.0–10.09m
Colour: Hue 5YR 3/3 (dark reddish brown)
Components: Ld 2; Dg 2; Dh ++; Ag +; Th +

10.09–10.10m
Colour: Hue 2.5YR 2.5/0 (black)
Components: Dg 2; Ld 1; Dh 1; Ag +; Th +

10.10–10.55m
Colour: Hue 5YR 3/3 (dark reddish brown)
Components: Ld 2; Dg 2; Ag +; Dh +

10.55–11.0m
Colour: Hue 5YR 3/3 (dark reddish brown)
Components: Ld 2; Dg 2; Tb +

11.0–11.58m
Colour: Hue 5YR 3/2 (dark reddish brown)
Components: Ld 2; Dg 1; Tb 1; Dh +

11.58–11.62m
Colour: Hue 5YR 4/3 (reddish brown)
Components: As 2; Ag 1; Ga 1

11.62–11.78m
Colour: Hue 2.5YR 3/2 (very dark greyish brown)
Components: Th 2; Dh 1; Dg 1; Tb +

11.78–11.84m
Colour: Hue 5YR 4/2 (dark reddish grey)
Components: As 2; Ag 1; Ga 1

VEGETATION–ENVIRONMENT INTERACTIONS ON CLARE ISLAND

Timothy Ryle

ABSTRACT

Clare Island lies at the mouth of Clew Bay, in County Mayo (H27), some 5km from the Irish mainland. It was intensively studied in the early part of this century, when a team of naturalists led by Robert Lloyd Praeger conducted the first Clare Island Survey (1909–1911). The research described here, as part of the New Survey of Clare Island, was undertaken in the years 1992–1996 to update the baseline information provided by the original multidisciplinary project. The aims of this study were to survey the flora and vegetation of Clare Island and map the major vegetation types. The key environmental factors that have influenced the island's vegetation are also discussed. A complete inventory of all higher plant species was made. Three hundred and eighty four species were located in the present survey, many of which were included among the four hundred and thirteen species listed in the original survey (Praeger 1911). Fifty-nine species from the original list were not recorded, and eight new records for Clare Island were discovered. A detailed survey of the native and semi-natural plant communities is presented. A modern approach to community classification did not prevent comparisons with the original vegetation descriptions. The utilisation and management of seventeen vegetation classes (and their related subcommunities) over 85 years is discussed, and related to recent research from similar habitats in Ireland and Great Britain. The distribution of the vegetation units has been mapped, and comparison between the current vegetation distribution and the distribution of units shown on Praeger's map (Praeger 1911) is made, with discussion of the individual environmental factors that have facilitated these changes. While grazing, particularly by sheep, is recognised as a major causal agent for vegetation change, historical land use, land management and climate are all recognised as having influenced the complex vegetation mosaic that occurs on Clare Island. The new map and vegetation data provide a baseline against which future changes may be assessed and will facilitate the development of environmentally sensitive management strategies for Clare Island.

Background—The Clare Island Survey, 1909–1911

Robert Lloyd Praeger had a passion for natural history and was constantly surveying wherever he travelled. Islands and the biogeography of island floras (e.g. Krakatoa, Christmas Island and the Faeroes) held a particular fascination for naturalists of his period. Praeger and his wife spent a vacation on Clare Island in 1903. Over the course of a week, Praeger studied the island vegetation, listing 393 higher plant species (this figure was later increased to 413 in 1911). Clare Island seemed to provide an interesting case, as it was located in Clew Bay, 5km from the Irish mainland at Roonagh Quay. It was felt that the island was sufficiently removed from the mainland to raise interesting questions as to the immigration of its flora and fauna. The flora was rich by comparison to the adjacent mainland (Praeger 1903), as it had an abundance of habitats that were not found on other islands along the west coast of Ireland (summarised in Table 1). The island's north-facing cliffs with their alpine habitats were a particular example.

On 13 April 1908 a meeting of Irish naturalists 'likely to be interested' in a major survey of an island was held in the National Museum, Dublin. Clare Island was selected for study because it contained a variety of habitats and was one of the only islands on the western seaboard that had not been extensively studied, apart from its flora (Praeger 1903). Praeger was instrumental in bringing together many of the great naturalists of the time and took on many of the administrative duties of the Clare Island Survey.

Over the course of three field seasons (1909–1911) Praeger's naturalists made many trips to Clare Island. Many went beyond their remit and included areas of the mainland for comparative investigation. The results of these investigations were prepared for publication—sixty-seven papers were published in the *Proceedings of the Royal Irish Academy*: one paper on peat deposits was never published. This original Clare Island Survey was one of the earliest baseline studies to include a comprehensive account of the major aspects of an island's natural history.

Table 1

Comparative details of topography and higher plant floras from several islands off the Irish west coast. The Belmullet Peninsula, with its narrow landbridge, is included for comparison. These data are based on Praeger 1903, 1904, 1905 and 1907

Region	Area (km^2)	Highest elevation (m)	Total higher plants
Achill Island	148.0	672	414
Clare Island	16.8	462	393
Inishbofin	11.7	89	379
Belmullet	117.0	132	348
Inishturk	5.8	192	327

Introduction—the present survey

Clare Island's landscape has undergone much change in the last ten thousand years. Far from being a natural landscape, it is the product of years of human management, providing, as it did, a sustainable resource for the survival of man and livestock (Usher and Thompson 1993). In 1988 the Praeger Committee of the Royal Irish Academy agreed that a resurvey of Clare Island should be undertaken. The background for such a suggestion was that (a) the baseline material presented in the first survey was of sufficient quality and depth to justify comparison with a modern survey, (b) considerable change has been observed in the island flora (Doyle and Foss 1986a) and (c) the island represented an unpolluted environment on the fringe of Europe (Evans 1957; Cullen and Gill 1991; Doyle and Whelan 1991).

Shortly before the time of the original survey there were significant changes in land-use practices on Clare Island. Social and agricultural practices (rundale farming) led to a problem of recurring land subdivision. Each son got a portion of the land upon marriage, and the farms got progressively smaller and comprised scattered fields. Upon purchasing the island in 1895 for £5472, the Congested Districts Board implemented a five-year plan (1895 to 1901) and set about rearranging the layout of the landscape in an effort to improve the social and economic structure. This they did by 'striping' the land and consolidating farm holdings. Although the land pattern was drastically altered, redistributing the land more equally (fifteen to twenty acres per household), the practice of booleying, or summer farming on the uplands, was left undisturbed. Further work included construction of 56km of clay boundary fences and an 11km dry stone wall (Congested Districts Board Wall) extending nearly the length of the island. The works completed, the island was sold back to the native population in 1901, establishing them once again as landowners.

Thereafter, there was a notable change from tillage to pastoral farming, which was in part due to the decline in the island's population (Mac Cárthaigh 1999). Praeger's initial survey of the island in 1903 noted that changes in farming practices and increases in livestock, mainly sheep, was beginning to have an effect on the upland vegetation (Praeger 1903). These changes would appear to have accelerated since that time. The heather community that once dominated the commonage has been almost entirely replaced by severely grazed rough grassland. A quarter of the plant species were not re-recorded after an interval of seventy-five years (Doyle and Foss 1986a). Several new species have been recorded (Doyle and Foss 1986a; Brodie 1991), reflecting the dynamic nature of the Clare Island flora.

Aims

Offshore islands, by their very nature, can be satisfactorily distant from the mainland to excite interest as to the possible migration pathways of plants and animals onto the island as well as the causal impacts that shape the island's ecology, e.g. Pembrokeshire islands (Gillham 1953; 1955; 1956a; 1956b; Goodman and Gillham 1954) and the Isle of Skye (Birks 1973). One of the aims of this vegetation survey was to resurvey the flora of Clare Island, building up a complete modern species list, allowing comparisons with those lists compiled by Praeger (1911) and Doyle and Foss (1986a). Secondly, the result of this survey allows direct comparison with vegetation described elsewhere in Ireland and places the communities in a European context (cf. White and Doyle 1982). Finally, the mapping of the vegetation units on the island allows comparisons with the distribution of vegetation types presented in Praeger's (1911) map and serves as a baseline for monitoring future changes in the vegetation (see fold-out map 1).

Clare Island

Clare Island (L8076) is situated in the mouth of Clew Bay, Co. Mayo (H27), 5km from the west coast of the Irish mainland (Fig. 1). The island measures approximately 7km along its east to west axis and 2.5km along its north to south axis (Fig. 2). Its area covers in the region of 1600ha. In addition to the precipitous cliffs along the northern shore, the twin peaks of Knockmore (463m) and Knocknaveen (223m), with the ridge that runs between them, are the most conspicuous landmarks on the island (Fig. 2).

Geology

The bedrock geology of Clare Island is complex, its arrangement controlled by some major faults that run through the Clew Bay area (Phillips 1965; 1973; Graham 1994; 2001). The stratigraphy includes Dalradian, Ordovician, Silurian, Devonian and Lower Carboniferous rocks, composed mainly of grits, schists, conglomerates, sandstones and some limited limestones (Cole *et al.* 1914; Hallissy 1914; Phillips 1965; Whittow 1974). The bulk of the rocks, however, are of Silurian age (see Fig. 3, p. 32).

Precambrian and early Palaeozoic rocks, like those on Clare Island, generally give rise to acid and infertile soils. On weathering, the rocks give rise to podzols on high ground, while in low-lying areas, impermeable clays are formed. In some areas, these clays are overlain with peat that can reach up to eleven metres in depth (see Coxon *et al.*, this volume, p. 11).

The pattern of glaciation in Ireland during the Pleistocene period is predominantly one of extensive ice masses pushing westwards along 'structural corridors' from the centres of ice accumulation—Achill Island and Nephin Beg in the case of Clare Island (Synge 1968; Coxon 1994; 2001). The earliest phase of glacial activity is recognised by moderately thick deposits of ice-moulded till in the south of the island. The glaciers, however, were reaching their westernmost limit on Clare Island (Browne 1991), as evidenced by the undulating nature of the

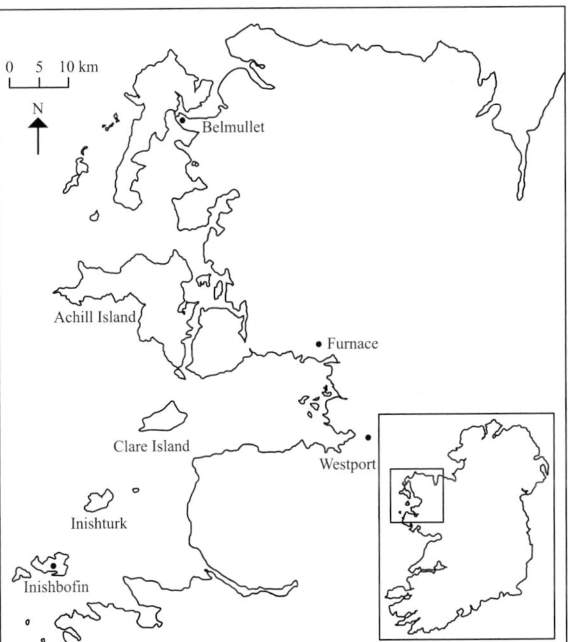

Fig. 1 Location of Clare Island (and adjacent islands), off the west coast of Ireland. The locations of the metereorological stations at Belmullet, Newport (Furnace) and Inishbofin are indicated, as is Westport, the principal town serving the island.

landscape surface in the north-east of the island, where the till surface is indicative of abrupt deposition of sediment at the extremity of the ice mass, resulting in a poorly sorted hummocky moraine.

Climate

There is no meteorological station located on Clare Island (see Fig. 2). The climatic character of the island must be based on data from the nearest meteorological stations at Inishbofin in County Galway, Furnace in County Mayo and Belmullet, again in County Mayo (Rohan 1986). The location of Clare Island is such that data from any one of these stations does not accurately reflect its climatic regime. Data on annual rainfall (for the years 1990 to 1995, when the majority of fieldwork was carried out) from the three climatological stations (Table 2) show considerable variation. The Belmullet data (Table 3) are considered the best approximation of the Clare Island climate, fluctuating less than 10%, based on the data from the other sites (D. Fitzgerald, Irish Meteorological Service, pers. comm.).

Although Belmullet lies further north than Clare Island and is located 5km inland, it has little high ground that provides shelter from incoming Atlantic weather. Both locations are influenced by the westerly Gulf Stream and its associated weather patterns (Keane 1986). The climate of the general area is temperate and oceanic (*sensu* Lamb 1977). Data for the principal climatic parameters for Belmullet for the period 1951 to 1980 are presented in Table 3 and are taken as a reasonable approximation of the Clare Island climate.

Annual rainfall on Clare Island has been estimated at 1200mm (averaged over a thirty-year period), following the suggestion of a 10% factorial increase of the Belmullet rainfall data (D. Fitzgerald, Irish Meteorological Service, pers. comm.). Clare Island experiences similar monthly rainfall patterns as Belmullet, with the driest periods occurring from April to July (less than 80mm of rain). Despite this, prolonged dry periods are rarely encountered.

Exceptional weather patterns were experienced throughout Ireland in 1995 and 1996 (Irish Meteorological Service 1995; 1996). The first three months were extremely wet, with temperatures below average. This pattern soon changed, yielding prolonged sunshine and higher temperatures that continued into the autumn.

Mean monthly air temperature ranges from 5.7°C to 14.1°C. During the summer months the mean maximum temperature can reach 17.2°C, although an extreme high of 28.0°C was recorded from Belmullet in 1995. Overall the range of temperatures is rarely severe.

Lamb (1977) has argued that the most significant climatic fluctuations affecting vegetation have been in temperature. Within the past few hundred years there has been a general trend of rising mean temperatures in Europe. Since the 1950s, however, there has been a gradual decline, although recent exceptional weather patterns have led to the perception of increasing temperatures. Long-term averaging of the data reveals no significant change in temperature in the past 100 years (D. Fitzgerald, Irish Meteorological Service, pers. comm.).

Some recent research has focused on the effects of precipitation on vegetation and the growing concern over the increased deposition of airborne pollutants in upland areas (Heil *et al.* 1987; Lee *et al.* 1992; Whitehead *et al.* 1997; Lee and Caporn 1998). Clare Island's location, removed from any major populated or industrial area, and the prevailing westerly frontal patterns mean that the effects of significant airborne inputs on the vegetation could be dismissed (Bailey 1984; Farrell *et al.* 1993; Jordan 1997).

Farming on Clare Island

Changes in farming practices on the west coast, from tillage to pastoral farming, were dictated by the thin peaty soils and the declining rural depopulation after the Famine of 1845. Samuel Lewis's *Topographical Dictionary* (Lewis 1837) records 1600 people on Clare Island. By 1911, this number had fallen to 300, and the population currently stands at around 160 inhabitants. This was a common feature of western areas, i.e. Connemara, northern Mayo and Donegal. The soils on Clare Island are relatively poor owing to the underlying geology and are often shallow, occurring as they do on rocky or sloping ground. Deeper blanket peats occur patchily in valleys and topographical depressions where water gathers.

Clare Island's isolation from the mainland meant that there was little economic future in tillage. Land formerly cropped was allowed to

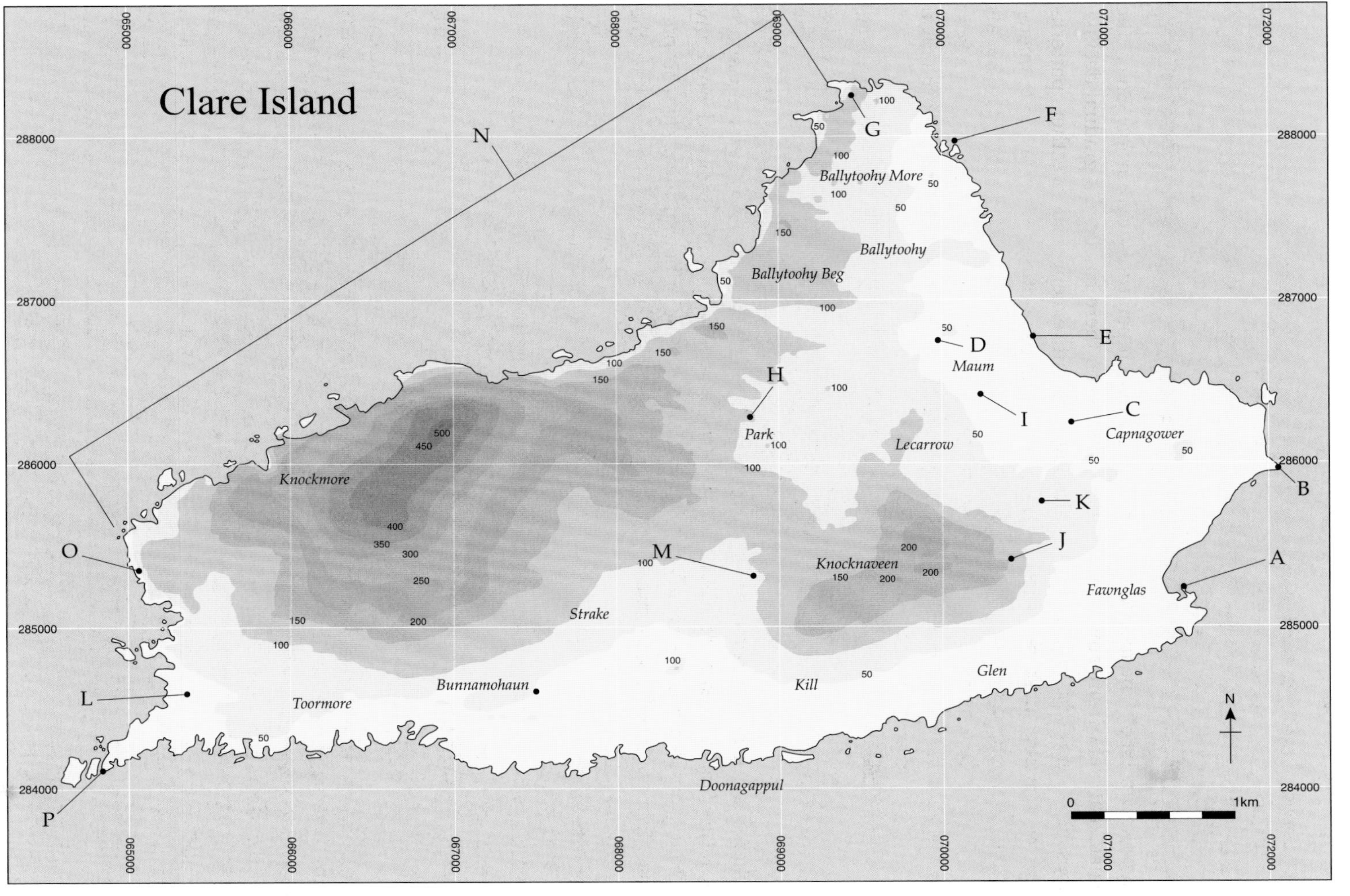

Fig. 2 Topographical map of Clare Island (based on Coxon and O'Connell 1994) featuring the principal townlands. Also included are sites/landmarks mentioned in the text.

A) The quay/harbour
B) Kinnacorra
C) The mill
D) Lassau
E) Portlea
F) Lighthouse cove
G) Lighthouse
H) Park
I) Lough Avullin
J) Leck
K) Poirtín Fuinch
L) Gorteen
M) Lough Leinapollbauty, Lough Merrignagh and Creggan Lough
N) Northern cliffs
O) Signal Tower
P) Kinatevdilla (Beetle Head)

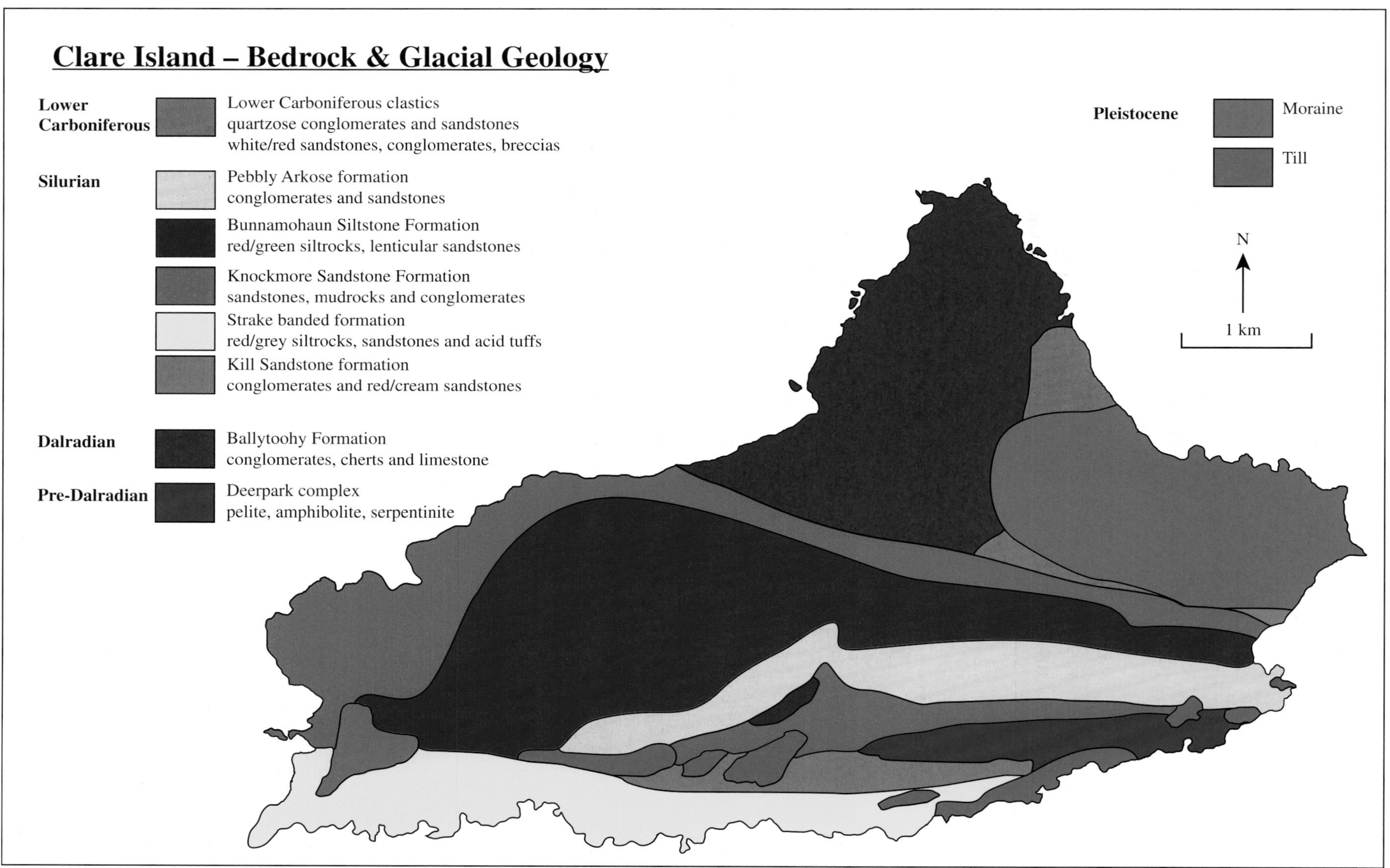

Fig. 3 Simplified geological map of Clare Island based on Graham (1994) and Browne (1991). Bunnamohaun and Strake geological formations do not correspond with similarly named townlands on the island.

Table 2
Annual rainfall data from Belmullet, Inishbofin and Furnace (Newport) for the years 1990 through 1995

Year	Belmullet	Inishbofin	Furnace
	Total annual rainfall (mm)		
1990	1447	1490	1805
1991	1154	1287	1550
1992	1448	1492	1771
1993	1188	1354	1544
1994	1316	1456	1734
1995	1203	1177	1373

revert to poor pasture. Sheep farming became the mainstay of the community, as in many other western regions, since climate, poor soils and hilly terrain rendered other methods of farming uneconomic.

The exploitation of the natural resources from marginalised land was a necessity for a self-sufficient island population such as on Clare Island. The demographic pressures meant that there was a heavy dependence on local materials, which included building materials, fuel and fertilisers for food production. The Congested Districts Board (see, for example, Ruttledge-Fair 1892) and other commentators of the time (C. Mac Cárthaigh, pers. comm.) condemned practices of extracting scraw turf, as the bare soils were more liable to erosion when unvegetated.

Six rabbits and six hares were introduced onto the island in 1906 for the purposes of shooting, and populations have since proliferated. Several large rabbit warrens exist along coastal grassland habitats along the northern cliffs from the lighthouse (northernmost point on Clare Island) to the base of the steep ascent of Knockmore, and on the extreme western end of the island below the signal tower. While the rabbits have undoubtedly contributed to the characteristic nature of the grass sward in places, the introduced lagomorphs have not had as significant an impact on the vegetation as domestic livestock.

Livestock include sheep, cattle, horses and donkeys. Cattle are generally kept infield, as are some donkeys, but the majority wander unattended. A number of 'feral' horses roamed the commonage, among the predominant sheep herd.

Table 3
Climate data from Belmullet meteorological station, Co. Mayo. Data are averaged for a thirty-year period: 1961–1990. (Data courtesy of Met Éireann)

	J	F	M	A	M	J	J	A	S	O	N	D	Annual
Temperature (°C)													
Mean max.	8.2	8.3	9.7	11.6	13.7	15.7	16.8	17.2	15.7	13.4	10.3	9.0	12.5
Mean min.	3.1	2.9	3.9	4.9	7.0	9.5	11.1	11.1	9.8	8.2	5.1	4.3	6.7
Mean	5.7	5.6	6.8	8.2	10.3	12.6	14	14.1	12.8	10.8	7.7	6.6	9.6
Rainfall (mm)													
Mean	123.5	80.1	95.8	58.1	68.0	67.3	67.6	93.7	108.0	132.9	127.7	119.9	1142.7
Rain days	23	19	23	19	18	18	19	20	21	24	23	24	249
Wet days	20	15	18	13	14	12	12	15	16	19	19	19	193
Sunshine (hours)													
Daily mean	1.47	2.41	3.29	5.27	6.14	5.36	4.29	4.63	3.65	2.63	1.74	1.08	3.50
Other (days)													
Snow or sleet	4.6	4.4	4.0	1.6	0.2	0	0	0	0	0	0.7	2.6	18.1
Snow at 9 a.m.	0.8	0.6	0.3	0.1	0	0	0	0	0	0	0.0	0.4	2.3
Fog	1.1	0.6	0.7	1.7	1.3	2.0	3.3	2.4	1.1	1.2	0.6	0.6	16.6
Air frost	5.0	4.3	2.1	0.9	0.0	0.0	0.0	0.0	0.0	0.1	1.3	3.1	16.7
Ground frost	10.5	9.5	7.3	5.4	1.9	0.1	0.0	0.0	0.5	1.7	5.5	7.8	50.3

At the time of this survey, the last two remaining goats that once inhabited the cliffs had not been sighted for several years.

Statistics prepared for the Congested Districts Board in 1892 (Ruttledge-Fair 1892) show that sheep were the preferred livestock and that Clare Island had 3000 sheep, 160 cattle, 74 horses and 16 asses. Praeger noted that between his first excursion in 1903 and his return for the Clare Island Survey (Praeger 1911) the 'appreciable numbers of' sheep were having a clear impact on the vegetation. Sheep numbers remained at these levels for the first half of the twentieth century and remained the dominant grazer on Clare Island.

Methods

Field survey

In updating the vegetation data on Clare Island, the field methods of the Braun-Blanquet school of phytosociology were used during this project (Westhoff and Van der Maarel 1973; 1978; White and Doyle 1982). While it is not directly comparable to Praeger's earlier descriptions on vegetation formation, the quality of the earlier information is such that it serves as an impressive baseline from which to draw comparisons.

The Braun-Blanquet approach was chosen as it is a relatively straightforward and widespread method of field description that is suitable for use in numerical analysis (Moore *et al.* 1970; Kent and Coker 1992). The results of this survey allow direct comparison with vegetation described elsewhere in Ireland and place the communities in a European context (cf. White and Doyle 1982).

The field work was carried out over a number of field seasons from 1993 to 1996.

Vegetation descriptions

The field descriptions for each quadrat were listed on a standard relevé (vegetation description/quadrat) card that allowed for the following information to be recorded:

- relevé number, date, location and grid reference of the quadrat;
- topographical features of the surrounding landscape and habitat, including aspect, slope, altitude, drainage and land-use management;
- soil type, degree of soil wetness and pH;
- total vegetation cover, height and cover of individual layers within a quadrat;
- cover abundance for all species of higher plants and bryophytes present in the quadrat (Table 4).

The regional soil classification scheme for Ireland prepared by An Foras Talúntais (1974) only records the occurrence of peaty gleys on Clare Island. A full soil survey was undertaken, as part of the New Survey of Clare Island (Vullings *et al.* 2013).

The names of two hundred of the commonest species were printed on the reverse of the field card to enable rapid recording. Any bryophyte species not identified in the field was collected and classified on return to the laboratory. Nomenclature for higher plants follows Tutin *et al.* (1964–1980). Nomenclature was later updated to take into account changes in characterisation and classification in plant systematics (Stace 1997). Nomenclature for bryophytes and lichens follows Smith (1990) and Grolle (1976) respectively.

Species lists

The changes in the flora and possible explanations as to the floral diversity, including a comprehensive species list of the plant species recorded over time, is presented elsewhere in this volume (see Jebb, pp 179–208). Further reports in this volume concentrate on the bryophytes (see Synnott, pp 209–232), lichens (see Seaward and Richardson, pp 233–251) and fungi (see Mitchel, pp 253–318).

Table 4
Braun-Blanquet cover abundance values used in the relevés

Cover abundance value	% cover	Frequency of individuals
5	76–100	Any number
4	51–75	Any number
3	26–50	Any number
2	6–25	Any number
1	<5	Numerous
+	Small cover	Few individuals
r		Solitary plant

Vegetation analysis
The phytosociological investigation relies on the collection of numerous relevés and subsequent rearrangement of the relevé tables to characterise the plant communities. Traditionally, the Braun-Blanquet style analysis relied on the physical reworking of tables, which allowed identification of patterns in the relevé data that were indicative of floristic communities or syntaxa. The syntaxa became apparent after successive rearrangements of the table of relevés. Each syntaxon was distinguished by a group of species that consistently occurred together.

With the advent of specialised computer packages, the time taken in analysing the raw data has been greatly reduced. The basis on which these programs work is that relevé tables are matrix tables and can be analysed mathematically, with patterns/similarities in groups of numbers being highlighted. Often the user can define the type of arrangement needed in the final output.

By consulting descriptive texts on European vegetation it is possible to classify the communities on floristic grounds following the Braun-Blanquet approach. These vegetation units are defined by *character*, *differential* and *companion* species, which are used in the construction of a hierarchical syntaxonomy. Even so, difficulties remain, arising from the depauperate nature of the Irish flora. An additional designation of indeterminate rank, *diagnostic species*, suggested by White and Doyle (1982), was used to circumvent floristic difficulties that arose due to the inadequate investigation of some Irish vegetation classes.

The association is the fundamental unit in this system (Westhoff and Van Der Maarel 1978; White and Doyle 1982), although further subdivisions, including subassociations and units, have been recognised. Associations are grouped together successively into alliances, orders and classes. Association names are composed of the names of one or two of the most typical species, and the generic and specific are latinised according to the *Code of Phytosociological Nomenclature* (Barkmann *et al.* 1986). The appending of the ending '-etum' to the name of the more dominant of the species completes the association name. There are now universally accepted rules included in a formalised nomenclatural system, governed by the code. The rules for the classification, priority of naming, etc., are analogous to the code of species nomenclature adapted in plant taxonomy. The endings for the hierarchical divisions are presented in Table 5. Thus an example of a correctly designated association is Pleurozio purpureae-Ericetum tetralicis Br.-Bl. et Tx. 1952 em. Moore 1968, originally described by Braun-Blanquet and Tüxen in 1952 and later revised by Moore in 1968: this describes the vegetation of Atlantic blanket bog in which both *Pleurozia purpurea* and *Erica tetralix* are typical species.

Table 5
Phytosociological endings employed in naming the vegetation communities

Rank	Ending
Class	-etea
Order	-etalia
Alliance	-ion
Association	-etum
Sub-Association	-etosum

Construction of phytosociological tables
Diagnostic species, based on phytosociological criteria identified by the user, are used in designing the table. All other species are ordered automatically in decreasing order of occurrence within the set of relevés analysed. The final vegetation tables were constructed using a spreadsheet, as the current computer package was not designed to produce fully edited phytosociological tables. The information on the phytosociological tables can be further summarised in the form of synoptic tables. This uses presence class data for each species. The presence classes used in the Braun-Blanquet approach are given in Table 6. In the current survey these values are used only as an indication of the frequency of occurrence of an individual species within the overall vegetation, rather than in each grouping recognised within a particular table.

Distribution of sites
The vegetation of the entire island was described in the present survey. Because many community types were present in the landscape, the approach taken in making the vegetation descriptions (relevés) remained constant. Four hundred and seven relevés were collected during the survey. As far as was

Table 6
Presence class values used in the Braun-Blanquet approach

Presence class	% occurrence in relevés assigned to a syntaxon
V	81–100
IV	61–80
III	41–60
II	21–40
I	11–20
+	6–10
R	0–5

possible, relevés were taken from homogenous stands of the particular community being described. Relevé sizes of 2m × 2m were generally used throughout the study: these are considered adequate by a number of authors for investigation of non-forest systems (Westhoff and Van Der Maarel 1973; 1978; Mueller-Dombois and Ellenberg 1974). In the case of the wet scrub woodland, 5m × 5m quadrats were used, reflecting the more complex nature of the vegetation type.

The locations and grid references for the quadrats are listed in Appendix 1. The grid references correspond to the map prepared for the new survey, which is lodged at the Royal Irish Academy in Dublin. The survey map, which was made at a scale of 1:7500, is based on a set of four six-inch Ordnance Survey maps that cover Clare Island (84, 84a, 85 and 75 of the 1920 edition).

Vegetation map

A topographical map at the scale of 1:7500 was constructed for the project (University of Glasgow 1992). A copy of this map served as the basis for the vegetation map, with field notes and boundaries transcribed onto it (see fold-out map 2). In addition, oblique still aerial images from television footage of Clare Island were also used, as was Ordnance Survey of Ireland aerial photography (1995 series).

Other maps that appear in the paper have been constructed primarily from data in Coxon and O'Connell (1994), supplemented by information from a variety of sources, which are referenced in the appropriate legends.

Soil analysis

A limited soil nutrient survey was undertaken to examine the possible correlation among soil parameters and the corresponding vegetation community types. Soil samples were collected from the centre of selected quadrats (258 samples from the 407 sites). The parameters that were investigated included pH (every relevé), extractable sodium (Na), calcium (Ca), magnesium (Mg), potassium (K), total available nitrogen (N), available phosphorous (P) and loss on ignition of organic content. These data do not form part of this paper as a comprehensive study into soil environment conditions was undertaken separately as part of the New Survey of Clare Island (Vullings *et al.* 2013.).

Vegetation of Clare Island

Postglacial vegetation

Praeger (1911) felt that remnants of the Arctic-Alpine plant communities found on the steep northern cliffs of Knockmore could not alone have been responsible for the colonisation of Clare Island. It was his belief that seeds floated across the narrow channel or those migrating animals coming across the fast melting land-bridge transported the seeds of many of the plants (Praeger 1911). There was little information on the vegetation history of Clare Island, as the work conducted by A.G. More on the postglacial sediments in the original survey was never published.

Prior to the commencement of the new Clare Island survey, Coxon (1987) produced a postglacial pollen diagram from a peat hollow containing five metres of sediment, on the north-eastern side of Clare Island. The results, which are included in volume 2 of the *New Survey of Clare Island* (Coxon 2001) describe a local area in Lecarrow. More recently, an eleven-metre sediment core from Lough Avullin was radiocarbon dated. The results suggest that the island had a 'typical' vegetational sequence, with woodland predominating until 5000 years BP, when anthropogenic or human-induced disturbance began (see Coxon *et al.*, this volume, pp 1–25). It provides a valuable insight into how the early Holocene vegetation developed and the effects of the human interference on the Neolithic and Bronze Age landscape prior to the original Clare Island Survey and Praeger's detailed botanical descriptions (1903; 1911).

The new survey and the botany

Despite its relatively compact size (1600ha), Clare Island is remarkable for the diversity of habitats and vegetation types that it supports. This makes it an interesting study area, since the variety is not repeated in any of the other offshore islands along the western coast of Ireland (Praeger 1903; Doyle and Foss 1986a; Mitchell 1986; McCarthy 1988; Brodie and Sheehy-Skeffington 1990). In presenting a complete phytosociological analysis of the vegetation of Clare Island, it is worth emphasising that the sociological interrelationship of the various syntaxa is complex, a feature that is related to local disturbances and land-use history.

The bulk of this chapter comprises a number of separate sections detailing a separate vegetation type. Within each section, the presentation of the results follows a similar format as far as is possible:

- Each section commences with an outline description of the character and general history of the vegetation and includes such details as its origins or genesis from previous vegetation types. The historical impacts on the land that have influenced the composition and structure of the vegetation are discussed. Particular emphasis is placed on the changes that have occurred in the time since the vegetation formations were originally described from Clare Island (Praeger 1911).
- The floristic assemblage of the vegetation includes a list of the diagnostic species that are useful in characterising each vegetation type. While particular combinations of species are frequently recorded from certain vegetation types on Clare Island, few species are exclusively associated with any habitat, as the composition of the vegetation has been heavily modified since farming communities were first established.
- The description and classification of the vegetation on Clare Island is discussed in relation to published counterparts from Ireland and north-western Britain. Because the land on Clare Island has been intensively utilised throughout the centuries, there is, understandably, quite a degree of local variation evident in the structure and composition of the vegetation, and overlap in syntaxa is not uncommon in the complex vegetation mosaic.

Despite the paucity of certain habitats and/or vegetation descriptions (particularly in upland regions), the regional climatic differences and the impoverished nature of the Irish flora, there is an intrinsic similarity between the vegetation of Ireland and Britain. The British National Vegetation Classification (BNVC) (Rodwell 1991–2000) has assembled a standardised descriptive scheme based on the evaluation of some 33,000 plus relevés, and has put the findings into context with phytosociological literature through landmark vegetation descriptions (e.g. Tansley 1911; 1939; McVean and Ratcliffe 1962; Birks 1973). This modern synthesis, published as a series of five volumes comprising detailed descriptions and phytosociological tables, allows comparisons with many of the vegetation communities presented in this paper. This has been of most assistance for vegetation that has been little investigated and/or where there are no analogues in the European literature.

The vegetation descriptions in this paper follow strict guidelines and as a result are academic in nature. For this reason, an additional classification heading has been included where possible at the start of each section within the vegetation descriptions. The Heritage Council's Habitats Classification system (Fossitt 2000) was introduced shortly after the completion of the research, but in the intervening years has become an important and widely used document for all those interested in habitats in Ireland. This classification scheme cannot convey the subtle complexities of Clare Island's fascinating vegetation and its floristic diversity. For this reason the data follow the conventions of the European school of phytosociology, as outlined earlier in this paper. However, links with the Heritage Council's Habitat Classification system are included where possible at the start of each section on phytosociology, so that the reader may easily contextualise some of the academic classification.

Compendium of the plant communities

The results of the vegetation analysis are summarised in the following compendium, giving an insight into the complexity of the vegetation recorded from Clare Island. A total of 34 syntaxa are described (although some syntaxon classified to the level of the order are repeated, as they describe vegetation from two different habitats). In the following compendium, some of the communities are classified to alliance or order level only, as the classification of fragmentary or disturbed communities poses many problems. Furthermore, there is a paucity of information for many of the smaller vegetation types from Ireland. Owing to the depauperate nature of the Irish flora, some communities are only tentatively described, as there is an absence of definite character or differential species.

The syntaxonomic classification of the vegetation units follows the scheme presented by White and Doyle (1982), and is only modified where recent research on Irish vegetation had been published at the time of this survey. Owing to the variability of the vegetation on Clare Island, it was necessary to assign a relevé to a syntaxon, even though some or all of the diagnostic species themselves are absent. This may result in a repetition of some syntaxa in different habitats, particularly in more disturbed or anthropogenically altered areas. For this reason, the order of the syntaxa does not follow the sociological progression of vegetation classes outlined for Ireland by White and Doyle (1982); rather they are arranged according to habitat.

Beach Complex (p. 40)

Cakiletea maritimae Tx. et Prsg. 1950
Cakiletalia maritimae Tx. in Oberd. 1949
Salsolo-Honkenyion peploidis Tx. 1950

Ammophiletea Br.-Bl. et Tx. 1943
Elymetalia arenarii Br.-Bl. et Tx. 1943
Agropyro-Honkenyion peploidis Tx. 1945 in Br.-Bl. et Tx. 1952

Koelerio-Corynephoretea Klika in Klika et Novak 1941
Festuco-Sedatalia Tx. 1951
Galio-Koelerion (Tx. 1937) Den Held et Westhoff 1969

Saltmarsh Community (p. 44)

Phragmitetea Tx. et Prsg. 1942
Phragmitetalia Koch 1926 em. Pignatti 1953 denuo em. Segal et Westhoff in Westhoff et Den Held 1969
Phragmition Koch 1926 em. Balátová-Tuláckovă 1963
Scirpetum maritimi (Christiansen 1934) Tx. 1937

Plantain Sward (p. 46)

Saginetea maritimae Westhoff, Van Leeuwen et Adriani 1962
Saginetalia maritimae Westhoff, Van Leeuwen et Adriani 1962
Saginion maritimae Westhoff, Van Leeuwen et Adriani 1962
Plantaginetum coronopodo-maritimi Praeger ex. White 1982

Lake Vegetation (p. 51)

Potametea Tx. et Prsg. 1942
Magnopotametalia Den Hartog et Segal 1964
Magnopotamion Den Hartog et Segal 1964
Nymphaeion Oberd. 1957 em. Neuhausl 1959

Littorelletea uniflorae Br.-Bl. et Tx. 1943
Littorelletalia uniflorae Koch 1926

Oxycocco-Sphagnetea Br.-Bl. et Tx. 1943
Scheuchzerietalia palustris Nordh. 1936
Rhynchosporion albae Koch 1926
Scheuchzerietum Paul 1910

Phragmitetea Tx. et Prsg. 1942
Phragmitetalia Koch 1926 em. Pignatti 1953 denuo em. Segal et Westhoff in Westhoff et Den Held 1969
Phragmition Koch 1926 em. Balátová-Tuláckovă 1963
Scirpo-Phragmitetum Koch 1926 em. Pignatti 1953 denuo em. Segal et Westhoff in Westhoff et Den Held 1969

Littorelletea uniflorae Br.-Bl. et Tx. 1943
Littorelletalia uniflorae Koch 1926
Hydrocotylo-Baldellion Tx. et Dierßen 1972
Scorpidio-Eleocharitetum multicaulis Ivimey-Cook et Proctor 1966

Littorelletea uniflorae Br.-Bl. et Tx. 1943
Littorelletalia uniflorae Koch 1926
Hydrocotylo-Baldellion Tx. et Dierßen 1972
Baldellio-Littorelletum Ivimey-Cook et Proctor 1966

Calluna Heathlands (p. 59)

Calluno-Ulicetea Br.-Bl. et Tx. 1934
Vaccinio-Genistetalia Schubert 1960

Genisto-Callunion (Duvign. 1944) Tx. in Prsg. 1949
Calluno-Ericetum cinereae Lemée 1937

Calluno-Ulicetea Br.-Bl. et Tx. 1934
Ulicetalia minoris (Duvign. 1944) Géhu 1973
Ulici-Ericion cinereae Géhu 1973
Potentillo erecti-Ericetum erigenae Foss 1986

Oxycocco-Sphagnetea Br.-Bl. et Tx. 1943
Sphagnetalia compactii Tx., Miyawki et Fugiwara 1970
Ericion tetralicis Schwick. 1933
Narthecio-Ericetum tetralicis Moore 1968

Grasslands (p. 68)

Nardetea Rivas Goday et Borja Carbonell 1961
Nardetalia Prsg. 1949
Nardo-Galion saxatilis Prsg. 1949
Achilleo-Festucetum tenuifoliae Birse et Robertson 1976

Nardetea Rivas Goday et Borja Carbonell 1961
Nardetalia Prsg. 1949
Nardo-Galion saxatilis Prsg. 1949
Nardo-Caricetum binervis (Pethybridge et Praeger 1905) Br.-Bl. et Tx. 1952

Molinio-Arrhenatheretea Tx. 1937
Molinietalia Koch 1926
Junco-conglomerati-Molinion Westhoff 1968

Molinio-Arrhenatheretea Tx. 1937
Molinietalia Koch 1926
Filipendulion (Duvigneaud 1946) Segal 1966

Molinio-Arrhenatheretea Tx. 1937
Arrhenatheretalia Pawlowski 1928
Arrhenatherion elatioris Br.-Bl. 1925

Molinio-Arrhenatheretea Tx. 1937
Arrhenatheretalia Pawlowski 1928
Cynosurion cristati Tx. 1937

Bog Communities (p. 117)

Oxycocco-Sphagnetea Br.-Bl. et Tx. 1943
Eriophoro vaginati-Sphagnetalia papillosi Tx. 1970
Calluno-Sphagnion papillosi (Schwick. 1940) Tx. 1970
Pleurozio purpureae-Ericetum tetralicis Br.-Bl. et Tx. 1952 em. Moore 1968

Oxycocco-Sphagnetea Br.-Bl. et Tx. 1943
Scheuchzerietalia palustris Nordh. 1936
Rhynchosporion albae Koch 1926
Sphagno tenelli-Rhynchosporetum albae (Osvald 1923) Koch 1926
Littorelletea uniflorae Br.-Bl. et Tx. 1943
Littorelletalia uniflorae Koch 1926
Hydrocotylo-Baldellion Tx. et Dierßen 1972
Hyperico-Potametum oblongi (Allorge 1926) Br.-Bl. et Tx. 1952

Oxycocco-Sphagnetea Br.-Bl. et Tx. 1943
Scheuchzerietalia palustris Nordh. 1936
Rhynchosporion albae Koch 1926
Sphagnum cuspidatum–Eriophorum angustifolium **community** Tx. 1958

Vegetation of Springs and Seepage Zones (p. 140)

Montio-Cardaminetea Br.-Bl. et Tx. 1943
Montio-Cardaminetalia Pawlowski 1928
Cratoneurion Koch 1928
Cratoneuretum filicino-commutati (Kuhn 1937) Oberd. 1977

Scheuchzerio-Caricetea nigrae (Nordh. 1936) Tx. 1937
Caricetalia davallianae Br.-Bl. 1949
Caricion davallianae Klika 1934

Littorelletea uniflorae Br.-Bl. et Tx 1943
Littorelletalia uniflorae Koch 1926

Fen Vegetation (p. 143)

Scheuchzerio-Caricetea nigrae (Nordh. 1936) Tx. 1937 em. Tx. 1980
Caricetalia davallianae Br.-Bl. 1949
Caricion davallianae Klika 1934
Schoenetum nigricantis (Allorge 1922) Koch 1926

Cliff Community (p. 146)

Betulo-Adenostyletea Br.-Bl. 1948
Adenostyletalia Br.-Bl. 1931
Dryoptero-Calamagrostidion purpureae Nordhagen 1943
Luzula sylvatica–Vaccinium myrtillus **association** Birks 1973

Calluno-Ulicetea Br.-Bl. et Tx. 1943
Vaccinio-Genistetalia Schubert 1960
Vaccinio-Callunion Moore in Mhic Daeid 1979
Herberteto-Polytrichetum alpini Mhic Daeid 1979

Montio-Cardaminetea Br.-Bl. et Tx. 1943
Montio-Cardaminetalia Pawlowski 1928
Cratoneurion Koch 1928
Saxifragetum aizoidis McVean et Ratcliffe 1962

Woodland (p. 152)

Franguletea Doing 1962 em. Westhoff 1968
Salicetalia auritae Doing 1962 em. Westhoff 1968
Salicion cinereae Th. Müller et Görs 1958

Franguletea Doing 1962 em. Westhoff 1968
Salicetalia auritae Doing 1962 em. Westhoff 1968
Salicion cinereae Th. Müller et Görs 1958
Salici-Betuletum pubescens Görs 1961

Querco-Fagetea Br.-Bl. et Vlieger 1937
Fagatalia sylvaticae Pawlowski 1928
Circaeo-Alnenion (Oberd. 1953) Doing 1962
Corylo-Fraxinetum Br.-Bl. et Tx. 1952 em. Kelly et Kirby 1981

Alnetea glutinosae Br.-Bl. et Tx. 1943 em. Th. Müller et Görs 1958
Alnetalia glutinosae Vlieger 1937 em. Th. Müller et Görs 1958
Alnion glutinosae (Malcuit 1929) Meijer Drees 1936 em. Th. Müller et Görs 1958
Osmundo-Salicetum atrocinereae Br.-Bl. et Tx. 1952

BEACH COMPLEX

The only appreciable sand deposit found on Clare Island lies at the mouth of the harbour on the eastern side of the island (Pl. I). Here a beach 300m long has developed in the curving bay, which is relatively sheltered owing to its eastern location and the impact of the harbour wall, which protects the area from the ravages of the Atlantic Ocean. The beach complex is small in extent, occupying a fraction of the land cover, approximately 0.5% of the whole island (Doyle and Foss 1986a). Three separate zones in a condensed succession are recognised: the strand-line, foredune and dune grassland communities. The development of a classic dune sequence is interrupted owing to the impacts of climatic and anthropogenic factors, with shifting sand regularly inundating the vegetation. In all, seven relevés are presented for the beach communities (Table 7).

Phytosociology

Strand-line

Cakiletea maritimae Tx. et Prsg. 1950
Cakiletea maritimae Tx. in Oberd. 1949
Salsolo-Honkenyion peploidis Tx. 1950
Table 7, column 1

Salsolo-Honkenyion peploidis differential species: *Honckenya peploides*
Cakiletea maritimae character species: *Cakile maritima*

Heritage Council classification: Sandy shores—LS2

As the sand around drift-lines is stabilised and consolidated, vegetation can establish, occurring some fifteen metres up from the low water mark. This species-poor community (Table 7, column 1) represents the pioneer vegetation on the strand-line. This community occurs on beaches with fine sand where accretion rates may be high but where wind movements may also lead to considerable reworking of the sediment (Gaynor 2008).

The severity of the environment is such that strand-lines are rarely colonised by more than a

Pl. I The extent of the only substantial sand deposits on Clare Island, which occur alongside the harbour.

Table 7
Relevé table for the beach complex on Clare Island

Column	1	2	3	4	5	6	7	Synoptic value
Relevé	156	157	159	158	161	160	201	
Salsolo-Honkenyion peploidis								
Cakile maritima	1				+			II
Honckenya peploides	+		1	1	1			III
Atriplex patula		+		+				II
Agropyro-Honkenyion								
Elymus juncea		4	2	3	3	2		IV
Ammophilion borealis								
Ammophila arenaria				2				I
Galio-Koelerion species								
Festuca rubra					4	3	3	III
Carex arenaria			+		1	2	+	III
Koeleria macrantha					1			I
Galium verum						1		I
Violo curtisii-Tortuletum ruraliformis								
Tortula ruralis ssp. *ruraliformis*						1		I
Cynosurion cristatus								
Lotus corniculatus				+	2	1	2	III
Daucus carota					+	1	+	III
Plantago lanceolata					+	1	2	III
Taraxacum officinale				+	1	2		III
Senecio jacobaea				r		+	1	III
Trifolium repens				+		2		II
Cynosurus cristatus							3	I
Anthoxanthum odoratum							2	I
Elymus repens							2	I
Trifolium pratense							2	I
Holcus lanatus							1	I
Leontodon autumnalis							1	I
Arrhenatherum elatius						1		I
Companion species								
Agrostis stolonifera			2		2			II
Rumex crispus			1		2			II
Bellis perennis						+	+	II
Poa pratensis						2		I
Potentilla anserina					2			I
Atriplex prostrata					1			I
Tripleurospermum maritimum					1			I
Achillea millefolium							+	I
Cerastium fontanum							+	I
Heracleum sphondylium					+			I
Isothecium myosuroides						+		I
Leontodon taraxacoides					+			I
Plantago maritima				r				I
Number of species	2	2	5	9	17	15	15	

few plant species, a finding also reported from around Scottish coasts by Dargie (1993). On Clare Island, only *Cakile maritima* and *Honckenya peploides* are recorded, with some *Atriplex patula* in areas associated with human disturbance. The plants are scattered on the strand, emerging only where tidal litter gathers. Although not found during the present survey, *Salsola kali* was previously recorded (Praeger 1903; 1911; Doyle and Foss 1986a). Praeger (1911) noted that this plant was capricious in its appearance, disappearing for several years before re-emerging on the strand-line with some vigour. This community is assigned to the Salsolo-Honkenyion peploidis of the class Cakiletea maritimae, and is analogous to the *Honckenya peploides–Cakile maritima* SD2 strand-line community in the BNVC classification scheme (Rodwell 2000).

FOREDUNES

Ammophiletea Br.-Bl. et Tx. 1943
Elymetalia arenarii Br.-Bl. et Tx. 1943
Agropyro-Honkenyion peploidis Tx. 1945 in Br.-Bl. et Tx. 1952

Table 7, columns 2 and 3

Agropyro-Honkenyion peploidis character species: *Elymus juncea, Honckenya peploides*

Heritage Council classification: Embryonic dunes—CD1

There is a subtle topographical elevation directly behind the strand-line vegetation, which supports a small discontinuous band of embryonic dunes characterised by *Elymus juncea* and occasional *Honckenya peploides* (Table 7, columns 2–4). The embryonic foredune community described by Praeger (1911) also contained *Salsola kali*. Despite the absence of the character species *Calystegia soldanella, Euphorbia paralias* and *Eryngium maritimum*, the occurrence of *Honckenya peploides* and *Elymus juncea* allows reference to the Agropyro-Honkenyion peploidis, an alliance that is recorded on the landward side of strand-line vegetation (White and Doyle 1982; Cooper *et al.* 1992; Gaynor 2008). The Clare Island community is similar in composition and structure to the species-poor *Elymus juncea* SD4 foredune community described from Britain (Rodwell 2000).

MARRAM DUNES

Ammophiletea Br.-Bl. et Tx. 1943
Elymetalia arenarii Br.-Bl. et Tx. 1943
Ammophilion borealis Tx. (1945) 1952

Table 7, column 4

Ammophilion borealis character species: *Ammophila arenaria*

Heritage Council classification: Marram dunes—CD2

Deposition of eroded sand often covers the foredune vegetation. The elevated sand deposits are colonised by *Ammophila arenaria* (Table 7, column 4), first described on the island by Doyle and Foss (1986a). The environmental conditions and the relatively narrow extent of the beach on Clare Island results in a complex overlapping mosaic of foredune vegetation of the Agropyro-Honkenyion peploidis and the Ammophilion borealis, which is typical of yellow or building dune systems. Similar observations have been made in western Scotland, where a number of dune associations overlap and can occur simultaneously in the foredune sequences (Gimingham 1964; Birks 1973). The marram grass growing on the relatively low dunes on Clare Island is not as robust as that found on typical yellow dunes elsewhere in Ireland, reaching only 40cm in height, compared with about one metre in many other areas. It is unlikely that this precursor yellow dune community will become fully established on Clare Island or will develop into a habitat of considerable height, such as might be expected for mature marram dune habitat in other more extensive coastal sites (Gaynor 2008), given the frequency with which hostile environmental conditions affect the beach.

The vegetation has affinities with the Ammophilion borealis, analogous to the *Ammophila arenaria* SD6 mobile dune community (Rodwell 2000) that is generally recorded on the landward side of *Elymus juncea* foredunes (Braun-Blanquet and Tüxen 1952; Schouten and Nooren 1977; White and Doyle 1982).

Praeger (1911) also described a second minor community from the foredune habitat that included *Potentilla anserina, Senecio jacobaea, Plantago lanceolata* and *Taraxacum officinale*. These species were occasionally recorded on the eroding faces of the banks at the front of the dune grassland, but given the nature of the habitat they are highly ephemeral and do not constantly occur from year to year.

DUNE GRASSLAND

Koelerio-Corynephoretea Klika in Klika et Novak 1941
Festuco-Sedetalia Tx. 1951
Galio-Koelerion (Tx. 1937) Den Held et Westhoff 1969

Table 7, columns 5–7

Galio-Koelerion character species: *Festuca rubra, Galium verum*

Festuco-Sedetalia differential species: *Koeleria macrantha*

Koelerio-Corynephoretea diagnostic species: *Carex arenaria*

Companion species: *Daucus carota, Senecio jacobaea, Lotus corniculatus, Plantago lanceolata, Taraxacum officinale* agg.

Heritage Council classification: Fixed dunes—CD3

Dune grasslands occupy the greater part of the beach complex and are quite distinct from the previous communities. In general, dune grasslands are considered more stable than the communities found on seaward side or bare sand (Gaynor 2008). They are less dynamic owing to the infrequency of sand redistribution and the nature of the vegetation that stabilises the community. This does not always follow for the Clare Island community, given the condensed nature of the beach habitats from the seaward to landward side (at most twenty metres but usually less then ten metres).

The low growing sward is characterised by grasses and dicotyledonous herbs (Pl. II A). Grazing pressure is light, created by a few cattle that occasionally wander over the dunes. Anthropogenic features influencing the physiognomic structure of the community include visiting tourists that camp there, the storage of material and equipment by the local authority and occasional use as a car-park during busy periods at the harbour.

Table 7, columns 5–7 describe this species-rich community. The vegetation is assigned to the Galio-Koelerion (class Koelerio-Corynephoretea) based on the dominance of *Festuca rubra* and the occasional occurrence of *Galium verum*, *Carex arenaria* and *Koeleria macrantha*. The Galio-Koelerion is well represented in Ireland (Gaynor 2008), comprising coastal grasslands that are subjected to grazing by rabbits or have been converted to golf-links. Cattle are the only grazers of the Clare Island community. Other common species include *Lotus corniculatus*, *Daucus carota*, *Senecio jacobaea* and *Plantago lanceolata*. These species are more heavily concentrated on subsiding dune faces (column 5).

The seaward sides of the dunes are subject to blowouts and removal of sand (Pl. II B). These disturbed areas support vegetation that shows some similarities with the Festuco-Galietum maritimi. Under the BNVC classification scheme, the *Festuca rubra–Galium verum* fixed dune SD8 grassland encompasses the Galio-Koelerion and the Festuco-Galietum maritimi vegetation, despite the absence of the diagnostic species listed by White and Doyle (1982), which include *Bromus hordeaceus* and *Polygala vulgaris*. These areas are frequently colonised by *Carex arenaria*, *Galium verum* and *Thymus praecox*.

Pl. II A Dune grassland located on the landward side of the foredunes. The community is typified by several species including *Festuca rubra*, *Daucus carota*, *Taraxacum officinale* agg., *Carex arenaria* and *Rumex crispus*.

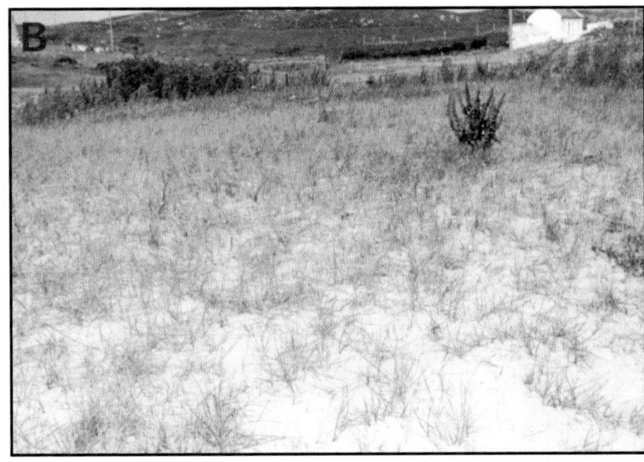

Pl. II B Disturbed area at the seaward part of the dunes where wind-blown sand is stabilised by *Carex arenaria* and *Elymus juncea*.

Elements of the Violo curtisii-Tortuletum ruraliformis association are found in open, sandy conditions. While *Viola tricolor* ssp. *curtisii*, a character species of the association, was not found on Clare Island, the moss *Tortula ruraliformis*, another character species, was found. This association, first described from Irish stands (Braun-Blanquet and Tüxen 1952) and since confirmed from other locations (Ivimey-Cook and Proctor 1966; Beckers *et al.* 1976; Ní Lamhna 1982), is a common feature of the seaward boundary of dune grassland.

As the vegetation cover increases and open sand becomes less prevalent, elements of mesotrophic grasslands characterised by the increased abundance of Cynosurion cristati species occur (Table 7, columns 6 and particularly 7). The meadow species are all quite stunted and only grow tall at the landward side of the dunes, where tourists and cattle seldom venture. The meadow species include *Arrhenatherum elatius*,

Cynosurus cristatus and *Anthoxanthum odoratum*. Further back the mature dune grassland merges into Cynosurion grasslands, comparable to the *Arrhenatherum elatius* MG1 grassland and the *Cynosurus cristatus–Centaurea nigra* MG5 grassland (Rodwell 1992). This under-utilised area of dune grassland is similar in composition and appearance to the few remaining examples of hay meadows left on Clare Island, described under the grasslands section.

SALTMARSH COMMUNITIES

Saltmarshes generally develop on alluvial sediments bordering saline water bodies whose water table fluctuates, usually tidally. During the original survey on Clare Island the only area of saltmarsh vegetation described by Praeger (1911) was located on the eastern tip of the island at Kinnacorra behind a distinctive V-shaped boulder beach.

The small saltmarsh community at Kinnacorra remains and is regularly affected by sea spray and inundated with salt water. There are no obvious flows of freshwater through the saltmarsh area. Despite the relatively short distance from the front of the marsh to the back, the soil water content decreases from 87% to 54%. An analogous trend is seen in organic content, which varies from 56.5% at the relatively wet front of the marsh to 10% in the drier upper marsh. The organic debris is derived from both rotting plant material and tidal debris trapped by the tall plants at the front of the marsh.

Other saltmarsh vegetation has since been noted on Clare Island (Doyle and Foss 1986a): *Puccinellia maritima*, newly recorded on Clare Island in 1984, was found forming a proto-Puccinellietum community on the mudflats behind the beach complex at the harbour. This species was recorded in the current survey.

Floristics

The saltmarsh vegetation described by Praeger, which was dominated by *Bolboschoenus maritimus* (*Scirpus maritimus*), is still found behind the boulder beach at Kinnacorra. The tall-growing (approx. 90cm) *Bolboschoenus* provides dense cover. The only other tall-growing plant found there is *Juncus maritimus*, which is confined to areas of standing water at the seaward side of the marsh. Species typical of the marsh vegetation include *Triglochin maritima* and *Plantago maritima*, although they are poorly represented. The mud rush, *Juncus gerardii*, was located in the area in 1994, but was not seen again during the project, reflecting the severity of impact of storms in this area. More commonly encountered are herbaceous species of both damp grasslands and small sedge communities. With the exception of *Agrostis stolonifera*, which is occasionally found in the undergrowth, most of the other plants are confined to the sheltered margins at the landward side of the community. These include *Trifolium repens*, *Sagina procumbens*, *Hydrocotyle vulgaris*, *Galium palustre* and *Ranunculus flammula*. Nardetalia species, including *Festuca ovina*, occur where the vegetation grades into severely grazed rough agricultural land. No bryophytes were recorded from this community.

Phytosociology

While saltmarshes found on western coasts form distinctive communities that differ from the surrounding vegetation, they are not always readily classified (White and Doyle 1982; Rodwell 2000). Many of the species found on western saltmarshes are not confined to this habitat. A further difficulty arises because saltmarsh communities are not fully described for Ireland (White and Doyle 1982).

SALTMARSH VEGETATION

Phragmitetea Tx. et Prsg. 1942

Phragmitetalia Koch 1926 em. Pignatti 1953 denuo em. Segal et Westhoff in Westhoff et Den Held 1969

Phragmition Koch 1926 em. Balátová-Tuláckovà 1963

Scirpetum maritimi (Christiansen 1934) Tx. 1937

Table 8

Scirpetum maritimi character species: *Bolboschoenus maritimus*

Diagnostic species: *Juncus maritimus*, *Juncus gerardii*

Heritage Council analogue: Upper salt marsh—CM2

Three relevés were taken in this vegetation (Table 8): one was taken in a pure stand of the vegetation (column 3), while the remaining two relevés (columns 1 and 2) represent the floristically richer margins of the saltmarsh vegetation. The Clare

Island vegetation dominated by *Bolboschoenus maritimus* is classified in the Scirpetum maritimi, analogous to the typical and *Agrostis stolonifera* variants of the *Scirpus maritimus* S21 swamp (Rodwell 2000). First recorded from Wexford by Braun-Blanquet and Tüxen (1952), the widespread distribution of this association has since been confirmed by several authors (Ivimey-Cook and Proctor 1966; Brock *et al.* 1978).

Table 8

Relevé table for the Scirpetum maritimi

Column	1	2	3	
Relevé	96	95	200	
				Synoptic value
Scirpetum maritimi				
Bolboschoenus maritimus	1	3	5	V
Juncus maritimus	2			II
Companion species				
Agrostis stolonifera	1		+	IV
Potentilla palustris			+	II
Sagina procumbens	+	2		IV
Hydrocotyle vulgaris	+	1		IV
Ranunculus flammula	1	+		IV
Trifolium repens	+	+		IV
Galium palustre	+	+		IV
Juncus articulatus	2			II
Eleocharis palustris	2			II
Iris pseudacorus	1			II
Carex nigra	1			II
Festuca ovina		3		II
Carex panicea		2		II
Potentilla erecta		1		II
Calliergonella cuspidata		+		II
Carex viridula		+		II
Eleocharis multicaulis		+		II
Plantago coronopus		+		II
Plantago maritima		+		II
Taraxacum officinale		+		II
Triglochin maritima		r		II
Number of species	12	16	3	

The margins of the Clare Island community would appear to be related to the Junco maritimi-Oenanthetum lachenalii (class Asteretea tripolii), since there is a distinct lack of tall growing halophytes typical of the Phragmitetea. Under the BNVC scheme, it is referable to the *Juncus maritimus* SM18 saltmarsh community (Rodwell 2000). The Junco maritimi-Oenanthetum lachenalii has been described in detail from British saltmarshes from supra-littoral zones growing on silty substrates in estuaries (Adam 1977; 1978). A community assignable to this association was recorded by Schouten and Nooren (1977) from counties Wexford and Waterford. Adam (1977) placed the association in the Armerion maritimae, with vegetation of upper saltmarshes, based not only on species assemblage but also on the level of grazing. Diagnostic species for that saltmarsh community include *Juncus maritimus* and *Bolboschoenus maritimus* along with *Oenanthe lachenalii, Juncus gerardii* and *Apium graveolens* (this last species was only recorded from one other habitat on Clare Island).

SALINE FLATS (NO RELEVÉS PRESENTED)

Heritage Council analogue: Lower salt marsh—CM1

A small saline flat was described from the harbour area. This lay behind the drainage channel that flows onto the beach (G. Doyle, pers. comm.). The saline flat is surrounded by damp grassland that is occasionally grazed by cattle. Doyle and Foss (1986a) recorded *Puccinellia maritima* and *Apium graveolens* growing there. These were new records for Clare Island at that time. Later examination of the site has shown it to be significantly degraded. The saline flat was hardly recognisable, as the area was inundated by sand deposited there during storms in 1993 and 1994 (P. Gill, pers. comm.). A small amount of *Puccinellia maritima* remains, occurring patchily on bare muddy ground among damp dune grassland along a stream bank at the back of the beach.

SHINGLE BEACHES (NO RELEVÉS PRESENTED)

Heritage Council analogue: Shingle and gravel banks—CB1

While a number of species can withstand the salt concentrations experienced in marine marsh vegetation, fewer still manage to successfully establish on shingle or boulder beaches (Randall 1989; Packham and Willis 1997), where constant movement of the substrate alters the habitat with nearly every tidal inundation. Species known to occur in such habitats include *Rumex crispus, Atriplex prostrata, Tripleurospermum maritimum, Sedum anglicum* and *Polygonum maritimum*, especially on or just above any drift lines that may occur. Only *Rumex* and *Atriplex* are commonly encountered on shingle beaches on Clare Island, occurring only as widely scattered individuals on such habitats.

During the summer of 1993, two small straggling specimens of *Lathyrus japonicus* were discovered on a shingle beach at a small cove near Portnakilly. The species is included in the *Red Data Book for Ireland* (Curtis and McGough 1988) and is considered a differential species of the Elymo-Ammophiletum on mobile coastal dunes in Europe (White and Doyle 1982). The BNVC classification lists this habitat as *Rumex crispus–Glaucium flavum* SD1 shingle community (Rodwell 2000).

The stability of shingle beaches is strongly affected by the hydrological activity and the nature of the substrate and its particle size (Packham and Willis 1997). The transitory nature of sand-based habitats meant that the species has been generally under-recorded (Minchin and Minchin 1996; T. Curtis, pers. comm.). Within a week of the discovery of *Lathyrus* at Portnakilly, the sea had scoured the site, and despite regular searching of that site and other similar habitats, this elusive plant was not refound (Ryle and Doyle 1998).

PLANTAIN SWARD

This distinctive community is found on Clare Island at elevations ranging from sea level to 100 metres. The greatest expanse of the sward is located at the extreme western end of the island, near Kinatevdilla (Beetle Head). Elements of the sward can be found at higher elevations, but these are transitional with coastal grasslands, located above the cliffs, at the lighthouse at the northernmost tip of the island (Pl. III).

The plantain community develops on peaty or sandy peaty soils (approx. 10cm deep), derived from coarsely textured sandstone. The shallow soils that have developed over these permeable rocks are subject to constant maritime inputs with the result that only a small number of plant species can survive in the extreme nature of this exposed coastal habitat. Standing water rarely gathers, dissipating rapidly through the vegetation and underlying lithologies. The vegetation is of low stature, about 1–3cm tall, owing to the constant exposure to sea spray, persistent strong winds and grazing.

Pl. III Plantain sward at extreme western end of the island among extensively grazed acid grassland. Much of the ground in the central parts of the photograph is dominated by *Plantago coronopus* and *P. maritima*, with minor contributions from other vascular plants.

The plantain sward is not as heavily stocked as other areas of the surrounding commonage, since grazing for already sparse grasses is difficult in the dwarf turf. Personal observations have revealed that sheep tend to gather and shelter from inclement weather behind the many abandoned turf clamps that are located on the plantain sward. The faecal material that accumulates must have a significant impact on the nutrient regime within the community.

Floristics

The community consists almost entirely of *Plantago maritima* and *P. coronopus*, with *Festuca rubra* (sometimes *Festuca ovina*) and *Plantago lanceolata* the only other constantly occurring species of the sward. Other species such as *Radiola linoides, Carex viridula* and *Sagina procumbens* are regular components of the vegetation. Within the community there is considerable species diversity with a variety of short-lived annuals and dwarf therophytes—Praeger listed some 28 such species for the community. Many of these species are not exclusive to the plantain sward. Most are grassland species, but some heathland elements occur at higher elevations. Grasses that increase in frequency at greater distances from the sea include

Agrostis capillaris, Holcus lanatus, Aira praecox and *Koeleria macrantha*. A considerable number of small herbaceous species are important in characterising the community, although few occur with any great abundance or regularity. Such species include *Anagallis tenella, Carex panicea, Hypochaeris radicata, Hydrocotyle vulgaris, Jasione montana* and *Sedum anglicum*. Bryophytes and lichens may form an important feature of the sward, although their total cover is variable too. The more common species include *Hypnum jutlandicum* and *Mnium hornum*, with *Frullania tamarisci* growing through the higher plants.

Phytosociology

Saginetea maritimae Westhoff, Van Leeuwen et Adriani 1962
Saginetalia maritimae Westhoff, Van Leeuwen et Adriani 1962
Saginion maritimae Westhoff, Van Leeuwen et Adriaini 1962
Plantaginetum coronopodo-maritimi Praeger ex. White 1982 Table 9

Plantaginetum coronopodo-maritimi character species: *Plantago coronopus, Plantago maritima, Plantago lanceolata*

No analogue in Heritage Council Classification

Praeger (1903; 1911) first described this plantain-dominated maritime community from Clare Island. Gimingham (1964) described analogous sea cliff vegetation from Atlantic coasts of Scotland, where it was confined to more steeply dipping cliffs. While characterised by the presence of *Festuca rubra* and the three *Plantago* species, that community was dominated by *Armeria maritima*.

The Festuco-Armerietum rupestris subassociation with *Plantago coronopus* was recorded from ungrazed or inaccessible maritime cliffs of the Lizard Peninsula in Cornwall (Malloch 1971). Although cattle were prevented from grazing in that community, Malloch noted that the increase in *Plantago* cover was due to grazing by sheep. Ivimey-Cook and Proctor (1966) defined two separate *Plantago*-dominated associations from the Burren in County Clare: their *Carex distans–Plantago maritima* and the *Cerastium diffusum–Plantago coronopus* associations lacked the species diversity of Praeger's plantain community. Other authors have described different coastal communities that have one or other of the *Plantago* species and have usually referred their communities to the Asteretea tripolii. For example, Braun-Blanquet and Tüxen (1952) recorded a *Plantago coronopus–Cerastium diffusum* association. This is quite different from the *Plantago* sward since *Plantago maritima* was poorly represented and *Armeria maritima* and *Carex distans* have high cover. The latter species are more typical of the Armerion communities from sea cliffs. The closest BNVC counterpart is *Festuca rubra–Plantago* spp MC10 maritime grassland with elements of both the *Armeria maritima* and *Carex panicea* subcommunities.

None of these descriptions seems to provide a strict comparison with Praeger's *Plantago* sward, although Gillham in an elegant series of papers (Gillham 1953; 1954; 1955; 1956a; 1956b; Goodman and Gillham 1954) described what she considered an unusual type of maritime grassland on Grassholm island, south Pembrokeshire, which had at the time only appeared to have been previously recorded on Worm's Head, south Wales, and Clare Island, western Ireland. However, the underlying geology on the Pembrokeshire islands is basaltic—a basic rock type not found on Clare Island.

The habitat on Clare Island was syntaxonomically validated and defined as a community of exposed or low-lying coastal cliffs in western Ireland (White 1982), classified as the Plantaginetum coronopodo-maritimi, in the Saginion maritimae and Saginetalia maritimae of the class Saginetea maritimae. The inclusion of this association in the Saginetea maritimae, as opposed to the Armerion maritimae, vegetation of upper salt-marsh zone, seems justified based on the floristic composition and the general physiognomy of the vegetation. In general, *Armeria maritima* is infrequently recorded, increasing in abundance where fencing prevents grazing or on cliff ledges.

The plantain sward, as found on Clare Island, is a low-growing community characterised by *Plantago maritima* and *P. coronopus* (Table 9). These species dominate the vegetation, providing at least 90% of the ground cover. Praeger recognised two forms of the plantain sward (Praeger 1911; White 1982). The first was the 'pure' plantain sward with only a few other constant species, while the second had a greater diversity of grassland

Table 9
Relevé table for the Plantaginetum coronopodo-maritimi

Column	1	2	3	4	5	6	7	8	9	10	11	12	13	14	15	16	17	18	19	20	21	Synoptic value
Relevé	22	114	378	167	23	181	168	126	175	377	112	113	35	375	115	128	133	376	180	125	178	
Plantaginetum coronopodo-maritimi																						
Plantago maritima	4	4	4	4	5	4	4	4	4	5	3	4	4	2	2	2	5	3	4	2	3	V
Plantago coronopus	3	3	+	3	3	2	2	3	1	2	2	2	3	+	2	2	+	1	+	2	+	V
Plantago lanceolata			+			2	+	1	1	+	+		+			+	1	+	1		1	IV
Radiola linoides						+		+		+		1	1	1	1	+	+	+	1	+	+	IV
Carex viridula					+	1	1	2		+	1		+			1	+		2	2	1	III
Sagina procumbens		+				1					+	1	+			+						II
Sedum anglicum			+									+	+		2	+	r					II
Mnium hornum		+										+		+	1		+	+				II
Aira praecox											+	+		2	2	+	1	1				II
Potentilla erecta								1						+		+	+	+	1	+	2	II
Anagallis tenella											1		+		+	2		1	1	+	1	II
Hypnum jutlandicum													1	1		3	2	2	1			II
Luzula campestris/multiflora														+	+		+	+	+			II
Cerastium fontanum						+			+			r										I
Euphrasia officinalis			+							+								+				I
Sagina maritima	2																					R
Centaurium erythraea																					r	R
Constant species																						
Festuca ovina/rubra	2	2	1	3	2	3	2	3	4	1		1	2	4	2	3	2	2	3	2	4	V
Agrostis capillaris				1	2			2		+	2		2	3	3	2	2	3	2	2	2	IV
Armerion maritimae																						
Armeria maritima	2	2	2	1		1						1										III
Montio-Cardaminetea species																						
Bryum pseudotriquetrum											2				+							+
Juncus acutiflorus											2											R
Pellia epiphylla											2											R
Philonotis fontana											2											R
Campylium elodes											1											R
Cratoneuron filicinum											1											R
Nardetalia species																						
Danthonia decumbens								2												2	1	I
Galium saxatile														+						+		+
Nardus stricta																				3		R
Companion species																						
Leontodon autumnalis									2	+	+							r	+	+	1	II
Hydrocotyle vulgaris							+				+					+			+	+	+	II

(Continued)

Table 9 *(Continued)*

Column	1	2	3	4	5	6	7	8	9	10	11	12	13	14	15	16	17	18	19	20	21	
Relevé	22	114	378	167	23	181	168	126	175	377	112	113	35	375	115	128	133	376	180	125	178	**Synoptic value**
Frullania tamarisci										1				+			2	+	+			II
Carex panicea								+						+			+	+		+	+	II
Carex nigra						1			+	+						1			+			II
Koeleria macrantha			+	+		1													+		1	II
Agrostis stolonifera		1		1			1				1											I
Bellis perennis								+			+								+		2	I
Hypochaeris radicata				+			+						+			+					+	I
Thymus praecox								1							+					+	2	I
Trifolium repens			+			+			1											+		I
Holcus lanatus						2			1										2			I
Juncus bulbosus											1					1		+				I
Rhytidiadelphus squarrosus											1								+		+	I
Jasione montana														+	+		+					I
Lotus corniculatus									2				1									+
Glaux maritima				2			1															+
Sphagnum subnitens											+									2		+
Calluna vulgaris														+				+				+
Cladonia portentosa													+				+					+
Eleocharis multicaulis																		+		+		+
Eurhynchium praelongum																			+		+	+
Prunella vulgaris											+								+			+
Ranunculus flammula										+										+		+
Viola riviniana																				r	+	+
Campylopus introflexus																				2		R
Diplophyllum albicans																	2					R
Campylopus paradoxus								1														R
Isolepis setacea											1											R
Cirsium dissectum																					1	R
Cladonia floerkeana															1							R
Cladonia uncialis															1							R
Cochlearia officinalis									1													R
Hypnum cupressiforme										1												R
Campylium stellatum							+															R
Campylopus atrovirens																		+				R
Campylopus fragilis										+												R
Cardamine flexuosa											+											R
Cerastium glomeratum			+																			R
Cladonia pyxidata															+							R
Dicranella heteromalla							+															R
Dicranum scoparium																	+					R
Fissidens adianthoides											+											R

(Continued)

Table 9 *(Continued)*

Column	1	2	3	4	5	6	7	8	9	10	11	12	13	14	15	16	17	18	19	20	21	Synoptic value
Relevé	22	114	378	167	23	181	168	126	175	377	112	113	35	375	115	128	133	376	180	125	178	
Peltigera canina										+												R
Pohlia nutans															+							R
Polytrichum commune														+								R
Rumex acetosa			+																			R
Samolus valerandi											+											R
Thuidium tamariscinum																				+		R
Number of species	5	7	11	9	5	13	10	13	11	15	27	10	15	16	17	17	19	20	21	21	22	

species. White (1982) regarded *Sagina maritima* as a character species of the community as his research showed it to be more commonly distributed than *S. procumbens*. On Clare Island *Sagina procumbens* is the more common species, although *S. maritima* appears to be the preferential species in stands that develop at sea level.

ARMERIA-RICH SUBTYPE

An extreme form of the sward is located on the seaward side of the community. It is not widespread, existing only as a narrow fringe at sea level. This depauperate community has on average just five species (Table 9, columns 1–3). Both of the plantains occur, together with *Festuca rubra,* and some minor cover is contributed by *Agrostis capillaris*. The only other constant occurrence is the dwarf form of *Armeria maritima,* which has up to 25% ground cover. This depauperate form of the community is similar to that recorded by Gimingham (1964) on coastal cliff tops and may have closer affinities with the Armerion rather than the Saginetea maritimae.

TYPICAL SUBTYPE

Therophytes become more conspicuous within this Clare Island community with distance from the sea. Typically, ten species per square metre are found within any stand of the 'pure' sward (Table 9, columns 4–10). Other than *Plantago coronopus* and *P. maritima,* constant occurrences include *Festuca rubra, Plantago lanceolata* and *Carex viridula,* with *Radiola linoides* and *Sedum anglicum* occasionally found in the sward. Some of the species listed as frequent occurrences by White (1982), such as *Euphrasia* spp and *Cerastium fontanum,* are less abundant in the community on Clare Island. Their restricted distribution there may be linked with the removal of turf from these areas in earlier times.

FLUSH SUBTYPE

Relevé 112 (Table 9, column 11) was made in a flush, located within the plantain sward. The vegetation was distinct from the plantain sward, occurring only where there was obvious seepage of water through layers within the sedimentary rocks. The vegetation is still classified within the plantain sward (total plantain cover was 50%), although it has clear affinities with the Montio-Cardaminetea since the vegetation is dominated by bryophytes, including *Philonotis fontana, Campylium stellatum, Cratoneuron filicinum, Bryum pseudotriquetrum* and *Pellia epiphylla*. Vegetation belonging to the Montio-Cardaminetea is represented in other spring habitats on Clare Island where bryophyte-dominated vegetation develops.

SPECIES-RICH VEGETATION

The species-rich form of the plantain sward (Table 9, columns 12–19) remains dwarfed, owing in part to sea-spray, but also to the impacts of grazing.

Sea plantain (*Plantago maritima*) remains constant, with 80% ground cover. There is a decline in *P. coronopus* with competition from grassland species and increased grazing pressure. Although never far-removed from the sea, *Festuca rubra* is replaced by *F. ovina* and *Agrostis capillaris* the further inland that the vegetation develops. This species-rich form of the community is found on sites that are slightly more sloping than those that support the 'pure sward'. There is a greater abundance of *Radiola linoides*, *Anagallis tenella*, *Carex viridula* and *Potentilla erecta*, while *Aira praecox*, *Luzula multiflora*, *Hypnum jutlandicum* and *Mnium hornum* are common elements of the community.

White (1982) concludes that such grassland admixtures in the plantain sward occur in situations less exposed to sea-spray. Malloch however, describing his *Plantago maritima* nodum, felt that the rate of spray deposition was lower in this nodum than the Festuco-Armerietum sward and that impeded drainage encouraged a greater species richness in this community. From the evidence on Clare Island, drainage does appear to be a factor as there is lateral runoff of excess water into the shallow stream basin that drains out near Kinatevdilla.

Surprisingly on Clare Island, small stands of a transitional plantain/coastal grassland community can occur up to 125 metres above sea level in gullies along the northern cliffs from the lighthouse to the base of Knockmore. Species numbers average fifteen per square metre (Table 9, columns 12–19,), although in stands that encroach upon heathland communities (columns 20 and 21) this rises to nineteen species.

NARDETALIA SUBTYPE

This final subcommunity (Table 9, columns 20 and 21) is confined to the western slopes of Knockmore, above the elevations at which the 'pure sward' is located. It is rich in Nardetalia species including *Nardus stricta* and *Danthonia decumbens*, and there is also a greater frequency of species such as *Koeleria macrantha*, *Hydrocotyle vulgaris*, *Bellis perennis* and some *Leontodon autumnalis* than in this last group. The community is lacking many of the constant species typically associated with the plantain sward on Clare Island, including *Hypnum jutlandicum*, *Mnium hornum* and *Sedum anglicum*.

LAKE VEGETATION

There are surprisingly few water bodies on Clare Island, despite the underlying impervious, siliceous geology. These include five permanent lakes and numerous smaller, ephemeral water bodies that only appear during wet spells. Notwithstanding this fact, the vegetation surrounding the open water bodies on Clare Island frequently displays zonation that reflects the transition from emergent aquatic communities through to colonising communities of terrestrial lakeshores. The composition and structure of the vegetation is controlled by a number of factors: (a) water depth and the nature of the lake bottom, (b) water circulation resulting from wind action, (c) groundwater fluctuations and water recharge from surrounding areas, (d) nutrient content of the water and (e) anthropogenic influences.

The vegetation of the permanent lakes on Clare Island is highly modified owing to the impact of man: the lakeshores are heavily grazed. As a consequence, the vegetation manifests itself as incomplete sequences that rarely extend around the entire perimeter of the lakes.

The five permanent lakes described by Praeger (1911) have undergone considerable change in their appearance and vegetation. Of the three small lakes originally described in the centre of the island, two are now completely infilled by a floating mat of vegetation dominated by bryophytes (Lough Merrignagh and Lough Leinapollbauty). The remaining lake, Creggan Lough, is a small open lake that displays a vegetation zonation ranging from the floating plants of relatively deep waters through to emergent vegetation.

The fourth lake, Poirtín Fuinch, is the largest body of water on the island and is structurally different from all other lakes (Pl. IV A). It occurs in a depression in the drumlin landscape and is located in commonage between Maum, Capnagower and Glen townlands, in an area rich in *Rhynchospora*-dominated blanket bog. No account of this water body was presented in Praeger's (1911) report of the island's vegetation. The lake is characterised by numerous small inaccessible peat islands. There are marked vegetational differences between the lake islands and the adjacent bog areas, a finding that was also reported from other peat islands in County Mayo (Doyle *et al.* 1987). These authors recorded a species-rich heather-dominated vegetation, with *Juniperus communis*, from the islands, in

Pl. IV Two of the five water bodies on Clare Island: A) Poirtín Fuinch, with numerous small peat islands. The marginal vegetation surrounding the lake is dominated by soft rush. The most conspicuous plants include yellow flag iris on the islands, while broad-leaved pondweed occurs in relatively sheltered corners of the lake; B) reed swamp at Lough Avullin. Over the past 80 years, these pale- coloured reeds, 3m high, have completely infilled this lake.

a marked contrast to the species-poor graminoid-dominated bog vegetation of the surrounding open areas. The peat islands at Poirtín Fuinch support vegetation that is completely different to that described by Doyle *et al.* (1987), and have moderately well-developed *Agrostis capillaris-Festuca ovina* swards (Achilleo-Festucetum tenuifoliae). Despite the lack of obvious grazing, the development of the grass sward suggests that either (a) access to the islands was previously possible or (b) that the water body has increased in area owing to peat extraction. Other obvious features of the peat islands include the occurrence of well-developed taller herbs such as *Iris pseudacorus* and *Lythrum salicaria*.

At the time of the original survey, Lough Avullin was an open body of water as deep as 2.5 metres in places and surrounded by a fringe of *Phragmites australis* and *Schoenoplectus lacustris* (Praeger 1911). Today the lake is completely infilled with an extensive reed bed and is inundated with water only when flooded by the encircling river. It is likely that a reduction in the water velocity occurred with the abandonment of the mill-race that is located downstream from the lake, and this resulted in accelerated accumulation of sediment in the lake.

The distinctive appearance of the reed swamp makes it stand out from the surrounding landscape and vegetation (Pl. IV B). Another smaller stand of *Phragmites australis*, measuring five square metres, occurs in a damp meadow overlooking the harbour on Clare Island, where a depauperate community, composed entirely of reeds, is found growing on a mound of boulders. It is possible that these reeds are a remnant of vegetation that once occupied a wetter area that was infilled by boulders. There is a suggestion that the stone mound found in this disturbed field is of archaeological significance (P. Gosling, pers. comm.).

The ephemeral lakes that are widely distributed over the island generally support vegetation that cannot be distinguished phytosociologically from that of the surrounding areas, despite the increase in the height of the water table with its concomitant flushing and localised nutrient enrichment. Several of the ephemeral lakes near Toormore, however, undergo considerable fluctuations in water level and support a characteristic vegetation assigned to the Baldellio-Littorelletum (relevé 139, presented separately).

Phytosociology

The classification of the vegetation of the lakes reflects the pattern of zonation of the five permanent lakes on Clare Island (Table 10). The relevés are separated into several groups reflecting differences in floristic composition, hydrological regime and structural ecology of the lake vegetation zonation. A feature common to all the lakes on Clare Island is the discontinuous zonation. Some lakes support a single syntaxon, while others have an arrested successional sequence reflecting the degree of water inundation, the amount of shelter, the peat depth and the level of grazing that occurs down to the water's edge.

While there are clear transitions between separate zones, the presence of transgressive species occurring in other vegetation units often leads to overlap in phytosociological communities. This concurs with other vegetation descriptions from shallow bog lakes in Ireland (Beckers *et al.* 1976; Brock *et al.* 1978; Van Groenendael *et al.* 1979; Doyle 1982a), where there seems to be little consensus as to the classification owing to the local edaphic and management conditions.

Table 10
Relevé table for the lakeside communities on Clare Island

Column	1	2	3	4	5	6	7	8	Synoptic value
Relevé	145	144	142	143	330	331	337	338	
Nymphaeion									
Nymphaea alba	3								I
Littorelletalia uniflorae									
Potamogeton natans	2	2							II
Myriophyllum alterniflorum	3	4							II
Littorella uniflora		2							I
Apium inundatum		1							I
Caricetum rostratae									
Carex rostrata	1	2	1	1	2				IV
Scheuchzerio-Caricetea nigrae									
Eleocharis multicaulis	2	1	1	2	2	2	1	+	V
Carex limosa		2	2	2	1		+	2	V
Juncus bulbosus		1	2	2	+	+	2	1	V
Hydrocotyle vulgaris		1	1	1	+	+		+	V
Ranunculus flammula	1	1	1	2	1				IV
Scheuchzerietum									
Sphagnum cuspidatum		1	2	3	5	5	5	5	V
Companion species									
Hypericum elodes	2	1	3	2	3	r			V
Scorpidium scorpioides	3	2	2						III
Agrostis stolonifera	1	1					2		III
Anagallis tenella		+	1	2					III
Mentha aquatica		+	+	+	r				III
Carex echinata		1	+			1	1	2	IV
Menyanthes trifoliata		1			+	+	+	+	IV
Drosera rotundifolia				+	+	+	+		III
Nardus stricta				1		1	2		III
Sphagnum palustre					+	1	2		III
Sphagnum auriculatum				2				2	II
Callitriche stagnalis			1	2					II
Drosera anglica				1	1				II
Anthoxanthum odoratum				1		1			II
Juncus effusus			1					+	II
Sphagnum capillifolium					1			+	II
Polytrichum commune						+	1		II
Aneura pinguis				1	+				II
Epilobium palustre			+	+					II
Aulacomnium palustre					+		+		II
Carex viridula				+	r				II
Calliergonella cuspidata		2							I
Agrostis capillaris						2			I
Sphagnum subnitens								2	I
Potamogeton polygonifolius			1						I
Sphagnum tenellum			1						I
Galium palustre				+					I
Eriophorum angustifolium					+				I
Odontoschisma sphagni					+				I
Potentilla erecta					+				I
Juncus squarrosus							+		I
Dicranum majus							+		I
Centaurea nigra					+				I
Juncus acutiflorus								+	I
Carex panicea						+			I
Viola palustris						+			I
Carex nigra						+			I
Narthecium ossifragum						+			I
Eurhynchium praelongum						+			I
Festuca ovina						+			I
Number of species	9	18	16	21	21	20	14	12	

The main classes include the Potametea, Littorelletea and Scheuchzerio-Caricetea nigrae. The Littorelletea and Scheuchzerio-Caricetea are common components on flooded soils in a variety of habitats, while the Potametea is readily distinguishable, even though it is poorly investigated in Ireland (White and Doyle 1982).

FLOATING VEGETATION

Potametea Tx. et Prsg. 1942
Magnopotametalia Den Hartog et Segal 1964
Magnopotamion Den Hartog et Segal 1964
Nymphaeion Oberd. 1957 em. Neuhausl 1959

Table 10

Nymphaeion character species: *Nymphaea alba, Potamogeton natans*

Heritage Council analogue: Mesotrophic lakes—FL4

The Potametea is poorly represented on Clare Island and is nearly always associated with infilling lakes. This vegetation corresponds with the species-poor subcommunity, the *Nymphaea alba* A7 community (Rodwell 1995). Occasional gaps and tears in the quaking *Sphagnum*-dominated mat are colonised by *Nymphaea alba* and, rarely, by *Nuphar lutea*. The water-lilies are also recorded from shallow open waters on Poirtín Fuinch, where the sheltered site allows other species such as *Potamogeton natans* and *Myriophyllum alterniflorum* to become established. Both *Juncus bulbosus* and *Scorpidium scorpioides* are confined to the shallow edges of the community (Table 10, column 1).

VEGETATION ROOTED IN THE SUBSTRATE OF DEEP TO SHALLOW WATERS

Littorelletea uniflorae Br.-Bl. et Tx. 1943
Littorelletalia uniflorae Koch 1926

Table 10, column 2

Diagnostic species: *Myriophyllum alterniflorum, Potamogeton natans, Littorella uniflora, Hypericum elodes*

Heritage Council analogue: Acid oligotrophic lakes—FL2

No relevés are presented for vegetation from lakes with water depths over 0.75m—the area occupied by this community represents a negligible fraction of total vegetation cover for the entire island. The vegetation of these areas is characterised by the occurrence of *Sparganium angustifolium, Potamogeton natans* and *Menyanthes trifoliata* with occasional *Nymphaea alba*.

In shallower waters, *Utricularia minor, Myriophyllum alterniflorum* and *Littorella uniflora* occur, species that are clearly characteristic of the Littorelletalia uniflorae (Table 10, column 2). Owing to the fragmented nature of the zonation, however, it seems unwise to assign the vegetation to a syntaxonomic status below that of the order. Doyle (1990) has recorded similar species combinations in conjunction with *Lobelia dortmanna* and *Eriocaulon aquaticum* (a species of restricted distribution in Ireland) from bog pools in north-west Mayo. The vegetation of such lakelets was classified in the Isoeto-Lobelietum, which belongs to the Lobelion dortmannae, in the Littorelletalia uniflorae, which corresponds with the *Littorella uniflora* sub-community of the *Littorella uniflora-Lobelia dortmanna* A22 community in the BNVC scheme (Rodwell 1995).

Marginal lake vegetation of the Littorelletalia from Clare Island shows considerable floristic overlap with the Scheuchzerio-Caricetea nigrae, particularly where the gently sloping, muddy substrates are exposed. However, a community characterised by *Carex rostrata* typically demarcates transitional areas from aquatic to terrestrial situations (Table 10, column 2). Similar communities from lake zonations in Ireland have been recorded by several authors (Braun-Blanquet and Tüxen 1952; Brock *et al.* 1978; Dierßen 1978; Van Groenendael *et al.* 1979; 1983a; O'Connell 1980; 1981). The bottle sedge, *Carex rostrata*, presents syntaxonomical problems as its broad ecological amplitude in Ireland complicates classification, particularly at the association level (Westhoff and Den Held 1969; Ó Críodáin 1988; Ó Críodáin and Doyle 1994). Van Groenendael *et al.* (1979) considered that the species was exclusively confined to vegetation assignable to the Scheuchzerio-Caricetea in the west of Ireland. It can occur rooted in water at depths of up 50cm, as in relevé 144, although it is more abundant in shallow, clearer waters. It is usually the dominant species in the vegetation. Other species recorded in close proximity with *Carex rostrata* include *Scorpidium scorpioides, Potamogeton natans* and *P. polygonifolius, Carex limosa, Myriophyllum alterniflorum* and *Littorella uniflora*. These generally tend to develop behind *Carex rostrata* stands, on the landward side of the lake margins.

A greater proportion of species characteristic of the Scheuchzerio-Caricetea nigrae (Table 10, columns 3 and 4) marks the transition towards terrestrial conditions. However, where the substrate is subject to alterations in the water table or is periodically flooded, there is an increase in the abundance of differential species of the Littorelletea, particularly *Hypericum elodes*, which is common in oceanic regions of Ireland (G. Doyle, pers. comm.). There is a reduction in the abundance of *Carex rostrata*, although *Carex limosa* still provides at least 10% of groundcover. Other species associated with this zone include *Callitriche stagnalis*, *Mentha aquatica* and *Drosera anglica*. They occur in small puddles that remain once the water levels decrease. On regularly exposed waterlogged soils *Sphagnum cuspidatum* may cover up to 25% of the ground. The common species are *Juncus bulbosus*, *Anagallis tenella*, *Eleocharis multicaulis*, *Ranunculus flammula* and *Hydrocotyle vulgaris*.

QUAKING VEGETATION

Oxycocco-Sphagnetea Br.-Bl. et Tx. 1943
Scheuchzerietalia palustris Nordh. 1936
Rhynchosporion albae Koch 1926
Scheuchzerietum Paul 1910

Table 10, columns 3 and 4

Scheuchzerietum diagnostic species: *Sphagnum cuspidatum*

Heritage Council analogue: Transition mire and quaking bog—PF3

The first record of the Scheuchzerietum in Ireland coincided with the only known location of *Scheuchzeria palustris* from a wet hollow on Pollagh Bog, in County Offaly (Moore 1955). This site was later destroyed by commercial peat extraction. Later Doyle (1990) assigned some of the vegetation from deep hollows within blanket bog complexes to this association. The community develops on a thick *Sphagnum*-dominated quaking vegetation mat and represents the final stages of an infilling lake. Continuity with the marginal aquatic vegetation is maintained, with reduced abundance of *Carex limosa* and *Menyanthes trifoliata* recorded, particularly on the periphery of the floating vegetation. While no direct equivalent for the community on Clare Island is described from the BNVC, it is clear that it has some affinities with the M2 *Sphagnum cuspidatum/recurvum* bog pool communities, which can consist of floating mats or carpets of vegetation, and to some degree the M1 *Sphagnum auriculatum* bog pool.

The vegetation recorded on Clare Island is characterised by well-developed floating mats of *Sphagnum cuspidatum*, which provide almost 100% cover (Table 10, columns 5–8; Pl. V A, V B). Other bryophytes, including *Aulacomnium palustre*, *Sphagnum capillifolium* and *S. palustre*, are an intrinsic, though infrequent, feature of the vegetation mat. Several vascular plants occasionally protrude through the quaking vegetation, including *Drosera rotundifolia*, *Menyanthes trifoliata* and *Carex echinata*. Species typical of the Scheuchzerio-Caricetea nigrae that are commonly associated with other waterlogged habitats include *Eleocharis multicaulis*, *Juncus bulbosus* and *Carex limosa*.

REED SWAMP

Phragmitetea Tx. et Prsg. 1942
Phragmetalia Koch 1926 em. Pignatti 1953 denuo em. Segal et Westhoff in Westhoff et Den Held 1969

BALBRIGGAN LIBRARY
PH: 8704401

Pl. V A Apart from some white water-lilies at top right corner of Lough Leinapollbauty, the remainder of the lake is completely infilled with *Sphagnum cuspidatum*.

Pl. V B Stand of bottle sedge in shallow waters of Creggan Lough.

Phragmition Koch 1926 em. Balátová-Tulácková 1963
Scirpo-Phragmitetum Koch 1926 em. Segal et Westhoff in Westhoff et Den Held 1969 (Table 11)

Scirpo-Phragmitetum character species:
Phragmites australis

Heritage Council analogue: Reed and large sedge swamps—FS1

The reed *Phragmites australis* is characteristic of the class Phragmitetea, which includes monodominant vegetation of tall clonal perennial grasses and rushes found in swampy areas near lakes, rivers and streams and in some fens. Praeger's original description of the reed community on Lough Avullin noted the occurrence of *Schoenoplectus lacustris*, which, in conjunction with *Phragmites*, suggests that the vegetation corresponds with the Scirpo-Phragmitetum. This second species, *Schoenoplectus lacustris*, recorded by Praeger (1911) and more recently by Doyle and Foss (1986a), was not relocated during this survey. Stands of reeds are commonly found in transitions from open water to infilled lakes (White and Doyle 1982), although few vegetation descriptions of Irish communities have been made (Duff 1930; Braun-Blanquet and Tüxen 1952; Ivimey-Cook and Proctor 1966).

On Lough Avullin, the monodominant canopy of *Phragmites australis*, reaching three metres tall, is broken only by the occurrence of some *Alnus* and *Salix* trees (Pl. IV B). The field layer is poorly developed with some straggling *Galium palustre, Epilobium palustre, Ranunculus flammula* and *Angelica sylvestris* (Table 11, columns 1–3). Other less abundant species include *Lythrum salicaria, Mentha aquatica* and *Iris pseudacorus*. It appears that *Iris pseudacorus* is gradually spreading upstream along the banks of a neglected mill-race that is located downstream, and penetrating into the reed swamp.

Few species can compete with the reeds and those that do are usually confined to the perimeter of the reed swamp. Where the canopy is more open, on the fringes of the reed swamp, herbaceous species increase in abundance (Table 11, columns 4 and 5). Constant species include *Menyanthes trifoliata, Galium palustre, Epilobium palustre* and *Carex rostrata*. Other occasional species include *Mentha aquatica, Ranunculus flammula, Hydrocotyle vulgaris, Cardamine pratensis, Lythrum salicaria, Angelica sylvestris* and *Myosotis scorpioides*. Syntaxonomically, these two relevés are not strictly related to the Scirpo-Phragmitetum and are more correctly assigned to the Scheuchzerio-Caricetea nigrae and Littorelletea uniflorae.

Van Groenendael *et al.* (1979; 1983a) considered that the herbaceous vegetation represented the transitional stage from the Phragmitetea to the Scheuchzerio-Caricetea nigrae, typically found behind the tall-growing helophytes. Satisfactory classification of this vegetation presents difficulties, since this vegetation assemblage was only recorded in a narrow

Table 11
Relevé table for the Scirpo-Phragmitetum

Column	1	2	3	4	5	Synoptic value
Relevé	4	4	4	4	4	
	0	0	0	0	0	
	3	4	5	6	7	
Scirpo-Phragmitetum						
Phragmites australis	5	5	3	1		IV
Iris pseudacorus		+	r			II
Myosotis scorpioides			+			I
Equisetum fluviatile				+		I
Companion species						
Lythrum salicaria	+	+				II
Angelica sylvestris	r	r				II
Cardamine pratensis	+					I
Alnus glutinosa	r					I
Mentha aquatica		+	+		+	III
Littorelletea uniflorae						
Galium palustre		+	1			II
Potamogeton polygonifolius				1	1	II
Juncus bulbosus				+	+	II
Scheuchzerio-Caricetea nigrae						
Epilobium palustre	+		+	+		III
Menyanthes trifoliata			2	3	3	III
Ranunculus flammula			+	r	+	III
Carex rostrata				1	1	II
Potentilla palustris				1	+	II
Number of species	6	6	8	9	7	

fringe from the edges of the reed swamp. These herbs are typical of the Littorelletea uniflorae and occur on marshy soils that are regularly inundated by fresh water.

The marginal areas surrounding the reed swamp, where there is no *Phragmites* cover, are dominated by *Menyanthes trifoliata* (Table 11, column 5). This species is typical of the Scheuchzerio-Caricetea nigrae, as are *Equisetum fluviatile, Potentilla palustris* and *Carex rostrata*. The pondweed *Potamogeton polygonifolius* is found floating on the water surface, while *Juncus bulbosus* is rooted in the soils beneath the shallow water.

The BNVC classification refers the reed-dominated community to the Phragmitetum australis S4 reed swamp found in sheltered areas fed by gently flowing waters. Rodwell's (1994) classification recognises two subcommunities that correspond with the gradual decrease in canopy cover provided by the reeds: the *Galium palustre* and *Menyanthes trifoliata* variants. These are comparable with the marginal vegetation characterised by the Littorelletea and Scheuchzerio-Caricetea species at the Lough Avullin site. The community found abutting the reed swamp is classified as the *Galium palustre* subcommunity of the Phragmitetum australis S4 swamp. Closely allied to this subcommunity is the *Menyanthes* variant. It also occurs in open waters (depth <50cm) alongside reed vegetation.

VEGETATION DOMINATED BY ELEOCHARIS MULTICAULIS AND SCORPIDIUM SCORPIOIDES

Littorelletea uniflorae Br.-Bl. et Tx. 1943
Littorelletalia uniflorae Koch 1926
Hydrocotylo-Baldellion Tx. et Dierßen 1972
Scorpidio-Eleocharitetum multicaulis Ivimey-Cook and Proctor 1966 Table 12

Scorpidio-Eleocharitetum multicaulis character species: *Eleocharis multicaulis, Scorpidium scorpioides*

Heritage Council analogue: Acid oligotrophic lakes—FL2

This vegetation is commonly recorded from oligotrophic lake margins that undergo regular

Table 12
Relevé table for the Scorpidio-Eleocharitetum multicaulis

Column	1	2	
Relevé	367	194	Synoptic value
Scorpidio-Eleocharitetum			
Scorpidium scorpioides	4	5	V
Eleocharis multicaulis	2	+	V
Carex viridula		1	III
Constant species			
Agrostis stolonifera	2	1	V
Philonotis fontana	1	2	V
Anagallis tenella	1	2	V
Juncus bulbosus	1	1	V
Festuca ovina	+	1	V
Isolepis setacea	+	1	V
Carex panicea	+	+	V
Companion species			
Equisetum palustre	2		III
Hydrocotyle vulgaris	1		III
Potamogeton polygonifolius	1		III
Mnium hornum		2	III
Carex nigra		1	III
Koeleria macrantha		1	III
Mentha aquatica	+		III
Aneura pinguis	+		III
Bellis perennis	+		III
Cardamine pratensis	+		III
Sagina procumbens	+		III
Selaginella selaginoides	+		III
Ranunculus repens	+		III
Galium palustre	+		III
Holcus lanatus	+		III
Drosera rotundifolia		+	III
Molinia caerulea		+	III
Narthecium ossifragum		+	III
Pellia epiphylla		+	III
Plantago lanceolata	r		III
Number of species	22	17	

fluctuations in the water table. It was rarely encountered on Clare Island, and only then on peripheral locations of *Schoenus*-dominated fen vegetation on the western end of the island. It is characterised by the overwhelming dominance attained by *Scorpidium scorpioides*. Other companion species include *Philonotis fontana, Agrostis stolonifera, Anagallis tenella, Juncus bulbosus, Festuca ovina, Carex panicea* and *Isolepis setacea* (Table 12).

This vegetation type was previously defined from the muddy surfaces of some gently sloping lake margins in the Burren (Ivimey-Cook and Proctor 1966). These authors considered the vegetation sufficiently distinctive to separate it from the closely related Eleocharitetum multicaulis, which is widely recorded in Ireland on lake margins that undergo periodic fluctuations in the water table (Braun-Blanquet and Tüxen 1952; Schoof van Pelt 1973; Dierßen 1978; Brock *et al.* 1978 and Van Groenendael *et al.* 1979). While the Scorpidio-Eleocharitetum community has been recognised in Ireland, Rodwell (1995) suggests that it be included in the *Littorella uniflora-Lobelia dortmanna* A22 community, which includes many of the characteristic elements that are found in moderately fertile stretches of calm, open water.

While the vegetation is assigned to the Littorelletea, on Clare Island it is spatially affiliated to the Caricetalia davallianae, since it was always found in close proximity with the Schoenetum nigricantis (Table 25, see p. 144). This vegetation is related to the *Schoenus nigricans–Scorpidium scorpioides* nodum described from Connemara (Van Groenendael *et al.* 1979). Along with the eponymous species, the vegetation contained a mixture of calcicole and calcifuge species, such as *Carex dioica* and *Eriophorum angustifolium*.

VEGETATION OF EPHEMERAL CALCICOLOUS LAKES

Littorelletea uniflorae Br.-Bl. et Tx. 1943
Littorelletalia uniflorae Koch 1926
Hydrocotylo-Baldellion Tx. et Dierßen 1972
Baldellio-Littorelletum Ivimey-Cook et Proctor 1966 relevé 139

Baldellio-Littorelletum character species: *Baldellia ranunculoides, Littorella uniflora*

Heritage Council analogue: Acid oligotrophic lakes—FL2

Relevé 139 (14 species)	
Littorella uniflora	2
Apium inundatum	3
Juncus bulbosus	2
Eleocharis multicaulis	2
Hypericum elodes	2
Ranunculus flammula	2
Agrostis stolonifera	1
Baldellia ranunculoides	1
Sphagnum cuspidatum	3
Eleogiton fluitans	2
Eleocharis palustris	2
Carex rostrata	2
Hydrocotyle vulgaris	1
Glyceria fluitans	1

Vegetation characterised by *Baldellia ranunculoides* and *Littorella uniflora* is assigned to the Baldellio-Littorelletum, originally described from certain ephemeral lakes in the Burren that were situated on calcareous marl (Ivimey-Cook and Proctor 1966). It may have some affinities with the as yet poorly sampled Hydrocotylo-Baldellion M30 vegetation of seasonally inundated habitats (Rodwell 1991). On Clare Island this vegetation is mainly located around Toormore, at the western end of the island (relevé 139) and is confined to fluctuating water bodies in coastal areas situated on moderately deep peaty substrates. The vegetation is found in small, naturally occurring hollows, never occupying more than a couple of square metres in any area.

Toormore once supported the small-scale linen industry that was carried out on the island (Mac Cárthaigh 1999, p. 49), since the small lakes were ideal for preparing the flax. The surrounding vegetation is composed of rough grassland with a significant *Calluna* component and several species diagnostic of wet heaths. Other than *Baldellia* and *Littorella*, both *Apium inundatum* and *Sphagnum cuspidatum* may also be locally prominent features of the community. The distinctive vegetation also includes *Eleogiton fluitans, Eleocharis palustris* and *Glyceria fluitans*, together with species commonly occurring in other wetland habitats, such as *Eleocharis multicaulis, Agrostis stolonifera, Hydrocotyle vulgaris* and diminutive plants of *Carex rostrata*.

HEATHLANDS

Heather-dominated vegetation is widespread in the British Isles, often occupying large tracts of land. The plant species composition of much of the vegetation here is unique in Europe (Averis *et al.* 2004). There is clear evidence that heather moorland has declined in extent in the British Isles with a considerable increase in the grassland component of the vegetation (Birse 1980; Ratcliffe and Thompson 1988; Usher and Thompson 1993; Bunce 1989; Thompson *et al.* 1995a). These uplands are increasingly subjected to pressure from recreational and agricultural forces that interact with climate, leading to fragmentation or loss of habitat. Agricultural management of upland areas through afforestation, burning of heather (muirburn), application of fertiliser, soil drainage and, particularly, grazing affect the vegetation in a variety of ways (Miles 1988). There is considerable evidence in Britain linking grazing pressure by sheep with changes in species diversity and structure of heather-dominated communities, often creating mosaics of shrub-dominated and grass-dominated communities (Ball *et al.* 1982; Welch 1984a; 1984b; Miles 1988; Stevenson and Thompson 1993).

Heathlands in Ireland are not as extensive as their British counterparts (Thompson *et al.* 1995a; 1995b; 1995c; Bleasdale 1998), nor are they as actively managed. Irish heathlands, however, are rarely free from human interference. Most heathland is located in commonages, especially in the west of Ireland, where grassland mosaics are found in intimate association with dwarf shrub vegetation. With Ireland's entry into the European Union in 1973 and the implementation of the agricultural subsidies schemes, sheep numbers further increased. Land may be heavily overstocked, causing significant changes to natural and semi-natural vegetation in these areas (Bleasdale and Sheehy-Skeffington 1992; 1995; MacGowan and Doyle 1996; 1997; Bleasdale 1998; McKee *et al.* 1998). Other threats to these habitats include burning, land reclamation and improvement, and increasingly, recreation and tourism pressures.

Heathlands with dominant *Calluna vulgaris* once covered the slopes of Knockmore and Knocknaveen and much of the drier areas of the commonage on Clare Island (Praeger 1911). Doyle and Foss (1986a) subjected Praeger's Clare Island vegetation map (see fold-out map 1) to digital analysis and calculated the percentage area that had been covered by various vegetation types in 1911. The heather formation accounted for 49% of the total land area, making it the most extensive of the vegetation types on the island. Indeed anecdotal evidence describes the tall, shrubby foliage of *Calluna* that was extensive in the upland areas, often obscuring sheep from the casual observer (Doyle and Foss 1986a). The heather community remained relatively intact until the 1950s but was gradually transformed to grassland as the heather community could not provide sufficient fodder to support the numbers of sheep.

The heather community is now drastically reduced in extent, compared with its distribution in the early part of the century (Pl. VI A). Remnants of the original community, which form distinct stands and are distributed at lower altitudes around the island, are often found in close proximity to *Molinia*-dominated blanket peats. The transitions from blanket peat to heathland and the increased abundance of *Calluna* generally reflect changes in topography and soil characteristics. Sloping topography, above 15°, precludes the accumulation of deep, wet peat except in some terraced landscapes and favours heathland development.

The majority of the heather-dominated vegetation is recorded from dry, relatively shallow peaty podzols, mainly along the eastern coast of Clare Island, but also in some inland areas. The community is noticeable where fencing prevents access by sheep, such as at a site located above Portlea. Less extensive stands of heathland vegetation are located on inaccessible overhanging ledges along the northern coastline and in some riverbeds on Knockmore.

In the late 1980s, some land that once supported low-growing heather was fenced off and sheep excluded (C. Cullen, pers. comm.; Pl. VI B). Since livestock has been excluded, *Calluna* regeneration has been rapid, with significant growth occurring over three years. As yet, there has been little development of the herbaceous and bryophyte ground flora that is usually associated with heather communities.

Floristics

The heathland community is generally dominated by *Calluna vulgaris*, providing from 50% to 95% cover. In heavily grazed situations the distribution of *Calluna* may be lower (<25%) (Table 13, column 8). On deeper peats, *Molinia caerulea* may be co-dominant (Table 14, columns 2 and 3, p. 64). The cover, height

Pl. VIA Comparison of the impacts of management regimes of closely cropped agricultural land in Strake townland and adjacent commonage. These areas once supported heather-dominated vegetation. In the foreground, agricultural land that once supported crop production but reverted to pastoral farming at the time of Praeger's survey, is now less intensively grazed, allowing low-growing heather to become re-established in peripheral areas. Beyond the Congested Districts Board wall, seen running across the centre of the photograph, lies the more heavily grazed commonage.

Pl. VIB Regenerating heathland at Ballytoohy More. A fence, barely visible, runs along the valley bottom. On the left hand side of the fence, large numbers of sheep have been excluded from grazing, allowing regeneration of the heather community. On the far side of the fence, extensive peat cutting and continual grazing result in degraded grassland community.

and structure of the subcanopy can vary quite markedly, with the lower-growing shrubs, such as *Erica tetralix* and *Erica cinerea* frequently present. Another ericaceous shrub, *Erica erigena*, is confined to a boulder cliff at one location in Portlea, where it forms a distinctive community (Praeger 1911; Doyle and Foss 1986a; Foss 1986; Foss *et al.* 1987).

Grasses are another important component of the vegetation, although their overall cover is variable. The commonest of the grasses encountered is *Anthoxanthum odoratum*, although *Agrostis capillaris* and *Festuca ovina* are also present. In higher altitude areas *Festuca vivipara* is commonly found in the vegetation, predominantly along streams where the vegetation is largely inaccessible to sheep. In general, grasses have a patchy distribution, growing through small gaps in the heather canopy. Other important graminoids include *Eriophorum angustifolium* and *Molinia caerulea*, while small tussocks of *Trichophorum caespitosum* are common on shallow soils and in open patches in the heather canopy. Common sedges include *C. panicea*, *C. echinata* and *C. viridula*. The taller *Carex binervis* is typical of stands near coastal areas and is commonly found where fences prevent access by sheep.

Herbs are infrequent in the vegetation, with only *Potentilla erecta* regularly recorded. Occasionally *Polygala serpyllifolia*, *Succisa pratensis* and *Anagallis tenella* are found. Where the heather canopy becomes less dominant, particularly in marginal areas near the coastal grasslands, smaller herbs are encountered, such as *Lotus corniculatus*, *Trifolium repens* and *Cerastium fontanum*.

A number of bryophytes and lichens are associated with the heathlands, although they rarely contribute significantly to ground cover. Those consistently recorded include *Pseudoscleropodium purum*, *Hypnum jutlandicum*, *Hylocomium splendens*, *Rhytidiadelphus squarrosus* and *Plagiothecium undulatum*. They are found growing through moist litter in the understorey of the heather canopy. Apart from the moderately abundant *Sphagnum subnitens*, other *Sphagnum* species such as *S. capillifolium* and *S. papillosum* are confined to damper substrates, in transitions to blanket peats. The commonest of the lichens are *Cladonia uncialis* and *C. portentosa*, but again they are not an abundant component of the community.

Phytosociology

Notwithstanding the characteristic appearance and physiognomy of heathlands, no satisfactory classification exists for stands of such vegetation in Ireland (Cotton 1975; O'Sullivan 1976; 1982; White and Doyle 1982). This is due, in part, to the rather catholic nature of *Calluna vulgaris*, which is found in wet and dry heathlands as well as in each of the major Irish peatland types. The classification is further complicated by considerable species overlap with bogs and to a lesser degree with grassland communities. The scheme adopted in this survey follows that outlined in White and Doyle (1982),

Table 13
Relevé table for the Calluno-Ericetum cinereae

Column	1	2	3	4	5	6	7	8	9	10	11	12	13	14	15	16	17	18	19	20	Synoptic value
Relevé	122	145	250	127	304	208	71	90	86	277	37	382	233	348	349	350	319	352	41	210	
Calluna vulgaris	5	5	4	5	5	4	4	2	4	3	4	5	5	5	4	4	5	3	4	5	V
Calluno-Ericetum cinereae																					
Erica cinerea		1		2	+	1		+	+	1	+	2	2	1	1	1	+	2	1	+	IV
Thuidium tamariscinum	1	2		2	+	1	1	+		+	1	+		+			+	+	1		IV
Hylocomium splendens	1	1	3		1	2				1			1	2	1	2	+			2	III
Rhytidiadelphus squarrosus			1			2		2						1	+	+	1	+			III
Pseudoscleropodium purum	2	2	1	2							1	+			+	+					II
Dactylorhiza maculata	r	r			r												r	r			II
Nardetalia species																					
Agrostis capillaris			1						2	2	2	1	1		1	2		2	2	1	III
Festuca ovina										3	2	1	1	+	2	1	1	3		3	III
Nardus stricta		2																	2	2	I
Festuca vivipara																			2	1	+
Hypericum pulchrum														1	+	1		+			II
Carex hostiana																	+	+			+
Pteridium aquilinum					r																R
Myrica gale			1																		R
Cirsium dissectum																			+		R
Blechnum spicant																+					R
Constant species																					
Anthoxanthum odoratum	2	1		1	1		2	2	2		1	2	1	2	2	3	2	3	2	1	V
Potentilla erecta	+	1	1	+	+	2	1	+	+	2	1	+	1	2	+	1	2	+	2	2	V
Molinia caerulea	3	2	3	3	2	1						+	1	1					2	2	IV
Erica tetralix	1	2	2		+	2	2			1	2	+							2	2	III
Hypnum jutlandicum			1		1		1	2	1			+	1		+	1		1	2		III
Carex panicea				+			+	1	+			1					+		+		III
Sphagnum subnitens			1	1		2								+		1		1			II
Carex binervis	+		1	1	+	1	1													1	II
Eriophorum angustifolium	2	2	3			3										+			2	2	II
Plagiothecium undulatum			+		+	1								+	+	+		+			II
Lophocolea bidentata			+		+		1				+				+	+		+			II
Cladonia portentosa		1	1	+			+		+	+		+	+								II
Carex nigra			1			1	+		+		1		+								II
Leucobryum glaucum				+			+	1	+	2		+	2						+		II
Frullania tamarisci			+				+			2			+	+	+					1	II
Peltigera canina							+	+	+	+		+	+							+	II
Rhytidiadelphus triquetrus					+					2		+	1	+	1	+				2	II
Mnium hornum				+					+	+			+		+	+	+	+			II

(Continued)

Table 13 *(Continued)*

Column	1	2	3	4	5	6	7	8	9	10	11	12	13	14	15	16	17	18	19	20	
Relevé	122	145	250	127	304	208	71	90	86	277	37	382	233	348	349	350	319	352	41	210	**Synoptic value**
Euphrasia officinalis													2	+	+	+					II
Nardia scalaris									1		1						1	+			II
Succisa pratensis			1					+	+			r									II
Companion species																					
Pedicularis sylvatica			+										+						+		I
Polygala serpyllifolia									r	+	+										I
Eurhynchium praelongum					+		1										+			1	I
Carex viridula								1		1									1		I
Trichophorum caespitosum	1	2				1															I
Lotus corniculatus								+	+			2	+								I
Salix repens				+				+	+												I
Luzula multiflora									+	r								+			I
Eurhynchium striatum														1		2					+
Lepidozia reptans				1											+						+
Isothecium myosuroides											1			+							+
Dicranum majus		+																+			+
Hypnum cupressiforme														+				+			+
Aira praecox								+	+												+
Hypochaeris radicata																+	+				+
Galium saxatile																+		+			+
Trifolium repens								+	+												+
Cerastium fontanum								+	+												+
Juncus squarrosus													1								R
Eleocharis multicaulis											1										R
Aulacomnium palustre										1											R
Pleurozia purpurea				1																	R
Sphagnum recurvum						2															R
Salix aurita				1																	R
Rhytidiadelphus loreus						1															R
Diplophyllum albicans																				2	R
Carex flacca																				1	R
Campylopus paradoxus													+								R
Cladonia uncialis								+													R
Sphagnum capillifolium												+									R
Juncus bulbosus																		+			R
Dicranum scoparium									+												R
Odontoschisma sphagni												+									R
Viola riviniana																		+			R
Carex echinata																+					R

(Continued)

Table 13 *(Continued)*

Column	1	2	3	4	5	6	7	8	9	10	11	12	13	14	15	16	17	18	19	20	Synoptic value
Relevé	122	145	250	127	304	208	71	90	86	277	37	382	233	348	349	350	319	352	41	210	
Rumex acetosa																		+			R
Mylia taylorii									+												R
Pleurozium schreberi			+																		R
Holcus lanatus									+												R
Scapania gracilis							+														R
Sagina procumbens								+													R
Cladopodiella fluitans			+																		R
Pellia epiphylla															+						R
Plagiochila punctata																		+			R
Saxifraga spathularis																		+			R
Bryum pseudotriquetrum																		+			R
Sphagnum compactum																+					R
Plagiochila spinulosa												+									R
Lophozia sp.													+								R
Number of species	12	15	22	17	16	16	17	22	27	22	17	25	26	20	22	26	18	28	17	19	

after Géhu's (1975) suggestion that shrub and grass heathlands should be assigned to different classes, the Calluno-Ulicetea and Nardetea respectively.

These syntaxonomic difficulties are further illustrated by the shrub-dominated heathlands found on mountain slopes of intermediate altitudes, which are classified in the alliance Genisto-Callunion, order Vaccinio-Genistetalia, in the class Calluno-Ulicetea. While this classification does not usually account for *Calluna*-dominated vegetation that occurs below 300m, the character of low-altitude heathlands in many places in Ireland suggests that in extreme climatic conditions the heathland vegetation associated with higher altitudes elsewhere in Europe descends down to sea level in Ireland (personal observation).

On Clare Island the heathland vegetation that once clothed much of the landscape is diminished in extent, with the majority located in the eastern half of the island. Some isolated patches occur in inaccessible areas at higher elevations, while wet heathland is associated with the margins of *Molinia*-dominated blanket bog and is transitional between blanket bog and dry heath vegetation.

The heathland vegetation is referred to three distinct associations based on floristic criteria, with local influences reflecting edaphic and management conditions:

Calluno-Ericetum cinereae (dry heathlands) (Table 13)

Potentillo erecti-Ericetum erigenae (*Erica erigena* heathland) (relevé 151)

Narthecio-Ericetum tetralicis (wet heathlands) (Table 14)

DRY HEATHLANDS

Calluno-Ulicetea Br.-Bl. et Tx. 1943
Vaccinio-Genistetalia Schubert 1960
Genisto-Callunion (Duvign. 1944) Tx. in Prsg. 1949
Calluno-Ericetum cinereae Lemée 1937 Table 13

Calluno-Ericetum cinereae diagnostic species: *Calluna vulgaris, Erica tetralix, Erica cinerea*

Constant species: *Molinia caerulea, Anthoxanthum odoratum, Potentilla erecta*

Table 14
Relevé table for the Narthecio-Ericetum tetralicis

Column	1	2	3	4	Synoptic value
Relevé	249	353	40	6	
Narthecio-Ericetum tetralicis					
Erica tetralix	2	1	1		IV
Carex panicea	1	+	+		IV
Potentilla erecta		+	1	1	IV
Polygala serpyllifolia		+	+		III
Pedicularis sylvatica	+				II
Succisa pratensis		1			II
Constant species					
Calluna vulgaris	5	2	3	2	V
Molinia caerulea	2	3	3	1	V
Hypnum jutlandicum	1	1	1		IV
Sphagnum subnitens		1	1	2	IV
Nardetalia species					
Festuca ovina		1	1		III
Agrostis capillaris			3		II
Festuca vivipara			1		II
Nardus stricta		1			II
Companion species					
Eriophorum angustifolium	2	3			III
Sphagnum capillifolium	2	2			III
Anthoxanthum odoratum			2	1	III
Juncus bulbosus		+	1		III
Anagallis tenella		+	1		III
Cladonia uncialis	+		+		III
Lophocolea bidentata		+	+		III
Juncus squarrosus				4	II
Cladonia portentosa	2				II
Eleocharis multicaulis		2			II
Trichophorum caespitosum	2				II
Sphagnum papillosum			2		II
Schoenus nigricans		1			II
Carex nigra			1		II
Plagiothecium undulatum			1		II
Polytrichum commune				1	II
Carex viridula			1		II
Aulacomnium palustre			1		II
Carex binervis	+				II
Pleurozia purpurea	+				II
Campylopus paradoxus		+			II
Galium saxatile				+	II
Dicranum scoparium	+				II
Odontoschisma sphagni	+				II
Rhytidiadelphus squarrosus				+	II
Campylopus atrovirens	+				II
Selaginella selaginoides			+		II
Calypogeia spp	+				II
Narthecium ossifragum		+			II
Aneura pinguis		+			II
Pinguicula vulgaris		+			II
Carex pilulifera				+	II
Number of species	17	24	22	12	

Heritage Council analogue: Dry siliceous heath—HH1

The dry heathland community (Table 13) is referred to the Calluno-Ericetum cinereae, an association that includes much of the *Calluna*-dominated dry heathland elsewhere on Irish mountain slopes at elevations of 300m to 800m (White and Doyle 1982). On Clare Island, however, the lower limits of this community are at 50m. The community is floristically and structurally analogous with the *Calluna vulgaris–Erica cinerea* (H10) heath of the BNVC classification (Rodwell 1991), which includes much of the heathland vegetation that occurs in many of the oceanic parts of Scotland and western England. The heathland community on Clare Island is physiognomically and floristically homogenous, with sixteen species, on average, per square metre. The overwhelming dominant is *Calluna vulgaris* (other than Table 13, column 8), and this species forms a dense, closed canopy, typically up to 50cm tall, with *Erica tetralix* as a common component. Diagnostic species confined to drier habitats include *Erica cinerea, Anthoxanthum odoratum, Hylocomium splendens, Thuidium tamariscinum* and *Pseudoscleropodium purum*. These species always remain subordinate to *Calluna*. Other species recorded on these heathlands include *Molinia caerulea, Eriophorum angustifolium, Carex binervis, Potentilla erecta, Hypnum jutlandicum* and *Sphagnum subnitens*: these are also constantly recorded in wetter situations.

There is a paucity of this vegetation at higher elevations of the commonage where grazing has eradicated the heather. Despite this fact, isolated patches of heather-dominated vegetation are occasionally found in mountain riverbeds and on some scree slopes along the northern cliffs that are inaccessible to grazing animals (Table 13, columns 9–20). This vegetation represents the last fragments of heath vegetation that once covered the hills.

Unlike the lowland heath vegetation with which it shares many floristic similarities, this upland heather community has an open canopy, allowing the establishment of other plants, notably graminoid species. There is a Nardetalia element, with *Agrostis capillaris* and *Festuca ovina* present, while *Nardus stricta, Festuca vivipara* and *Molinia caerulea* are typically found at the higher elevations (Table 13, columns 19 and 20). In stands of older heather there is a greater presence of *Molinia* (at least 25%) and *Eriophorum angustifolium*. One diagnostic species for the alliance, *Vaccinium myrtillus*, is absent from the upland heather community but is recorded in acid grasslands at similar altitudes on Clare Island. Heather physiognomy is variable, although *Calluna* remains overwhelmingly dominant, particularly in pioneer stands of the vegetation, a finding recorded by other authors (McVean and Ratcliffe 1962; Birse and Robertson 1976; Birse 1980). The bushy nature of *Calluna* in these situations allows a taller mixed shrub and forb vegetation to develop. There is a higher proportion of *Erica cinerea*, though this is less abundant than *Erica tetralix*. Some shade tolerant species such as *Carex sylvatica* (no relevés presented) are confined to these habitats.

ERICA ERIGENA HEATHLANDS

Calluno-Ulicetea Br.-Bl. et Tx. 1943
Ulicetalia minoris (Duvign. 1944) Géhu 1973
Ulici-Ericion cinereae Géhu 1973
Potentillo erecti-Ericetum erigenae Foss 1986
relevé 151

Potentillo erecti-Ericetum erigenae character species: *Erica erigena*
Class, Order and Alliance character species: *Calluna vulgaris, Erica cinerea, Potentilla erecta*

Heritage Council analogue: Dry Calcareous Heath—HH2

Relevé 151 (28 species)	
Erica erigena	3
Molinia caerulea	3
Festuca ovina	3
Agrostis capillaris	2
Myrica gale	2
Anthoxanthum odoratum	2
Rhytidiadelphus squarrosus	2
Erica cinerea	1
Carex pulicaris	1
Hypericum pulchrum	1
Pseudoscleropodium purum	1
Carex panicea	+
Plagiothecium undulatum	+
Viola riviniana	+
Plantago lanceolata	+
Calluna vulgaris	2
Festuca vivipara	2
Pteridium aquilinum	2
Hylocomium splendens	2
Erica tetralix	1
Potentilla erecta	1
Carex hostiana	1
Thuidium tamariscinum	1
Blechnum spicant	1
Cirsium dissectum	+
Anagallis tenella	+
Trifolium repens	+
Succisa pratensis	r

At Portlea, bushes of *Erica erigena* are relatively tall, reaching 1m in some places, and form the dominant element in the canopy (Pl. VII). Other heathland ericoids such as *Calluna vulgaris, Erica tetralix* and *Erica cinerea* play a secondary role, together with scattered *Myrica gale*. Foss (1986) indicated that *Ulex europaeus* was a diagnostic species for *Erica erigena*-dominated vegetation elsewhere in Ireland, yet this shrub was not recorded in the Clare Island community. The species assemblage is diverse with a strong monocotyledonous ground flora, including *Molinia caerulea, Anthoxanthum odoratum, Agrostis capillaris, Festuca*

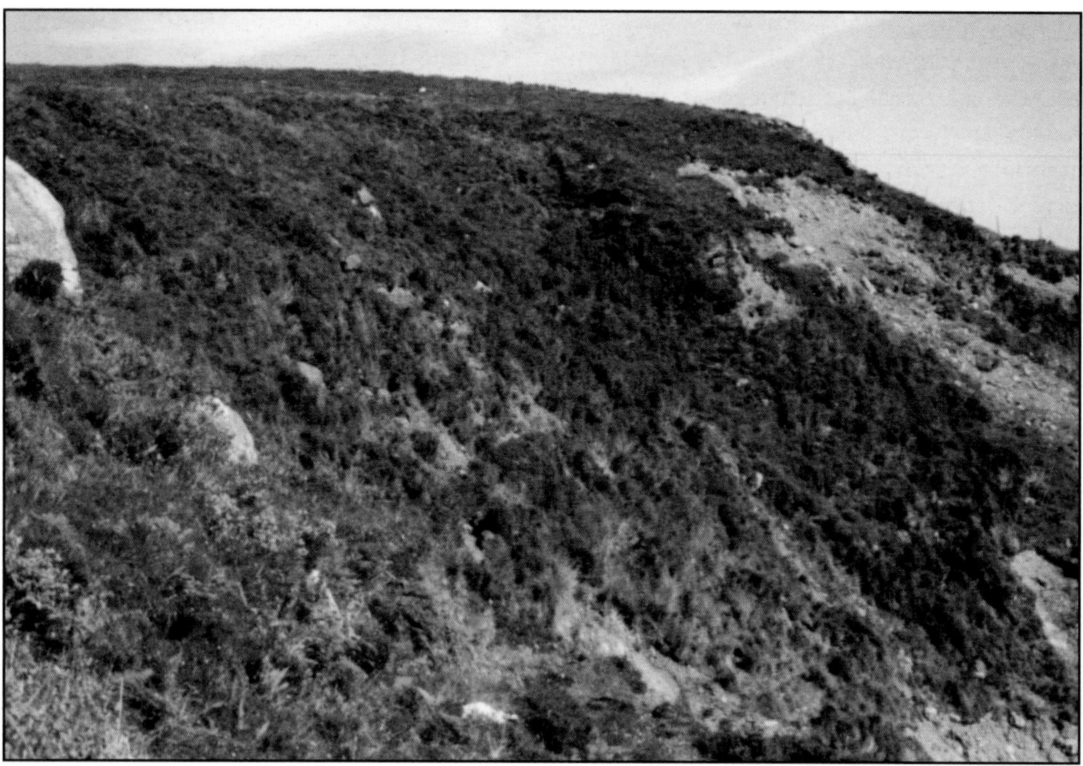

Pl. VII Mediterranean community, Potentillo erecti-Ericetum erigenae, occurring on the cliffs at Portlea. *Erica erigena* bushes are winter brown in this photograph.

ovina, *Festuca vivipara*, *Carex pulicaris* and *C. hostiana*. The presence of black bog rush, *Schoenus nigricans*, recorded in the Clare Island community (Foss 1986), was confined to open ground on the wetter margins of the community. Common bryophytes include *Rhytidiadelphus squarrosus* and *Hylocomium splendens*, although they provide only moderate coverage.

The shrub *Erica erigena* has a disjunct distribution in Europe, occurring in Galway and Mayo in the west of Ireland, and near Bordeaux and in several locations on the Iberian Peninsula (Foss 1986; Foss *et al.* 1987; Foss and Doyle 1988). Foss (1986) examined possible mechanisms for the dispersal of *Erica erigena* that might explain its distribution in Ireland. He examined the links between wave dispersal with distribution of the shrub along the western coasts of Portugal, Spain and Ireland. His results showed that seeds, fresh branches and dried branches were not buoyant (floating for 66 minutes, 6.5 days and 13.5 days respectively) and thus concluded that this method of dispersal was unlikely. Foss also considered seed dispersal by migratory birds that travel up along the western coast of Europe. Since the majority of birds that follow that route are passerine, they do not usually eat seeds, so this method of dispersal is again unlikely (Foss 1986; Foss and Doyle 1988).

Using palynological and scanning electron microscopy identification, Foss and Doyle (1988) examined the distribution of *Erica erigena* pollen in peat profiles in an area considered the centre of the shrub's distribution in Ireland, at Bellacragher Bay, near Mulranny and Claggan Mountain, in County Mayo (both sites are on the Irish mainland, some 15km to the north-east of Clare Island). The first positive identification of the pollen within the peat profile at Claggan was 477 radiocarbon years BP. This date coincided with the establishment of the Dominican Abbey at Borrishoole, 14km east of Claggan, prompting the speculation that the shrub had been introduced by man.

Despite the mechanisms suggested by these authors for the dispersal of the shrub, there is insufficient evidence on Clare Island that might explain its occurrence, except to note that Achill Island has a major colony of the shrub in Ireland. A distance of some 5km separates Portlea from Achill. It seems possible, therefore, that the introduction was due to fishermen coming ashore at the small harbour to the north of Portlea, carrying the seed either attached to their clothes or in bedding for livestock.

WET HEATHLANDS

Oxycocco-Sphagnetea Br.-Bl. et Tx. 1943
Sphagnetalia compactii Tx., Miyawki et Fugiwara 1970
Ericion tetralicis Schwick. 1933
Narthecio-Ericetum tetralicis Moore 1968
Table 14

Narthecio-Ericetum tetralicis character species: *Erica tetralix*
Ericion tetralicis differential species: *Potentilla erecta, Pedicularis sylvatica, Carex panicea, Polygala serpyllifolia, Succisa pratensis*
Constant species: *Calluna vulgaris, Molinia caerulea, Hypnum jutlandicum, Eriophorum angustifolium*

Heritage Council analogue: Wet heath—HH3

Wet heathlands (Table 14) are found in close proximity with *Molinia*-dominated blanket bog communities, in the low-lying areas in Ballytoohy (Pl. VIII). The land is within agricultural boundaries and has supported grazing animals. Grazing went unchecked until ten years ago (2001), when the area was fenced off in an effort to restore the heather (C. Cullen, pers. comm.). This has proven successful, with *Calluna* gradually re-establishing, although still less abundant than previously. The community is found on sloping ground (maximum slope 30°), on a range of soils varying from thin podzols to moderately deep peat. In certain situations the bedrock is exposed where vegetation cover has been completely removed.

Heathland vegetation of wetter situations, while retaining many diagnostic species of the Calluno-Ericetum cinereae (Table 14), is more closely allied to the Ericion tetralicis of the class Oxycocco-Sphagnetea, an alliance comprising wet heathland vegetation from Atlantic regions of north-western Europe (Moore 1968). Moore considered that the Irish communities were sufficiently distinct from the more continental Ericetum tetralicis to be given the revised classification. The floristically poor Narthecio-Ericetum tetralicis community on Clare Island has an average of eighteen species per square metre. Vegetation assignable to this association has been recorded by several authors and places, scattered from around the country (Braun-Blanquet and Tüxen 1952; Moore 1968; Vanden Bergen 1975; Dierßen 1978). Birse and Robertson (1976) and later Birse (1980) remarked that the

Pl. VIII Vegetation referred to the Narthecio-Ericetum tetralicis. Small shrubs of ling are largely interspersed among tussocks of purple moor grass (except for the small patch of heather where the rock outcrops).

vegetation was widely distributed but rarely extensive. This vegetation corresponds with the typical community of the *Scirpus cespitosus–Erica tetralix* M15 wet heath (Rodwell 1991), with the *Carex panicea* subcommunity representing vegetation occurring on shallow, bare peat.

Some of the character species listed by Moore (1968) are poorly represented in the wet heathland vegetation on Clare Island, being more abundant in wet hollows within blanket peats. Both *Calluna* and *Molinia* have a similar cover, with *Molinia* occurring between individual heather shrubs. Other constants include *Erica tetralix, Eriophorum angustifolium, Potentilla erecta* and *Hypnum jutlandicum*. Bryophytes are more numerous here than in *Calluna*-dominated vegetation on drier substrates, reflecting the relatively more moist conditions under the vegetation canopy in the wet heathland habitat. Blanket bog associates such as *Pleurozia purpurea, Odontoschisma sphagni* and *Sphagnum papillosum* are present. Other abundant species include *Hypnum jutlandicum, Hylocomium splendens* and *Thuidium tamariscinum*.

Heathland vegetation found in areas regularly irrigated by surface waters (Table 14, column 3) has an increased abundance of species indicative of wetter areas, including *Eleocharis multicaulis, Carex panicea, Carex nigra, Juncus bulbosus, Sphagnum papillosum* and *Aulacomnium palustre*. Some other areas that are severely damaged by sheep (Table 14, column 4) support a depauperate vegetation type that has, on average, only twelve species per square metre. In these situations *Juncus squarrosus* supplants *Calluna* as the major species.

GRASSLANDS

The establishment of early farming settlements on Clare Island initiated a process that led to the virtual destruction of woodland vegetation, the establishment of derived heathland and grassland communities, and the final replacement of heathland by acid grassland. Farming enclaves were established in any suitable land on the island, generally in sheltered areas in the lowlands, chiefly at elevations below 150 metres. Probably, from the time of the earliest settlements, tillage was concentrated close to the lowland farm settlements, while the upland areas were probably used for summer grazing. This pattern of farming was consolidated in the recent historical past, when the Congested Districts Board's walls and associated earthworks (built in 1895) finally demarcated farmland from the commonage on the upland areas.

When the Clare Island population numbered several thousand, during the period from 1800 to 1850, the upland commonage supported many activities. Peatland commonage provided turf for domestic use and rushes for thatching houses. Some of the drier slopes were cultivated for crops, and livestock were brought onto the hills during the summer months. The post-Famine decline in population reduced the pressure for tillage on the commonage, with farming practice favouring grazing. By the time of the original Clare Island Survey in 1911, there had been a noticeable increase in sheep grazing on the commonage (Praeger 1911). Since that time, the upland commonage area has become increasingly important for grazing, prompting a decline in heathland and an increase in grass-dominated communities, which now also occupy abandoned upland tillage areas.

Very little of the commonage and abandoned pastures on Clare Island are either species-rich or of good quality. These lands, which were generally reclaimed from *Calluna*-dominated heathlands, now lie derelict. Much of the grass-dominated vegetation on Clare Island can be referred to the upland grassland class Nardetea or included in the 'rough upland grazings' described by several authors (Tansley 1939; McVean and Ratcliffe 1962; O'Sullivan 1982; Rodwell 1992). Vegetation belonging to two Nardetea associations can be distinguished, the Achilleo-Festucetum tenuifoliae and the Nardo-Caricetum binervis. Although usually considered upland communities, these are found on Clare Island from the highest point on Knockmore (465m) down to sea level.

On Clare Island, fourteen subtypes can be distinguished within the Achilleo-Festucetum, including acid grasslands on podzolic and mineral soils, acid grasslands in coastal areas and a replacement community dominated by *Pteridium aquilinum*. A further four subtypes are distinguished from the Nardo-Caricetum binervis and a further four subtypes characterise derelict Nardo-Caricetum vegetation. Grassland vegetation of abandoned and extant meadows, and of some abandoned tillage areas, has a different floristic character and is assigned to the syntaxa of the Molinio-Arrhenatheretea. Classification of these different grasslands has proven difficult, owing to the fragmentary nature of the stands, the profound impact of grazing animals and the complex history of land use. Five subtypes have been distinguished in the case of Clare Island.

Achilleo-Festucetum Tenuifoliae

Subtypes occurring on podzolic soils

Severely grazed stands (Table 15, columns 1–38)
Racomitrium-rich stands (Table 15, columns 39–49)
Stands containing *Ulex europaeus* (Table 15, columns 50–51)

Subtypes occurring on non-podzolic soils

Typical subtype occurring on mineral soils (Table 15, columns 52–59)
Achilleo-Festucetum with Arrhenatheretalia elements (Table 15, columns 60–67)
Achilleo-Festucetum with *Festuca rubra* (Table 15, columns 68–70)
Achilleo-Festucetum with *Plantago maritima* (Table 15, columns 71–73)
Impoverished subtype occurring on mineral soils (Table 15, columns 74–79)

Coastal subtypes

Upland coastal subtype (Table 16, columns 1–5)
Upland coastal vegetation transitional to the Cynosurion cristati (Table 16, columns 6–8)
Cynosurion-rich subtype (Table 16, columns 9–17)
Vegetation transitional between the Cynosurion and Koelerio-Corynephoretea (Table 16, columns 18–22)
Koelerio-Corynephoretea rich subtype (Table 16, columns 23–29)

INVASIVE BRACKEN COMMUNITY

Bracken-dominated subtype (Table 5, column 11)

Nardo-Caricetum Binervis

Typical subtype (Table 18, columns 1–31)
Wet subtype (Table 18, columns 32–45)
Agricultural subtype (Table 18, columns 46–66)
Degraded agricultural subtype (Table 18, columns 67–72)

DERELICT SUBTYPES

Derelict Nardo-Caricetum binervis vegetation (Table 19, columns 1–4)
Vegetation with Montio-Cardaminetea elements (Table 19, columns 5–8)
Derelict vegetation of abandoned peat cuttings (Table 19, columns 9–16)
Species-poor vegetation of abandoned peat cuttings (Table 19, columns 17–20)

Meadow and Pasture Vegetation of the Molinio-Arrhenatheretea

Junco-Molinietum (Table 20, columns 1–5)
Degraded Junco-Molinietum (Table 20, column 6)
Filipendulion (Table 20, columns 7–9)
Arrhenatherion (Table 20, column 10)
Cynosurion (Table 20, columns 11–15)

ACID GRASSLANDS

Upland areas in Ireland and Great Britain are characterised by the occurrence of a wide range of vegetation types, ranging from grasslands and shrubby heathland to mire and small sedge communities (O'Sullivan 1982; Rodwell 1991; 1992; McFerran *et al.* 1994a; 1994b). The floristic and structural composition of the vegetation mosaic is thought to be related to gradients in soil moisture, nutrient supply and grazing intensity (Birse and Robertson 1976; Rodwell 1992).

The effect of the altered management practices on Clare Island has left its imprint on the landscape, resulting in the replacement of the *Calluna*-dominated vegetation in upland areas by a vegetation mosaic reflecting a combination of edaphic, climatic and anthropogenic influences (Pl. IX). Several communities now characterise the uplands of Clare Island. Shrub-dominated heathlands are very much diminished in extent, and generally confined to inaccessible cliff ledges, riverbanks or areas where sheep have been prevented from grazing. Peatland vegetation of the Oxycocco-Sphagnetea forms part of the mosaic in areas of impeded drainage, in wet hollows of drumlin areas and in sites where shallow peat has been cut away. Vegetation of the small sedge class (Scheuchzerio-Caricetea nigrae) is commonly found along the streambanks and floodplains, in waterlogged cutaways and along drainage channels.

The acid grassland communities encompass the majority of the vegetation on Clare Island,

Table 15

Relevé table for the Achilleo-Festucetum tenuifoliae

Column										1										2										3					
	1	2	3	4	5	6	7	8	9	0	1	2	3	4	5	6	7	8	9	0	1	2	3	4	5	6	7	8	9	0	1	2	3	4	5
Relevé	1	3	3			1	1	1	2	2		2			2	2		1	2	3		3			2	3	1	1	3	1	1	2	1		2
	0	0	1	9	4	0	6	9	6	9	5	4	3	2	6	6	5	9	0	9	1	7	2	2	6	1	7	3	1	0	0	1	6	3	9
	6	7	0	7	7	2	4	7	2	1	6	0	1	9	6	8	5	5	5	2	0	1	7	8	0	1	3	4	3	9	3	8	5	0	5
Achilleo-Festucetum tenuifoliae																																			
Agrostis capillaris																																			
Festuca ovina	2	2	2	3	2	2	4	5	3	3	4	2	2	2	2	4	3	3	2	2	3	3	3	3	2	3	4	2	3	1	2	2	1	3	2
Polygala serpyllifolia				r	r		r		r	+	+	r				+	+	+				+				+		+		+	r				
Viola riviniana																								+	+	r									
Nardetalia																																			
Potentilla erecta	+	+	+	1	+	+	1	+	1	+	2	1	1	1	2	1	2	+	+	+	1	+	2		1	1	3	1	+	2	+		1	1	+
Galium saxatile	1	+	+	+	+	+	2	1	1	1		2	2	1	+	1		1	2	1	1		2	1	1	+	1	1	1		1	+		2	1
Anthoxanthum odoratum		1	1	1	2	1	2		2	2		1	2	1	1	2	+	2	2	+	2	1	1		1	1	2		2			1	1		1
Holcus lanatus			+			1			1					+	1							+	1		1	+	1		1	1	1	2			
Rhytidiadelphus squarrosus	+	2	1			+				+	+	+	+		1	1	3	+	2	+			2		1		1	2	+	1	+	1	1		+
Aira praecox	1	1	+	+	+	+	1	1	1	+					+			1	1	+		1	+		+	1	1		2		2			2	
Carex viridula				+		+	+	+	+	+	1	+	1		+		1	1		+		+					+	+		+				1	
Luzula multiflora																				+															
Constant species																																			
Calluna vulgaris			+	+										1			3	+	+	2		+	2				1		+	2					
Hypnum jutlandicum	2	1	2			1	1			2					2	+		+	2	+		+				2		1	1		1	3	3		1
Peltigera canina		+		+		+	+	r		+								+	+	+		+			r	+	+	+	+	+					+
Polytrichum commune										+				1			1		2	+		+	1		2							3		1	3
Thuidium tamariscinum	+		+	+		+				1	1	1	2		+		2	+	2		2			+		1	1		+			1		1	
Mnium hornum	+	1	+	+	+		2	+	+	1		1						1	+	1		+				+		1	+		2	1	1		+
Frullania tamarisci	+	+	+	+			2					+						+	1	+				+	+	1		1	1	+		+			
Sphagnum subnitens												1	2				2		1		3	+				2				1			+		
Cladonia portentosa				+											+	+			+	+		+							+	+	+				
Leucobryum glaucum		+		2	1					1		+				1				+		+		1			1	1	+	+					
Racomitrium community																																			
Racomitrium lanuginosum																																			
Ulex community																																			
Ulex europaeus																																			
Agricultural land																																			
Prunella vulgaris																						+		+	+	+									
Plantago lanceolata																													+						
Lotus corniculatus																										+	2	1		1					
Trifolium repens																								1	1		1		+						
Thymus praecox																								1											
Danthonia decumbens																																			

				4										5										6										7										Synoptic value
6	7	8	9	0	1	2	3	4	5	6	7	8	9	0	1	2	3	4	5	6	7	8	9	0	1	2	3	4	5	6	7	8	9	0	1	2	3	4	5	6	7	8	9	
	1			2	2	2	2	1		3		3			1	1		3	1	1	2	2	2			2	3	3	3		2	3	3	3		1	1						1	
1	9	7	6	8	0	7	5	0	5	0	4	7	1	1	8	6	8	1	0	8	1	2	2	6	9	5	8	8	9		0	4	2	4		6	3		6	1	9	8	5	
7	9	5	1	6	3	6	6	7	8	9	6	4	9	4	6	9	0	4	1	4	3	1	3	3	3	2	3	6	4	8	7	5	1	4	3	8	5	9	4	1	8	7	2	
			2	1	2	4	3	3	5	2	2	+	4	5	4	2	4	5	3	5	4	2	2	2	2	3	3	+	1	2	3	2	2	1	3	2	2	2	3	5	2	3	2	V
2	1		3	2	2	2	2		4	1	2	2	2	3	3	4	3	4	2	3	3	2	3	1	3	2	2	+	2	2	2	2	+	1	1	2	4	2		2	2	3	4	V
					+	+	r	r	+		+	+						+						+		r											r		+					II
																	+	+				1	+								+		+											I
		+	1	+	+	2	2	+	2	+	1	+	1	1	1	1	+	+			1	1	1		+	2	+	2		+	2		+		1		2		1	+	+	+	3	V
1	+	+	2		1	1	1	+	2	+	1	+	2	1	1		+	+	1		1	2		+		1	+				+								1	1				IV
		1	1	1	1	2	1	1			1		2		2	1	2	1	1		1	1	1	3	1	2	2	4	2		3	3		3				2	1		3	2	3	IV
																+						1		3	1	4	4	2	3		2	4	2	4					1		1	2	1	III
	3	1	+		+	2	1	+	2						1					+	+		1				+				+	+									+	2		III
+		+					+	1						1	1		+	1	+	2	1		+																					III
				+	+	1	+		1		1				1	+			+				1							+			+		3	1	1							III
																r	+	+	+		+	1	+	+		+	+	+			+								1				+	I
1		2	1	2	+	2	3			+		1	1			+			+		2	3	1					+	+		3		1				+		+		+	2	2	III
2	1		2	2	1	3	1	1					2		3			1					1										+									+		III
				1		+	r										+	+	+		+	+	+				+												+			+	1	II
3	2	2	2		+				+				1	2			1				+							1		1	+							1						II
	1	1							1				+										+			+								+							+	+		II
		+		+	+		1	+		2	1																																	II
					+	2	2	+					1			+		+			+	+	1																			+		II
		3	1		2		1				2												+																3					II
		+		+	+	1	+	+			+	+	+																															II
				2	1	1	2			4	3							+			+																							+
			2	3	+	+	1	1	2	+	2	2	1																															I
														1	2																													R
																+	+	+	+		r	+	+			+	+		+		+		+		r		+							II
																+	+	1	+				+	1	+	+	+		1	+			+		+	+	1		r					II
																+	+	1	+		+	1		1		+	1	+	+		1		+		1				+		+	+		II
														+			+	+		1	+	1	+	1	+	1	1	1	1		2	1		+				1	1	1	+			II
																+		+	+		+	+	+										1				+							I
																+		+	1		1		1				+	+			1						1							I

(Continued)

Table 15 *(Continued)*

Column										1										2										3					
	1	2	3	4	5	6	7	8	9	0	1	2	3	4	5	6	7	8	9	0	1	2	3	4	5	6	7	8	9	0	1	2	3	4	5
Relevé	1	3	3			1	1	1	2	2		2			2	2		1	2	3		3			2	3	1	1	3	1	1	2	1		2
	0	0	1	9	4	0	6	9	6	9	5	4	3	2	6	6	5	9	0	9	1	7	2	2	6	1	7	3	1	0	0	1	6	3	9
	6	7	0	7	7	2	4	7	2	1	6	0	1	9	6	8	5	5	5	2	0	1	7	8	0	1	3	4	3	9	3	8	5	0	5
Bellis perennis																								r											
Cerastium fontanum																									+		+			+				+	
Cirsium dissectum																																			
Ranunculus repens																																			
Hypochaeris radicata																											1								
Rumex acetosa																							1		+										
Cynosurus cristatus																											1								
Festuca rubra																																			
Oxalis acetosella																																			
Plantago maritima																																			
Plantago coronopus																								+											
Hydrocotyle vulgaris																																			
Companion species																																			
Hylocomium splendens												+			+				3	+		+							+	1			+		
Carex panicea						1			+		1				+				1											1			1		+
Rhytidiadelphus triquetrus	1											1				+			2	2	2				+				2						
Erica tetralix														1			1			+										+					
Sedum anglicum	+	+			+	+	r	1																				+			+	+			
Polytrichum juniperinum						+	2				+	+								+									2						
Epilobium brunnescens																																			
Scapania gracilis				+	+																											+			1
Juncus effusus			1								1																		+				2	1	2
Juncus squarrosus															+																				+
Juncus bulbosus		+	+																													+	1		+
Rumex acetosella		+			+	+														+						+	1		+		1				
Dicranum scoparium	+							+						1				+	2			+						1					+		
Anagalis tenella			+								+														1			1	+	2					
Jasione montana				r											r							r													
Succisa pratensis																						r								+	1				
Carex nigra							1						+														2			+					
Pseudoscleropodium purum																								+								+		1	
Sagina procumbens										+			+											+											
Eurhynchium praelongum																																			
Sphagnum papillosum																																4			1
Aulacomnium palustre																													+						+
Hypnum cupressiforme					+	+					1													1											
Carex echinata																																	+		1
Radiola linoides																						+					1	+							

				4										5										6										7										
6	7	8	9	0	1	2	3	4	5	6	7	8	9	0	1	2	3	4	5	6	7	8	9	0	1	2	3	4	5	6	7	8	9	0	1	2	3	4	5	6	7	8	9	
	1			2	2	2	2	1		3		3			1	1		3	1	1	2	2	2			2	3	3	3		2	3	3	3		1	1						1	
1	9	7	6	8	0	7	5	0	5	0	4	7	1	1	8	6	8	1	0	8	1	2	2	6	9	5	8	8	9		0	4	2	4		6	3		6	1	9	8	5	
7	9	5	1	6	3	6	6	7	8	9	6	4	9	4	6	9	0	4	1	4	3	1	3	3	3	2	3	6	4	8	7	5	1	4	3	8	5	9	4	1	8	7	2	
																																												Synoptic value
																+		+						+	1				+		+		+		+		1			+				**I**
																r	+			+		r							+		+	+		+									+	**I**
																1				r	+										+				1		2				2			**I**
																	+	+						1		1	+	+	1					+				1						**I**
																									+	+			+				r			+	1							**I**
																								2		+			+			+		+									+	**I**
																								2	2	2	2	2	2	1	1								+					**II**
																																2	3	3										**R**
																																+		+				+					+	**+**
								2								1							1												3	4	2							**+**
																+		+	+	+															3	2								**+**
																+																			1	+	+		+					**I**
				+		+																					+	2											2		+			**I**
						+										1					+				1														1			+		**I**
							+				+									+																								**I**
							+		1							1																					1					+	+	**I**
											r	+							+		r																							**I**
						+	+	+		+					1				+				+																					**I**
	+																																											**I**
	1		1	+	1		1	+			+																																	**I**
1	+	+											1		1														+	2								3					1	**I**
1	2	1	+																																									**I**
1																													+										1					**I**
													+	+						1							+																	**I**
														1																														**I**
																+		+								+				1			+		3				1			+		**I**
								+															r										r	+									1	**I**
			+													+																					1					+	+	**I**
			1							+																		+	+								1					+		**I**
	1																					+						+			+			+									+	**I**
																	1			1		r			+													+		+				**I**
														1									1				+				+	+		+										**+**
					2						2																																	**+**
			1		+	+																																	2			+		**+**
															3				+																									**+**
	+																											+		1									1					**+**
												+				+																					+							**+**

(Continued)

Table 15 *(Continued)*

Column										1										2										3					
	1	2	3	4	5	6	7	8	9	0	1	2	3	4	5	6	7	8	9	0	1	2	3	4	5	6	7	8	9	0	1	2	3	4	5
Relevé	1	3	3			1	1	1	2	2		2			2	2		1	2	3		3			2	3	1	1	3	1	1	2	1		2
	0	0	1	9	4	0	6	9	6	9	5	4	3	2	6	6	5	9	0	9	1	7	2	2	6	1	7	3	1	0	0	1	6	3	9
	6	7	0	7	7	2	4	7	2	1	6	0	1	9	6	8	5	5	5	2	0	1	7	8	0	1	3	4	3	9	3	8	5	0	5
Plagiothecium undulatum			+																										+				1		
Lophocolea bidentata										+																+			+			+			+
Euphrasia officinalis																					+					+			+						
Vaccinium myrtillus																							+												
Erica cinerea											1			+																					
Campylopus paradoxus				+		+	+					+															+								
Polytrichum formosum	1	+	+		+								1																				+		
Cirsium palustre																									+										
Leontodon autumnalis																																			
Pteridium aquilinum																								1	+						2				
Achillea millefolium																																			
Rhytidiadelphus loreus																						+													
Carex pulicaris																																			
Viola palustris																																			
Salix repens																																			
Dicranum majus	+	+																																	
Campylopus introflexus										1																+									
Campylopus pyriformis															1								1												
Pleurozium schreberi																																	1	2	
Polytrichum piliferum									1						1	1																			
Mylia taylorii																																			
Blechnum spicant																		+																	
Carex pilulifera																																			
Molinia caerulea																																			
Nardia scalaris																																			+
Dactylorhiza maculata																																			
Eleocharis multicaulis											1																								
Hymenophyllum wilsonii																					1											2			
Isothecium myosuroides								+																											
Sphagnum capillifolium																																			1
Lepidozia reptans																																			
Pellia epiphylla																									+										
Plagiochila punctata																														+					
Pleurozia purpurea																																			
Cladonia pyxidata															+	+																			
Cladonia uncialis																																			
Saxifraga spathularis				r																															
Carex sp.																								+											

				4										5										6										7										
6	7	8	9	0	1	2	3	4	5	6	7	8	9	0	1	2	3	4	5	6	7	8	9	0	1	2	3	4	5	6	7	8	9	0	1	2	3	4	5	6	7	8	9	
	1			2	2	2	2	1		3		3			1	1		3	1	1	2	2	2			2	3	3	3		2	3	3	3		1	1						1	
1	9	7	6	8	0	7	5	0	5	0	4	7	1	1	8	6	8	1	0	8	1	2	2	6	9	5	8	8	9		0	4	2	4		6	3		6	1	9	8	5	
7	9	5	1	6	3	6	6	7	8	9	6	4	9	4	6	9	0	4	1	4	3	1	3	3	3	2	3	6	4	8	7	5	1	4	3	8	5	9	4	1	8	7	2	**Synoptic value**
		1																								+								+									2	**+**
																																1		2										**+**
																	+	+									+																	**+**
	+		+		+																																							**+**
						+							+									+									2													**+**
								+																																				**+**
																																							+					**+**
																										2			r		1													**+**
												r													+	+											+		+					**+**
																		2																				1					1	**+**
																		+		+				+			+																	**+**
																						1		+										1										**+**
																+							+								+		+											**+**
																												1		1		+							+					**+**
																+																										+		**R**
		+																																										**R**
2																																												**R**
						+																																						**R**
									+																																			**R**
																																												R
						+																																						**R**
					+																										+													**R**
			+																																									**R**
	1																																										2	**R**
																																												R
																									+																			**R**
																																												R
																																												R
																+																											1	**R**
																																												R
																																												R
																																						+						**R**
																																												R
							+																																					**R**
																																												R
							+														+		1																					**R**
																																		r										**R**
																																												R

(Continued)

Table 15 *(Continued)*

Column									1										2										3					
	1	2	3	4	5	6	7	8	9	0	1	2	3	4	5	6	7	8	9	0	1	2	3	4	5	6	7	8	9	0	1	2	3	4	5
Relevé	1	3	3			1	1	1	2	2		2			2	2		1	2	3		3			2	3	1	1	3	1	1	2	1		2
	0	0	1	9	4	0	6	9	6	9	5	4	3	2	6	6	5	9	0	9	1	7	2	2	6	1	7	3	1	0	0	1	6	3	9
	6	7	0	7	7	2	4	7	2	1	6	0	1	9	6	8	5	5	5	2	0	1	7	8	0	1	3	4	3	9	3	8	5	0	5
Cerastium diffusum																																			
Pedicularis sylvatica																																			
Poa annua																																		2	
Poa trivialis																								1											
Polypodium vulgare																																			
Atrichum undulatum																																			
Bryum pseudotriqeutrum																																			
Campylopus atrovirens																																	+		
Dicranella heteromalla				+																															
Polytrichum alpestre																	1																		
Sphagnum molle																																			
Sphagnum palustre													1																						
Radula complanata																+																			
Scapania irrigua																																			
Cladonia floerkeana																																			
Cladonia sp.	+																																		
Sphagnum auriculatum																																			3
Lophozia sp.																																			
Plagiochila asplenioides																																			
Campylopus pyriformis																																			
Polytrichum piluliferum																																			
Viola sp.																																			
Ranunculus flammula																																			
Carex binervis																																			
Agrostis stolonifera																																			
Carex flacca																																			
Drosera rotundifolia																																			
Lysimachia nemorum																																			
Calliergon cuspidatum																																			
Agrostis canina																																			
Angelica sylvestris																																			
Carex ovalis																																			
Centaurium erythraea																																			
Primula vulgaris																																			
Selaginella selaginoides																																			
Iris pseudacorus																																			
Lemna minor																																			
Mentha aquatica																																			
Sphagnum palustre																																			
Eleocharis palustris																																			

36	37	38	39	40	41	42	43	44	45	46	47	48	49	50	51	52	53	54	55	56	57	58	59	60	61	62	63	64	65	66	67	68	69	70	71	72	73	74	75	76	77	78	79	
17	199	75	61	286	203	276	256	107	58	309	46	374	19	14	186	169	80	314	101	184	213	221	223	63	93	252	383	386	394	8	207	345	321	344	3	168	135	9	64	11	98	87	152	Synoptic value
														+																														R
																														1														R
																													+											2				R
																				3					2																			R
		+																																										R
				+																																								R
																	+																											R
																						+																						R
																																				+								R
																																												R
				3																																								R
																																												R
																																												R
					1																																							R
+																																												R
																																												R
																																												R
																																												R
										+																																		R
																																	+											R
																						+																						R
																			+		1																+							R
																									2					1									+					R
																										1		+			1													R
																																				1								R
																									1														2					R
																														+									+					R
																													+	1				+										R
																									2								+											R
																												2						1										R
																																r	r											R
																												2																R
																+										1																		R
																																r	+											R
																							+										+											R
																														4														R
																																						2						R
																														2														R
																												1																R
																														1														R

(Continued)

Table 15 *(Continued)*

Column										**1**										**2**										**3**					
	1	**2**	**3**	**4**	**5**	**6**	**7**	**8**	**9**	**0**	**1**	**2**	**3**	**4**	**5**	**6**	**7**	**8**	**9**	**0**	**1**	**2**	**3**	**4**	**5**	**6**	**7**	**8**	**9**	**0**	**1**	**2**	**3**	**4**	**5**
Relevé	**1**	**3**	**3**			**1**	**1**	**1**	**2**	**2**		**2**			**2**	**2**		**1**	**2**	**3**		**3**			**2**	**3**	**1**	**1**	**3**	**1**	**1**	**2**	**1**		**2**
	0	**0**	**1**	**9**	**4**	**0**	**6**	**9**	**6**	**9**	**5**	**4**	**3**	**2**	**6**	**6**	**5**	**9**	**0**	**9**	**1**	**7**	**2**	**2**	**6**	**1**	**7**	**3**	**1**	**0**	**0**	**1**	**6**	**3**	**9**
	6	**7**	**0**	**7**	**7**	**2**	**4**	**7**	**2**	**1**	**6**	**0**	**1**	**9**	**6**	**8**	**5**	**5**	**5**	**2**	**0**	**1**	**7**	**8**	**0**	**1**	**3**	**4**	**3**	**9**	**3**	**8**	**5**	**0**	**5**
Callitriche stagnalis																																			
Carex hostiana																																			
Myosotis secunda																																			
Glaux maritima																																			
Arrhenatherum elatius																																			
Carex distans																																			
Cerastium glomeratum																																			
Digitalis purpurea																																			
Galium palustre																																			
Juncus acutiflorus																																			
Leontodon taraxacoides																																			
Luzula sylvatica																																			
Oxyria digyna																																			
Trichophorum caespitosum																																			
Senecio aquaticus																																			
Stellaria alsine																																			
Trifolium pratense																																			
Triglochin maritima																																			
Campylium stellatum																																			
Ulota crispa																																			
Taraxacum officinale																																			
Carex sp.																																			
Hypericum pulchrum																																			
Schoenus nigricans																																			
Koeleria macrantha																																			
Narthecium ossifragum																																			
Fissidens adianthoides																																			
Number of species	**1**	**1**	**2**	**2**	**1**	**2**	**1**	**1**	**1**	**2**	**1**	**1**	**1**	**1**	**2**	**1**	**1**	**2**	**2**	**2**	**1**	**2**	**1**	**2**	**2**	**2**	**2**	**1**	**3**	**2**	**1**	**2**	**2**	**1**	**2**
	7	**9**	**1**	**3**	**8**	**0**	**9**	**4**	**4**	**0**	**6**	**9**	**5**	**3**	**2**	**5**	**5**	**0**	**2**	**2**	**1**	**6**	**3**	**0**	**4**	**3**	**4**	**9**	**0**	**4**	**6**	**0**	**0**	**4**	**3**

				4										5										6										7										
6	7	8	9	0	1	2	3	4	5	6	7	8	9	0	1	2	3	4	5	6	7	8	9	0	1	2	3	4	5	6	7	8	9	0	1	2	3	4	5	6	7	8	9	
	1			2	2	2	2	1		3		3			1	1		3	1	1	2	2	2			2	3	3	3		2	3	3	3		1	1						1	
1	9	7	6	8	0	7	5	0	5	0	4	7	1	1	8	6	8	1	0	8	1	2	2	6	9	5	8	8	9		0	4	2	4		6	3		6	1	9	8	5	
7	9	5	1	6	3	6	6	7	8	9	6	4	9	4	6	9	0	4	1	4	3	1	3	3	3	2	3	6	4	8	7	5	1	4	3	8	5	9	4	1	8	7	2	Synoptic value
																																						1						R
																																												1	R
																																						1						R
																																				1								R
																												+																R
																														+														R
																																	+											R
																																						+						R
																																+												R
																									+																			R
																																	+											R
																																		+										R
																																	+											R
																														+														R
																												+																R
																																						+						R
																												+																R
																																								+					R
																																				+								R
																																	+											R
																									+																			R
																																								+					R
																																	r											R
																																			2									R
																	+																												R
																	+																												R
																	+																												R
1	1	1	2	1	2	2	2	1	1	1	1	1	1	1	1	3	1	2	1	1	2	2	2	1	1	2	2	2	2	1	2	1	3	2	1	1	2	1	2		1	2	2	
4	8	9	1	5	6	5	6	7	4	4	8	6	9	3	6	2	8	7	9	2	3	2	7	6	2	3	1	4	1	8	6	7	1	1	4	2	1	6	9	8	4	3	1	

Table 16

Relevé table for the coastal subtypes of the Achilleo-Festucetum tenuifoliae

Column									1										2										
	1	2	3	4	5	6	7	8	9	0	1	2	3	4	5	6	7	8	9	0	1	2	3	4	5	6	7	8	9	
Relevé							1		1	1	4			3				2	3	1	3	2			1		1	1		
	2	2	2	2	4	4	4	5	5	7	0	3		8	8	5	8	2	2	2	8	2		2	3	8	1	2		
	4	6	5	1	9	2	0	0	3	2	0	9	2	1	5	3	9	5	0	7	0	4	5	0	2	4	9	0	1	**Synoptic value**
Achilleo-Festucetum tenuifoliae																														
Festuca ovina/rubra	3	3	3	1	3	3	4	3	4	3	5	2	2	2	4	2	4	4	5	4	2	4	4	3	3	3	3	4	2	V
Agrostis capillaris	3	3	3	2	3	4	2	4	2	2	2	2	2	1	3	3	5	2	3	2	+	2	2	5	4		3	3	2	V
Potentilla erecta	1	1	1	2	+	2	1	1	+					+	+		+	+	+	1	+		1		1		1	+		IV
Luzula multiflora	1		+			+	+	+					+				+		+		+	+	+	+	1		1	+		III
Galium saxatile	2	1	1	2	2		+								1		+		+					1	1		+			III
Anthoxanthum odoratum			2		2	2			1		+			2	2	2	+	1	+		2									III
Lotus corniculatus					+	+	+		+	+					1	1	+			+	1		1		+	+	+			III
Nardetalia species																														
Nardus stricta	2	1	3	1	4	2	2	4									1							2	2					II
Danthonia decumbens		1	2	2	1	2	1		1								+						1							II
Polytrichum commune	2		1	+	1																			+			2			I
Juncus squarrosus	2	2		4																										+
Plantaginetum coronopodo-maritimi																														
Plantago lanceolata			+			+	1		1	1	1	+	2	+	+	+	+	1	1	1	+	1	+	1	1	+	2			IV
Plantago maritima						2	2	2		+	+	4	2		1	1		2	+		1		1	2	+		+	1	4	III
Plantago coronopus						+	+	1	+	+	+	2	4	1			+							1		1	2	2	2	III
Anagallis tenella						1	+	+	1					+	5	1	+		+	1	+						+			III
Hypochaeris radicata						+	2		3		+					+		1			+	r		+		+	+			II
Cerastium fontanum						+			+	+					+			+	+				+		+				+	II
Thymus praecox						+	+		+		+		+						+			+		1		+				II
Sagina procumbens													2		+	+	r	1				1	2				+	+		II
Radiola linoides							+	+								+	+										+	2		II
Cynosurion cristati species																														
Holcus lanatus							1	1	2		1	1	1	3	2	2	1	3	2	3	2	2								III
Trifolium repens						1	+		1	2		1	1	+	+	1	+	2	+			+	+							III
Prunella vulgaris						+	1	r	+		r				+	r	+	1	+		+	1	+							III
Carex panicea					1	1	+	1	1		+	+		1			+		+		+									II
Leontodon autumnalis						+	+		1	+	+		+				r													II
Cirsium dissectum						1	1	+	r				+	1			+				2									II
Carex viridula					1	2	+	1				1	1				1		2						+			+		II
Hydrocotyle vulgaris					1	+	+	+				+	1																	II
Cynosurus cristatus							2			4				1	1	2														I
Trifolium pratense									1	1	1												+							I
Ranunculus repens										2					1	2		1												I
Daucus carota									1	+	+				+											+				I
Bellis perennis							+			1	+		+		+															I

(Continued)

Table 16 *(Continued)*

	1	2	3	4	5	6	7	8	9	10	11	12	13	14	15	16	17	18	19	20	21	22	23	24	25	26	27	28	29	
Column										**1**										**2**										
	1	**2**	**3**	**4**	**5**	**6**	**7**	**8**	**9**	**0**	**1**	**2**	**3**	**4**	**5**	**6**	**7**	**8**	**9**	**0**	**1**	**2**	**3**	**4**	**5**	**6**	**7**	**8**	**9**	
Relevé							**1**		**1**	**1**	**4**			**3**				**2**	**3**	**1**	**3**	**2**			**1**		**1**	**1**		
	2	**2**	**2**	**2**	**4**	**4**	**4**	**5**	**5**	**7**	**0**	**3**		**8**	**8**	**5**	**8**	**2**	**2**	**2**	**8**	**2**		**2**	**3**	**8**	**1**	**2**		
	4	**6**	**5**	**1**	**9**	**2**	**0**	**0**	**3**	**2**	**0**	**9**	**2**	**1**	**5**	**3**	**9**	**5**	**0**	**7**	**0**	**4**	**5**	**0**	**2**	**4**	**9**	**0**	**1**	**Synoptic value**
Koelerio-Corynephoretea species																														
Sedum anglicum	+		+		r				+													2	+	+	+	+	+	+	+	III
Jasione montana									+									+	r	+	+	2	+				+	+		II
Aira praecox			1													+								2	2		2	2	1	II
Koeleria macrantha																	2	3	+	3		2				+				II
Hypnum cupressiforme				1															1					+				2		I
Companion species																														
Calluna vulgaris		+																	1	+	2	2	+				+	+		II
Euphrasia officinalis						+		+			+	r							+				r	+					+	II
Rumex acetosa										2					+		+	+				1					+			II
Viola riviniana								+			+								+		+	+		r						II
Cladonia portentosa		+	+		+							+															+		+	II
Peltigera canina															+		+		+	+					+		+			II
Hypnum jutlandicum		1															2				+				1		1			I
Sphagnum subnitens		2			2			+											+					1						I
Rumex acetosella	+		+																+				2					1		I
Carex nigra			+	+		+								+						2										I
Armeria maritima													1									1				2			3	I
Angelica sylvestris									+	1								1		1										I
Thuidium tamariscinum					+		+	1									1													I
Dicranum scoparium	1							+												+							1			I
Mnium hornum										+															+		+	1		I
Centaurium erythraea						+	r		1		+																			I
Succisa pratensis																		1	r	+	+									I
Festuca vivipara				2													1											1		I
Juncus bulbosus				1	1											+														I
Carex pulicaris									1					+							+									I
Erica tetralix														+							1		+							I
Ophioglossum vulgatum												r			r		+													I
Polygala serpyllifolia				r	r				+																					I
Pseudoscleropodium purum														+					+		+									I
Rhytidiadelphus squarrosus		1																	+						+					I
Frullania tamarisci			+																+									+		I
Carex binervis															1				2											+
Sphagnum palustre					2									1																+
Campylopus pyriformis		1			1																									+
Carex flacca																				2							+			+
Salix repens																	+				2									+
Juncus articulatus														+		2														+
Narthecium ossifragum		+												1																+

(Continued)

Table 16 *(Continued)*

Column	1	2	3	4	5	6	7	8	9	10	11	12	13	14	15	16	17	18	19	20	21	22	23	24	25	26	27	28	29	
Relevé							1		1	1	4			3				2	3	1	3	2			1		1	1		
	2	2	2	2	4	4	4	5	5	7	0	3		8	8	5	8	2	2	2	8	2		2	3	8	1	2		
	4	6	5	1	9	2	0	0	3	2	0	9	2	1	5	3	9	5	0	7	0	4	5	0	2	4	9	0	1	**Synoptic value**
Pteridium aquilinum									+														1							+
Aulacomnium palustre					1									+																+
Leucobryum glaucum																								+			1			+
Polytrichum juniperinum																						1			+					+
Tortula ruralis ssp. *ruraliformis*											+															1				+
Ulota crispa									+																				1	+
Rhytidiadelphus loreus					+									+																+
Primula vulgaris											+							+												+
Ranunculus flammula							+						+																	+
Viola palustris						+								+																+
Eurhynchium praelongum							+																		+					+
Viola sp.							r		+																					+
Elymus repens										2																				**R**
Arrhenatherum elatius																		2												**R**
Eleocharis multicaulis														2																**R**
Juncus effusus																2														**R**
Plantago major																				2										**R**
Silene maritima																										2				**R**
Equisetum arvense																					1									**R**
Erica cinerea																						1								**R**
Eriophorum angustifolium														1																**R**
Molinia caerulea														1																**R**
Poa annua																							1							**R**
Vaccinium myrtillus		1																												**R**
Vaccinium oxycoccos	1																													**R**
Calliergon cuspidatum														1																**R**
Isothecium myosuroides																				1										**R**
Agrostis stolonifera											1																			**R**
Carex pilulifera																								1						**R**
Carex echinata														+																**R**
Carex paniculata																			+											**R**
Sagina subulata												+																		**R**
Schoenus nigricans														+																**R**
Selaginella selaginoides																			+											**R**
Senecio aquaticus																					+									**R**
Spergularia rupicola																										+				**R**
Stellaria media																													+	**R**
Tussilago farfara																					+									**R**
Hypericum pulchrum																		+												**R**
Pleurozium schreberi	+																													**R**
Sphagnum capillifolium														+																**R**
Aneura pinguis														+																**R**
Scapania gracilis																											+			**R**

(Continued)

Table 16 *(Continued)*

Column									1										2										
	1	2	3	4	5	6	7	8	9	0	1	2	3	4	5	6	7	8	9	0	1	2	3	4	5	6	7	8	9	
Relevé							1		1	1	4			3				2	3	1	3	2			1		1	1		
	2	2	2	2	4	4	4	5	5	7	0	3		8	8	5	8	2	2	2	8	2		2	3	8	1	2		
	4	6	5	1	9	2	0	0	3	2	0	9	2	1	5	3	9	5	0	7	0	4	5	0	2	4	9	0	1	**Synoptic value**
Bryum capillare																			+											R
Centaurea nigra															+															R
Dactylorhiza purpurella																					+									R
Cladonia uncialis																									+					R
Cladonia sp.																										+				R
Cirsium sp.																			+											R
Taraxacum officinale																		+												R
Cerastium sp.																						+								R
Cirsium palustre																					r									R
Achillea millefolium																		r												R
Sonchus arvensis																		r												R
Dactylorhiza majalis														r																R
Number of species	1	1	1	1	2	2	3	2	2	1	2	1	1	3	2	1	2	2	3	1	2	2	2	1	2	1	2	1	1	
	3	5	6	3	2	6	0	0	9	8	1	4	7	1	3	9	9	4	4	9	8	0	1	9	0	4	7	9	2	

both in commonage regions and agricultural land (Pl. IX). By far the greatest proportion of the grassland mosaic is now covered by vegetation of the Nardetea. The shallow, impoverished podzols are ideal for the development of grass-dominated communities and rough grazing pasture. These impoverished grassland communities are distinctive, since they are generally dwarfed as a result of the significant grazing intensity that is almost universally applied in the uplands throughout the island. Some authors believe that in other locations such low-growing grassland swards are partly maintained by climate (Bleasdale and Sheehy-Skeffington 1992; 1995; Bleasdale 1998).

Floristics

The acid grasslands are characterised by *Festuca ovina*, *Agrostis capillaris*, *Nardus stricta* and *Anthoxanthum odoratum*, which are the most constant and generally most extensive components. At higher altitudes, *Festuca vivipara* becomes more abundant, while *Holcus lanatus* is locally abundant in drier stands of the vegetation, often accompanied by *Aira praecox* or *Aira caryophyllea*. Sheep's bent, *Nardus stricta*, is frequent and sometimes co-dominant with *Agrostis* and *Festuca*. The small, distinctive tussocks of *Nardus* are unpalatable to sheep (Grant *et al.* 1985). The animals often uproot the grass, helping it to further extend its distribution by dislodging the seeds, so that it generally achieves its greatest dominance in severely grazed areas, along with *Danthonia decumbens* and *Juncus squarrosus*.

Other graminoid species found in the upland grassland systems are generally associated with mineral-rich soils and are indicative of the agricultural influences in abandoned fields or farm enclosures. *Holcus lanatus* is occasionally found in drier stands, but it attains its greatest abundance either where cattle or sheep gather or in relict field systems. Other species indicative of past agricultural influence include *Cynosurus cristatus*, *Poa trivialis*, *Poa pratensis* and, rarely, *Lolium perenne*. Both of the *Poa* species are especially common in vegetation at the base of walls near gates, while *Cynosurus cristatus* is a regular component of agricultural areas along the southern coastline. Along coastal areas *Festuca rubra* and *Koeleria macrantha* are occasional components

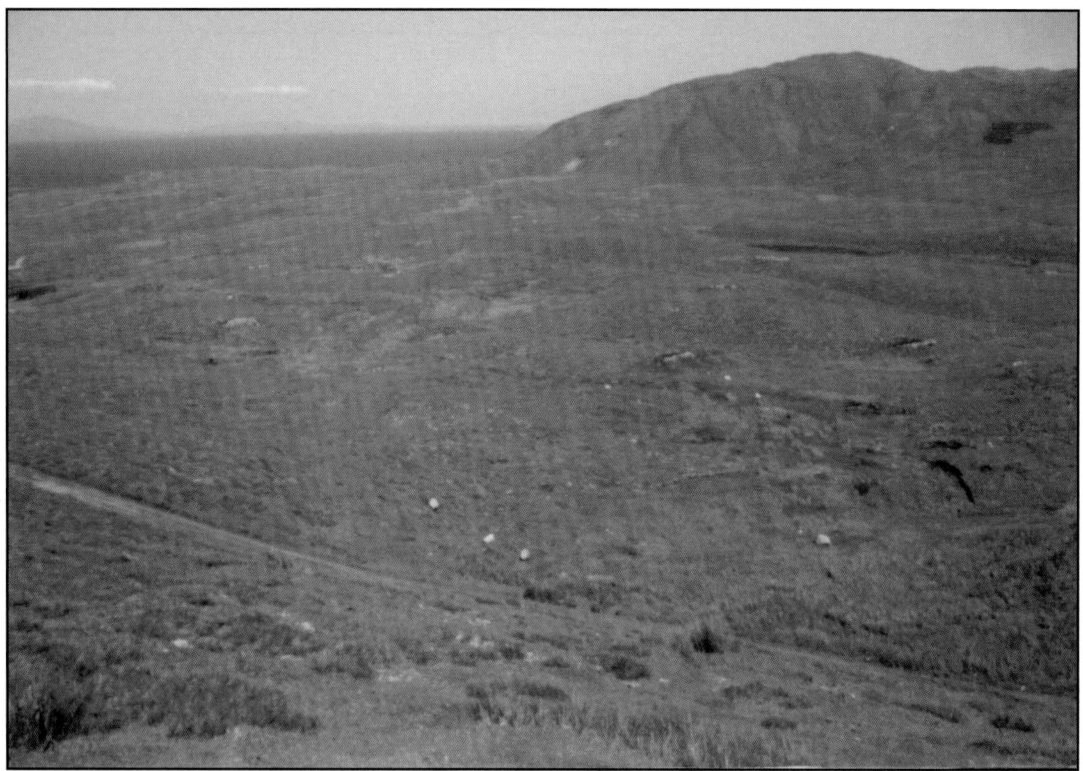

Pl. IX Park, in the centre of the island, was characterised by heathland in 1911. Remnants of heather persisted up until the middle of the twentieth century, when drastic changes in the sheep population resulted in the replacement of the heather community by a vegetation mosaic of acid grassland communities.

in the grasslands. The red fescue, *F. rubra*, is abundant in areas where grazing pressure is reduced, especially along cliff margins where fencing has been erected to safeguard sheep. Although frequently recorded from damp upland habitats in Britain, *Agrostis canina* is not frequently recorded from the grassland commonage on Clare Island. Stunted forms of *Molinia caerulea*, typically found on deeper peats and blanket bogs, are sometimes encountered in upland grasslands where pockets of peat have accumulated in topographical depressions. In a number of places, *Deschampsia flexuosa* forms a tall-growing sward: this type is generally confined to places where the main grazing is by cattle.

The impact of various forms of management and utilisation are often reflected by subtle variations within the dicotyledonous flora of these grasslands. Most of the dicots are only locally abundant and then not very frequent. The only constantly occurring species is *Potentilla erecta*, accompanied by occasional *Galium saxatile, Viola riviniana, Polygala serpyllifolia, Anagallis tenella, Rumex acetosa* and *Euphrasia officinalis*. At higher elevations, *Epilobium brunnescens* and *Vaccinium myrtillus* are recorded. Vegetation that has developed over dry, stony and somewhat less acidic soil often has *Thymus praecox, Jasione montana* and *Bellis perennis* present. Some species, such as *Lotus corniculatus, Prunella vulgaris* and *Plantago lanceolata*, are recorded in transitional stands associated with agricultural land at elevations up to 200 metres. Other species, such as *Trifolium repens, Succisa pratensis, Cerastium fontanum, Achillea millefolium* and *Hypochaeris radicata*, are generally confined to relatively richer mineral soils.

While grazing has almost eradicated the shrub-dominated heathland vegetation, impoverished shrubs of *Calluna vulgaris* and *Erica tetralix* are often found among the grassland swards. The denuded remains of the heathers may have on the underside of the woody branches small green shoots that remain inaccessible to grazing sheep. Where grazers are excluded over the summer months, more robust *Calluna* with a prostrate growth form is found. Such heather is typically found in some under-utilised agricultural areas, rather than in commonage areas (Pl. X).

Rushes and sedges are an integral part of the grassland vegetation, although these remain only locally abundant. Constant occurrences include

Pl. X In under-utilised land, the development of prostrate heather can occur, as in this photograph taken in Strake. Note also the distinctive fairy-ring growth form that is typical of the *Juncus squarrosus*, a species that is unpalatable to sheep.

Carex viridula, *Carex panicea* and *Luzula multiflora*, while in damper situations *Carex echinata* and *Juncus effusus* are found. Other common rushes include *Juncus articulatus*, *J. acutiflorus* or *J. bulbosus*, which are found on bare soil where percolating water occurs and in ephemeral mountain streambeds. Rosettes of *Juncus squarrosus* are distinctive in wet acid grasslands and attain dominance in areas that have been severely grazed.

In the short close turf, bryophytes are usually only locally abundant. Their cover is quite variable, depending on the structure of the sward and the proportion of bare peat. On terraced areas, on bare patches of soil or where soils are wet, *Sphagnum* species are common components of the grassland community. Where there is a gradation to wet heath or bog vegetation, *Sphagnum* species can become locally dominant. Common species include *S. subnitens* and *S. palustre*, with *S. papillosum* in damper areas and on terraced slopes. The common true mosses include *Rhytidiadelphus squarrosus* and *Thuidium tamariscinum*, while *Mnium hornum*, *Polytrichum commune* (*P. formosum* on mineral soils), *Eurhynchium* spp, *Hylocomium splendens*, *Pseudoscleropodium purum* and *Aulacomnium palustre* occur, but with reduced cover. Patches of *Hypnum jutlandicum* are found in very dwarf turf, while cushions of *Leucobryum glaucum* are especially common on ridges. In some dry areas *Racomitrium lanuginosum* is locally abundant.

Lichens and liverworts are not especially characteristic or abundant in the grassland vegetation: *Cladonia uncialis*, *C. portentosa* and *Frullania tamarisci* are the most common, while the dog lichen, *Peltigera canina*, is particularly obvious on the dry, eastern slopes of Knockmore.

Phytosociology

All of the acid grasslands on Clare Island are related to the widely distributed and easily identifiable *Agrostis–Festuca* grassland or rough grazings (Tansley 1939; McVean and Ratcliffe 1962; Birse and Robertson 1976; Birse 1980; O'Sullivan 1982; Rodwell 1992), classified within the Nardetea. Attempts to produce a satisfactory classification of Irish upland grassland vegetation have not been successful (O'Sullivan 1976; 1982; Cotton 1975). Grassland vegetation mosaics are extensive over hill slopes where grazing of stock hinders reversion to heath vegetation. Continual grazing has been shown to reduce the species diversity of similar habitats in Britain (Rawes 1981; Ball *et al.* 1982; Welch 1984a; 1984b; 1986; Miles 1988). However, Rawes (1981) emphasised the heterogeneity of these grasslands, which were subject to constant species compositional changes. He considered it unwise to offer conclusions, based on short-term exclusion experiments, such as those at Moor House, since for other than grasses a decline in the number of species was recorded, even after the exclusion experiments had ceased.

The acid grassland vegetation on Clare Island is assigned to two separate associations within the Nardetea: the Achilleo-Festucetum tenuifoliae (Tables 15, 16, and 17) and the Nardo-Caricetum binervis (Tables 18 and 19). There is considerable variation in the physiognomy and overall appearance of the vegetation. This is particularly true of commonage areas where the signs of historical management/habitation are imprinted on the landscape, complicating the pattern of the vegetation mosaic. Despite the syntaxonomic differentiation, both associations are closely related with very few clearly defined vegetation boundaries and with many plant species in common. Many of these species have wide ecological amplitudes, characterising other vegetation types such as lowland grasslands, heathlands and mires. The Nardo-Caricetum binervis is distinguished primarily from the Achilleo-Festucetum tenuifoliae on the occurrence of *Nardus stricta* in the vegetation.

Subtypes Occurring on Podzolic Soils

SEVERELY GRAZED
ACHILLEO-FESTUCETUM VEGETATION

Nardetea Rivas Goday et Borja Carbonell 1961
Nardetalia Prsg. 1949
Nardo-Galion saxatilis Prsg.1949
Achilleo-Festucetum tenuifoliae Birse et Robertson 1976 Table 15

Nardetea diagnostic species: *Polygala serpyllifolia, Galium saxatile, Juncus squarrosus, Danthonia decumbens, Festuca vivipara*
Achilleo-Festucetum tenuifoliae diagnostic species: *Festuca ovina, Agrostis capillaris*
Constant species: *Potentilla erecta, Anthoxanthum odoratum, Rhytidiadelphus squarrosus, Calluna vulgaris, Hypnum jutlandicum*

Heritage Council analogue: Dry-humid acid grassland—GS3

The occurrence of the Achilleo-Festucetum tenuifoliae is well documented in the British Isles (e.g. McVean and Ratcliffe 1962; Birks 1973; Birse and Robertson 1976; Birse 1980; Rodwell 1992), occurring on dry soils in the sub-montane zone between agricultural land and more heathy vegetation. In the BNVC scheme, the species-poor community is classed in the *Festuca ovina–Agrostis capillaris–Galium saxatile* U4 grassland (Rodwell 1992). O'Sullivan (1982) considers that the Achilleo-Festucetum tenuifoliae is the only adequately defined association in the class in Ireland, based upon a review of Irish literature and research. It is widespread on the lower slopes of hills in the eastern part of Ireland and found at elevations ranging from 200m to 400m. In the west of Ireland the vegetation is floristically less diverse than that recorded elsewhere and is rarely free from the impacts of grazing animals. It generally occurs at lower altitudes and is often bounded by blanket bog. Eight subcommunities of the association, representing the transition from podzolic to mineral soils, are recognised from Clare Island. These groups are found in elevations ranging from sea level to the summit of Knockmore (465m).

The first group (Table 15, columns 1–38) represents a severely grazed form of the original hill vegetation at intermediate elevations ranging from 100m to 250m. It is generally confined to drier soils and is most noticeable along old green roads and tracks that intersect much of the commonage. This has a basic complement of 30 species per square metre, with an average of 21 species in any one stand. Broad-leaved grasses such as *Agrostis capillaris* and *Anthoxanthum odoratum*, along with the ubiquitous *Festuca ovina*, dominate the community, with occasional species of more mesophytic pasture present. Associated species include *Potentilla erecta, Galium saxatile, Rhytidiadelphus squarrosus, Hypnum jutlandicum, Thuidium tamariscinum, Polytrichum commune, Mnium hornum, Frullania tamarisci* and *Cladonia portentosa*. Dwarf shrubs are generally a minor component of the vegetation but may have a cover of up to 25% in places. The distinctive physiognomy of *Calluna* in this grassland highlights the dramatic changes that have occurred in the commonage regions where significant sheep numbers have eradicated all large bushes of heather, leaving a landscape littered with dwarfed woody remains, with small leafy shoots confined to the underside of old wooden shoots. At higher elevations (200m to 350m), the order of the dominant species is altered, with an increase in the abundance of *Festuca ovina*. On the drier areas *Aira praecox, Plantago lanceolata, Lotus corniculatus* and *Trifolium repens* are also present, while in wetter locations *Juncus effusus* is more common.

RACOMITRIUM-RICH
ACHILLEO-FESTUCETUM

The vegetation represented in columns 39–49 (Table 15) is not very widespread and is generally found at intermediate altitudes on Clare Island. It has similarities with some of the *Racomitrium* heathlands found on similar substrates throughout the British Isles (Braun-Blanquet and Tüxen 1952; Mhic Daeid 1976), since *Racomitrium lanuginosum, Cladonia portentosa, Campylopus paradoxus, Polytrichum juniperinum* and *P. formosum* are a conspicuous component of the vegetation. Again the dominant grassland elements are *Agrostis capillaris, Festuca ovina* and *Anthoxanthum odoratum*. Apart from *Galium saxatile* though, some Nardetalia species are less abundant than in the previous group. It is characteristically located in level areas underlain by rock fragments, such as old tracks and areas of scree accumulation on the commonage slopes. Its closest analogue from the BNVC is the *Racomitrium lanuginosum* sub-community of the *Nardus stricta–Galium saxatile* U5 grasslands

(Rodwell 1992), although Birse (1980) and Rodwell (1992) note several *Racomitrium*-rich communities occurring in upland situations in Britain, which seem to be controlled by the surrounding vegetation as much as the underlying soil type and slope.

ACHILLEO-FESTUCETUM WITH *ULEX EUROPAEUS*

This community (Table 15, columns 50 and 51) is found confined to several isolated locations on the island, on soils that are glacially derived that are relatively drier and more freely draining than those supporting the surrounding vegetation. Floristically, the vegetation is composed of a depauperate Achilleo-Festucetum tenuifoliae species assemblage with the ground flora dominated by *Agrostis capillaris* and *Festuca ovina*, together with isolated bushes of *Ulex europaeus* that are up to one metre high. The impact of grazing, however, has altered the vegetation structure so that the grass sward is now the predominant feature of the vegetation. This disturbed community may represent the remnants of vegetation once dominated by *Ulex*, which would have been referred to the Calluno-Ulicetea.

ACHILLEO-FESTUCETUM OCCURRING ON NON-PODZOLIC SOILS

The remaining five subcommunities are also classified in the Achilleo-Festucetum tenuifoliae (analogous with the typical subcommunity of the *Festuca ovina–Agrostis capillaris–Galium saxatile* U4 grassland of the BNVC scheme). In general the vegetation is richer in species, and contains a number typical of the Molinio-Arrhenatheretea and might overlap with the more calcicolous *Festuca ovina–Agrostis capillaris–Thymus praecox* CG10 grassland. These include several grassland elements indicative of agricultural influence. The vegetation is found on better draining mineral soils (10cm to 40cm deep) on the majority of agricultural land and some of the steeper slopes of the commonage. It is also recognised from derelict summer settlements, booley sites and abandoned pre-Famine sites located on the commonage. The vegetation reflects the earlier agricultural utilisation of these areas, and includes grasslands that are not given over to crop or hay production. It also includes some of the abandoned sites, which are frequently invaded by bracken, *Pteridium aquilinum* (Pls XIVA and XIVB, see p. 94). An impressive and extensive network of ancient crop drills underlies most of the vegetation.

The concentration of the majority of the agricultural land in comparatively sheltered regions is along the south coast of Clare Island, which coincides with better draining, shallow loam soils. The vegetation is visually distinct from the hill sward in that it is generally less closely cropped. The soils retain some of the effects of the sand and seaweed that were applied in historical times to the farmland to enhance the soil fertility, so the vegetation is generally more healthy than similar vegetation on the poorer soils that are widespread in commonage areas. There is a considerable reduction in stocking rates on these areas, since sheep are kept off much of the land during the summer months. Sheep are only herded infield when the grazing on the commonage becomes exhausted.

From an agricultural viewpoint this community is an important type, generally supplying the best quality herbage for livestock on Clare Island. Again, *Agrostis capillaris* and *Festuca ovina* are the major graminoids, although *Anthoxanthum odoratum* and *Holcus lanatus* provide significant cover. Associated constant species include *Potentilla erecta, Prunella vulgaris, Galium saxatile, Plantago lanceolata, Lotus corniculatus, Luzula multiflora* and *Bellis perennis*. In certain situations on commonage, particularly on land that is under-utilised or not readily accessed by sheep, *Calluna vulgaris* may become re-established, with its prostrate growth forming a distinctive and abundant component of the vegetation (Pl. X)

Typical Achilleo-Festucetum occurring on mineral soils

The typical community of the Achilleo-Festucetum, recorded from mineral soils on Clare Island, is characterised by the occurrence of *Agrostis capillaris* and *Festuca ovina*, with several constantly recorded species including *Anthoxanthum odoratum, Potentilla erecta* and *Luzula multiflora* (Table 15, columns 52–59). However, it has significantly more herbaceous dicots than vegetation described from podzolic soils and whose presence on Clare Island is typical of dry areas/slopes (Pl. XI A, XI B). For this reason, it would seem that the community is related to the *Festuca ovina–Agrostis capillaris–Thymus praecox* CG10 grassland (Rodwell 1992), as many of the herbaceous species are usually indicative of agricultural influences or drier soils, although there may be no obvious current agricultural activity. The species include *Prunella vulgaris, Plantago*

Pl. XIA Fence in Capnagower, erected to prevent sheep from straying onto cliffs, clearly showing the vegetation differences associated with varying grazing pressure on Nardetea grasslands. Closely cropped commonage is on right side of fence, while heather-rich sward on the left-hand side has been ungrazed for some time.

Pl. XIB Impact of management on the Achilleo-Festucetum tenuifoliae. On the left is a typical derelict field. The field lying to the right of the fence with its lush vegetation was used for grazing in the winter months only.

Pl. XII Two views of Achilleo-Festucetum tenuifoliae from agricultural land in Glen townland: A) Closely cropped sward over lazy beds; B) Rush-dominated hollows in lazy beds in a neglected field.

lanceolata, Thymus praecox, Trifolium repens and *Lotus corniculatus*. Interestingly, *Danthonia decumbens*, a species usually associated with severely grazed areas, is commonly recorded from this community, although it does not provide any significant contribution to the overall ground cover (<1%). In general, heather occurs only where there has been a cessation in grazing that has allowed the gradual reversion to the former heathland.

Achilleo-Festucetum with Arrhenatheretalia influences

In certain situations the grassland vegetation resembles stunted Arrhenatheretalia meadows (Table 15, columns 60–67). This community characterises grassland vegetation of distinctly dry areas within agricultural boundaries. The appearance of the community is varied, depending on the type of management to which the vegetation is subjected. Typically, the vegetation is composed of a closely cropped sward, although it can become quite luxuriant and tall (approx. 25cm) where sheep are excluded from grazing (Pl. XII).

Along with the ubiquitous *Agrostis capillaris* and *Festuca ovina*, species characteristic of mesotrophic grasslands include *Holcus lanatus, Plantago lanceolata, Prunella vulgaris, Lotus corniculatus, Thymus praecox, Danthonia decumbens* and *Bellis perennis*. Other regular components of the vegetation that reflect this community's agricultural heritage include *Cynosurus cristatus, Rumex acetosa, Hypochaeris radicata* and *Ranunculus repens*. This assemblage corresponds with the *Holcus lanatus–Trifolium repens* subcommunity of the *Festuca–Agrostis–Galium* U4 grassland, with some influence of agricultural improvement (Rodwell 1992). O'Sullivan (1982) considered that *Cynosurus cristatus* was a clear indicator of human management and grassland improvement. However, variations in soil characteristics, e.g. depth and drainage, result in minor differences in the botanical composition of these stands with other species such as *Festuca rubra* or *Plantago maritima* replacing *Cynosurus* as the diagnostic species. They often reflect influences of other, vegetation types as well as historical, human influences.

ACHILLEO-FESTUCETUM WITH FESTUCA RUBRA

This vegetation (Table 15, columns 68–70) is typically associated with turf clamps, whether unstacked or stacked with turf. For much of the year, land surrounding the turf clamps is not grazed as the majority of the sheep are left on the commonage. As with much of the Achilleo-Festucetum, *Agrostis capillaris, Festuca ovina, Holcus lanatus* and *Rhytidiadelphus squarrosus* are the most abundant components of the vegetation. Other species located within this community include *Rumex acetosa, Oxalis acetosella* and *Carex nigra,* while *Festuca rubra* is recorded in conjunction with *Festuca ovina,* indicating the influence of increasingly maritime conditions.

ACHILLEO-FESTUCETUM WITH PLANTAGO MARITIMA

This subcommunity (Table 15, columns 71–73) is not widespread, occurring sparingly on small earthen field boundaries in agricultural land abutting low-lying coastal areas. The vegetation is characterised by *Agrostis capillaris* and *Festuca ovina,* with some minor amounts of *Carex viridula* and *Potentilla erecta.* Of note is the absence of constant species that typify the acid grassland vegetation such as *Galium saxatile, Holcus lanatus, Anthoxanthum odoratum* and bryophyte species, including *Hypnum jutlandicum, Polytrichum commune* and *Thuidium tamariscinum,* and the lichen species *Peltigera canina* and *Cladonia portentosa.*

The primary difference in the composition of the vegetation is the abundance of *Plantago maritima.* The community is distinguished from the grass-rich plantain sward since it is located in areas far removed from the usual distribution of the Plantain Sward, described elsewhere in the text. Many of the diagnostic species from the plantain sward community, including *Sedum anglicum, Aira praecox, Anagallis tenella* and *Luzula multiflora,* are not recorded.

IMPOVERISHED SUBTYPE OF THE ACHILLEO-FESTUCETUM OCCURRING ON MINERAL SOILS

The least floristically diverse community of the acid grassland vegetation occurring, on mineral soils on Clare Island, primarily characterises the vegetation recorded from dry, sheltered areas bordering on agricultural land, where sheep are gathered before being brought infield (Table 15, columns 74–79). The vegetation is characterised by *Agrostis capillaris* and *Festuca ovina,* with several constantly recorded species including *Anthoxanthum odoratum, Potentilla erecta, Trifolium repens* and *Rhytidiadelphus squarrosus.* Many of the constant species commonly associated with the Achilleo-Festucetum are reduced in abundance and occur only sparingly, particularly the bryophytes such as *Thuidium tamariscinum, Mnium hornum, Frullania tamarisci, Sphagnum subnitens* and *Leucobryum glaucum.* Species that typify agricultural land and are absent or much reduced from this subtype include *Prunella vulgaris, Plantago lanceolata, Bellis perennis, Trifolium repens, Rumex acetosa* and *Cynosurus cristatus.*

COASTAL SUBTYPES

The coastal subtypes comprise the dwarf turf grasslands found in coastal areas, the greatest extent of which are confined to a narrow fringe bordering agricultural land on the southern shore of Clare Island with the remainder located in sheltered hollows on the northern cliffs, and correspond with the 'Natural Grassland Formation' described from Clare Island by Praeger (1911). The vegetation is transitional between the acid grasslands of the Nardetalia and the vegetation of the Plantain sward. The vegetation rarely extends beyond 100m inland and does not occur at elevations greater than 150m above sea level.

In contrast to the other grassland communities on Clare Island, the vegetation included in the Coastal Grasslands occurs in areas with distinguishing soil characteristics and grazing regimes (predominantly lagomorphs). These shallow soils (<0.5m) are typically less podzolised than most of the other soils on Clare Island, and are composed of freely draining loams with a pH range of 5 to 6. They are associated with shales and sandstone conglomerates and have numerous rock fragments incorporated in the soil matrix.

Species composition and physiognomy are largely influenced by the grazing regime and soil characteristics of the habitat (Malloch 1971; 1972; Birks 1973; Carter 1988). This coastal vegetation has been highly modified since Praeger's time. Lagomorphs (six rabbits and six hares) were intentionally introduced onto Clare Island in 1906 to provide populations for hunting. Their populations have since multiplied, particularly along areas of the northern coast. Rabbits are now the chief grazers of the Coastal Grassland, since the dwarf vegetation with its fine-leaved graminoid component provides an ideal food source (Rodwell 1992). Some grazing by cattle and sheep also occurs near agricultural holdings.

Floristics

The coastal grasslands are characterised and dominated by grass mixtures in which *Festuca rubra/ ovina* and *Agrostis capillaris* are the most abundant species, accounting for at least 90% cover. Unlike the acid grasslands of the Achilleo-Festucetum tenuifoliae found in most areas on Clare Island, the dwarf sward of these coastal grasslands are homogenous in appearance. Distinctive rosettes of *Plantago lanceolata* and *P. maritima* are another consistent component of the vegetation at elevations from 50m to 150m above sea level. Additional prominent graminoid elements include *Anthoxanthum odoratum*, *Aira praecox* and *Holcus lanatus*, while *Koeleria macrantha* and *Cynosurus cristatus* attain local prominence in places. Other grasses are not quite so frequent, although *Nardus stricta*, *Danthonia decumbens* and *Festuca vivipara* are occasionally encountered in vegetation transitional with the upland Achilleo-Festucetum tenuifoliae grassland.

Dwarf annual and perennial herbs, although present, rarely flower owing to the constant grazing and wind pruning in the coastal climate. Subcommunities of the coastal grasslands are distinguished by means of species assemblages from either the Cynosurion or Galio-Koelerion. Hemicryptophytes and putative, nitrogen fixing plants are an abundant component of the vegetation, although the overall cover that they provide is low (<10%). Many species are indicative of dry grasslands of both upland and agricultural pastures. These species include *Potentilla erecta*, *Lotus corniculatus*, *Cerastium fontanum*, *Trifolium repens*, *Prunella vulgaris* and *Euphrasia officinalis*. Rushes, carices and low-growing broad-leaved grasses typical of agriculturally improved areas are also a common component. The occurrence of *Thymus praecox*, *Jasione montana* and *Sedum anglicum* along with *Koeleria macrantha* while they do occur in several communities, particularly on freely-draining soils, typically on Clare Island, they occur in close proximity with maritime situations.

Stands of this vegetation are homogenous in appearance. Where sheep have been excluded from grazing on maritime cliffs, *Armeria maritima* and *Silene maritima* can occur. Stunted bushes of *Calluna vulgaris* establish at the margins, particularly near heathland areas where heather is regenerating.

Bryophytes, though numerous, provide little of the cover. Species consistently recorded include *Mnium hornum*, *Polytrichum commune* and *Thuidium tamariscinum*. Along with minute specimens of *Cladonia* species, the only lichen recorded is *Peltigera canina*, which is found in damp hollows.

Phytosociology

While the Coastal subtypes (Table 16) recorded on Clare Island are included in the Achilleo-Festucetum tenuifoliae, there are important structural and ecological differences that set it apart from other acid grassland communities (also classified as the Achilleo-Festucetum tenuifoliae). In the BNVC classification scheme, these moderately species-rich acid grasslands are transitional between the *Festuca ovina–Agrostis capillaris–Galium saxatile* U4 grassland and the more basiphilous *Festuca-Agrostis-Thymus* grassland CG10 (Rodwell 1992). The diminutive sward is dominated by *Festuca ovina/rubra* and *Agrostis capillaris*, providing at least 90% of the cover (Table 16). Constant species include *Potentilla erecta*, *Luzula multiflora* and *Lotus corniculatus*. Three distinct communities of the coastal grassland are identified, reflecting the influence of species characteristic of oceanic or agricultural habitats.

UPLAND SUBTYPE

At higher altitudes on Clare Island, a species-poor community occurs (Table 16, columns 1–5), which includes a collection of Nardetalia species such as *Nardus stricta*, *Danthonia decumbens* and *Juncus squarrosus*. This community is not floristically diverse, and other than those already listed has few additional species.

The vegetation represented by the first five relevés is separated from the Achilleo-Festucetum community on Clare Island, to which it is syntaxonomically analogous, because of the reduction in mean number of species recorded per quadrat (mean species number thirteen as opposed to seventeen). The pedogenic characteristics of this depauperate community are also dissimilar to those encountered under typical upland stands of the Nardetalia, as the soils are dry and loamy rather than podzolic.

The influence of species characteristic of the Plantaginetum coronopodo-maritimi is widespread in the remaining coastal grasslands on Clare Island, since *Plantago lanceolata* and *P. maritima* contribute 20% of ground cover on average (Table 16, columns 8–29). Leguminous herbs including *Trifolium* sp. and *Lotus corniculatus* and other small plants such as *Thymus praecox*, *Anagallis tenella* and *Hypochaeris radicata* are scattered in the short dense turf.

Malloch (1971) describes a somewhat similar community from stable cliff tops grazed by cattle, the *Festuca*-Armerietum rupestris nodum. The structure of that community is relatively similar to that described for Clare Island, with the vegetation occurring on a 'turf mattress', but differs primarily in the floristic composition with higher abundances of certain species such as *Armeria maritima* and *Daucus carota*. Malloch's (1971) research showed that successional sequences were not uncommon with the maritime grasslands grading into maritime heaths where grazing has been excluded. Similar vegetation has been recorded elsewhere in the British Isles by several authors (Ivimey-Cook and Proctor 1966; Birks 1973; Cotton 1975; Rodwell 1992). Four subcommunities within the plantain-rich community are distinguished, based on overlapping assemblage of herbaceous species of the Cynosurion cristati and Koelerio-Corynephoretea.

Transitional upland vegetation with Cynosurion cristati influence

In this subcommunity (Table 16, columns 6–8) the vegetation is transitional between the upland subtype and the remainder of the coastal subtypes, which occur at lower altitudes. It is the most species-rich community with floristic influences of several grassland classes. The bulk of relevés have a strong Cynosurion cristati element, a token of the agricultural background. The floristic differences between this community and the next are characterised by the occurrence of *Nardus stricta* and *Danthonia decumbens*.

Cynosurion-rich subtype

The third subcommunity of the plantain-rich coastal grassland (Table 16, columns 9–17) is found on the low cliffs adjacent to agricultural land, often demarcated from farmland by low earthen walls. These barriers are poorly maintained, resulting in cattle and sheep roaming along the coast. The soils are quite dry, and are of a loam composition.

The basic composition of the vegetation is derived from the Achilleo-Festucetum element with *Festuca ovina* and *Festuca rubra*, along with *Agrostis capillaris* and *Potentilla erecta*. Of the plantain species, only *Plantago lanceolata* and *Plantago maritima* are recorded, while some of the associated herbaceous flora is less abundant. Grasses such as *Holcus lanatus* and *Cynosurus cristatus*, and herbaceous species such as *Prunella vulgaris*, *Leontodon autumnalis* and *Cirsium dissectum* are also common. These Cynosurion elements are diagnostic of agricultural land. When extremely dry patches are encountered, species such as *Ranunculus repens*, *Bellis perennis* and *Daucus carota* are a noticeable component of the vegetation.

Coastal grassland vegetation transitional between the Cynosurion and Koelerio-Corynephoretea

This subcommunity (Table 16, columns 18–22) of the plantain-rich grassland contains species characteristic of the Koelerio-Corynephoretea, vegetation typical of sandy, dry grasslands (White and Doyle 1982). Apart from *Agrostis capillaris* and *Festuca ovina* and associated herbaceous flora previously listed, this subcommunity is characterised by the occurrence of *Koeleria macrantha*. Several authors (Braun-Blanquet and Tüxen 1952; Ní Lamhna 1982; White and Doyle 1982) regard *Festuca rubra* as a class character species of the Irish stands of the Koelerio-Corynephoretea. However, the short dense sward of this community on Clare Island prevented the separation of *Festuca rubra* and *F. ovina* with certainty. It is worth noting that stands of comparable, although taller vegetation located behind fences further along the cliffs suggest that both species occur in equal proportion.

Essentially this subtype (Table 16, columns 18–22) is found where favourable edaphic conditions exist, typically in a narrow band near agricultural boundaries along the northern cliffs. It is transitional between coastal grassland located in agricultural areas and vegetation of the next subtype, which has a greater proportion of Koelerio-Corynephoretea diagnostic species. Apart from *Holcus lanatus*, species diagnostic of the Cynosurion are of minor significance and other than sheep, there is no obvious or current agricultural influence and/or settlement.

Koelerio-Corynephoretea rich subtype

This subcommunity (Table 16, columns 23–29) is confined mainly to topographical depressions along the northern cliffs. These sheltered hollows coupled with the dry soils allow vegetation distinct from the surrounding unevenly textured acid grassland and heathland vegetation to develop. The vegetation also supports the majority of the island's rabbit populations (Pl. XIII).

The vegetation is again characterised by a closely cropped sward dominated by *Festuca rubra* and/or *F. ovina*, *Agrostis capillaris* and *Plantago coronopus*. Of the Koelerio-Corynephoretea character species, only *Aira praecox* and *Sedum anglicum* occur with any frequency, although with diminished abundance. The other diagnostic species including *Jasione montana*, *Hypnum cupressiforme* and *Koeleria macrantha* are less abundant. The occurrence of this vegetation some distance away from the any agricultural influence manifests itself in the general absence, or at least reduction, in species typical of the Cynosurion including *Holcus lanatus*, *Cynosurus cristatus*, *Trifolium pratense* and *T. repens*. A notable feature is that *Armeria maritima* was locally abundant in certain situations. Malloch (1971) considered that *Armeria* could rapidly spread in some of his maritime communities, if grazing pressures were reduced, thus re-establishing the Festuco-Armerietum rupestris. It was not possible to prove this hypothesis.

Bracken community

Bracken (*Pteridium aquilinum*) is a characteristic species of a variety of vegetation types and is especially important in oak woodland. It is an invasive species of dry upland pastures, heathlands and marginalised farmland, areas that were originally occupied by woodland (Birks 1973; Cotton 1975; Jermy and Crabbe 1978; O'Sullivan 1982; Thompson *et al.* 1986; Marrs *et al.* 1998a). Bracken is a clonal perennial that proliferates through rhizomes that penetrate down to 2m into the soil (Sheffield *et al.* 1989). It is an aggressive

Pl. XIII Closely cropped sward of coastal grassland. Although grazed by sheep, this habitat is the preferred site for the larger populations of the island rabbits, as evidenced by the numerous burrows in this photograph.

weed, overcoming other plants through the shading provided by its dense foliage (Whitehead *et al.* 1997). Taylor (1986) has estimated that bracken is extending its range in upland areas of Great Britain at an annual rate of *c.* 2%.

In Ireland, bracken is most commonly found encroaching onto *Calluna*-dominated heathland and abandoned pastures or lazy beds (O'Sullivan 1982). Deforestation, changes in land management of upland and marginal agricultural areas and overgrazing by sheep have all been cited for the spread of bracken (Marrs and Hicks 1986; Marrs 1987; Whitehead *et al.* 1997). More recently, it has been suggested that climate changes benefit the growth and spread of bracken rather than *Calluna* and that the success of *Pteridium* relates to (a) deep penetration of its rhizomes and (b) the tolerance of bracken to water deficits (Marrs 1993; Pakeman and Marrs 1996).

Once established, eradication of bracken is difficult. It was probably cut under less intensive farming regimes. Currently, removal is achieved by continual application of weedkillers (Thompson *et al.* 1995b) and/or cutting of the fronds (Marrs 1987a). Studies have shown that once cut, cattle can limit the spread by mechanical pressure of trampling and faecal and urine deposition, encouraging other species to outgrow the bracken (Lowday 1984; Lowday and Marrs 1992). Eradication is not guaranteed and the effectiveness of these methods does not last indefinitely and cessation of the treatment can result in rapid re-establishment of bracken (Marrs *et al.* 1998a; 1998b; 1998c).

In the original Clare Island Survey, bracken was found as a frequent component of vegetation confined to marginal areas of farmland, although no distinct bracken community was described (Praeger 1911). Following 'striping' of the island by the Congested Districts Board, and the decline in the population, land previously cultivated fell into neglect. Since then bracken has spread into abandoned fields and on to lazy beds. The decline in booleying, the transfer of people and animals to upland summer grazing sites in the commonage, has led to a major expansion of bracken outside established farming settlements, making it a striking component of the vegetation throughout the island.

Bracken is a common and often extensive element in upland grassland mosaics. The community is easily recognised among the stunted turf of rough grazings on western Irish hills where it is an invasive species (O'Sullivan 1982; White and Doyle 1982). Indeed Moore (1960) considered the bracken community as a variant of either *Ulex* or *Calluna* heathlands or of the *Agrostis-Festuca* grassland rather than an independent vegetation type. The vegetation on Clare Island corresponds with the Achilleo-Festucetum tenuifoliae, analogous with the *Festuca ovina-Agrostis capillaris-Galium saxatile* U4 grassland (Rodwell 1992). Unlike Irish accounts, the BNVC classification separates the bracken community from the *Festuca-Agrostis* (U4) grasslands. This is assigned to the *Pteridium aquilinum-Galium saxatile* U20 community (Rodwell 1992), although it should be noted that a second *Quercus-Rubus-Pteridium* community is described from fertile soils in or around Carpinion forest in Britain (Rodwell 1991). This community is widespread in its distribution throughout the British Isles, especially in lowland areas up to elevations of 200m.

Very little of the commonage and abandoned pastures on Clare Island are either species-rich or of good quality. These lands were reclaimed from *Calluna*-dominated heathlands but now lie derelict. The majority of the bracken community is found on lazy beds, with their distinctive pattern of parallel ridges. These lands, which previously supported potato and oat crops, have since fallen into neglect. The soils, though shallow, remain dry and provide an ideal habitat for the spread of the *Pteridium aquilinum* (Little and Collins 1995).

Phytosociology

No applicable phytosociological community described for Ireland (Table 17).

Heritage Council analogue: Dense bracken—HD1

The structure and composition of the bracken community is relatively simple, with *Pteridium aquilinum* as the sole dominant. The overwhelming dominance of *Pteridium* is such that it has few competitors. Hence the bracken community has a relatively depauperate ground flora. While the major associated grass species is *Festuca ovina*, occasionally *F. rubra* near coastal sites, (Table 17) other grasses such as *Holcus lanatus* and *Poa pratensis*, develop where gaps and trails occur.

There are occasional occurrences of *Agrostis capillaris* and *Anthoxanthum odoratum*. Herbaceous dicotyledons such as *Galium saxatile*, *Oxalis acetosella*

Table 17
Relevé table for the Bracken community

Column	1	2	3	4	
Relevé	3	3	3	3	
	1	1	1	1	
	8	6	5	7	
					Synoptic value
Pteridium aquilinum	5	5	5	5	V
Nardetalia species					
Festuca ovina	2	1	4	3	V
Galium saxatile	+	2	+	+	V
Oxalis acetosella	2	2		2	IV
Holcus lanatus	1		3	1	IV
Potentilla erecta	+		+		III
Anthoxanthum odoratum	2				II
Agrostis capillaris			2		II
Companion species					
Thuidium tamariscinum	2	1	+		IV
Hylocomium splendens	2	2			III
Juncus effusus	1	+			III
Rhytidiadelphus triquetrus	3		+		III
Rhytidiadelphus squarrosus		2	+		III
Poa pratensis			2	3	III
Viola riviniana			+	+	III
Mnium hornum		+		+	III
Peltigera canina			+	+	III
Polytrichum commune	+			+	III
Cerastium fontanum			r	+	III
Cirsium arvense			+	r	III
Sphagnum recurvum	+				II
Pseudoscleropodium purum		2			II
Sphagnum subnitens		2			II
Dicranum scoparium		1			II
Bryum pseudotriquetrum		+			II
Leucobryum glaucum		+			II
Sphagnum papillosum		+			II
Ophioglossum vulgatum			+		II
Rumex acetosa			+		II
Digitalis purpurea				1	II
Sagina procumbens				1	II
Trifolium repens				+	II
Luzula multiflora				+	II
Anagallis tenella				+	II
Rumex acetosella				+	II
Number of species	1	1	1	1	
	3	5	6	8	

and *Potentilla erecta* are low-growing and generally lack vigour. Occasionally, *Digitalis purpurea* breaks through the canopy. The bryophyte flora is poorly-developed with only *Thuidium tamariscinum* and *Hylocomium splendens* occurring with any regularity. Other mosses that occur occasionally include *Pseudoscleropodium purum*, *Rhytidiadelphus squarrosus* and *Mnium hornum*. Lichen cover is sparse, with *Peltigera canina* sometimes found on bare ground.

The spread of bracken appears to continue unchecked in some places on Clare Island. No concerted efforts have been made to clear the bracken since it does not have a significant impact in many farms. Occasionally small tracts of bracken are cleared, freeing space for cattle pens. In the long term, it is doubtful if bracken will colonise large swathes of commonage, as the damp peaty soils in many places inhibit successful establishment (Pl. XIV A, below, and Pl. XV B, p. 104).

Pl. XIV A Late summer view of agricultural land in Gorteen. Derelict lazy beds with their dwarf grassland sward is the most distinctive habitat on the island in which bracken (pale brown) readily establishes itself.

Pl. XIV B Bracken rapidly expands in neglected fields. It occurs on the drier slopes seen in the background and on drier parts of the rush-dominated plain in the topographical depression in Maum.

One of the relevés presented (Table 17, column 1) was recorded in a bracken community covering an ancient habitation site lying in a bog in Maum townland. The area is surrounded by waterlogged soils supporting small-sedge and bog communities. The site comprises a large gathering of fulachta fiadh with angular pebbles and stones incorporated into the gley soil where the temporary cooking places were built. These sites, with their characteristic soils, allow water to drain quickly off the terraced slopes of the surrounding drumlin topography into the valleys below. An abundance of bryophytes colonise these sites owing to the moist conditions. The bracken canopy, although providing nearly 100% cover, is not as dense as that encountered at drier sites and allows more light to penetrate beneath the canopy. Along with the very common *Oxalis acetosella* and *Galium saxatile*, there is a greater abundance of small herbs (18 species compared with an average of 14 per quadrat), although the contribution towards ground cover is minimal.

NARDO-CARICETUM GRASSLANDS

Nardetea Rivas Goday et Borja Carbonell 1961
Nardetalia Prsg. 1949
Nardo-Galion saxatilis Prsg.1949
Nardo-Caricetum binervis (Pethybridge et Praeger 1905) Br.-Bl. et Tx. 1952 (Table 18)

Nardetea diagnostic species: *Polygala serpyllifolia, Galium saxatile, Juncus squarrosus, Danthonia decumbens, Festuca vivipara*
Nardo-Caricetum binervis diagnostic species: *Nardus stricta, Luzula multiflora*
Constant species: *Agrostis capillaris, Festuca ovina, Potentilla erecta*

Heritage Council analogue: Dry humid acid grassland—GS3 (and Wet grassland—GS4)

Typical subtype

The bulk of Praeger's (1911) heather community has been replaced with grassland vegetation, referred to the Nardo-Caricetum binervis (Table 18, columns 1–31), an impoverished form of the *Agrostis–Festuca* grassland (O'Sullivan 1982; White and Doyle 1982). The Nardo-Caricetum binervis has been recorded in Ireland by several authors (Braun-Blanquet and Tüxen 1952; Cotton 1975; Brock *et al.* 1978), and is thought to be the main constituent of grassy 'heathlands', particularly in the west of Ireland (White and Doyle 1982). Interestingly *Carex binervis,* defined as a character species of the Irish vegetation (Cotton 1975), is on Clare Island generally confined to heather communities along cliff ledges. More recently, Bleasdale and Sheehy-Skeffington (1992; 1995) have examined species-poor grassland vegetation that occurs on podzolic and mineral-rich soils in Connemara, County Galway. These authors did not make use of the Nardo-Caricetum binervis appellation in their findings, merely placing the vegetation in various categories of the *Agrostis-Festuca* rough grassland. This highlights the phytosociological difficulties of dealing with the acid grasslands that are rarely intact or undisturbed.

There is a general diminution in the cover provided by *Agrostis capillaris* and *Festuca ovina,* compared with the Achilleo-Festucetum vegetation. This is compensated for by the increased cover of *Nardus stricta* and *Juncus squarrosus*, except for the agricultural subtype in which *Agrostis* and *Festuca* are still dominant over *Nardus.* Constant species including *Potentilla erecta, Galium saxatile, Luzula multiflora, Carex panicea, C. viridula* and *C. nigra* are also common. Bryophytes and lichens such as *Sphagnum subnitens, Hypnum jutlandicum, Rhytidiadelphus squarrosus, Racomitrium lanuginosum, Leucobryum glaucum, Cladonia portentosa, C. uncialis* and *Frullania tamarisci* are common components of the ground layer. The heather species, *Calluna vulgaris* and *Erica tetralix,* are recorded in the vegetation, although they are dwarfed in stature and confined to depressions in the degraded vegetation, where they are overlooked by sheep. The distinctive appearance of *Calluna,* with its bare wooden shoots is the only reminder of the heather community that once dominated the hills on Clare Island.

Wet subtype

Towards the opposite edaphic extreme, particularly in the wetter conditions that exist on higher ground, the Nardo-Caricetum binervis grassland becomes less productive (Table 18, columns 32–45). Uncontrolled grazing by sheep of these infertile habitats allows *Nardus stricta* and *Juncus squarrosus* to become firmly established in such grassland (Grant *et al.* 1978). Since the vegetation provides nutrient-poor herbage, the sheep, although not selective grazers, generally avoid such areas.

Table 18

Relevé table for the Nardo-Caricetum binervis

Column										**1**										**2**										**3**									
	1	**2**	**3**	**4**	**5**	**6**	**7**	**8**	**9**	**0**	**1**	**2**	**3**	**4**	**5**	**6**	**7**	**8**	**9**	**0**	**1**	**2**	**3**	**4**	**5**	**6**	**7**	**8**	**9**	**0**	**1**	**2**	**3**	**4**	**5**	**6**	**7**	**8**	**9**
Relevé	**1**	**1**	**1**	**2**	**2**	**2**	**2**	**2**	**3**	**1**	**1**	**2**	**1**	**2**				**2**	**2**	**3**	**1**	**3**	**3**		**2**	**2**		**2**	**2**	**2**	**2**			**2**			**1**	**1**	
	0	**2**	**6**	**3**	**3**	**3**	**5**	**7**	**8**	**6**	**8**	**0**	**9**	**8**	**3**	**3**	**4**	**1**	**5**	**9**	**8**	**6**	**6**	**3**	**1**	**1**	**1**	**3**	**5**	**6**	**7**	**4**	**5**	**7**	**4**	**4**	**3**	**9**	**5**
	8	**4**	**2**	**1**	**4**	**5**	**7**	**5**	**7**	**3**	**8**	**4**	**0**	**7**	**2**	**3**	**8**	**7**	**4**	**1**	**9**	**8**	**9**	**4**	**4**	**6**	**8**	**9**	**8**	**9**	**1**	**5**	**7**	**4**	**3**	**4**	**0**	**8**	**2**
Nardo-Caricetum binervis																																							
Nardus stricta	2	4	3	3	4	4	2	2	3	2	4	3	4	1	3	2	2	1	2	2	3	2	3	1	2	2	1	4	3	4	3	2	1	3	2	3	4	1	2
Juncus squarrosus	2	2	2	3	2	2	1			2	2	2	1		1	1	2	1		2		3	1	2	2		2	2	3		3	2	1	4	2	3	2	3	1
Danthonia decumbens		2	1								2	1	2					1				+	+	1				+									2		
Luzula multiflora		+											+				+	+																	+				
Achilleo-Festucetum tenuifoliae																																							
Agrostis capillaris	2	2	1	1			2	2	+	1	1	2	2	1	3	4	3	4	1	1	2	2	2	4	1	2	4	2	2	3	1	3	2	3	3	4	2	3	3
Festuca ovina		2	2	1	1	1	2	2		1		1	3	3	2	2	2	2	1	+	2	+	1	2	1	1	3	2	2	3	2	1	3	2	2	3	2		3
Polygala serpyllifolia		r	+	+	+	+	+		+		+	+	+	+		+	+		+			+	+		+	+	r	+	+	+	+	+	+	r		+	+		
Festuca vivipara	1				1	+					2						+						+												+		1		
Nardetalia species																																							
Potentilla erecta	+	+	2	1	1	1	1	1	+		1	1	1	1	2	1	2	1	1	1	+	+	1	1	+	+	1	1	2	1	2	1	1	2	1	1	1	+	1
Rhytidiadelphus squarrosus	1	+	2	2		+	1	1		+	+	1	2				1	+		+		2	+					2		1	2			3	1				2
Carex viridula		+	2	+					+	+		+	+	+	1	1			2				+	1	+		+	+	+	1	1	1							1
Carex panicea	+	+	+		+	+	+	+	+	+		1	1	+		1	+	+		+	1		+		+	1			+	+	+		1				+		+
Galium saxatile												+					1													1		2	2	+			+		+
Agricultural land																																							
Cynosurus cristatus																																							
Holcus lanatus																																							
Trifolium repens																																							
Prunella vulgaris																																							
Anagallis tenella	+	1																																			1		
Plantago lanceolata																																							
Aira praecox																																			1				
Viola riviniana																																							
Peltigera canina														1			+																						
Thymus praecox																																							
Hydrocotyle vulgaris																																							
Companion species																																							
Calluna vulgaris	1	+	3	2	1	2	2	1	1	3	1	1	+	1	2	2		2	2	2	1	2	+	2	1	2	3	1	2	+	3		3	2		1			1
Sphagnum subnitens	2	1	2		2	3		2		+	2	2	1		2		3	1	3		2	2	2	2		3	2	2	3		1	3	3		3	2	3	+	2
Hypnum jutlandicum	3	2	2	1			2	2	+	+		+	2					3	1	+		2	2		1	3		2	2	2	2		2	2			2	+	2
Cladonia portentosa	+	+		1	+	1	1	+	+	1	1	+	+		1	+	+	+	2	+		+	+		+	+	2	+	+	+	2	+	+	+					
Erica tetralix	+		+	+	+	1	1		+	+	+	+			+	1		1	+	1	+	+	+	1	+		2	1	1	+	2	+		1	+	+	+		
Anthoxanthum odoratum															+			+					1	1		1	2	1	2	1			+	2			1		
Racomitrium lanuginosum		1		+	3	2	3	1	+	3	4		2	3			3		1	1	+	+	2			+		2				1	2						

40	41	42	43	44	45	46	47	48	49	50	51	52	53	54	55	56	57	58	59	60	61	62	63	64	65	66	67	68	69	70	71	72	
265	373	399	59	211	398	51	116	263	237	193	202	212	123	196	264	70	185	131	187	306	370	372	4	36	261	285	259	91	94	384	222	92	**Synoptic value**
2	3	+	2	2	1	2	3	1	4	+	1	+	2	1	2	2	2	4	2	4	1	1	2	1	2	3	+	1	2	+	2	2	V
4	2	3	2	2	3															1													III
2	2					2						1			1			1	1		1	+				1		+		1			II
		+				+		+		1	+	+	+	1	+		+				+			+	+	+	+	+	+		+		II
3	2	2	3	2	2	4	2	5	3	5	5	3	4	5	5	3	4	2	5	2	5	4	3	4	4	2	1	4	3	2	3	2	V
2	2	+	1	2	2	4	2	4	2	4	4	4	3	4	3	2	3	2	4	2	2	2	2	3	4	3		2	3	3	4		V
+	+					+	+		+			r											+										III
2	2		2		+											3																	I
1	+	+	+	1	+	1	1	1	1	+	+	1	1	1	1	1	1	1	1	+	+	+	1		1	1		+	+	+	+	+	V
1	+	2		2	2			+	1			1	1		+	+		1		2		+							+			+	III
	+	+		1		1		+	1					1		2	1	1	1	+	+	+	+		1	1		1					III
	+	+				1									+	1	+	1		+	+	+		+	+	+			1				III
+	+	+	1	+	+	2	+	1	1	2	1	1	1	2	1	+	1	+	1	+	+			1	1	+							III
																												1		2	3	2	+
	1											+	2	1	1			+		+							1	2	2	2	2	1	I
										+	+	+		1			+	+			+	+		1	+		+	+	+	2	1	+	II
						r				+	r	+		+	r	r	+	1	+	+	+	+	+		1			r		+	+		II
							1		1				1	+		2		1	1	+	+		2		+	+	2	1		+		+	II
						+				+	r	+		2		r	+		+		1	1	1	+	1	r		+		+			II
						2	1	1		1	2	2	+	1			1	+	1		+	+						1					II
						+		+	+	+		+		+	+		r		+	r	+	+			+								I
						+			r	+	+	+			+			r	+		+								+			+	I
								1					+	+			+		+		+	+	+		1								I
									+							1	2	1		+	+	+	1				1						I
			1			1	+		1										+		+				+				+				III
+	2	+		3													+	+		1	+		1										III
	+	3	2		1		2				+	1	1				+		1	+									1				III
						+	1		+										1														III
							+		+												+		1										III
1	1	+		1	2			1				1			1	2				2		+	1					2	1	4	1	3	III
	+				+		3		2										+	+						1							II

(Continued)

Table 18 *(Continued)*

Column										1										2										3									
	1	2	3	4	5	6	7	8	9	0	1	2	3	4	5	6	7	8	9	0	1	2	3	4	5	6	7	8	9	0	1	2	3	4	5	6	7	8	9
Relevé	1	1	1	2	2	2	2	2	3	1	1	2	1	2				2	2	3	1	3	3		2	2		2	2	2	2			2			1	1	
	0	2	6	3	3	3	5	7	8	6	8	0	9	8	3	3	4	1	5	9	8	6	6	3	1	1	1	3	5	6	7	4	5	7	4	4	3	9	5
	8	4	2	1	4	5	7	5	7	3	8	4	0	7	2	3	8	7	4	1	9	8	9	4	4	6	8	9	8	9	1	5	7	4	3	4	0	8	2
Dicranum scoparium						+			+			1							1	1		+		1	+	1		1	1	+	1	+		1			+	+	1
Cladonia uncialis	+			1	+	+		+		+	1	+	+	+							+			+	+	+	1		+		+			+		+			+
Polytrichum commune								1					+					+														2	1	+	2	1			
Sphagnum papillosum																													1			3	3	2	3	2	2		3
Frullania tamarisci		2									+		+		+				2	1	+	+	+						1	+	+			1					1
Plagiothecium undulatum			1	+														+	+	+		+			+				2	+	1			1				1	
Carex nigra		1							+		+	+	+					r	+	+			+		+														
Leucobryum glaucum							2		2	+	+	+	1						2	+		1		1	1		1		1	1									
Scapania gracilis							2							1	+		+	+		2	+	+	+	+	1				2					1					
Thuidium tamariscinum															1					+					2		1		+		1			2					1
Campylopus paradoxus	+	2		+		2	2			+		2						+					+														1		
Eleocharis multicaulis	1	1			1				1							2				+	1																+		
Sphagnum capillifolium							2	3	3										2						3				3					3					
Erica cinerea		+									+		+	+	2	+						+			+														
Juncus bulbosus	+	+	+					2	3																														
Mnium hornum														+								+	+							1				1			1		
Ranunculus repens																																							
Lotus corniculatus																																							
Plantago coronopus																																							
Plantago maritima		+																																					
Sedum anglicum																																r				r			
Hylocomium splendens	2																							1	2	2				+	1				1				
Juncus effusus																		+	+																1				
Rhytidiadelphus triquetrus							+								2										3				+			2					1		
Zygogonium ericetorum			2						3					3							4		+																
Campylopus atrovirens			+		1			2			1										1																		
Breutelia chrysocoma																												+		1									
Polytrichum juniperinum													+													+											+	+	
Cirsium dissectum																																							
Radiola linoides																																							
Rhytidiadelphus loreus																2			2						1		1									1			
Sphagnum palustre																3	2									+													
Eriophorum angustifolium							1													+																		2	
Aulacomium palustre																																		1		+			
Campylopus pyriformis													+		1									+															
Pleurozium schreberi																	+															+							
Blechnum spicant																				+										+									
Jasione montana																																							

4										5										6										7			
0	1	2	3	4	5	6	7	8	9	0	1	2	3	4	5	6	7	8	9	0	1	2	3	4	5	6	7	8	9	0	1	2	
2	3	3		2	3		1	2	2	1	2	2	1	1	2		1	1	1	3	3	3			2	2	2			3	2		
6	7	9	5	1	9	5	1	6	3	9	0	1	2	9	6	7	8	3	8	0	7	7		3	6	8	5	9	9	8	2	9	
5	3	9	9	1	8	1	6	3	7	3	2	2	3	6	4	0	5	1	7	6	0	2	4	6	1	5	9	1	4	4	2	2	
																																	Synoptic value
					+	+	1										+												+	+			II
							+																										II
2			2	2		2			+									+		+									+				II
2			2	3					1																		1						I
				2		2	2		+			+							1	+					+								II
+		+		2	+					+	+	+																					II
			+		+												+	+	+								3	1			1	+	II
											1								+		+					2							II
			1								+															1							II
									1	1		1	+	2	+			+		+	+							1	1			1	II
									1							1	+																I
									+									2															I
					3				2																								I
																							+										I
	+								1									+		+				1			3				1		I
									+		+	1									+			+								+	I
												+									+	+	1					1			+	2	+
						+					+	+	+								+							+		2	+	+	+
																1	1		+		+	+		2	2								+
							1									+	+	+				+	2										+
						+					+		+				+							+									+
	+	+																															+
													2					1									1	2				3	+
			2																														+
							1																										+
																		1								+							+
	+			+					+																								+
					2									+																			+
														r			r			+								+					+
													1					+	+		+			+	+								+
																					+												+
	3																																+
																															2		+
3																											1						+
															+																		+
1																													1				+
													+							+													+
				r	r						+		+				+																+

(Continued)

Table 18 *(Continued)*

Column									1										2										3									
	1	2	3	4	5	6	7	8	9	0	1	2	3	4	5	6	7	8	9	0	1	2	3	4	5	6	7	8	9	0	1	2	3	4	5	6	7	8	9
Relevé	1	1	1	2	2	2	2	2	3	1	1	2	1	2				2	2	3	1	3	3		2	2		2	2	2	2			2			1	1	
	0	2	6	3	3	3	5	7	8	6	8	0	9	8	3	3	4	1	5	9	8	6	6	3	1	1	1	3	5	6	7	4	5	7	4	4	3	9	5
	8	4	2	1	4	5	7	5	7	3	8	4	0	7	2	3	8	7	4	1	9	8	9	4	4	6	8	9	8	9	1	5	7	4	3	4	0	8	2
Polytrichum piliferum												+																											
Hypochaeris radicata																																							
Leontodon autumnalis																																							
Cerastium fontanum																																							
Sagina procumbens																																							
Bellis perennis																																							
Carex pulicaris																																							
Euphrasia officinalis																																							
Narthecium ossifragum				+	1		1																																
Hymenophyllum wilsonii																														1									
Vaccinium myrtillus																																							
Mylia taylorii											+	+																		1									
Pleurozia purpurea				2																+	1																		
Carex echinata																+									+														
Molinia caerulea																				+								1											
Dicranum majus																																			+	+			
Diplophyllum albicans																																							
Leiocolea alpestris																														1	1								
Nardia scalaris									2	+																													
Plagiochila spinulosa										+																													
Drosera rotundifolia																			+																				
Campylopus fragilis																										1													
Campylopus introflexus			+																																				
Drepanocladus fluitans																																							
Scorpidium scorpioides																																							
Sphagnum compactum																												1											
Sphagnum molle																						+																	
Sphagnum tenellum																							+																
Mylia anomala																																							
Odontoschisma sphagni																											+				+								
Cladonia sp.														+																									
Calypogeia sp.					+																																		
Pedicularis sylvatica			r																																				
Pteridium aquilinum														+																									
Succisa pratensis													r																										
Sagina subulata																																							
Achillea millefolium																											r												
Eurhynchium praelongum																																							
Viola sp.																																							

4										5										6										7			
0	1	2	3	4	5	6	7	8	9	0	1	2	3	4	5	6	7	8	9	0	1	2	3	4	5	6	7	8	9	0	1	2	
2	3	3		2	3		1	2	2	1	2	2	1	1	2		1	1	1	3	3	3			2	2	2			3	2		
6	7	9	5	1	9	5	1	6	3	9	0	1	2	9	6	7	8	3	8	0	7	7		3	6	8	5	9	9	8	2	9	
5	3	9	9	1	8	1	6	3	7	3	2	2	3	6	4	0	5	1	7	6	0	2	4	6	1	5	9	1	4	4	2	2	**Synoptic value**
								+		+	+	+					+																+
													1			+	+								r					+		2	+
																		+				+		r						r	1	+	+
													+											+				+			+		+
													+	+		+						+					+	+					+
																						1	+	+	1								+
																+		+		1			+				+						+
						+						1									+	+								+			+
									1																								R
			1	2																													R
		+	1		1																												R
																																	R
																																	R
																											1						R
																																	R
											+																						R
					+				1																								R
																																	R
																																	R
				2																													R
																											+						R
																																	R
																+																	R
									+																								R
									+																								R
																										2							R
																																	R
																																	R
					+																												R
																																	R
																																	R
																																	R
																																	R
										+																							R
																																	R
																								+									R
																																	R
										+																				+			R
													+			+		+															R

(Continued)

Table 18 *(Continued)*

Column										1										2										3									
	1	2	3	4	5	6	7	8	9	0	1	2	3	4	5	6	7	8	9	0	1	2	3	4	5	6	7	8	9	0	1	2	3	4	5	6	7	8	9
Relevé	1	1	1	2	2	2	2	2	3	1	1	2	1	2				2	2	3	1	3	3		2	2		2	2	2	2			2			1	1	
	0	2	6	3	3	3	5	7	8	6	8	0	9	8	3	3	4	1	5	9	8	6	6	3	1	1	1	3	5	6	7	4	5	7	4	4	3	9	5
	8	4	2	1	4	5	7	5	7	3	8	4	0	7	2	3	8	7	4	1	9	8	9	4	4	6	8	9	8	9	1	5	7	4	3	4	0	8	2
Ranunculus flammula																																							
Rumex acetosella																																							
Hypnum cupressiforme																																							
Pellia epiphylla																																							
Agrostis stolonifera																																							
Bryum pseudotriqeutrum																																							
Carex pilulifera																																							
Bryum capillare																																							
Polytrichum formosum																																							
Lepidozia reptans																																							
Fissidens taxifolius																																							
Scapania undulata																																							
Campylium elodes																																							
Eurhynchium striatum																																							
Alchemilla vulgaris agg.																																							
Philonotis fontana																																							
Osmunda regalis																																							
Cirsium palustre																																							
Viola palustris																																							
Carex binervis																																							
Carex flacca																																							
Calliergon cuspidatum																																							
Potamogeton polygonifolius																																							
Hypericum elodes																																							
Juncus articulatus																																							
Bromus hordeaceus																																							
Epilobium palustre																																							
Plagiochila punctata																																							
Ophioglossum vulgatum																																							
Ranunculus acris																																							
Carex ovalis																																							
Number of Species	1	2	2	1	1	1	1	1	1	1	2	2	2	1	1	1	1	2	2	2	1	2	2	1	2	1	1	2	2	2	2	1	1	2	1	1	2	1	1
	8	4	0	6	8	7	8	8	8	9	0	3	5	7	7	6	8	1	1	4	6	2	6	7	5	8	8	0	4	5	2	7	6	4	6	6	2	0	7

4										5										6										7			
0	1	2	3	4	5	6	7	8	9	0	1	2	3	4	5	6	7	8	9	0	1	2	3	4	5	6	7	8	9	0	1	2	
2	3	3		2	3		1	2	2	1	2	2	1	1	2		1	1	1	3	3	3			2	2	2			3	2		
6	7	9	5	1	9	5	1	6	3	9	0	1	2	9	6	7	8	3	8	0	7	7		3	6	8	5	9	9	8	2	9	
5	3	9	9	1	8	1	6	3	7	3	2	2	3	6	4	0	5	1	7	6	0	2	4	6	1	5	9	1	4	4	2	2	
																																	Synoptic value
																		+									1						**R**
											+	+	+																				**R**
													1									+			+								**R**
													1							+							2						**R**
													1															1					**R**
													+														1						**R**
								+																				+					**R**
															+	+																	**R**
												+																					**R**
											+																					+	**R**
																						1											**R**
																		1															**R**
																					+												**R**
																						+											**R**
																						+											**R**
																									1								**R**
													r																				**R**
																											r	+		r			**R**
																											+						**R**
																															2		**R**
																																1	**R**
																											+						**R**
																											3						**R**
																											2						**R**
																											1						**R**
																													+				**R**
																															+		**R**
																														+			**R**
																													r				**R**
																														+			**R**
																																+	**R**
1	2	1	1	1	2	2	1	1	3	1	2	3	2	1	1	2	2	3	2	2	3	2	1	1	2	1	2	2	1	2	1	2	
7	1	5	6	8	0	3	8	4	0	7	1	0	9	9	7	3	7	1	3	6	1	7	9	7	2	6	7	8	9	3	7	4	

The community is readily identifiable in the north-west of Clare Island, among the extensive peat cuttings at Ballytoohy More (Pl. XV A). The community is the most seriously degraded form of the Nardo-Caricetum binervis and phytosociologically, has some similarities with the *Nardus-Scirpus cespitosus* heath described by Bleasdale and Sheehy-Skeffington (1992; 1995), although *Trichophorum caespitosum* is not a regular component of the Clare Island vegetation. The nearest analogue in the BNVC scheme is the *Nardus stricta-Galium saxatile* U5 grassland although it would grade into the *Juncus squarrosus-Festuca ovina* U6 community in extremely wet peaty situations, where *Juncus* can be the overall dominant species.

Sheep's bent, *Nardus stricta*, and *Juncus squarrosus* are the most conspicuous component of the vegetation. Both attain their greatest dominance in these severely grazed sites and on terraced slopes, occupying upwards of 50% of ground cover (Curran *et al.* 1983). The distinctive 'fairy rings' (Pl. XV B) of *Juncus squarrosus* are locally dominant on deeper, moist peaty soils especially where runoff from surrounding slopes is a notable feature. The herbaceous flora commonly associated with the Nardetalia, such as *Galium saxatile* and *Viola riviniana*, is reduced in abundance. The bryophyte assemblage of this degraded hill grassland is similar to that recorded in the previous subcommunity (Table 18, columns 1–31), although the wetter ground is characterised by an increase in the abundance of *Sphagnum papillosum* (average cover 20%–40%). In these wetter situations, species typical of the Oxycocco-Sphagnetea indicate gradations to wet-heath or bog communities.

Pl. XV A Extensive network of peat cutting in Ballytoohy More. The surrounding grassland vegetation is dominated by *Nardus stricta* and *Juncus squarrosus*. Note the presence of fairy rings in the foreground.

Pl. XV B Severely degraded grassland. The distinctive fairy rings of hard rush are evident towards the right side of the photograph. The *Zygogonium ericetorum* algal mat develops in situations where water scouring is a feature of the landscape.

Agricultural subtype

The remaining two groups (Table 18, columns 46–66 and 67–72 respectively), describe agricultural land where *Nardus stricta* is frequent in the grassland vegetation (average cover of 20%). It occurs in peripheral areas of abandoned fields, and is characteristic of much marginal land in Ballytoohy. The soils supporting the vegetation are siliceous soils, although less podzolized than usual. Floristically, the vegetation represents a succession from severely grazed hill grassland through to marginal agricultural land where sheep are enclosed for short periods at regular intervals throughout the year.

The agricultural grassland differs only slightly from the typical Nardo-Caricetum community. These differences are based on the presence of species characterising anthropogenic influences. The vegetation is transitional between the enclosed fields of farming areas in Ballytoohy that border onto the commonage (Table 18, columns 46–66). The common species include *Agrostis capillaris*, *Festuca ovina*, *Juncus effusus*, *Prunella vulgaris*, *Plantago lanceolata*, *Trifolium repens* and *Cirsium palustre*. Both *Prunella vulgaris* and *Plantago lanceolata* are listed as differential species for the Hylocomio-Centaureetum nigrae, an ill-defined association that is closely related to the Nardo-Caricetum binervis in Ireland (Cotton 1975; White and Doyle 1982). For this reason and the fact that these two species are constant companions in a variety of plant communities described from Clare Island, the syntaxon has not been employed in the current research.

Degraded agricultural subtype

The final subcommunity is distinguished by the overwhelming dominance of graminoid species typical for the Nardetalia and an absence of many of the small herbs diagnostic of the Nardetalia (Table 18, columns 67–72). The community is not widespread and is recorded from a limited number of degraded areas in agricultural land. The sward comprises *Agrostis capillaris, Festuca ovina, Holcus lanatus, Anthoxanthum odoratum, Cynosurus cristatus* and *Nardus stricta*. There is a general reduction in the herbaceous species typical of the previous group, with only *Ranunculus repens, Trifolium repens* and *Anagallis tenella* recorded. In damper situations *Juncus effusus* is occasionally recorded. Other species of reduced cover and frequency include *Aira praecox, Thymus praecox, Plantago lanceolata* and *Peltigera canina*.

Derelict subtypes

The final collection of twenty relevés (Table 19) is separated from the description of the Nardo-Caricetum binervis, as the vegetation is seriously degraded and may represent a mosaic of Caricion davallianae and Littorellion vegetation. The closely grazed vegetation is heavily poached and there is a considerable amount of bare soil, generally accounting for between 20% to 40% of all quadrats. The soils rarely exceed 0.5m in depth. The vegetation is typically recorded from peripheral areas of the upland mosaic, from both acid grassland and bog vegetation.

Four subtypes are recognised, related to the continuous irrigation and localised nutrient enrichment, which alters the basic character of the vegetation. The overall composition of the vegetation reflects its acid grassland ancestry, classified in the Nardo-Caricetum binervis. However, there is a strong influence of Littorelletea species, in particular species diagnostic of the Hydrocotylo-Baldellion. In the wetter habitats, where scouring is an obvious feature of the soil surface, there is an increased incidence of species diagnostic of the Scheuchzerio-Caricetea.

The continuous seepage of nutrient-enriched waters results in the development of vegetation that has some similarities with the Carici nigrae-Juncetum articulati. This vegetation, previously assigned to the Caricion curto-nigrae by other authors (Braun-Blanquet and Tüxen 1952; Ivimey-Cook and Proctor 1966) was later reassigned to the Caricion davallianae (Ó Críodáin 1988), owing to the occurrence of species typically confined to base-rich areas. The association is distinguished by the occurrence of *Juncus articulatus*, a character species of the association, while *Ranunculus flammula, Carex nigra* and *Hydrocotyle vulgaris* are transgressive class character species with a clear preference for this vegetation. It has been recorded from a variety of waterlogged habitats in low-lying areas in Ireland, typically neglected wet acid grassland, particularly on deeper soils (White and Doyle 1982; Ó Críodáin 1988). Doyle (1990) recorded vegetation referable to the Carici nigrae-Juncetum articulati from some deep drainage channels within blanket bog.

Unlike classic stands of the association, many of the characteristic species are reduced in abundance, while constant species characteristic of the Caricetalia davallianae are absent. This may be due to the level of grazing suffered by the vegetation but also in part to the level of irrigation by base-poor waters. The occurrence, also, of several peat-forming species characteristic of the Oxycocco-Sphagnetea highlights the historical pressures to which the commonage was subject.

Typical subtype

The first set of relevés (Table 19, columns 1–4) was recorded from neglected pastoral land in Maum townland. The vegetation, recorded in areas where bare soil accounts for <25% of any quadrat, is characteristic of places subject to fluctuating water levels. This vegetation, which is related to both the Littorelletea and the Scheuchzerio-Caricetea, is rather variable in its composition, often showing site-specific differences.

Other than the primary graminoid species including *Nardus stricta, Agrostis capillaris* and *Festuca ovina*, the associated flora includes *Juncus bulbosus, Anagallis tenella, Potamogeton polygonifolius, Carex nigra, Ranunculus flammula* and *Hydrocotyle vulgaris*. In areas where accumulated water does not immediately dissipate, there is an increased incidence of *Juncus effusus* and *Sphagnum cuspidatum*, which appears to be at the expense of *Nardus stricta*.

Vegetation with Montio-Cardaminetea elements

Four relevés (Table 19, columns 5–8) describe vegetation recorded from severely grazed, wet

Table 19
Relevé table for the derelict vegetation assigned to the Nardo-Caricetum binervis

Column	1	2	3	4	5	6	7	8	9	10	11	12	13	14	15	16	17	18	19	20	Synoptic value
Relevé	100	105	13	15	166	118	179	111	141	366	83	65	73	292	296	270	360	332	77	297	
Nardetalia																					
Nardus stricta	1	4			2	3	3	2	2	1	2	2	2	2	2	4	+	2	2	+	V
Potentilla erecta	+	+	2	1	+		1			+	+	2	+	+	+	1	+	+	1	+	V
Carex panicea	1	+			+		1		2		1	2	1	1		1		+	+	+	IV
Festuca ovina	2		1	1		2	3	1	2	+				3		2	+	+			IV
Agrostis capillaris	3	2	2	3	1									2		1	1				III
Rhytidiadelphus squarrosus	1	+		1	+							1	+					+	2		III
Littorelletea uniflorae																					
Juncus bulbosus	+	+			2	+	2	1	2	1			1	2	+	2	+	+	+	+	IV
Eleocharis multicaulis			1		3	2	2		4	2	1	1	1		2	1		2	+	+	IV
Anagallis tenella	+	+	1	1	1	1	3		1	1	+	2		+	1	2					IV
Potamogeton polygonifolius	1		1		1	1		3	2		2		2		2	3					III
Carex viridula	1	1			+	2	2		1		1		+			1					III
Hypericum elodes	2									1					1	+					II
Carici nigrae-Juncetum articulati																					
Juncus articulatus				2	1	1	3	1													II
Carex nigra	+		1	1	2	1	1	3		3	+			+	1	+					IV
Ranunculus flammula	+	1	1	1	1	1	2	1	3	+		+				1					IV
Hydrocotyle vulgaris	+	+	1	2	1	1	2	1	2	+		+	+	+	+	2	1				V
Montio-Cardaminetea																					
Pellia epiphylla					1	1	1		+	1	+	+	+								III
Calliergon cuspidatum		+			1	2	2	1								2					II
Caricetalia nigrae																					
Carex echinata	2			2	1				+	2	+	1	+	+	1		+		+	+	IV
Viola palustris	+									+	1	+	1		1						II
Sphagnum palustre	2								2	1										3	II
Oxycocco-Sphagnetea																					
Drosera rotundifolia			+						+		+		+	1	1	+	+	+	+	r	III
Erica tetralix			1						+		+	1	+			1	+	1			III
Sphagnum papillosum					2						1		1	1	2	1	3	3			III
Sphagnum subnitens			3								2			2		2	3		3		II
Molinia caerulea							2		2			2					+	+		1	II
Narthecium ossifragum					+				+		+					+		+	+	+	II
Eriophorum angustifolium											3	2			1		2	+		1	II
Polygala serpyllifolia											r		+	+		r	+	r			II
Aulacomnium palustre										+					1		+	+		3	II

(Continued)

Table 19 *(Continued)*

Column										1										2	
	1	2	3	4	5	6	7	8	9	0	1	2	3	4	5	6	7	8	9	0	
Relevé	1	1			1	1	1	1	1	3				2	2	2	3	3		2	
	0	0	1	1	6	1	7	1	4	6	8	6	7	9	9	7	6	3	7	9	
	0	5	3	5	6	8	9	1	1	6	3	5	3	2	6	0	0	2	7	7	
																					Synoptic value
Companion species																					
Sphagnum cuspidatum	2		3	2	2								2								II
Juncus effusus	+			3	2					2			+	+							II
Anthoxanthum odoratum	2			2						1			1	1							II
Galium saxatile	+	+											+	+		+					II
Trifolium repens		+		1		+	+	+													II
Bellis perennis		1			r	+	1														II
Prunella vulgaris	r	1				+		+								+					II
Agrostis stolonifera							+	1		1						+					II
Aneura pinguis					1				+	+	+						+				II
Succisa pratensis						+					+	2				+		+			II
Hypnum jutlandicum														+		1	+	+	+		II
Polytrichum commune					+								+	+	1		2				II
Calluna vulgaris																2	+	1	2		II
Leontodon autumnalis						1	1									r					I
Bryum pseudotriquetrum		+			+				1												I
Carex pulicaris		1										+	+								I
Galium palustre				+				+		+											I
Holcus lanatus							1		1	+											I
Equisetum palustre	+												+			r					I
Plantago lanceolata		+				2						+									I
Plantago maritima		+				+	+														I
Sphagnum capillifolium															2		+			1	I
Festuca vivipara														1			1				+
Mnium hornum														2					2		+
Juncus squarrosus																			2	1	+
Breutelia chrysocoma					2														+		+
Cirsium dissectum						+			1												+
Cirsium palustre	+									+											+
Dicranum majus												1						+			+
Fissidens adianthoides		1				1															+
Hylocomium splendens												1				1					+
Isothecium myosuroides									1							2					+
Leucobryum glaucum											+						+				+
Lophocolea bidentata												+				+					+
Ranunculus repens		+				+															+
Selaginella selaginoides												+				+					+
Taraxacum officinale	r	+																			+

(Continued)

Table 19 *(Continued)*

Column										**1**										**2**	
	1	**2**	**3**	**4**	**5**	**6**	**7**	**8**	**9**	**0**	**1**	**2**	**3**	**4**	**5**	**6**	**7**	**8**	**9**	**0**	
Relevé	**1**	**1**			**1**	**1**	**1**	**1**	**1**	**3**				**2**	**2**	**2**	**3**	**3**		**2**	
	0	**0**	**1**	**1**	**6**	**1**	**7**	**1**	**4**	**6**	**8**	**6**	**7**	**9**	**9**	**7**	**6**	**3**	**7**	**9**	
	0	**5**	**3**	**5**	**6**	**8**	**9**	**1**	**1**	**6**	**3**	**5**	**3**	**2**	**6**	**0**	**0**	**2**	**7**	**7**	**Synoptic value**
Juncus acutiflorus								2													**R**
Scorpidium scorpioides	2																				**R**
Sphagnum auriculatum										2											**R**
Sphagnum tenellum																			2		**R**
Trichophorum caespitosum						2															**R**
Campylium stellatum					1																**R**
Eurhynchium striatum				1																	**R**
Pseudoscleropodium purum												1									**R**
Hypnum cupressiforme																	1				**R**
Isolepis setacea								1													**R**
Lotus corniculatus				1																	**R**
Rhytidiadelphus triquetrus																1					**R**
Odontoschisma sphagni																		1			**R**
Plagiothecium undulatum	1																				**R**
Dactylorhiza maculata												+									**R**
Danthonia decumbens																	+				**R**
Dicranum scoparium																	+				**R**
Empetrum nigrum																				+	**R**
Grimmia trichophylla																	+				**R**
Aira praecox														+							**R**
Chiloscyphus sp.												+									**R**
Cladonia portentosa																		+			**R**
Cratoneuron commutatum					+																**R**
Mentha aquatica								+													**R**
Mylia anomala																	+				**R**
Pedicularis palustris												+									**R**
Pinguicula vulgaris					+																**R**
Plantago coronopus		+																			**R**
Poa annua			+																		**R**
Pohlia nutans									+												**R**
Racomitrium lanuginosum														+							**R**
Rhynchospora alba																	+				**R**
Rhytidiadelphus loreus													+								**R**
Scapania gracilis															+						**R**
Pinguicula lusitanica											r										**R**
Epilobium sp.										r											**R**
Vaccinium oxycoccos																				+	**R**
Number of species	2	2	1	1	3	2	2	1	2	2	2	2	2	2	1	3	2	2	1	1	
	7	3	4	7	0	3	3	6	3	4	2	5	2	3	8	4	7	2	6	6	

grassland. Overall, the vegetation is similar with that described in columns 1–4 of the typical subtype described above, with a basic composition including *Nardus stricta*, *Festuca ovina*, *Eleocharis multicaulis*, *Juncus bulbosus*, *J. articulatus*, *Carex nigra*, *Ranunculus flammula* and *Hydrocotyle vulgaris*. These species, which distinguish the Carici nigrae-Juncetum articulati, occur in greater abundance than the previous subtype.

However, the vegetation is typified by the complete absence of *Agrostis capillaris*, while *Calliergonella cuspidata* and *Pellia epiphylla/neesiana*, species that are characteristic of spring-fed bryophyte-rich vegetation are common. In this case, these last two species are found among the constantly recharging water pools on the bare, poached ground.

Derelict vegetation of abandoned peat cuttings

The remaining two groups (Table 19, columns 9–16 and 17–20) have increased incidences of species typically associated with peatland habitats. The more abundant species of the Oxycocco-Sphagnetea include *Sphagnum papillosum*, *Drosera rotundifolia* and depauperate *Erica tetralix*. Others that are recorded with less frequency include *Molinia caerulea*, *Narthecium ossifragum*, *Polygala serpyllifolia* and *Aulacomnium palustre*. Both *Sphagnum subnitens* and *Eriophorum angustifolium* are generally abundant where recorded. The necessity for turf for domestic purposes in historical times, resulted in the exploitation of most suitable land not under tillage.

The vegetation of this derelict subtype (Table 19, columns 9–16) contains elements of the Caricetalia nigrae. The species here include *Carex echinata* and *Viola palustris*, with *Sphagnum palustre* occasionally recorded where standing water occurs. The occurrence of these species is indicative of areas that are subject to occasional flushing acid water. This is an important feature of the subtype and occurs as fragmentary stands in peat cuttings associated with derelict agricultural land.

Species-poor vegetation of abandoned peat cuttings

This subtype (Table 19, columns 17–20) is generally the most depauperate subtype of this derelict vegetation, with an average of nineteen species per quadrat. It is confined to small, peat-filled depressions in the commonage where harvesting for peat occurred at a time when the island's human population was significantly greater.

The basic Nardetalia element is greatly reduced in frequency and abundance, as is the Littorelletea uniflorae component, with just scattered *Juncus bulbosus* and *Eleocharis multicaulis* a feature. There is a complete absence of Scheuchzerio-Caricetea species, although this is not surprising since the community is not subject to water run-off. The vegetation bears little resemblance to the original vegetation that would have preferred the undulating peat. The bare peaty soil still supports remnants of the Oxycocco-Sphagnetea, including *Drosera rotundifolia*, *Narthecium ossifragum* and *Sphagnum papillosum*, however, these species are not abundant. Both *Drosera* and *Narthecium* favour the bare soil situations in these habitats.

Meadow and pasture vegetation of the Molinio-Arrhenatheretea

Several plant communities reflect the variable management history of the agricultural land. A considerable area was given over to crop production, while pastures and meadows were abundant where soils were deeper. Cattle pastures were located on marshy and alluvial soils that could not support crop production, mainly in the eastern half of the island. Arable crops were concentrated on the dry loamy soils common along the sheltered southern coast.

Agricultural land was developed where forest clearance had taken place and on land reclaimed along the sheltered coast. As the population diminished in the last century, these agricultural lands fell into disuse. The impoverished soils could not sustain crop production, without the major input of sand and seaweed that had been applied by earlier, larger farming populations. The traditional ridge and furrow systems (lazy beds) used for crops, particularly potatoes, remain as relicts of this mode of farming on much of the land enclosed by the dry stone wall that separates the farmland from the commonage on Clare Island. The majority of the relict lazy beds are now vegetated with a closely cropped, species-rich, Nardetalia sward.

Some stands that are free from grazing during the summer develop a taller sward often rich in

Arrhenatheretalia species. Some of these stands are left unused while others are grazed later in the year. Meadows are used for hay production and are still treated in a traditional manner, whereas the application of artificial fertilisers is confined to a few fields. Cattle are excluded from hay meadows in the earlier part of the year, but later allowed to graze when the hay harvest has been completed.

Many of the wetter field systems are neglected and have developed into rush-dominated wet grasslands. Cattle are rarely allowed onto these areas, except in situations where no other suitable land is available for grazing. Unmanaged roadside verges support vegetation that is rather similar to the tall grassland of wet pastures. These unmanaged grassland habitats are gradually extending their range along ditches and support many plants of restricted distribution on Clare Island.

Floristics

Many of the species associated with wet grasslands of the Molinio-Arrhenatheretea are commonly found in other vegetation classes such as the Oxycocco-Sphagnetea, Scheuchzerio-Caricetea nigrae, Nardetea and Plantaginetea majoris. Wet grasslands differ mainly in the structural arrangement and the abundance of individual species in the vegetation. By far the greatest constituents of the vegetation are graminoids, although tall forbs are prominent on certain substrates. Among the grasses, *Anthoxanthum odoratum*, *Holcus lanatus* and *Agrostis capillaris* are the most abundant, with *Anthoxanthum* often the most dominant. Regularly, *Festuca ovina* is found beneath the canopy of taller graminoids. Species that occasionally occur include *Cynosurus cristatus*, *Dactylis glomerata*, *Arrhenatherum elatius* and *Lolium perenne*, especially in vegetation that develops in ditches and in meadows given over to hay production.

Tall forbs such as *Lythrum salicaria*, *Filipendula ulmaria* and *Iris pseudacorus* are locally abundant, and when in flower, form a conspicuous feature of the vegetation. The forbs are the tallest component of the vegetation along with the rushes *Juncus effusus*, *J. articulatus* and *J. conglomeratus*. The last species, *J. conglomeratus*, is sometimes found growing together with *Gunnera tinctoria*, an invasive species that was first recorded on the island by Doyle and Foss (1986a). The gradual spread of *Gunnera* into neglected, damp fields and ditches along the south coast of Clare Island has been noted in recent years (Michael O'Malley, pers. comm.). Sedges are abundant in wetter stands, with *Carex nigra*, *C. echinata* and *C. ovalis* the most common. Other less frequently recorded sedges include *Carex dioica*, *C. panicea*, *C. viridula* and *C. pulicaris*.

Among the low-growing dicotyledonous herbs recorded in the wet meadows are leguminous, low-growing rosette plants and creeping herbs. These include *Ranunculus repens* and *Trifolium repens*, which are relatively frequent, with *Plantago lanceolata*, *Trifolium pratense*, *Galium palustre* and *Epilobium palustre* less common. On drier sites *Lotus corniculatus* and *Centaurea nigra* are obvious features. A few bryophytes are found growing within the moist litter, although total cover rarely exceeds 5%. In stands near watercourses, there are increases in bryophyte cover and species diversity. In such conditions, *Rhytidiadelphus squarrosus*, *Sphagnum palustre* and *Aulacomnium palustre* are the commonest bryophytes, occurring together with occasional *Sphagnum* species and members of the Brachytheciaceae, including *Rhytidiadelphus squarrosus*, *Brachythecium rutabulum* and *Eurhynchium praelongum*.

Phytosociology

The class Molinio-Arrhenatheretea includes the majority of anthropogenic lowland grasslands in Ireland (O'Sullivan 1965; 1982). The vegetation is found in isolated patches on Clare Island, within farmland boundaries and is quite distinct from the Nardetea vegetation typical of lazy beds. According to O'Sullivan (1982), species that separate vegetation of the Molinio-Arrhenatheretea from the Nardetea include *Holcus lanatus*, *Rumex acetosa* and *Trifolium pratense*. A number of these species that are characteristic of Molinio-Arrhenatheretea communities occasionally occur with low constancy in Nardetea grasslands, such as *Holcus lanatus*, while Nardetea species generally occur as constant companions in the meadow community on Clare Island.

Two orders, the Molinietalia and Arrhenatheretalia are used to separate low quality wet meadows of the former order from relatively higher quality grasslands assigned to the latter. Many of the characteristic species of the Molinio-Arrhenatheretea that have been defined from Central European vegetation are missing from the Irish examples. In western Irish counties, vegetation of the Molinio-Arrhenatheretea often borders on blanket bog. Rodwell (1992) considers that the differences between the Molinietalia and the Arrhenatheretalia are somewhat more complex and have been variously related to wetness, acidity and even peatiness of the soil. O'Sullivan (1965; 1982) considered that these features may not be directly applicable to Irish conditions, and felt that the distinction between each of the two orders is based on the absence or at least the reduced abundance of character species from the other order. Thus species characteristic of the Molinietalia, such as *Juncus effusus*, *J. conglomeratus* and *Filipendula ulmaria*, occur frequently in neglected hay meadows, whereas the reduced abundance of species including *Dactylis glomerata*, *Holcus lanatus* and *Trifolium* species are often the only differences between the orders. Four syntaxa have been distinguished on Clare Island (Table 20), referred respectively to the (a) Junco acutiflori-Molinietum, (b) the Filipendulion, (c) the Arrhenatherion, and (d) the Cynosurion. Both the Junco-Molinietum and the Filipendulion belong to the Molinietalia, while the last two syntaxa belong to the Arrhenatheretalia. The identification of these four syntaxa reflects vegetation differences related to soil composition, groundwater fluxes and management.

While the Arrhenatheretalia contains a number of reasonably well-developed associations such as the Centaureo-Cynosuretum, a widespread association characterised by the presence of *Cynosurus cristatus* and *Centaurea nigra*, the Clare Island communities present difficulties, since there is considerable overlap of diagnostic species at the association level. The approach taken here is to assign three of the four syntaxa to alliance level only, since these problems have been encountered elsewhere in the west of Ireland, where there appears to be a lack of ecological separation in the distribution of some species that are useful diagnostics elsewhere in Ireland and Europe.

VEGETATION OF WET MEADOWS

Molinio-Arrhenatheretea Tx. 1937
Molinietalia Koch 1926
Junco conglomerati-Molinion Westhoff 1968
Junco acutiflori-Molinietum Tx. et O'Sullivan 1964 in O'Sullivan 1968 Table 20, columns 1–6

Molinietalia character species: *Juncus effusus*, *Lythrum salicaria*
Junco acutiflori-Molinietum character species: *Carex nigra*, *C. echinata*, *C. ovalis*
Diagnostic species: *Epilobium palustre*, *Galium palustre*

Heritage Council analogue: Wet grassland—GS4

The Junco acutiflori-Molinietum (Table 20, columns 1–5) comprises wet grassland that is widely distributed in the western Irish counties (O'Sullivan 1965; 1976; 1982). Such vegetation develops on poorly draining soils and cutover peat and is often confined to neglected perimeters of farmland. On Clare Island such grassland communities are confined to drumlin valleys and often occur on the edge of peatland areas. The best example of such vegetation on Clare Island is located close to the small woodland at Lassau, an isolated area that supported smallholdings of cattle in historical times. Analogous wet grassland vegetation is assigned to a species poor variant of the *Holcus lanatus-Juncus effusus* MG10 rush pasture of the British National Vegetation Classification (Rodwell 1992) and are probably best seen as part of the Calthion.

On Clare Island, there is an average of sixteen species per quadrat in this grassland community. The vegetation is characterised by *Anthoxanthum odoratum*, *Holcus lanatus* and *Agrostis capillaris*, which are the main grasses, and by *Juncus effusus*, the main rush. Although *J. conglomeratus* was only first recorded from Clare Island by Doyle and Foss (1986a), it appears to be extending its range outside ditches owing to the lack of management of many former meadow sites. In areas where water percolation is an obvious feature, *Lythrum salicaria* occurs. The dense grassland sward of the Junco acutiflori-Molinietum is generally 1m tall, with some obvious stratification of the vegetation. Elements of the Scheuchzerio-Caricetea nigrae are typically found in Irish wet grasslands, represented here by the high incidence of small sedges in the vegetation (O'Sullivan 1982; Ó Críodáin and Doyle 1994). The vegetation is dominated by *Carex echinata* and *C. nigra*, while *Rumex acetosa*, *Galium palustre*, *Epilobium palustre* and *Luzula campestris* are also commonly found.

DEGRADED JUNCO-MOLINIETUM

A single relevé taken from the fringes of blanket bog represents a degraded form of the Junco

Table 20

Relevé table for grassland vegetation assigned to the Molinio-Arrhenatheretea

Column	1	2	3	4	5	6	7	8	9	10	11	12	13	14	15	
Relevé	146	147	250	364	365	384	278	279	174	402	253	88	385	393	401	Synoptic value
Molinio-Arrhenatheretea																
Holcus lanatus	3	2	4	2	3		2	1	1	2	3	2	3	4	1	V
Rumex acetosa	2	1	+					1			+	+	+	+		III
Trifolium pratense						1	1	1				+			+	II
Plantago lanceolata		+					+				+		+			II
Festuca rubra							3	2								I
Ranunculus acris							1	+								I
Poa trivialis								2								+
Constant species																
Anthoxanthum odoratum	3	4	3	2	3	3	4	3		2	2	3	4	3	3	V
Agrostis capillaris	2	3	2	3	4					2	2	4	2	+		IV
Molinietalia																
Juncus effusus	4	1	2		2			2		1	3			1	2	IV
Epilobium palustre	1	+	r	r		+	r	+	+							III
Galium palustre	1	1	+				+	1	2							II
Lythrum salicaria	1	1	1							1	+				1	II
Junco acutiflori-Molinietum																
Potentilla erecta	2	2	+			1							+	+		III
Carex nigra	+	1	1	+				3		+		1				III
Carex echinata		1		2	+	2	2	1		+				+		III
Thuidium tamariscinum				+	+									1		II
Hydrocotyle vulgaris		1		+		2										II
Centaurea nigra					1	3										I
Sphagnum palustre		2				3										I
Aulacomium palustre				3		2										I
Ranunculus flammula						2	+		3							I
Viola palustris				+		2										I
Juncus articulatus						1			3		2					I
Molinia caerulea						2										+
Eriophorum angustifolium						2										+
Agrostis canina						2										+
Filipendulion																
Filipendula ulmaria	+						1	2	3							I
Arrhenatherion elatioris																
Arrhenatherum elatius			1							3					+	II
Dactylis glomerata										4						+

(Continued)

Table 20 *(Continued)*

Column	1	2	3	4	5	6	7	8	9	10	11	12	13	14	15	
Relevé	146	147	250	364	365	384	278	279	174	402	253	88	385	393	401	**Synoptic value**
Taraxacum officinale										r			+			I
Heracleum sphondylium										+						+
Veronica chamaedrys										+						+
Cynosurion cristati																
Trifolium repens			+	+	2						1	+	1	2	1	III
Cynosurus cristatus							2						2		2	II
Lolium perenne															2	+
Companion species																
Rhytidiadelphus squarrosus	1	2		+	+	+					+	+		+		III
Luzula campestris		+	1		+	+	1	+				+	+			III
Ranunculus repens		2						1	2		1		1			II
Festuca ovina							1			1			2	2		II
Lotus corniculatus							1					1	+			II
Anagallis tenella		1		+		1	+									II
Carex ovalis			1		+			+			2					II
Iris pseudacorus							+		1		1					I
Potentilla palustris		2					1									I
Carex dioica			1		+											I
Brachythecium rutabulum							+	1								I
Sphagnum subnitens		1		+												I
Lophocolea bidentata					+						1					I
Eurhynchium praelongum	+										+					I
Carex panicea				+		+										I
Drosera rotundifolia				+		+										I
Pellia epiphylla				+			+									I
Galium saxatile					+							+				I
Myosotis secunda							r		2							I
Potentilla anserina													+		+	I
Calluna vulgaris										+		+				I
Mentha aquatica							+		1							I
Eleocharis multicaulis									3							+
Oenanthe crocata									2							+
Nasturtium officinale									2							+
Callitriche stagnalis									2							+
Juncus acutiflorus		2														+
Carex pulicaris							2									+
Juncus bulbosus				2												+
Juncus conglomeratus											1					+

(Continued)

Table 20 *(Continued)*

Column	1	2	3	4	5	6	7	8	9	10	11	12	13	14	15	
Relevé	146	147	250	364	365	384	278	279	174	402	253	88	385	393	401	**Synoptic value**
Nardus stricta				1												+
Lysimachia nemorum								1								+
Polytrichum commune					1											+
Sphagnum papillosum				1												+
Lepidozia reptans		1														+
Agrostis stolonifera		1														+
Pseudoscleropodium purum					+											+
Carex viridula				+												+
Crepis capillaris		+														+
Rumex acetosella					+											+
Triglochin palustris				+					+							+
Drepanocladus exannulatus		+														+
Eurhynchium striatum					+											+
Plagiomnium affine					+											+
Aneura pinguis		+														+
Cirsium palustre			r													+
Dactylorhiza maculata							+									+
Galium aparine								+								+
Rhinanthus minor							+									+
Stellaria palustris								+								+
Succisa pratensis							+									+
Equisetum telmateia							+									+
Cardamine pratensis							r									+
Alopecurus geniculatus									1							+
Sphagnum capillifolium											1					+
Senecio aquaticus									+							+
Centaurium erythraea											+					+
Hypochaeris radicata															+	+
Cerastium fontanum													+			+
Rumex obtusifolius															+	+
Plantago major															+	+
Hylocomium splendens												+				+
Arctium minus													+			+
Mnium hornum														+		+
Number of species	13	26	16	22	19	20	28	20	18	13	18	13	16	12	13	

acutiflori-Molinietum (Table 20, column 6). This depauperate wet grassland vegetation lacks many of the species recorded in the other sites with only *Anthoxanthum odoratum* and *Carex echinata* well represented. Knapweed, *Centaurea nigra*, is an obvious component in this degraded community, providing approximately 30% cover, while *Molinia caerulea* and *Eriophorum angustifolium* are also abundant. Other common species that are generally typical of wet habitats include *Sphagnum palustre*, *Viola palustris*, *Hydrocotyle vulgaris*, *Ranunculus flammula* and *Juncus articulatus*.

TALL GRASSLAND VEGETATION OF CONTINUOUSLY WET AREAS

Molinio-Arrhenatheretea Tx. 1937
Molinietalia Koch 1926
Filipendulion (Duvigneaud 1946) Segal 1966
Table 20, columns 7–9

Filipendulion character species: *Filipendula ulmaria*

Heritage Council analogue: Marsh—GM1

Species-rich stands of the Junco acutiflori-Molinietum merge with Filipendulion vegetation (Table 20, columns 7–9) in wetter situations. This Filipendulion vegetation is widespread in Ireland, although it remains poorly documented (White and Doyle 1982). Klein (1975) records a community referable to the Filipendulion from the wooded shores of Lough Ree. Other authors have noted its occurrence in close proximity to Franguletea vegetation (Ivimey-Cook and Proctor 1966; Brock *et al.* 1978). An analogous vegetation type under the BNVC is the *Filipendula-Angelica* M27 tall-herb fen community. Interestingly, there are records in Scotland that this wet community was often mown for hay (Rodwell *et al.* 2000).

Two associations within the alliance are listed in White and Doyle (1982), although authors in Ireland rarely classify such vegetation beyond the rank of alliance, since a number of the characteristic species used to define the association in Continental Europe are not found here.

The Filipendulion is floristically the most diverse of the peripheral wet grasslands on Clare Island (averaging 22 species per quadrat) and occurs on alluvial sites and along some ditches where slow moving streams deposit organic material. Columns 7 and 8 (Table 20) were recorded from the banks of a river that winds its way through wet grassland and into the woodland at Lassau, while relevé 174 (column 9) occurs in a ditch that borders a roadside (Pl. XVI). While such tall forb communities elsewhere in Ireland are generally characterised by *Filipendula ulmaria* and *Lythrum salicaria*, the latter is not located in the current relevés. On Clare Island, *Lythrum salicaria*, is recorded in vegetation referred to the Junco acutiflori-Molinietum confirming the overlapping status of these two alliances in the eu-oceanic conditions typical of the west of Ireland.

HAY MEADOWS

Molinio-Arrhenatheretea Tx. 1937
Arrhenatheretalia Pawlowski 1928
Arrhenatherion elatioris Br.-Bl. 1925
Table 20, column 10

Arrhenatherion elatioris character species: *Dactylis glomerata*, *Arrhenatherum elatius*, *Taraxacum officinale* agg.

Heritage Council analogue: Dry meadows and grassy verges—GS2

The syntaxonomic allegiances of this Continental European alliance in Ireland are poorly defined, as the Irish examples are species poor (O'Sullivan 1965; 1982; White and Doyle 1982). Some meadow vegetation on Clare Island is referable to the Arrhenatherion elatioris (Table 20, column 10).

Pl. XVI Typical stand of Junco acutiflori-Molinietum, composed of tall grasses and forbs in a neglected field at Capnagower. When in flower, the vegetation is highly distinctive, particularly the purple loosestrife and meadowsweet among the rushes.

This vegetation represents good quality meadows found on well-drained soils. The highly productive grasses, such as *Arrhenatherum elatius* and *Dactylis glomerata* that form the bulk of the vegetation are typical of hay meadows and some fields that are left ungrazed during the summer (Pl. XVII). Beckers *et al.* (1976) record a community referable to the alliance from County Mayo, although O'Sullivan (1982) later suggested that the distinction between the Arrhenatherion elatioris and the Cynosurion cristati in Ireland was unclear, owing to the frequent overlap of character species. This classification difficulty is further compounded in that one BNVC analogue Arrhenatherum elatioris MG1 grassland is confined to lowland situations that are flooded in winter. Thus the current vegetation is more closely allied with the *Anthoxanthum odoratum-Geranium sylvaticum* MG3 grassland, though floristic affinities between the remnant habitat on Clare Island with British examples are not equal.

On Clare Island some differentiation between the two alliances of the Arrhenatheretalia is possible, based on species composition and habitat. In the absence of grazing, tall growing grasses predominate. A large proportion (95%) of the ground cover is provided by relatively few species: these include *Dactylis glomerata, Anthoxanthum odoratum, Holcus lanatus, Agrostis capillaris* and *Arrhenatherum elatius*. Other species include *Heracleum sphondylium, Taraxacum officinale* and *Veronica chamaedrys*, while minor amounts of *Juncus effusus* are usually confined to the edges of areas occupied by the vegetation. The diagnostic species considered necessary for characterisation of the association Arrhenatherum elatioris in Ireland (cf. White and Doyle 1982) are not recorded from Clare Island. These species include *Pimpinella major, Tragopogon pratensis, Trisetum flavescens, Crepis biennis* and *Knautia arvensis*.

Pl. XVII Meadow vegetation assigned to the Cynosurion cristati. This meadow was left ungrazed until late in the summer each year.

MODERATE QUALITY PASTURE

Molinio-Arrhenatheretea Tx. 1937
Arrhenatheretalia Pawlowski 1928
Cynosurion cristati Tx. 1937

Table 20, columns 11–15

Cynosurion cristati character species: *Cynosurus cristatus, Trifolium repens, Lolium perenne*

Heritage Council analogue: Dry Meadows and Grassy Verges—GS2

This community (Table 20, columns 11–15) is again classified just to the alliance level as from a European perspective its classification poses problems at the level of the association (Zuidhoff *et al.* 1995). It is referred to the Cynosurion cristati although the vegetation contains some species characteristic of the Centaureo-Cynosuretum (analogue *Cynosurus cristatus-Centaurea nigra* MG5 grassland). No equivalent vegetation type has been described from Central Europe (O'Sullivan 1982; Rodwell 1992). These moderate quality grasslands are of variable appearance and are generally dicot-rich (O'Sullivan 1982).

Originally described from Ireland by Braun-Blanquet and Tüxen in 1952, the widespread distribution of the Cynosurion vegetation in Ireland has since been confirmed by O'Sullivan (1965; 1982). Similar vegetation has been reported from Britain (Birks 1973; Rodwell 1992).

The vegetation occurs on mineral rich soils of some of the drier lazy beds and unmanaged roadside verges on Clare Island. The sward is typically 25cm high, although in the absence of grazing it can reach 75cm (Pl. XVIII A). Constant graminoid species include *Holcus lanatus, Anthoxanthum odoratum, Agrostis capillaris, Festuca ovina* and along unmanaged roadside verges, *Cynosurus cristatus*. Common species include *Trifolium pratense, T. repens, Potentilla anserina* and *Lotus corniculatus*. Other species include *Luzula campestris* and *Rhytidiadelphus squarrosus*. These species, which are characteristic of the Centaureo-Cynosuretum,

Pl. XVIII Two examples of Cynosurion cristati vegetation showing differences in management regimes: A) In the foreground a typical stand of species-rich grassland with an extensive layer of creeping buttercup; B) Intensively grazed meadow where many of the habitat character species are diminutive and there is a relative increase in the abundance of Nardetalia species. (Note also the spread of rushes in a neglected part of the field).

are also common components of the three grassland communities described previously. In wet derelict fields, *Juncus effusus* sometimes becomes the overall dominant, and may be assigned to the Centaureo-Cynosuretum-Juncetosum, a subassociation commonly found spreading along small streams and in disused corners of Cynosurion meadows (O'Sullivan 1965) (Pl. XVIII B).

VEGETATION OF BOG COMPLEXES

The development of blanket bogs in Ireland is well-documented (cf. Doyle 1990). The major peat system in the west of Ireland is Atlantic blanket bog. Its development, distribution and phytosociology has been discussed by Doyle and Moore (1980) and Doyle (1982b; 1990; 1997). Atlantic blanket bogs are formed under eu-oceanic climatic conditions, and are generally confined to coastal plains below 200m. The important climatic features of such areas may be summarised as mild and wet, with annual precipitation exceeding 1200mm. In general, Atlantic blanket bogs develop on acid lithologies, which are further influenced by the pattern of glacial deposition. However, the extensive tracts of blanket bog may be viewed as a complex of ombrotrophic vegetation together with a range of drainage features.

The general consensus is that peat formation within Atlantic blanket bog areas commenced in waterlogged hollows sometime after 8000 BP, although a number of radiocarbon dates from several locations in the west coast of Ireland suggest that initiation of peat formation was related to local conditions (cf. Doyle 1997). There are two obvious tree-laden strata (mainly pine) within the peat profile. In the Glenamoy region of northwest Mayo, initiation of blanket peat commenced around 7000 BP, based upon the dating of tree stumps, while a second layer of tree stumps, separated from the mineral layer by 20cm of peat, was dated at 4290 BP (Doyle 1982b; 1990).

The expansion of the blanket peats appears to have accelerated with the large-scale clearance of the forests by Neolithic farming communities. Some fossilised tree remains exposed on some peat surfaces at Maum townland, however, indicate that some peat-producing vegetation existed prior to forest clearance, since the woody subfossils are found above the peat layer (Pl. XIX A). Preliminary investigations of sediment cores from the eastern part of Clare Island reveal rises in *Calluna* and Cyperaceae pollen and suggest peat initiation at around 4000 years BP (Coxon 1994). This tentative chronology coincides with research conducted in areas adjacent to Clare Island, including the nearby island of Inishbofin (Ní Ghráinne 1993) and proximal mainland areas (Doyle 1982b; 1990; 1997; O'Connell 1990).

The timing of blanket peat initiation, however, is not synchronous throughout Ireland. In locations such as the Céide Fields in north-west Mayo it is clear that the expansion of blanket bog followed Neolithic farming settlement. Caulfield (1978; 1983) suggested that the widespread initiation of Atlantic blanket bog followed human interference in the fourth millennium BP. Recently, shallow peats have been discovered from the western end of Clare Island overlying old field systems (P. Gosling, pers. comm.). It has been suggested that the peat subsumed this field system, in a manner comparable with the prehistoric site described from the Céide Fields (Caulfield 1978).

Peat formation on Clare Island began in impervious clay basins within the drumlin landscape that was created during the last glaciation. Bog vegetation appears to have been extensive in its distribution on Clare Island, with many cutaway sites bearing testament to expansive bogs. Throughout the centuries, the commonage area has supplied much of the island's domestic turf (C. Mac Cárthaigh 1999). Although most of the island's soils are peaty, the peat that accumulated in shallow depressions and basins in the *Calluna*-dominated heathland yielded a maximum of four layers of scraw turf before a new site was exploited. Thus it is not uncommon to find a fragmented network of bare peat on the commonage.

The blanket bog on Clare Island exists as a complex, comprising ombrotrophic vegetation on level surfaces together with a series of drainage features such as shallow depressions, pools, lakes (with peat islands) and outliers of heathland vegetation and acid grassland vegetation that dissect the bog surface. The majority of the bog vegetation occurs in the extensive complex running from Maum townland to Ballytoohy Beg and bounded on the west by Knocknaveen (see fold-out map 2). It includes the majority of the deep peats and the only extensive area of extant blanket bog, situated in a plain below the lighthouse (Pl. XX, see p. 133).

One notable site is at Park, located on the commonage (Pl. XIX B). The site is easily distinguished both on the ground and from aerial photographs. It once supported a farming settlement during pre-Famine times, as evidenced by the lazy beds on its surface. Peat depths in excess of 6m have been measured. The abandoned peatland, now vegetated with species-poor grassland, lies derelict except for occasional peat-cutting.

Unlike comparable vegetation from Galway and Mayo, which averages 3.5m (Doyle 1990; 1997), peat depth on Clare Island averages less than 2m. In some depressions this figure may be closer to 7m, such as in the area around Poirtín Fuinch above Maum townland (P. Coxon, pers. comm.). However, most communities developed over fairly shallow peat that rarely exceeded two metres.

The peatland area, centred on Maum townland, now provides much of the island's present

Pl. XIX A Cutaway bog in Maum townland where peat removal has uncovered numerous tree stumps, indicative of a relict forest that was subsumed during peat formation.

Pl. XIX B Abandoned pre-Famine settlement in Park, which is located in the centre of the island. This remarkable site, which once supported crop production, as evidenced by the abundant lazy beds, occurs on peat measuring six metres deep. It now lies abandoned except for occasional peat cutting.

day turf needs, as it is easily accessed by road. Interestingly, many of the fulachtaí fia that have been uncovered during the present archaeological survey are also found here, confined to streams and waterlogged areas within the peatland.

The complex mosaic is generally fragmented. Description of these areas is divided into four sections, each comprising several syntaxa that characterise either native blanket bog or its various drainage features.

ATLANTIC BLANKET BOG

Relatively intact native blanket bog on Clare Island (Table 21, columns 1–14)
Schoenus-rich blanket bog (Table 21, columns 15–24)
Juncetosum subtype (Table 21, columns 25–47)
Recovering blanket bog (Table 21, columns 48–56)
Degraded blanket bog (Table 21, columns 57–60)

WET HOLLOWS

Rhynchospora-rich stands (Table 22, columns 1–4)
Rhynchospora-poor stands (Table 22, columns 5–7)

PEAT AND PEAT ISLANDS ON DEEP PEAT BASINS

Peat islands (Table 23, columns 1–2)
Ombrotrophic blanket peat surrounding the 'islands' (Table 23, columns 3–6)
Marginal vegetation on the peat basin (Table 23, columns 7–8)

DRAINAGE CHANNELS

Vegetation dominated by *Hypericum elodes* and *Potamogeton polygonifolius* (Table 24)
Vegetation dominated by *Sphagnum cuspidatum* and *Eriophorum angustifolium* (Table 24)

Floristics

Native Atlantic blanket bog vegetation is graminoid-dominated with *Molinia caerulea* and *Schoenus nigricans* as the major components, together with *Eriophorum vaginatum*, *Rhynchospora alba* and *Narthecium ossifragum*. In certain situations, however, *Eriophorum vaginatum* may be replaced by *E. angustifolium*. Ericoid shrubs present in depauperate form include *Calluna vulgaris* and *Erica*

Table 21
Relevé table for the Pleurozio purpureae-Ericetum tetralicis and its subcommunities

Column										1										2										3
	1	2	3	4	5	6	7	8	9	0	1	2	3	4	5	6	7	8	9	0	1	2	3	4	5	6	7	8	9	0
Relevé	2	2	2	2	1	2	2	1		2	2	1	2	1	1	2		1	3	2	1	3	3	1	3	3	3	1		2
	4	4	1	5	5	4	4	8	6	8	8	4	4	4	5	2	3	7	6	2	9	4	5	7	8	9	8	9	7	1
	3	4	5	1	5	7	8	2	6	1	2	8	2	6	4	7	8	7	4	8	2	7	4	6	8	0	9	1	8	2
Pleurozio purpureae-Ericetum tetralicis																														
Character species																														
Campylopus atrovirens				1							2							+	+								2	1		
Schoenus nigricans		1													2	1	2	2	3	3	2	2	4	2						
Pleurozia purpurea			2																										+	
Differential species																														
Potentilla erecta	1	1	1	1	1	2	1	2	2	1	1	2	2	2	1	1	1	1	+	+			+	+	+	+	+	1	1	1
Pedicularis sylvatica	+	+	+	+	+	+		+	1	+			1				+					+							r	
Polygala serpyllifolia	+	+	r	+						r	+		+				r		+		+	+		+					+	
Juncetosum																														
Carex panicea	2	+		+	+	2	2	1		r					1	2		1	+	1	+	+	+		+	+	+	2	+	
Eleocharis multicaulis												1						2	2		4		1	3	1	2	2	1	+	+
Juncus bulbosus						1										1			+		2				1	1	1		1	2
Calluno-Sphagnion papillosi																														
Molinia caerulea	2	3	2	3	2	3	2	3	3	2	2	5	3	4	3	2	3	2	3	3	3	4	2	4	1	1	1	2		+
Calluna vulgaris	4	3	3	3	4	3	2	3	2	3	3	1	3	3		2		3	+		+	1			1	1	+	1	1	1
Eriophorum angustifolium	3	3	+	3	3	2	2	2	3	2	3	3	3		2	2	1	3	1	2		2	3	3		+	+	+		2
Sphagnum subnitens	1		4		3			2	2	3	3	2	2					1	+		1		2		1	3			2	2
Cladonia portentosa	2		1	2	+	2	+	1		+	1							+					+				+			+
Cladonia uncialis	+	2	+	+																						+			+	
Campylopus paradoxus						1																3				2			1	1
Diplophyllum albicans					+			1					1														1			
Rhyncospora alba							2												2											
Mylia anomala																			+											
Order and Class species																														
Erica tetralix	3	2	+	2	2	3	1	2	1	2	2	2	2	2	1	2	1	2	+	2	1	2	+	2	+	+	2	+		1
Narthecium ossifragum	+	1		2		1	1			2	1	1	1	2	1	1	+	1	+	1	+	+	+	1	+	+	+	r	+	+
Drosera rotundifolia	+	+	+	r	+		+				+	r	+	+			+		+	+		+	+		+	+				+
Hypnum jutlandicum		2		1	2	1		+					2	1				2	+		1		+		1	+		1		2
Trichophorum caespitosum	1	3		2	3	3	3		1	2	2		2		2	2				2				+						
Sphagnum papillosum	3													1		1											3			
Plagiothecium undulatum	1								+		+				+										+					1
Hylocomium splendens	2								2				2	1			+					+								
Odontoschisma sphagni										1	1																			

1	2	3	4	5	6	7	8	9	40	1	2	3	4	5	6	7	8	9	50	1	2	3	4	5	6	7	8	9	60	
99	379	272	267	290	226	219	16	289	67	149	229	346	170	236	206	171	82	273	241	280	283	351	209	230	232	6	7	54	183	**Synoptic value**
				3		2							+						1											I
																														+
																														R
+	+	2	1	+	1	1				1	2	+	1	+	2	2	+	2	+	1	1	1	2	2	2	1	+	2	3	IV
	+						+					r			+	+		r			1				1	r		r		II
		+		+					+		r	+	r	+	1				r		+		+		+					II
1	+	1	1	1	2		1	1		+		+	+	1																III
1			2	+		1		+	2	1		1	+			1														III
		4			+	3		1	+					+	1												1			III
+	1	2	1		+		1	+	+	4	3	3	3	3	2	3	1	2	2	3	3	3	1	2	3	1	+	1	1	V
	3	1		+	1		4	2		2	1	2	1	1	1	3	1	1	2	3	3	3	3	3	3			1		V
2			+		2	1	2	+		3	2	1		1	1	3	2			1	1	+	1	2					3	IV
1	3	1	2		2			2	2	1	1	1	+		2		1	2			2		2	2	3	2		2		IV
+	+	+		+					+								+		+		+									II
		+				+		+	+					+			+		+											II
																														+
																														+
																														R
			+																											R
	+	1		1	1		1	1	1	3	3	2	+	+	2	2	+	1	2	2	1	+	2	2	2					V
1	+		1	1			+		1			+			1				+				+			+				III
		+		+		2		+	1	+		+					+													III
	1	2		+				+			1		3	1		1	+	2				1	1		3					III
					2					1									3	2	2	+								II
				2	1						1	1		1	2			+	2	+							1	3		II
	+										+			+	1	+		2					1	+	2					II
		2										+		+	2			2				1					3			II
	+																	1	1	+										+

(Continued)

Table 21 *(Continued)*

Column										**1**										**2**										**3**
	1	**2**	**3**	**4**	**5**	**6**	**7**	**8**	**9**	**0**	**1**	**2**	**3**	**4**	**5**	**6**	**7**	**8**	**9**	**0**	**1**	**2**	**3**	**4**	**5**	**6**	**7**	**8**	**9**	**0**
Relevé	**2**	**2**	**2**	**2**	**1**	**2**	**2**	**1**		**2**	**2**	**1**	**2**	**1**	**1**	**2**		**1**	**3**	**2**	**1**	**3**	**3**	**1**	**3**	**3**	**3**	**1**		**2**
	4	**4**	**1**	**5**	**5**	**4**	**4**	**8**	**6**	**8**	**8**	**4**	**4**	**4**	**5**	**2**	**3**	**7**	**6**	**2**	**9**	**4**	**5**	**7**	**8**	**9**	**8**	**9**	**7**	**1**
	3	**4**	**5**	**1**	**5**	**7**	**8**	**2**	**6**	**1**	**2**	**8**	**2**	**6**	**4**	**7**	**8**	**7**	**4**	**8**	**2**	**7**	**4**	**6**	**8**	**0**	**9**	**1**	**8**	**2**
Nardetalia species																														
Nardus stricta								2							2	2		3		2	2		+		2	2	3	4	2	1
Festuca ovina									2		1				2	3	3	2		2	2	1			+	+	+	2		
Agrostis capillaris			1									3		1	2	2	3		1		2			1	+	1				
Anthoxanthum odoratum									2			2	1	2								2		2						
Carex echinata												1	1	2	+	+	1			+		+		2			+			
Thuidium tamariscinum													1	1														+		
Festuca vivipara															1	+	1	3		+								1		
Rhytidiadelphus squarrosus	1		1						1					1										1	+					+
Companion species																														
Carex nigra	+		+	+	1							1	+	+			1		+		2			+		+	+	+		
Anagallis tenella								1							1	1	1	2	+	1	1	+	+	1				2		
Aulacomnium palustre	1	+							1			+		1							2			+	+	+		2		
Scapania gracilis	+		1		1	1				1	1														+					
Sphagnum capillifolium	1					2	1			2	+											2		1						
Juncus effusus																+									1					
Juncus squarrosus			3																			+			+	+		1	+	2
Succisa pratensis										+		1	1	+		+			+	1	+	+	+	1						
Hydrocotyle vulgaris																				+	+		+	1		+		+		
Sphagnum cuspidatum				2		3	4													1										2
Rhytidiadelphus triquetrus															+													1		
Carex binervis										1	+																			
Dicranum scoparium			1	+				1		+																				2
Pseudoscleropdium purum	1													1			1							2						
Carex viridula															1	1					3	+						1	+	
Cirsium dissectum								1							2	1		2	+	2	1			+						
Danthonia decumbens			+									1				+					1		1					2		
Carex pulicaris															+	+	+				1	+	+	+				1		
Sphagnum palustre											1						2		2				+	+	3					
Zygogonium ericetorum					1					3	1															2		1	1	
Polytrichum commune																														
Racomitrium lanuginosum				1						1																		+	3	1
Holcus lanatus															1			1												
Ranunculus flammula												+			+					+	1		+	1						
Lophocolea bidentata																		+						1	+					
Pleurozium schreberi										1			2																	
Frullania tamarisci				+																	1		1							
Viola palustris																+	+							+						
Dactylorhiza maculata												+																		

									4										5										6	
1	**2**	**3**	**4**	**5**	**6**	**7**	**8**	**9**	**0**	**1**	**2**	**3**	**4**	**5**	**6**	**7**	**8**	**9**	**0**	**1**	**2**	**3**	**4**	**5**	**6**	**7**	**8**	**9**	**0**	
	3	**2**	**2**	**2**	**2**	**2**		**2**		**1**	**2**	**3**	**1**	**2**	**2**	**1**		**2**	**2**	**2**	**2**	**3**	**2**	**2**	**2**				**1**	
9	**7**	**7**	**6**	**9**	**2**	**1**	**1**	**8**	**6**	**4**	**2**	**4**	**7**	**3**	**0**	**7**	**8**	**7**	**4**	**8**	**8**	**5**	**0**	**3**	**3**			**5**	**8**	
9	**9**	**2**	**7**	**0**	**6**	**9**	**6**	**9**	**7**	**9**	**9**	**6**	**0**	**6**	**6**	**1**	**2**	**3**	**1**	**0**	**3**	**1**	**9**	**0**	**2**	**6**	**7**	**4**	**3**	
																														Synoptic value
2	1	2	5	1	3	3					3		2	2	2	2		1	1			1	2	2	2		1	2		III
	1	3	2	2	4		1	3			3	1	3	3	2	2		1		2			2	2	2	2	2	3		III
	1	2	2	1	2	1		2			2		2	3	1			2			1	2	1	2	2	3	4	3	4	III
	2	2	1		2		1			2		2	+	1		1		1		2	3	2		1			1	1	3	III
		1			1				+			1		+	2									+			+	1	1	III
	2	1			1		1			2	+			2	2	+		2			1			1			2		2	III
	+				+						+			1	+	1						2	+	1	+					III
	+	2		+						3	1	+		+	2	+		3		+		+	2		2				2	II
				+		1	1			1	1		1	+	2	1			1		1		1	1				1	2	III
			+	+	1			+			1	+	+											1		1	2	2		III
		2			2					1								1							1			2		II
	+			+											+				1						+					II
	3			3					3					3						3		+	1							II
		+		2		+	1				1				+								1				2	2	2	II
			1														1	2					3	2	2					II
				+				r		+	2	+	+		1	+				+	+	+								II
				+		+		+				+	+													1	+	2	2	I
		3				2											1		4						3					I
			1				+			1				2				2			+			1						I
	+									1	+										1	+								I
	+	1					+								1						+				+					I
											+	+						+				+	2						3	I
			+								1	+	+	1													1			I
											1	+	1			+										1				I
		1						+			+		+	+	1															I
											1	+					+									1				I
			+			2	1	+														+								I
2				1				2																						I
		1					+							+	2									1	1		1		3	I
						+		2																						I
					2							1										+		1				2	2	I
													+																	I
		+																		+		+		+						I
										2										+	3			+						+
							+						+									+								+
				+																						1			1	+
										r						r				+	+					+				+

(Continued)

Table 21 *(Continued)*

Column										**1**										**2**										**3**
	1	**2**	**3**	**4**	**5**	**6**	**7**	**8**	**9**	**0**	**1**	**2**	**3**	**4**	**5**	**6**	**7**	**8**	**9**	**0**	**1**	**2**	**3**	**4**	**5**	**6**	**7**	**8**	**9**	**0**
Relevé	**2**	**2**	**2**	**2**	**1**	**2**	**2**	**1**		**2**	**2**	**1**	**2**	**1**	**1**	**2**		**1**	**3**	**2**	**1**	**3**	**3**	**1**	**3**	**3**	**3**	**1**		**2**
	4	**4**	**1**	**5**	**5**	**4**	**4**	**8**	**6**	**8**	**8**	**4**	**4**	**4**	**5**	**2**	**3**	**7**	**6**	**2**	**9**	**4**	**5**	**7**	**8**	**9**	**8**	**9**	**7**	**1**
	3	**4**	**5**	**1**	**5**	**7**	**8**	**2**	**6**	**1**	**2**	**8**	**2**	**6**	**4**	**7**	**8**	**7**	**4**	**8**	**2**	**7**	**4**	**6**	**8**	**0**	**9**	**1**	**8**	**2**
Leucobryum glaucum					+			2											+									+		
Erica cinerea								1																						
Luzula campestris									1																					
Viola riviniana																												+		
Sphagnum compactum			1										2															2		
Juncus articulatus															1					1		+	+	1						
Aneura pinguis																				1		+		1						
Mnium hornum								+																						+
Peltigera canina			r																					+						
Pellia epiphylla																							+							
Leontodon autumnalis																	r			1										
Lotus corniculatus																														
Plantago lanceolata																							r							
Campylopus introflexus																														
Isothecium myosuroides																														
Sphagnum tenellum																								1			+			
Calypogeia sp.											+			+																
Sphagnum auriculatum							2																							
Breutelia chrysocoma				1																			+	+						
Selaginella selaginoides																					+	+								
Salix repens																														
Nardia scalaris				1																										
Myrica gale																														
Trifolium repens																														
Dicranum majus																									+					
Rhytidiadelphus loreus																														
Juncus acutiflorus												1																		
Hypericum pulchrum																														
Eurhynchium praelongum																								2						
Viola sp.												1																		
Vaccinium oxycoccos																														
Blechnum spicant																														
Pinguicula vulgaris																														
Cirsium palustre																														
Equisetum palustre																														
Galium saxatile																														
Hypericum elodes																														
Hypochaeris radicata																								+						

									4										5										6	
1	2	3	4	5	6	7	8	9	0	1	2	3	4	5	6	7	8	9	0	1	2	3	4	5	6	7	8	9	0	
	3	2	2	2	2	2		2		1	2	3	1	2	2	1		2	2	2	2	3	2	2	2				1	
9	7	7	6	9	2	1	1	8	6	4	2	4	7	3	0	7	8	7	4	8	8	5	0	3	3			5	8	
9	9	2	7	0	6	9	6	9	7	9	9	6	0	6	6	1	2	3	1	0	3	1	9	0	2	6	7	4	3	
																														Synoptic value
																+														+
					+		2															+								+
										+											+							+		+
								r																			1	+		+
																			1											+
													+																	+
			r									+																		+
																		1							1					+
					+									+																+
						1		+																						R
													+																	R
							+									1					1									R
											r	+																		R
+						1		1																						R
			1				1									1														R
																			2											R
					+																									R
						4													2											R
																														R
																														R
													+			1														R
																								1						R
										+										2										R
										1																		1		R
																				+										R
			1																					1						R
																												1		R
																						+								R
																														R
																														R
							+																							R
																						+								R
									r																					R
																											1			R
		1																												R
																						+								R
												+																		R
								r																						R

(Continued)

Table 21 *(Continued)*

Column	1	2	3	4	5	6	7	8	9	10	11	12	13	14	15	16	17	18	19	20	21	22	23	24	25	26	27	28	29	30
Relevé	243	244	215	251	155	247	248	182	66	281	282	148	242	146	154	227	38	177	364	288	192	347	354	176	388	390	389	181	78	212
Jasione montana																														r
Huperzia selago																														
Plantago maritima																														
Radiola linoides																														
Isolepis setacea																														
Trifolium pratense																														
Campylopus fragilis																														
Eurhynchium striatum																														
Polytrichum formosum																														
Sphagnum molle																										+				
Sphagnum recurvum																														
Cladopodiella fluitans																														
Lepidozia reptans																														
Mylia taylorii																														
Drepanocladus fluitans																				1										
Bryum capillare																					2									
Agrostis canina																				1										
Potamogeton polygonifolius																				1										
Campylium stellatum																							+							
Prunella vulgaris																					+			+						
Agrostis stolonifera																					+									
Pinguicula lusitanica																							r							
Drepanocladus revolvens																				1										
Euphrasia officinalis																														
Number of species	23	15	21	22	17	16	14	19	15	21	21	20	21	19	22	24	21	20	24	28	30	25	24	33	23	23	18	28	17	22

tetralix. In general, the bryophyte component (and in particular *Sphagnum* species) is less dominant in its contribution to the vegetation than is usual on raised bogs. The important bryophyte species include *Pleurozia purpurea*, *Campylopus atrovirens*, *Sphagnum papillosum* and *Sphagnum subnitens*.

On Clare Island, the vegetation encountered reflects both the degraded nature of the habitat and the relative lack of extensive areas of natural blanket bog communities. It reveals the impact of human settlement on the habitat, as the peatland was vital in supplying domestic fuel for the largely self-sufficient islanders. The increasing levels of grazing over the past 50 years have exacerbated the impacts on bog vegetation.

Apart from vegetation that is typical of ombrotrophic surfaces, the blanket bog systems support a variety of drainage features—shallow depressions, pools, lakes (with peat islands) and outliers of heathland vegetation, which together form the characteristic and complex vegetation mosaic. The flora and vegetation reflect differences in soil depths and

									4										5										6	
1	2	3	4	5	6	7	8	9	0	1	2	3	4	5	6	7	8	9	0	1	2	3	4	5	6	7	8	9	0	
	3	2	2	2	2	2		2		1	2	3	1	2	2	1		2	2	2	2	3	2	2	2				1	
9	7	7	6	9	2	1	1	8	6	4	2	4	7	3	0	7	8	7	4	8	8	5	0	3	3			5	8	
9	9	2	7	0	6	9	6	9	7	9	9	6	0	6	6	1	2	3	1	0	3	1	9	0	2	6	7	4	3	**Synoptic value**
																														R
								2																						**R**
													1																	**R**
													+																	**R**
																											1			**R**
																										+				**R**
									+																					**R**
																						+								**R**
						2																								**R**
																														R
																														R
																														R
																													1	**R**
														+																**R**
																														R
																														R
																														R
																														R
																														R
																														R
																														R
																														R
																														R
							+																							**R**
1	2	2	2	2	2	1	2	2	1	2	2	2	2	2	2	2	1	2	1	1	2	2	2	2	2	1	1	2	1	
1	3	8	0	5	3	9	2	5	4	3	7	9	8	9	7	3	3	2	9	8	2	6	0	4	1	4	9	1	6	

hydrological gradients that occur in the bog habitats. Many of the plants characteristic of bogs are not confined to these habitats and are regularly found as part of the intricate vegetation mosaic.

Purple moor grass, *Molinia caerulea*, typical of Atlantic blanket bogs in the west of Ireland, is a regular component of the vegetation on Clare Island. The other major graminoid species, *Schoenus nigricans*, occurs infrequently on Clare Island and is confined to peats of moderate depths (0.5m to 2.5m) where flushing is an obvious feature of the habitat. Unusually, graminoid species diagnostic of the Nardetalia, such as *Anthoxanthum odoratum*, *Agrostis capillaris*, *Festuca ovina*, and in some instances *Nardus stricta* are recorded in situations where the bog habitat is seriously degraded. These grass species are less conspicuous in wetter situations, where there is a relative abundance of bryophytes.

Ericaceous species figure prominently in the vegetation, although with varying abundance depending on the situation. Ling heather, *Calluna vulgaris*, is the most common species, occurring

Table 22
Relevé table for the Sphagno tenelli-Rhynchosporetum albae

Column	1	2	3	4	5	6	7	Synoptic value
Relevé	12	356	355	388	361	357	359	
Sphagno tenelli-Rhynchosporetum albae								
Rhynchospora alba	2	3	3	1	+	+	+	V
Sphagnum cuspidatum	3	3	2	3	4	2	2	V
Eriophorum angustifolium		2	1	2		3	5	V
Pleurozio purpureae-Ericetum tetralicis								
Molinia caerulea		1	2	2	3	+	+	V
Potentilla erecta	1	+	+		+		+	V
Campylopus atrovirens		+	+					II
Pinguicula lusitanica	r				+			II
Polygala serpyllifolia					+			I
Eriophoro vaginati-Sphagnetalia papillosi								
Sphagnum papillosum		2	+		1	3		III
Odontoschisma sphagni		+	+					II
Narthecium ossifragum					+	+		II
Sphagnetalia compactii								
Drosera rotundifolia		+	+	+	+	+	+	V
Erica tetralix	1	+	+		+			IV
Companion species								
Eleocharis multicaulis	2	+	2	1	1	2		V
Sphagnum capillifolium		2	3	2	1	2		V
Nardus stricta		+	+	+	+			IV
Carex panicea			+	+		+		III
Calluna vulgaris	1	+			+			III
Agrostis canina			1	+	2			III
Agrostis capillaris	2	+						II
Sphagnum palustre				4			2	II
Hypericum elodes	1				1			II
Osmunda regalis		+			r			II
Festuca ovina	2							I
Carex echinata				1				I
Juncus bulbosus						+		I
Hypnum jutlandicum		+						I
Carex nigra				+				I
Scapania gracilis		+						I
Polytrichum commune				+				I
Zygogonium ericetorum		+						I
Potamogeton polygonifolius			+					I
Sphagnum tenellum				+				I
Racomitrium lanuginosum			+					I
Juncus effusus				+				I
Campylopus introflexus			+					I
Sphagnum compactum		+						I
Rhytidiadelphus squarrosus					+			I
Anthoxanthum odoratum					+			I
Anagallis tenella					+			I
Hydrocotyle vulgaris					+			I
Cladonia uncialis	+							I
Succisa pratensis			r					I
Number of species	11	20	19	16	21	11	7	

as stunted bushes (20–40cm), except where sheep have been excluded from grazing. On drier, undulating topographies, *Erica tetralix* is occasionally recorded alongside *Erica cinerea*. Two species that are recorded from small islands from a deep peat basin on Clare Island include *Vaccinium oxycoccos* and *Empetrum nigrum*. Neither of these two species is typical of Atlantic bog situations, with *Empetrum* generally restricted to montane blanket bog, while *Vaccinium*, generally confined to raised bogs in Ireland, is recorded from a few waterlogged depressions within the Atlantic blanket bog region (Doyle and Foss 1986b). Similar shrub-dominated communities characterised by *Calluna* and *Empetrum* together with *Juniperus communis* have been described from peat islands by Doyle *et al.* (1987).

The herbaceous flora is poorly developed in many of the bog communities on Clare Island. Species typical of bogs, which may attain local significance

Table 23
Relevé table for the peat islands and adjacent ombrotrophic peat

Column	1	2	3	4	5	6	7	8	Synoptic value
Relevé	322	324	325	323	327	326	328	329	
Pleurozio purpureae-Ericetum tetralicis									
Molinia caerulea	2	+	+	+		+	+	2	V
Drosera rotundifolia		+	+	+	+	+	+	1	V
Narthecium ossifragum	+		+		+	+	+	+	V
Potentilla erecta	+	+	+		+		+	+	IV
Schoenus nigricans		2	1	1					III
Scheuchzerietalia palustris									
Menyanthes trifoliata	+	+	+		+	+	+		V
Carex limosa	+	+							II
Sphagnum cuspidatum				2		4			II
Sphagnetalia compactii									
Erica tetralix	+		+		+			+	III
Sphagnum compactum					2				I
Juncus squarrosus				+					I
Oxycocco-Sphagnetea									
Vaccinium oxycoccos	+	+	+				+	+	IV
Empetrum nigrum	2	+							II
Mylia taylorii			1					+	II
Scheuchzerio-Caricetea nigrae									
Carex viridula		+	+	1	+		+		IV
Carex panicea			+	+	+			+	III
Carex nigra	+	2		1	+				III
Campylium stellatum		+	2	3	2				III
Aneura pinguis		+	+	+		+			III
Pellia epiphylla		+	+		+	+			III
Ranunculus flammula			+		+				II
Sphagnum palustre				+		1	5	3	III
Littorelletea uniflorae									
Eleocharis multicaulis	+	+	1	4	4	3	2	2	V
Juncus bulbosus			+	+	1	+	+	1	V
Potamogeton polygonifolius		+	+	1	1	+	+		V
Scorpidium scorpioides		2	2	2	3				III
Anagallis tenella	+	+	1	+	+				III
Hydrocotyle vulgaris				+	+	+			III
Hypericum elodes					1		+		II
Companion species									
Festuca ovina	2	2	2	+		2		2	V
Nardus stricta		2	3	+	1	2	+	3	V
Agrostis capillaris	1	2	+			+		+	IV
Carex echinata	1			+		+	+	1	IV
Eriophorum angustifolium	2			2		2	2		III
Sphagnum capillifolium			2		2			1	III
Sphagnum papillosum	+		+				+		III
Carex rostrata				2	+		+		III
Succisa pratensis		r	+		r				III
Calluna vulgaris	+				+			+	III
Carex paniculata	+	+	+						III
Pinguicula vulgaris		r		+	+				III
Aulacomnium palustre			+					2	II
Rhytidiadelphus squarrosus	2							+	II
Sphagnum subnitens						+	1		II
Juncus articulatus		+			+				II
Selaginella selaginoides			+		+				II
Lepidozia reptans			+			+			II
Pseudoscleropodium purum	2								+
Hylocomium splendens	3								+
Agrostis stolonifera			+						+
Scapania gracilis					+				+
Polytrichum commune							+		+
Calliergon cuspidatum				+					+
Hypnum jutlandicum								+	+
Rhytidiadelphus triquetrus	+								+
Lophocolea bidentata	+								+
Scapania undulata				+					+
Polytrichum juniperinum						+			+
Atrichum undulatum				+					+
Sagina nodosa				+					+
Ceratophyllum demursum		+							+
Mnium hornum	+								+
Anthoxanthum odoratum			+						+
Drosera anglica						r			+
Pinguicula lusitanica					r				+
Triglochin palustris					r				+
Dactylorhiza maculata								r	+
Number of species	24	25	32	27	30	21	19	21	

Table 24
Relevé table for the Hyperico-Potametum oblongi

Column	1	2	3	4	5	6	7	8	9	10	11	12	Synoptic value
Relevé	334	79	440	336	136	288	365	138	308	255	137	238	
Hyperico-Potametum oblongi													
Hypericum elodes	2	2	5	5	3	3	4	4	3	3	4	2	V
Potamogeton polygonifolius	+	2	2	+	2	2	2	1	3	2	1	2	V
Littorelletea uniflorae													
Eleocharis multicaulis	1		2	2	3	1		4	1	2	2	3	V
Hydrocotyle vulgaris	+	+		+	1	+	2	+	2		1	+	V
Anagallis tenella	+		+	+	+	+	3			1	1	1	IV
Juncus bulbosus	1	2	2		2	1	+			1		1	IV
Ranunculus flammula	+	+	+	1	2	+	+		1	2	3	1	V
Agrostis stolonifera	1			2	1			1			1		III
Carex echinata	2	2	1		+	1	2					+	III
Carex nigra				+	2	1	+		1	2	2	1	IV
Sphagnum cuspidatum-Eriophorum angustifolium													
Sphagnum cuspidatum	2			1	2						2	3	III
Eriophorum angustifolium		1				1			2	2		2	III
Scorpidium scorpioides	1			4						2	2	2	III
Carex panicea			+						2	+	2	+	III
Bryum pseudotriquetrum		1	2							+	1	2	III
Companion species													
Carex viridula							+		2	+		1	II
Juncus articulatus					+			2		+	2		II
Drosera rotundifolia	+					+	+					1	II
Aneura pinguis	+		+							+	+		II
Pellia epiphylla	+	+			+							+	II
Juncus effusus	+	1								1			II
Triglochin palustris		+			1						1		II
Galium palustre	+	+	+										II
Viola palustris		+	+				+						II
Carex limosa	+		1	3									II
Nardus stricta	+						+					2	II
Calliergon cuspidatum	1											1	I
Sphagnum subnitens			1			1							I
Aulacomnium palustre	+											1	I
Equisetum palustre		+								r			I
Sphagnum palustre	4												+
Festuca ovina									3				+
Apium inundatum								3					+
Baldellia ranunculoides											2		+
Carex flacca		2											+
Drepanocladus revolvens												2	+
Carex viridula		1											+
Juncus squarrosus												1	+
Pinguicula lusitanica												1	+
Agrostis capillaris			1										+
Mnium hornum			1										+
Sphagnum auriculatum							1						+
Agrostis capillaris							+						+
Aira praecox									+				+
Mentha aquatica		+											+
Fissidens adianthoides		+											+
Leontodon autumnalis		+											+
Campylopus introflexus						+							+
Cardamine pratensis	+												+
Pohlia nutans											+		+
Juncus acutiflorus			+										+
Menyanthes trifoliata	+												+
Potentilla erecta							+						+
Potentilla palustris				+							+		+
Thuidium tamariscinum			+										+
Cirsium dissectum											+		+
Erica tetralix												+	+
Equisetum telmateia				r									+
Myosotis secunda	r												+
Number of species	25	20	18	13	14	13	15	7	11	16	19	24	

in certain situations, include *Narthecium ossifragum*, *Drosera rotundifolia*, *Pinguicula lusitanica* (occasionally, the non-bog species *P. vulgaris*), *Potentilla erecta* and *Polygala serpyllifolia*. On Clare Island, a number of species, including *Anagallis tenella*, *Ranunculus flammula* and *Hydrocotyle vulgaris*, are consistently recorded that are not typical bog species elsewhere in Ireland. Both *Potamogeton polygonifolius* and *Hypericum elodes* are commonly found in drainage channels and revegetating cutaway areas.

Cyperaceous species such as *Carex nigra*, *C. panicea* and *C. echinata* are more abundant where there is significant soil waterlogging. In *Schoenus*-dominated bog vegetation, *Carex pulicaris* is common, while *Carex paniculata*, recorded from only two locations on Clare Island, is associated with *Sphagnum*-dominated vegetation on deep peat (>6m) that is subject to continuous irrigation. Rushes are variable in cover, with *Juncus bulbosus* the most frequent component, typical of waterlogged areas. Other juncaceous species include *Juncus articulatus* recorded on regularly flushed sites, while *J. effusus* and *J. acutiflorus* are usually confined to peripheral areas subjected to anthropogenic influences such as disturbance by drainage. Towards the boundaries of level peat masses or on shallow exposed peat (approx. 25%–50% bare ground cover), *Trichophorum caespitosum* is occasionally recorded.

Bryophytes form an important part of the vegetation and, in certain circumstances, are the main contributors to ground cover. Characteristic species include *Scorpidium scorpioides*, *Pleurozia purpurea* and *Odontoschisma sphagni*. The cover of these bryophyte species is much reduced here compared to undisturbed blanket bog elsewhere in Ireland, again reflecting the lack of undisturbed bog vegetation on Clare Island. Peat mosses are the most conspicuous and often the most abundant element of the bryophyte flora: *Sphagnum papillosum*, *S. palustre* and *S. subnitens* are the most important, with *Sphagnum cuspidatum* dominating bog pools and revegetating channels. Several other species that are recorded include *Sphagnum tenellum*, *S. auriculatum*, *S. compactum*, *S. squarrosum* and *S. recurvum*. Other mosses frequently recorded are *Aulacomnium palustre*, *Hylocomium splendens*, *Plagiothecium undulatum*, *Racomitrium lanuginosum*, *Polytrichum commune*, and unusually, *Campylium stellatum*, *Rhytidiadelphus squarrosus* and *R. triquetrus*. Other locally abundant species found mainly in marginal areas of bogs include *Bryum pseudotriquetrum*, *Isothecium myosuroides*, *Pseudoscleropodium purum* and *Calliergonella cuspidata*.

Liverworts are particularly important in defining certain associations within the bog class although, on Clare Island, most increase in frequency on deeper peat and/or increased soil saturation. These include *Scapania gracilis*, *Diplophyllum albicans*, *Lophocolea bidentata*, *Mylia anomala* and *M. taylorii*. Several species including *Aneura pinguis*, *Pellia epiphylla/neesiana* and the moss *Fissidens adianthoides* are characteristic of degraded peat subject to frequent irrigation.

Lichens are not a significant component of the vegetation of blanket bog mosaics on Clare Island. Only *Cladonia portentosa* and *C. uncialis* are occasionally found on drier hummocks or growing through taller tussock-forming plant species, such as *Molinia* and *Eriophorum*. Another component of the vegetation is the algal aggregate *Zygogonium ericetorum*, which plays an important role in maintaining the surface integrity of the peat in permanently waterlogged conditions (Doyle 1982b). On Clare Island, where it is found growing on and between tussocks and on denuded peat, its cover increases dramatically in frequently flushed areas with low vegetation cover and a higher proportion of exposed bare peat. The expansion in *Zygogonium* has been observed elsewhere in the blanket bog regions where intensive sheep grazing and pedestrian tourist traffic has produced extensive bared peat areas (MacGowan and Doyle 1996; 1997).

Phytosociology

Much of the vegetation of acid peatland ecosystems is assigned to the Oxycocco-Sphagnetea, which encompasses the ombrotrophic bog communities and wet heathlands of Europe. Considerable areas of western Ireland were covered by blanket bogs (Hammond 1979), and despite widespread utilisation for peat harvesting and grazing, it has persisted as the main vegetation in such areas. Recently, the uncontrolled expansion of sheep numbers, prompted by European Union headage payments for disadvantaged areas supporting blanket bog, has resulted in degradation of the vegetation in much of the west coast of Ireland (Bleasdale and Sheehy-Skeffington 1992; MacGowan and Doyle 1996; 1997). Since much of the Clare Island landscape has been intensively utilised, there remains very little undamaged bog vegetation. Species characteristic of the Scheuchzerio-Caricetea nigrae and Littorelletea uniflorae often form a major constituent of the degraded and fragmented bog vegetation complex on Clare Island, and it is not uncommon for elements of the Nardetalia to spread from the surrounding grassland areas into damaged sections of bog habitats. The vegetation description is presented under four headings:

Ombrotrophic blanket bog vegetation (Atlantic blanket bog)
Vegetation of wet hollows on bog surfaces
Vegetation of small islands found on deep ombrotrophic peats
Drainage channels

ATLANTIC BLANKET BOG

Oxycocco Sphagnetea Br. -Bl et Tx. 1943
Eriophoro vaginati-Sphagnetalia papillosi Tx. 1970
***Calluna*-Sphagnion papillosi** (Schwick. 1940) Tx. 1970
Pleurozio purpureae-Ericetum tetralicis Br.-Bl. et Tx. 1952 em. Moore 1968 Table 21

Pleurozio purpureae-Ericetum tetralicis character species: *Pleurozia purpurea, Campylopus atrovirens, Schoenus nigricans* (All reduced in abundance)
Differential species: *Molinia caerulea, Potentilla erecta, Pedicularis sylvatica, Polygala serpyllifolia*

Heritage Council analogue: Lowland blanket bog—PB3

The occurrence of large areas of lowland blanket peat, occurring up to elevations of 200m, is well documented for counties Galway and Mayo (Doyle and Moore 1980; Doyle 1982b; 1990; 1997; MacGowan and Doyle 1996) and Kerry (Mhic Daeid 1976). The characteristics of such western, or lowland Atlantic blanket bog (Pleurozio purpureae-Ericetum tetralicis) have been comprehensively discussed by Doyle and Moore (1980) and Doyle (1982b; 1990; 1997). Comparable vegetation in Britain has been assigned to the synonymous Trichoporeto-Eriophoretum (McVean and Ratcliffe 1962; Birks 1973) and the *Scirpus cespitosus-Eriophorum vaginatum* M17 blanket mire in the BNVC classification (Rodwell 1991).

The prominence of graminoid species, such as *Molinia caerulea* and *Schoenus nigricans*, the depauperate nature of the ericoid shrubs and the poor development of *Sphagnum* species, together with the occurrence of a distinct set of character and differential species, including *Pleurozia purpurea, Campylopus atrovirens, Pedicularis sylvatica, Pinguicula lusitanica, Polygala serpyllifolia* and *Potentilla erecta* differentiate Atlantic blanket bog from other Irish bog vegetation types. Alliance diagnostic species recorded from Clare Island include *Calluna vulgaris, Erica tetralix, Narthecium ossifragum* and *Cladonia portentosa*, while diagnostic species less frequently encountered include *Myrica gale, Diplophyllum albicans* and *Mylia anomala*, while *Eriophorum angustifolium* rather than *E. vaginatum* is a constant companion.

Five subtypes of the blanket bog community on Clare Island are presented, illustrating the gradations from 'native' blanket bog through peripheral and damaged vegetation, transitional with heathland and/or acid grassland vegetation.

Relatively intact native blanket bog on Clare Island

Columns 1–14 of Table 21 list relevés taken in relatively intact blanket bog, that is confined to a plain occurring below the lighthouse in the north-eastern part of the island (Pl. XXA). The vegetation is typified by the overall dominance of *Molinia caerulea* (Pl. XXB), which grows in a relatively depauperate form, rarely developing stout tussocks. The only other tall growing plant regularly found in the vegetation is *Eriophorum angustifolium*. The characteristic bryophytes *Campylopus atrovirens* and *Pleurozia purpurea* are present but not frequent. Constant occurrences include the association differential species *Potentilla erecta, Pedicularis sylvatica* and *Polygala serpyllifolia*. The ericoid shrubs *Calluna vulgaris* and *Erica tetralix* are well represented, although depauperate in their growth form, a consequence of constant waterlogging and the impacts of grazing. While *Trichophorum caespitosum* is recorded on intact blanket bog, it only becomes locally abundant where gaps exposing the underlying mineral soils occur in the moderately tall vegetation owing to the severity of grazing (Table 21, columns 50–52,).

Schoenus-rich blanket bog

Another species characteristic of western blanket bog, *Schoenus nigricans* (Doyle 1982b), is generally restricted in its distribution on blanket bog on Clare Island (Pl. XXI), and is mainly found on

Pl. XXA The only extensive area of Atlantic blanket bog on Clare Island is situated below the Lighthouse. The bog is confined to level ground seen in the lower part of the photograph. The slopes to the left-hand side of the picture are occupied by recovering heathland, whilst the remaining slopes support a closely-cropped Nardetalia sward.

Pl. XXB Bog in Ballytoohy More showing development of purple moor grass tussocks in the valleys with regenerating heather occupying the relatively drier slopes.

PL. XXI Black bog rush (*Schoenus nigricans*) is mainly confined to the wettest parts of blanket bog at Ballytoohy Beg. In the centre of the photograph, *Schoenus* is concentrated around the drainage areas (darkly coloured patches).

permanently waterlogged deep peats, subject to periodic flushing in Ballytoohy More (Table 21, columns 15–24). It is only rarely recorded from typical stands of the blanket bog vegetation (Table 21, column 2), which on Clare Island are dominated by *Molinia caerulea*.

The vegetation is characterised by the codominance of *Schoenus* and *Molinia*, while constant species include *Potentilla erecta*, *Eriophorum angustifolium*, *Erica tetralix*, *Narthecium ossifragum* and *Drosera rotundifolia*. Other diagnostic species of the Pleurozio purpureae-Ericetum tetralicis such as *Campylopus atrovirens*, *Pleurozia purpurea* and *Pedicularis sylvatica* are reduced in abundance. An obvious feature of the habitat is the presence of runnels, channelling water along the bog surface. These highly saturated situations favour the establishment of some Scheuchzerio-Caricetea species such as *Carex nigra*, *Eleocharis multicaulis* and *Anagallis tenella*, while the impact of sheep grazing increases the occurrence of Nardetalia species (columns 15 onwards), with dwarfed forms of *Agrostis capillaris* and *Festuca ovina* an obvious feature.

Juncetosum subtype

The remaining three subtypes describe degraded vegetation from peripheral areas of blanket bog, where there is often obvious water movement across the peat surface and disturbance through severe grazing. These subtypes each retain the basic vegetation composition typical of the intact bog on Clare Island, with similar contributions by *Molinia caerulea*, *Eriophorum angustifolium*, *Calluna vulgaris* and *Potentilla erecta*.

Columns 25–47 of Table 21 describe less intact stands of blanket bog vegetation that occur extensively as outliers in shallow depressions (<1.25m) among the commonage vegetation mosaic. In these situations the peat is regularly flushed by run-off from the surrounding slopes, so the vegetation is characterised by differential species of the Juncetosum subassociation (Doyle 1982b), which include *Eleocharis multicaulis*, *Carex panicea* and *Juncus bulbosus*. The occurrence of species diagnostic of the Nardetalia such as *Nardus stricta* and to a lesser extent *Festuca ovina*, *Agrostis capillaris* and *Anthoxanthum odoratum* reflects the degradation of the peatland habitat and the intensity of grazing on these peripheral areas on the blanket bog.

Recovering blanket bog

Some degraded blanket bog at Ballytoohy More was fenced off approximately twenty years ago, after many years of grazing by sheep. This vegetation (Table 21, columns 48–56) is found at an interface between *Molinia*-dominated blanket bog and wet heathland. The vegetation is readily distinguished from the relatively intact blanket bog, where the grazing impacts are still evident, with obvious patches of bare soil and a considerable reduction in the height of the vegetation, particularly in the graminoid component. These features are typical in intensively grazed or trampled areas, reported from blanket bogs in the west of Ireland (MacGowan and Doyle 1996; 1997).

Exclusion of grazing has allowed some recovery of the blanket bog vegetation. The main species providing ground cover include *Molinia caerulea*, *Eriophorum angustifolium*, *Nardus stricta*, *Calluna vulgaris*, *Sphagnum subnitens* and *Eleocharis multicaulis*. Some typical bog species, particularly smaller herbs such as *Pedicularis sylvatica* and *Polygala serpyllifolia*, are reduced in their abundance, while other species considered diagnostic of the Pleurozio purpureae-Ericetum tetralicis, such as *Pleurozia purpurea* and *Schoenus nigricans*, are completely absent.

Degraded blanket bog

This subtype (Table 21, columns 57–60) characterises the most seriously degraded blanket bog on Clare Island. Constant poaching and overgrazing by sheep has had dramatic effects, resulting in the almost total removal of the vegetation cover in some places. Where the vegetation has been stripped from the peat surface there has been subsequent soil erosion, with the underlying mineral soil exposed in some areas.

The overall cover of bog species is reduced with increasing numbers of sheep and the associated effects of poaching. Many of the species considered characteristic of the Oxycocco-Sphagnetea are absent, including *Erica tetralix*, *Drosera rotundifolia*, *Trichophorum caespitosum*, *Campylopus atrovirens*, *Pleurozia purpurea* and *Odontoschisma sphagni*. Bryophytes and small herbs are highly susceptible to the impacts of grazers and are virtually excluded in the most degraded sites. These observations coincide with the findings from severely grazed blanket bogs in Connemara and Mayo, where there are significant alterations in species cover, and in certain situations the eradication of entire suites of typical bog species (MacGowan and Doyle 1996).

In these disturbed sites there is an increased abundance of Nardetalia species, which appear to be establishing at the expense of *Molinia caerulea*. They include *Nardus stricta*, *Festuca ovina* and, particularly, *Agrostis capillaris*, which together may contribute up to 65% of total groundcover. Where water flows over the eroding bare peat surfaces, species of the Scheuchzerio-Caricetea occur. These include *Carex nigra*, *Hydrocotyle vulgaris* and *Anagallis tenella*. Interestingly, the mucilaginous algal aggregate *Zygogonium ericetorum*, which generally flourishes on bare, exposed peat in waterlogged conditions (Doyle 1982b; 1990; MacGowan and Doyle 1996; 1997), was not a major component of this subtype on Clare Island. Although not widespread, it occurs more frequently on gently sloping peat depressions situated in the commonage that are subject to run-off water from higher ground.

WET HOLLOWS ON BOG SURFACES

Oxycocco-Sphagnetea Br.-Bl. et Tx. 1943
Scheuchzerietalia palustris Nordh. 1936
Rhynchosporion albae Koch 1926
Sphagno tenelli-Rhynchosporetum albae
(Osvald 1923) Koch 1926 (Table 21)

Rhynchosporion albae character species:
Rhynchospora alba, Narthecium ossifragum
Diagnostic species: *Eriophorum angustifolium, Molinia caerulea, Eleocharis multicaulis, Sphagnum cuspidatum, S. papillosum*

Heritage Council analogue: Lowland blanket bog—PB3

The majority of the vegetation from wet hollows on flat bog surfaces is assigned to the Sphagno tenelli-Rhynchosporetum albae, which belongs to the Rhynchosporion albae, the Scheuchzerietalia palustris and the Oxycocco-Sphagnetea. The vegetation is included in the Scheuchzerietalia palustris, although the syntaxonomic status of this order is unclear in the west of Ireland. The classification is complicated by the overlap in the character species between the Scheuchzerietalia palustris and the Scheuchzerio-Caricetea nigrae. While the community has structural similarities with the *Narthecium ossifragum-Sphagnum papillosum* M21 valley mire described from low-lying areas in Britain, floristically it is more closely related to the Rhynchosporion assemblages of the BNVC, specifically the *Rhynchospora alba* subcommunity of the *Sphagnum cuspidatum/recurvum* M2 bog pool vegetation (Rodwell 1991).

The vegetation of wet hollows is common in situations where bare peat is much in evidence, with underlying peat depth ranging from 1.5m to 6m. It is characterised by a level peat surface with a series of waterlogged runnels on the flat peat surface, created after the abandonment of peat cutting. There is a paucity of extensive tracts except in low-lying areas of Maum townland on the eastern side of the island, where the bulk of this vegetation occurs. Small outliers referable to this association are centred on three small bog lakes at Lecarrow, in the centre of the island. The vegetation of wet hollows is assigned to two subtypes—*Rhynchospora*-rich and *Rhynchospora*-poor stands.

RHYNCHOSPORA-RICH STANDS

The considerable degree of structural variation within the vegetation is dependent on the depth of the ground water, the degree of flushing, the amount of bare peat present and the fragmentary nature of abandoned cutaway sites. Small tufts of *Rhynchospora alba* occur on the bare peat on permanently waterlogged hollows on blanket bogs.

Apart from *Rhynchospora alba*, constant species include *Eriophorum angustifolium, Molinia caerulea* and *Eleocharis multicaulis* (Table 22, columns 1–4). The character species *Sphagnum recurvum* was not recorded from this community on Clare Island. Other species with low cover include *Potentilla erecta, Carex panicea, Erica tetralix* and *Drosera rotundifolia*. In general, *Sphagnum* species flourish because the continuous irrigation inhibits successful colonisation of the bare peat surfaces by other bryophyte species. The relatively flat bog surface is only broken by low hummocks comprised of hardened peat, which are floristically richer and are home to typical upland species such as *Festuca ovina, Agrostis capillaris* and impoverished *Calluna vulgaris*. *Sphagnum cuspidatum* occurs in drainage channels adjoining this vegetation, where it withstands fluctuations in the water level. *Sphagnum papillosum* and *S. capillifolium* are found growing at the base of *Eriophorum angustifolium* clumps.

Rhynchospora-poor subtype

In certain situations on the flat bog surface, *Rhynchospora* cover is reduced, with *Eriophorum angustifolium* and *Sphagnum cuspidatum* growing in waterlogged cutaway peat (Table 22, columns 5–7). This impoverished form of the Sphagno tenelli-Rhynchosporetum albae has some similarities with the ill-defined *Sphagnum cuspidatum-Eriophorum angustifolium* community, which has been observed in drainage channels and waterlogged cutaway bogs in Ireland (White and Doyle 1982). It is also present in drainage channels on Clare Island.

The vegetation also has sparse growth of *Eleocharis multicaulis, Narthecium ossifragum* and *Drosera rotundifolia*, while *Molinia caerulea* and *Sphagnum capillifolium* are less adundant. Considerable variation in cover was recorded within this community, most noticeably for the bog cotton *Eriophorum angustifolium*, which ranged from 30% to nearly 90% (Table 22, column 7).

Despite extensive peat extraction and subsequent denudation, it is possible to find isolated stands of undisturbed vegetation. The general absence of grazing or recent peat cutting has resulted in some revegetation of this habitat. Species that increase in cover in these fully vegetated areas include *Molinia caerulea*, *Trichophorum caespitosum*, *Polygala serpyllifolia*, *Pedicularis sylvatica*, *Calluna vulgaris* and *Rhynchospora alba* (occasional occurrences) (Table 21, column 7). This subtype, typical of slightly drier habitats, may have been more extensive in the past, covering more of the bog surface, in Maum at least, before cutting altered the vegetation dynamics.

Pl. XXIIA Ombrotrophic peat basin located in the centre of the island, surrounded by severely grazed heather community.

Pl. XXIIB View of small raised peat islands. Grazed tufts of bottle sedge are indicative of the area's development from a small lake (see Table 12, column 4).

Peat and Peat Islands in Deep Peat Basins

Several raised hummocks or 'islands', averaging one metre in diameter, are found in a single basin in the centre of the island, at the junction of the three small lakes near Lecarrow (Pls XXIIA; XXIIB). The basin area is 100m by 75m. It seems that this peat basin is the last remnant of terrestrialisation of a former lake. Although this area on Clare Island is spatially separated from the blanket bog proper, it shares many floristic similarities. It is a permanently waterlogged ombrotrophic peat basin that occurs on deep humified peat (in excess of 6m deep). Similar vegetation has been described from deep, wet hollows on ombrotrophic bog surfaces where *Vaccinium oxycoccos* and *Empetrum nigrum*, two species of restricted distribution on Atlantic blanket bog, were recorded (Doyle and Foss 1986b; Doyle 1990). The community described by Doyle (1990) had a high percentage of bryophytes, with only localised collections of vascular plants, a reflection of severe grazing intensity.

The overall composition of the vegetation has an assemblage indicative of the Pleurozio purpureae-Ericetum tetralicis. The influence of grazing animals results in stunted vegetation on the small 'islands', although the vegetation is better developed than the vegetation on the surrounding peat basin and the severely grazed grassland vegetation (Nardetalia) on the adjacent slopes. The peat 'islands' support dwarf shrub vegetation, comprising *Calluna vulgaris* and *Erica tetralix*, together with *Agrostis capillaris* and *Festuca ovina*. Also found growing through the community are *Vaccinium oxycoccos* and *Empetrum nigrum* (this is one of only three stations for *Empetrum* on Clare Island). Other species, including *Menyanthes trifoliata* and *Narthecium ossifragum*, are recorded along the peripheral edges of these 'island' habitats.

Peat Islands

The island vegetation is referred to the Pleurozio purpureae-Ericetum tetralicis, although it has some allegiances with the Vaccinio-Ericetum tetralicis, based on the occurrence of *Vaccinium oxycoccos* and *Empetrum nigrum*. These species are generally characteristic of mountain blanket bog, although no suitable plateau that could support such a community occurs on Clare Island.

The 'peat islands' (Table 22, columns 1–2) are elevated, at most, 45cm above the height of the surrounding vegetation. As such they remain accessible to sheep that stray onto the surrounding treacherous habitat, unlike the peat islands described from other bog lakes by Doyle *et al.* (1987). The raised hummocks at the Clare Island site are floristically less diverse than the islands described by Doyle *et al.* (1987) because they are not protected by open water or the impacts of grazing and burning.

Apart from *Vaccinium oxycoccos* and *Empetrum nigrum*, the other main components of the small islands include *Molinia caerulea, Potentilla erecta, Festuca ovina, Agrostis capillaris, Eleocharis multicaulis* and *Anagallis tenella*, with minor contributions from other species considered typical of bog habitats, including *Narthecium ossifragum, Schoenus nigricans* and *Drosera rotundifolia*. Two other species, *Menyanthes trifoliata* and *Carex limosa*, are confined to the margins of the islands.

Ombrotrophic Blanket Peat Surrounding the 'Islands'

Bryophytes and non-bog species figure significantly in the ground layer in the next subtype, which bears little resemblance to any ombrotrophic bog communities described from Clare Island (Table 22, columns 3–6). Total vegetation cover is nearly 100%, the vascular plant cover varies between 20% and 75% and bryophyte cover less than 65% is rare. Occasionally, tears may occur in the vegetation surface after periods of extreme flooding or when sheep venture in search of food.

Phytosociologically, the vegetation is assigned to the Pleurozio purpureae-Ericetum tetralicis, although many of the character species, including *Molinia caerulea, Drosera rotundifolia, Narthecium ossifragum* and *Schoenus nigricans*, are less abundant. Mosses, in particular *Sphagnum* species, such as *Sphagnum palustre, S. capillifolium* and *S. subnitens*, are frequent components of the ground flora. Other constants include *Scorpidium scorpioides, Calliergonella cuspidata* and *Scapania gracilis*. The community also contains many elements that are diagnostic of both the Scheuchzerio-Caricetea and the Littorelletea, including *Eleocharis multicaulis, Juncus bulbosus* and *Potamogeton polygonifolius*. Many of the companion species include graminoid species diagnostic of the Nardetalia, such as *Festuca ovina, Agrostis capillaris* and *Nardus stricta*, that are found on the perimeter of the community.

It seems probable, given the proximity of this basin with the three water bodies, that the development of this bryophyte-rich vegetation represents the final stages of lake terrestrialisation and peat accumulation. Two of those water bodies (described previously) have developed a quaking bryophyte-dominated vegetation mat similar to schwingmoor vegetation assigned to the Scheuchzerietum, described from England (Tallis 1973) and more recently from Ireland (Doyle 1990).

Marginal vegetation on the peat basin

Columns 7 and 8 in Table 22, in particular, have a strong bryophyte component with a poorly developed vascular flora. The vegetation is located beside the small stream that flows into the peat basin at its northern tip. The heavily degraded and poached vegetation is a variant of the previous community, most noticeable by the reduced size of the grazed tussocks of *Molinia caerulea, Schoenus nigricans, Deschampsia caespitosa* and *Carex paniculata* (recorded from only one other location on Clare Island). Constant companion species include *Potamogeton polygonifolius, Eleocharis multicaulis, Drosera rotundifolia* and *Juncus bulbosus*. Elements of the Nardetalia, including *Nardus stricta* and *Festuca ovina*, are indicative of grazing pressure and the ensuing spread of acid grassland species.

Drainage channels

The vegetation of drainage channels is readily recognised where these features are part of intact ombrotrophic blanket bogs (Doyle 1990). Drainage channels on Clare Island are features of the degraded blanket bogs, so their vegetation is less clearly defined owing to the occurrence of diagnostic species typical of both the Ericion tetralicis and Calluno-Sphagnion papillosi. The difficulty in separating the common or diagnostic species characterising drainage features such as channels and flushes on blanket bogs was highlighted in studies in west Galway and north Mayo (Lockhart 1991). Apart from floristic difficulties, the overall structure of the vegetation is dependent on the depth of the peaty substrate, channel width, flow rate of the water that passes through the channel and, to some extent, the previous management regimes.

Two associations are recognised from Clare Island for the vegetation of drainage channels, the majority of which are associated with relict cutaway sites and are extensively distributed on the commonage. Some were created, or at least were perpetuated through anthropogenic influences, mainly through peat extraction.

The first association, the Hyperico-Potametum oblongi, includes vegetation of deeper channels occurring on ombrotrophic bog surfaces where peat extraction has left a network of waterlogged depressions. The second type, the *Sphagnum cuspidatum-Eriophorum angustifolium* community, comprises vegetation of shallower peats and is associated with abandoned cuttings at higher altitudes and occasionally with marginal areas of low-lying ombrotrophic blanket bogs.

VEGETATION DOMINATED BY HYPERICUM ELODES AND POTAMOGETON POLYGONIFOLIUS

Littorelletea uniflorae Br.-Bl. et Tx. 1943
Littorelletalia uniflorae Koch 1926
Hydrocotylo-Baldellion Tx. et Dierßen 1972
Hyperico-Potametum oblongi (Allorge 1926) Br.-Bl. et Tx. 1952 Table 23

Hyperico-Potametum oblongi character species:
Hypericum elodes, Potamogeton polygonifolius

Heritage Council analogue: Cutover bog—PB4

Vegetation assigned to the Hyperico-Potametum oblongi is dominated by *Hypericum elodes* and *Potamogeton polygonifolius* and is generally found in waterlogged hollows on a substantial depth of peat (over 1m deep) and occasionally found in roadside ditches along boggy areas. Such vegetation occurs elsewhere in Ireland in the wetter parts of blanket bogs (Braun-Blanquet and Tüxen 1952; Schoof van Pelt 1973; Dierßen 1978). The equivalent community in the BNVC scheme is the *Hypericum elodes-Potamogeton polygonifolius* M29 soakway (Rodwell 1991). More recently, Bleasdale and Conaghan (1995) describe a community from flushed blanket bog in Connemara that they regard as allied to the Hyperico-Potametum oblongi (= polygonifolii) but lacking *Eleogiton fluitans*, the character species of the community. This was also the case in the Clare Island vegetation where *E. fluitans* was not recorded.

The substrate in the drainage channels is of a treacherous nature, yet the ditches are heavily foraged by sheep. Bare peat is an obvious feature of the habitat supporting the vegetation on Clare Island. The surrounding hill vegetation, the majority of which supports vegetation assigned to the Nardetea, is closely cropped, facilitating the constant drainage into the hollows and derelict runnels and channels that were created to drain the moderately deep peats. Within this category two subtypes are described, reflecting flow rates and the age of the channels.

The first subtype (Table 23, columns 1–8) is characterised by a high constancy of *Hypericum elodes* and *Potamogeton polygonifolius*. Other species include Littorelletea diagnostic species such as *Eleocharis multicaulis, Ranunculus flammula, Hydrocotyle vulgaris* and *Anagallis tenella*. Species that are almost exclusive to this subtype include *Carex echinata, Juncus bulbosus, Galium palustre, Viola palustris* and *Drosera rotundifolia*. This subtype has a diverse species assemblage, the majority of species being unevenly distributed and are generally of low cover—typically less than 5%. Habitats supporting this subtype are subject to frequent scouring, which might account for the general reduction in the abundance of most species.

The occurrence of several species, all with relatively high cover values, is interesting. These include *Sphagnum palustre* (column 1) with a cover of 65%, *Scorpidium scorpioides* (column 4, 60% cover), *Carex flacca* (column 2, 15% cover) and *Apium inundatum* (column 8, cover 30%). While these species are characteristic of separate classes, they all show a preference for fluctuating water levels. Their occurrence in these channels reflects the anthropogenic impacts through domestic grazers and peat harvesting that influence the vegetation and facilitate spread of species from surrounding habitats.

The second subtype (Table 23, columns 9–12) is found on level peat and characterises older drainage channels. Bare or exposed peat is less common in these older drainage channels. However, during periods of extreme flooding, the integrity of the vegetation surface may be damaged, exposing fresh peat surfaces. In general, there is a higher constancy of *Potamogeton polygonifolius* relative to *Hypericum elodes*, with a greater constancy of species typical of the Scheuchzerio-Caricetea including *Ranunculus flammula* and *Carex nigra*. Diagnostic species of the Littorelletea common in this subtype include *Eleocharis multicaulis*, *Hydrocotyle vulgaris* and *Anagallis tenella*. Other important species include *Eriophorum angustifolium*, *Carex viridula* and *Carex panicea*, although they are typically recorded from infilling channels on deeper peats.

VEGETATION DOMINATED BY SPHAGNUM CUSPIDATUM AND ERIOPHORUM ANGUSTIFOLIUM

Oxycocco-Sphagnetea Br.-Bl. et Tx. 1943
Scheuchzerietalia palustris Nordh. 1936
Rhynchosporion albae Koch 1926
Sphagnum cuspidatum-Eriophorum angustifolium community Tx. 1958 (Table 24)

Diagnostic species: *Sphagnum cuspidatum*, *Eriophorum angustifolium*

Heritage Council analogue: Lowland blanket bog—PB3 / Cutover bog—PB4

Another vegetation type commonly recorded from drainage channels on Clare Island is referable to the *Sphagnum cuspidatum-Eriophorum angustifolium* community. This is one of two ill-defined but closely related communities of drainage channels that have been assigned to the Rhynchosporion albae. White and Doyle (1982) and Doyle (1990) considered that a second and closely allied *Sphagnum recurvum-Eriophorum angustifolium* community occurred in drainage channels both in mountain and Atlantic blanket bog areas in Ireland. No evidence of this distinctive community was recorded on Clare Island during the present survey, although it is possible that fragmentary stands occur among the degraded vegetation where *Sphagnum recurvum* is infrequently recorded.

The *Sphagnum cuspidatum-Eriophorum angustifolium* community is poorly documented in Irish literature and the only published reference to the community in Ireland is by Dierßen (1978). Rodwell's (1991) synthesis of British vegetation would see this vegetation classified in the *Sphagnum cuspidatum/recurvum* M2 bog pool. Unlike the community described by Dierßen, and other accounts of similar vegetation from Scotland (i.e. Birks 1973), *Eriophorum angustifolium* may be of little consequence to the general structure of the vegetation on Clare Island, as it is quite stunted owing to the intensity of grazing. It should be noted, however, that *Eriophorum angustifolium* is recorded as a major component of some nearby stands of the Sphagno tenelli-Rhynchosporetum albae. Other constant occurrences, that are more abundant than *Eriophorum* in the drainage channels, include *Juncus bulbosus*, *Eleocharis multicaulis*, *Nardus stricta*, *Potentilla erecta* and *Drosera rotundifolia*. Two subtypes of the community are distinguished, the occurrence of which appears to be related to the depth of the substrate and water flow rates. Slower flow rates favour the dominance of *Sphagnum cuspidatum* (≥75%).

The first subtype (Table 24, columns 1–3) is typical of deeper peat situations, where a significant portion of bare ground is present in and around the channel. In this situation, drainage is not as evident, except during periods of heavy rainfall, when the ensuing floodwaters from higher elevations swell the channels. The pale yellow-green mat of *Sphagnum cuspidatum* is distinctive, creating a stark physical contrast to the surrounding degraded vegetation. The other main components of the vegetation include *Eleocharis multicaulis*, *Hypericum elodes*, *Anagallis tenella* and *Nardus stricta*.

The second subtype (Table 24, columns 4–7) has both *Sphagnum cuspidatum* and *Eriophorum angustifolium* present, along with the constant species already listed. The subtype is confined to channels that are less saturated or prone to flooding. Indeed, the channels may only extend for a few metres, as they were harvested only briefly for turf when there was a greater need for domestic fuel. These areas formerly supported mixed grassland/heather vegetation. Species found here include *Calluna vulgaris*, *Carex panicea*, *Agrostis capillaris* and *Festuca ovina*; the occurrence of these species in the older channels is indicative of the length of time since they were used for harvesting peat. The subsequent consolidation of the substrate enabled the re-establishment of the more terrestrial species.

VEGETATION OF SPRINGS AND SEEPAGE ZONES

The communities are rarely extensive and occur sparingly as part of the varied mosaic on Clare Island. They are associated with the most severely degraded acid grasslands on the commonage and some peripheral areas of ombrotrophic bogs (Oxycocco-Sphagnetea). The syntaxonomic position of the Clare Island vegetation found in such situations is complicated because some stands are fragmented and often degraded, while some may have persisted due to human influence. The final diagnosis is difficult, with a distinction between the communities based primarily on species composition, and secondarily on vegetation structure and ecology. The approach taken here is to assess the diagnostic species of different classes, consider the ecology/habitat and to relate it to the most typical syntaxon.

The overall morphology of the vegetation is characterised by the presence of several low-growing cyperaceous and juncaceous species, while bryophytes are also an important feature of the ground flora. Spring vegetation is classified in the Montio-Cardaminetea. The majority of the vegetation units found near seepage zones have strong allegiances with the Littorelletea uniflorae, whose diagnostic species assemblage favours a fluctuating water table, although there is generally some water on the ground surface. The occurrence of several species characteristic of the Scheuchzerio-Caricetea nigrae further confuses the classification. This is a class related to vegetation from distinctly wet areas occurring in impoverished situations and recorded from a variety of wetter habitats, ranging from heathland to lake edge to damp scrub-forest to wet meadows. Its occurrence in Ireland has been comprehensively studied and described (Ó Críodáin 1988; Ó Críodáin and Doyle 1994; 1997). Two alliances within the class are recorded: the Caricetalia nigrae characterises vegetation of acid areas, while the Caricetalia davallianae is less well represented on Clare Island, since it encompasses more calcicole vegetation.

SPRING VEGETATION

Montio-Cardaminetea Br.-Bl. et Tx. 1943
Montio-Cardaminetalia Pawlowski 1928
Cratoneurion Koch 1928
Cratoneuretum filicino-commutati (Kuhn 1937) Oberd. 1977 relevé 335

Cratoneuretum character species: *Cratoneuron filicinum, Philonotis fontana*

Heritage Council analogue: Calcareous springs—FP1

Relevé 335 (24 species)	
Cratoneuron filicinum	3
Agrostis stolonifera	2
Sagina procumbens	2
Myosotis secunda	1
Juncus bulbosus	1
Hypericum elodes	+
Carex limosa	+
Pellia epiphylla/neesiana	+
Aneura pinguis	+
Bryum pseudotriquetrum	+
Epilobium palustre	+
Equisetum palustre	+
Calliergonella cuspidata	2
Philonotis fontana	2
Potamogeton polygonifolius	2
Hydrocotyle vulgaris	1
Rhizomnium punctatum	1
Eleocharis multicaulis	+
Cardamine pratensis	+
Aulacomnium palustre	+
Carex nigra	+
Juncus acutiflorus	+
Campylopus introflexus	+
Eurhynchium praelongum	+

The presence of spring communities generally reflects some degree of base-rich flushing, although this is not necessarily reflected by perceivable groundwater seepage. The only basiphilous spring community on Clare Island is assigned to the Cratoneuretum filicino-commutati, within the Montio-Cardaminetea (Relevé 335). The Cratoneurion is not well characterised in Ireland, with only two brief accounts of its occurrence here. Braun-Blanquet and Tüxen (1952) ascribed a single relevé from County Offaly to the association, while Ivimey-Cook and Proctor (1966) described a community from the Burren that White and Doyle (1982) considered was referable to this association. Neither of these descriptions is entirely analogous with the current vegetation type, which is occasionally found at elevations ranging between 350 and 450 metres on Knockmore, where there is a flow of base-rich waters derived from the band of calcareous rocks. The vegetation occurs locally in montane springs in the Pennines and the highlands of Scotland, where it is classified as the M38 *Cratoneuron commutatum-Carex nigra* spring (Rodwell 1991).

The spring vegetation is predominantly recognised by the lush bryophyte carpet, with species that are characteristic of calcareous areas (of which there is very little on Clare Island). The distinctive appearance of the spring vegetation is accentuated by the closely cropped and intensely grazed sward of the communities abutting the springs. Diagnostic species include *Cratoneuron filicinum*, *Calliergonella cuspidata* and *Philonotis fontana*. Other bryophytes include *Pellia epiphylla/neesiana*, *Aneura pinguis*, *Bryum pseudotriquetrum*, *Aulacomnium palustre*, *Campylopus introflexus* and *Eurhynchium praelongum*. The remaining ground flora of the spring vegetation is weakly developed. The only constant occurrences include species typical of both the Littorelletea and the Scheuchzerio-Caricetea, such as *Hydrocotyle vulgaris*, *Juncus bulbosus* and *Eleocharis multicaulis*. The contribution of these species varies depending on the topography and substrate composition, a feature noted by Ivimey-Cook and Proctor (1966).

VEGETATION OF SEEPAGE ZONES

Much of the upland area of Clare Island is subjected to irrigating water that promotes the development of vegetation with a number of species typical of flushed areas (*sensu* McVean and Ratcliffe 1962). Flush vegetation is dealt with elsewhere, since it is (a) an inherent feature of the vegetation mosaic of both acid grasslands and peatland ecosystems, and (b) bryophytes are rarely the major provider of ground cover in such flushes. However, vegetation located in and around areas of water seepage also display physiognomic differences from the surrounding vegetation. The wet soils are composed of clay and mud, although the matrix may include many rock fragments, or in some cases it may be found where water seeps along joints in the stratified lithologies.

While the seepage zone vegetation on Clare Island has similarities with that described from elsewhere in Ireland and Britain (McVean and Ratcliffe 1962; White and Doyle 1982; Rodwell 1991; Bleasdale and Conaghan 1995), the general paucity of detail in the literature prohibits a satisfactory overall classification scheme. Seepage zone vegetation on Clare Island is described by means of four relevés, each of which is assigned to a different alliance. The lack of sufficient relevé material and the rudimentary nature of the degraded vegetation prevents any classification to association level.

The first sample has obvious allegiances with the Scheuchzerio-Caricetea nigrae, although it contains elements of the Littorelletea uniflorae. The next three samples all have greater abundances of Littorelletea species, highlighting the difficulty in classifying the communities due to the admixture of various combinations of character species from separate syntaxa. The vegetation is typically associated with more derelict sites and abandoned fields that generally occur on moderately deep soils (predominantly peat-derived gleys).

VEGETATION DOMINATED BY CAMPYLIUM STELLATUM

Scheuchzerio-Caricetea nigrae (Nordh. 1936) Tx. 1937
Caricetalia davallianae Br.-Bl. 1949
Caricion davallianae Klika 1934

Relevé 294

Heritage Council analogue: Non–Calcareous Springs—FP2

Relevé 294 (13 species)	
Campylium stellatum	4
Eleocharis multicaulis	2
Potamogeton polygonifolius	1
Nardus stricta	1
Calliergonella cuspidata	1
Carex nigra	+
Bellis perennis	+
Juncus bulbosus	2
Carex viridula	2
Ranunculus flammula	1
Hypericum elodes	1
Pinguicula lusitanica	+
Carex panicea	+

The species-poor community is assigned to the Scheuchzerio-Caricetea nigrae, based on the occurrence of *Calliergonella cuspidata* and *Ranunculus flammula*. This is further referred to the Caricion davallianae, owing to the occurrence of a monodominant bryophyte layer comprising *Campylium stellatum*, *Pellia epiphylla/neesiana*, *Bryum pseudotriquetrum* and *Rhizomnium punctatum*. The occurrence of species characteristic of the Littorelletea uniflorae such as *Eleocharis multicaulis* and *Potamogeton polygonifolius*, with *Juncus bulbosus* and *Carex viridula* frequently emerging through the bryophyte carpet, are also characteristic features.

Unlike most typical stands of the Caricetalia davallianae, which can form extensive stands (Ó Críodáin 1988; Ó Críodáin and Doyle 1994), the vegetation described here is confined to a few areas surrounding rocky outcrops in upland areas that are subject to seepage by moderately enriched waters. The community is particularly common on the bare, podzolised soils that occur on the eastern slopes of Knockmore, although the occurrence of the vegetation is always fragmentary.

Other authors have also recorded similar seepage vegetation (McVean and Ratcliffe 1962; Birks 1973), although their descriptions are brief, as these communities are seldom extensive. Recently, Bleasdale and Conaghan (1995) described a comparable *Carex panicea-Campylium stellatum* flush community from the Connemara region of Ireland, which they considered to be a species-poor variant of the *Carex demissa-Juncus bulbosus* subcommunity of the *Carex dioica-Pinguicula vulgaris* (M10) mire (Rodwell 1991).

VEGETATION DOMINATED BY CALLIERGONELLA CUSPIDATA AND ISOLEPIS SETACEA

Scheuchzerio-Caricetea nigrae (Nordh. 1936) Tx. 1937
Caricetalia davallianae Br.-Bl. 1949
Caricion davallianae Klika 1934 relevé 110

Heritage Council analogue: Non-calcareous springs—FP2

Relevé 110 (19 species)

Calliergonella cuspidata	3
Potamogeton polygonifolius	2
Ranunculus flammula	2
Juncus acutiflorus	2
Agrostis stolonifera	1
Anagallis tenella	+
Carex nigra	+
Triglochin palustris	+
Sagina procumbens	+
Samolus valerandi	+
Isolepis setacea	3
Pellia epiphylla/neesiana	2
Hydrocotyle vulgaris	2
Eleocharis multicaulis	1
Bellis perennis	1
Carex viridula	+
Bryum pseudotriquetrum	+
Leontodon autumnalis	+
Mentha aquatica	+

Both *Calliergonella cuspidata* and *Isolepis setacea* dominate this moderately species-rich community, which is tentatively classified in the Caricion davallianae. It is worth noting that *Isolepis setacea* is a character species of the Isoeto-Nanojuncetea, which despite the widespread occurrence of its short-lived therophytic character species, remains virtually undescribed in Ireland. Braun-Blanquet and Tüxen (1952) described a community containing *Isolepis setacea* (previously called *Scirpus setaceus*) from a roadside in Connemara, which they considered was a subassociation of the Cicendietum filiformis in the Isoeto-Nanojuncetea. Birks (1973) in his account of the vegetation of Skye, recorded a rare, fragmentary community occurring on bare sand or gravel in areas that were prone to seasonal waterlogging, also related to the same class.

The present community was only recorded below the signal tower on the extreme western half of the island, where it is confined to a narrow drainage basin between two small streams. Water slowly seeping from stratified rocks feeds into one small stream above the drainage basin. The flow of water does not hinder the vegetation colonisation and subsequent growth. In addition to *Calliergonella* and *Isolepis*, each of which contributes about 30% ground cover, *Potamogeton polygonifolius, Ranunculus flammula, Hydrocotyle vulgaris, Juncus acutiflorus* and *Eleocharis multicaulis* are recorded. There are several other vascular plant and bryophyte species; they are generally of little cover including *Samolus valerandi.* This latter species is characteristic of the Samolo-Littorelletum within the Littorelletea uniflorae and is only recorded from a number of brackish sites on Clare Island.

VEGETATION DOMINATED BY MYOSOTIS SECUNDA AND FISSIDENS ADIANTHOIDES

Scheuchzerio-Caricetea nigrae (Nordh. 1936) Tx. 1937
Caricetalia davallianae Br.-Bl. 1949
Caricion davallianae Klika 1934 relevé 117

Heritage Council analogue: Non-calcareous springs—FP2

Relevé 117 (15 species)

Myosotis secunda	3
Callitriche stagnalis	2
Potamogeton polygonifolius	2
Mentha aquatica	2

Carex nigra	1
Juncus articulatus	1
Juncus bulbosus	+
Jungermannia sp.	+
Fissidens adianthoides	2
Eleogiton fluitans	2
Aneura pinguis	2
Agrostis stolonifera	1
Ranunculus flammula	1
Anagallis tenella	+
Hydrocotyle vulgaris	+

Relevé 117 describes vegetation found in a runnel in marshy grassland, from outside a gate at the western end of the Congested Districts Board Wall in Strake townland. The area is enriched by water draining off the slopes into a small stream that winds its way through the vegetation. Sheep heavily graze the surrounding vegetation, although larger herbivores, including feral horses and donkeys, congregate here, as evidenced by large concentrations of their faecal material.

The vegetation is characterised by the presence of *Fissidens adianthoides*, *Callitriche stagnalis*, *Potamogeton polygonifolius* and *Aneura pinguis*, each providing ground cover of between 15% and 25%, while the most significant cover is provided by *Myosotis secunda* (35%). The occurrence of this last species in upland areas of Clare Island is anomalous, since it is generally confined to drainage ditches along unmanaged road verges. Other species include *Carex nigra*, *Ranunculus flammula*, *Anagallis tenella*, *Juncus bulbosus*, *Juncus articulatus* and *Hydrocotyle vulgaris*, all of which are commonly associated with wet habitats.

VEGETATION DOMINATED BY ELEOGITON FLUITANS AND MYRIOPHYLLUM ALTERNIFLORUM

Littorelletea uniflorae Br.-Bl. et Tx 1943
Littorelletalia uniflorae Koch 1926

relevé 81

Analogue with Heritage Council scheme not present

Relevé 81 (9 species)	
Eleogiton fluitans	3
Eleocharis multicaulis	2
Juncus bulbosus	1
Eriophorum angustifolium	1
Narthecium ossifragum	+
Myriophyllum alterniflorum	3
Utricularia intermedia	2
Carex nigra	1
Drosera rotundifolia	+

Another community of uncertain status, with a depauperate and unusual species assemblage, was encountered in a man-made drainage channel along one of the smaller inter-drumlin bog hollows at Maum townland. These channels were constructed in the early 1980s to divert and control the water flow for the island's water scheme. The vegetation, characterised by *Eleogiton fluitans*, *Eleocharis multicaulis* and *Juncus bulbosus*, is allied to the Littorelletea uniflorae, while the presence of *Myriophyllum alterniflorum* and *Utricularia intermedia* is indicative of impeded water flow along the channel. The sparse vegetation also includes *Eriophorum angustifolium*, *Narthecium ossifragum* and *Drosera rotundifolia*.

FEN VEGETATION

In certain locations on Clare Island, notwithstanding the predominantly acid lithologies and the typically calcifuge vegetation, increases in moderately enriched ground and surface water benefits the establishment of calcicolous fen communities. Fen vegetation on Clare Island is characterised by *Schoenus nigricans*, although it is worth noting that *Schoenus*, a character species of Atlantic blanket bog, also occurs in some of the wetter parts of blanket bog.

On Clare Island, *Schoenus*-dominated fen exhibits marked physiognomic differences from the surrounding vegetation, which is typically a dwarf grassland sward. A second community, the Scorpidio-Eleocharitetum multicaulis, recorded only twice from Clare Island, is spatially allied with the *Schoenus*-dominated vegetation. It is more commonly recorded from oligotrophic lake margins, which undergo regular fluctuations in the water table.

FEN VEGETATION DOMINATED BY SCHOENUS NIGRICANS

Scheuchzerio-Caricetea nigrae (Nordh. 1936) Tx. 1937 em. Tx. 1980
Caricetalia davallianae Br.-Bl. 1949
Caricion davallianae Klika 1934
Schoenetum nigricantis (Allorge 1922) Koch 1926

(Table 25)

Schoenetum nigricantis character species:
Schoenus nigricans
Constant species: *Anagallis tenella*, *Eleocharis multicaulis*, *Carex viridula*

Table 25
Relevé table for the Schoenetum nigricantis

Column	1	2	3	Synoptic value
Relevé	72	129	104	
Schoenetum nigricantis				
Schoenus nigricans	3	2	3	V
Constant species				
Carex viridula	+	2	1	V
Anagallis tenella	1	1	+	V
Eleocharetosum multicaulis				
Eleocharis multicaulis	2	2	2	V
Potamogeton polygonifolius	2	2		III
Erica tetralix	+	+		III
Myrica gale	2			II
Cirsietosum dissecti				
Cirsium dissectum		1	+	III
Agrostis stolonifera		1	+	III
Scheuchzerio-Caricetea nigrae				
Pellia epiphylla		2	+	III
Ranunculus flammula		1	+	III
Hydrocotyle vulgaris		1	+	III
Plantago maritima		+	+	III
Juncus articulatus		+	+	III
Companion species				
Campylium stellatum	2			II
Hypericum elodes	1			II
Carex rostrata	1			II
Sphagnum papillosum	1			II
Eriophorum vaginatum	1			II
Nardus stricta		3		II
Carex panicea		1		II
Isolepis setacea		1		II
Sagina procumbens		1		II
Drepanocladus revolvens			2	II
Breutelia chrysocoma			2	II
Juncus bulbosus			2	II
Agrostis capillaris			2	II
Molinia caerulea			1	II
Scapania irrigua			1	II
Triglochin palustris	+			II
Drosera anglica	+			II
Pedicularis palustris	+			II
Leontodon autumnalis		+		II
Galium palustre		+		II
Succisa pratensis			+	II
Potentilla erecta			+	II
Mentha aquatica			+	II
Pinguicula vulgaris			+	II
Taraxacum officinale agg.			r	II
Centaurium erythraea			r	II
Number of species	15	19	22	

Heritage Council analogue: Rich fen and flush—PF1

Several authors have described fen and wet flush vegetation in Ireland, in which *Schoenus nigricans* is the physiognomic dominant. There was little consensus on the classification of such vegetation in Ireland (Braun-Blanquet and Tüxen 1952; Ivimey-Cook and Proctor 1966; O'Connell 1977; Van Groenendael *et al.* 1979), until Ó Críodáin and Doyle (1997) clarified the status of this community. The Schoenetum nigricantis is classified in the Caricion davallianae, order Caricetalia davallianae. The association does not include vegetation of ombrotrophic blanket bogs in which *Schoenus nigricans* may be co-dominant with *Molinia caerulea* (Doyle 1982b; 1990).

The BNVC, however, recognises a *Schoenus nigricans-Narthecium ossifragum* M14 mire that may not be easily integrated into the Schoenetum nigricantis M13 mire (Rodwell 1991). In general, *Schoenus* is usually the strong dominant species making it stand out at a distance from the usual acid grassland or bog vegetation.

In *Schoenus* fens and flush habitats in Ireland, *Schoenus* is the sole character species owing to its dominance, with cover generally above 25% (Ó Críodáin and Doyle 1997) and the structural impact of its pronounced tussock formation. Fen vegetation dominated by *Schoenus* is not extensive on Clare Island. Enriched waters heavily influence the composition and distribution of this vegetation type, which is recorded from only two locations. The greatest extent of the *Schoenus*-flush vegetation is found at the northern end of the island, below the lighthouse (Pl. XXIII). A smaller outlier is located in a peat basin at the extreme western end of the island. The vegetation occurs on moderately shallow peaty soils, where fragments of coarse-grained sandstone bedrock are an integral part of the soil matrix. Continuous irrigation and percolation of base-rich ions through the soils has resulted in less podzolized soils.

As presented in Table 25, the order of the relevés reflects the occurrence of fen vegetation on substrates ranging from shallow blanket peats to shallow loamy podzols. These habitat differences are reflected in the occurrence of species that are diagnostic of both the Eleocharetosum multicaulis and the Cirsietosum dissecti that are distinguished by Ó Críodáin and Doyle (1997). The stout tussocks of *Schoenus* are a conspicuous feature of the landscape standing out from the surrounding, closely cropped, grass sward (Pl. XXIII). The *Schoenus* tussocks are widely spaced, however, providing up to 35% cover at a maximum, compared with the usual lower figure for the Schoenetum of 50% (Ó Críodáin and Doyle 1997). The constant species assemblage for the association on Clare Island includes *Eleocharis multicaulis*, *Carex viridula* and *Anagallis tenella*. The sward supports varying abundances of *Agrostis capillaris*, *Festuca ovina* and *Nardus stricta*, with the vegetation gradient reflecting the level of grazing and enrichment due to water percolation.

Pl. XXIII One of two fen sites on Clare Island dominated by black bog rush, located below the lighthouse. The dark green vegetation behind the telegraph pole is located in gullies that are continually irrigated by water draining off the surrounding slopes.

The first relevé (Table 25, column 1) was taken from the margins of the blanket bog. Typically, the vegetation is found in more rheotrophic areas where there is a strong flow of water (Ó Críodáin and Doyle 1997). The peaty soils were harvested for scraw turf, leaving a small network of drainage channels.

Some soil erosion occurred and this was accelerated by drainage works that were carried out in the vicinity of the community. Water flows along small channels in the exposed peat surface and may form small pools after periods of rain. The vegetation is dominated by *Schoenus nigricans* (Table 25, column 1), although the high abundances of *Myrica gale, Eriophorum vaginatum, Sphagnum papillosum* and *Drosera anglica* reflect the relationship that the vegetation shares with the proximal blanket bog. Periodic standing water at the site favours the occurrence of species such as *Hypericum elodes, Carex rostrata* and *Campylium stellatum*.

The next two relevés (Table 25, columns 2 and 3) were recorded from gradually more elevated situations than the previous example and are found among runnels on a plain dominated by a grassland sward (classified in the Achilleo-Festucetum tenuifoliae, coastal subtype 2). This *Schoenus* vegetation has strong allegiances with the cirsietosum, another subassociation distinguished by Ó Críodáin and Doyle (1997). Floristically, column 2 represents transitional vegetation since it contains elements of both the eleocharetosum and cirsietosum subassociations. Vegetation described by means of the relevé in column 2 (Table 25) is the most extensive form of the *Schoenus* fen and in total comprises about 80% of the fen vegetation on Clare Island. The community is generally rich in species of vascular plants and includes *Carex panicea, Isolepis setacea, Sagina procumbens* and *Leontodon autumnalis*, while *Cirsium dissectum* is more prominent in regularly flushed swards. Structurally, it differs from vegetation described in the other relevés, since the *Schoenus* tussocks are more compactly grouped. They are also more stunted, however, as a result of severe cropping by sheep.

The cirsietosum is the most comprehensively described form of the Irish Schoenetum vegetation in Ireland, which occurs in the 'driest' *Schoenus* habitats (Ó Críodáin and Doyle 1997). The most typical example of the cirsietosum vegetation on Clare Island (Table 25, column 3) includes *Schoenus* and *Eleocharis multicaulis*, together with *Cirsium dissectum*, *Agrostis stolonifera* and a suite of Scheuchzerio species such as *Pellia epiphylla/neesiana*, *Ranunculus flammula* and *Hydrocotyle vulgaris*. Vegetation recorded from tussock interspaces, that was not recorded from less degraded fen vegetation, included *Drepanocladus revolvens*, *Breutelia chrysocoma*, *Juncus bulbosus* and *Scapania irrigua*, which are characteristic of areas subjected to water run-off.

CLIFF VEGETATION

Cliffs, by their very nature, are extreme habitats that support an impressive, albeit fragmentary, range of vegetation types. The vegetation is influenced chiefly by, (a) relief, (b) sea-spray (c) edaphic factors, and (d) grazing patterns. Much of the character of the cliff vegetation on Clare Island is composed of admixtures of an acid grassland-*Calluna* heathland, only differing in abundances of the species. Cliffs are generally free from intense grazing, since access is often impeded. Vegetation height varies on all cliffs, although the effect of wind often results in stunted vegetation. Sea spray deposition also plays an important role in determining the composition and distribution of the vegetation.

Much of the coastline on Clare Island is characterised by cliffs, which includes the smaller precipices along the south coastline that only rise a few metres above the sea, to the more rugged cliffs along the northern half of the island that ascend to 400 metres (Pl. XXIV). Owing to the severity of the slope, the majority of the sea cliffs are treacherous in nature. Scree is not an uncommon feature where the cliffs are sheer. The vegetation survives on thin, immature soils, consisting of minerals washed down the cliffs. They are liable to erosion as vegetation integrity is frequently breached. Another feature of the cliffs is the fragmentary nature of the vegetation. Some of the species found on the cliffs are localised, occurring only as widely scattered individuals. The influence of sea spray becomes increasingly important at lower levels on the larger sea cliffs, although some of the maritime species occur frequently at higher elevations owing to the upwelling air currents.

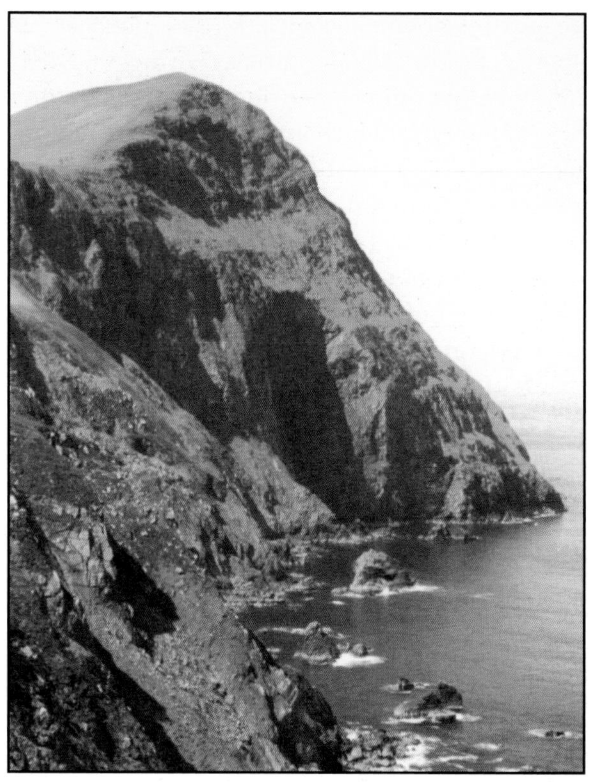

Pl. XXIV Northern cliffs of Knockmore, where much of the woodrush (*Luzula sylvatica*) community and calcicolous vegetation has developed. Vegetation from less severely sloped ground is covered by an undergrazed grassland community with some well-developed bushes of heather in sheltered/inaccessible areas.

There are also some impressive inland crags on Clare Island, such as those found at Leck (Pl. XXV B). Although occasional trees of willow and mountain ash occur on some rock-ledges, the greater part of the crags are treeless. These inaccessible stands probably represent the remnants of the ancient woodland that was once more abundant on Clare Island.

In the following description, three communities characterised by the woodrush, *Luzula sylvatica*, are distinguished on the cliffs on Clare Island, although other fragmentary and inaccessible cliff communities do occur. Outside of woodland communities, *Luzula sylvatica* occurs as an occasional species in a variety of upland communities in Ireland. It attains an overwhelming dominance in some cliff communities on mountain slopes and ledges, where it presents a marked contrast to the surrounding overgrazed hill vegetation (e.g. Mhic Daeid 1976). The Clare Island *Luzula* community is confined to ridges and cliffs on the north-eastern flanks of Knockmore (Pl. XXV A) and the inland cliffs at Leck (Pl. XXV B) and in the inaccessible crags of the Ballytoohy ridge system. The development of a calcicolous flora, within the

Pl. XXV A Woodrush-dominated vegetation on the ridge below Knockmore's summit.

Pl. XXV B Inland cliffs at Leck, which are dominated by *Luzula sylvatica* with occasional rowan and willow. These cliffs support vegetation that has some similarities with the precursor forest community that Braun-Blanquet felt gave rise to the extrasilvaticum subassociation.

Luzula community, is dependent on the seepage of nutrient-enriched waters from the only basic lithologies to outcrop on Clare Island.

Floristics

Woodrush, *Luzula sylvatica*, consistently characterises the three cliff communities, although in decreasing abundance, with only a few other species regularly encountered (Table 26) such as *Vaccinium myrtillus*, *Agrostis capillaris*, *Festuca ovina/vivipara* and *Anthoxanthum odoratum*. The continuity with Nardetalia communities is maintained since *Potentilla erecta*, *Holcus lanatus* and *Galium saxatile* are frequently recorded. Among the dense, luxuriant vegetation, other vascular plants are poorly represented. Occasionally, *Carex binervis*, *Deschampsia flexuosa*, *Eriophorum vaginatum*, *Oxalis acetosella* and *Saxifraga spathularis* are found. In north-west Britain, *Deschampsia caespitosa* is commonly recorded from similar vegetation (Rodwell 1992), but on Clare Island this species was only located at one location on an old sheep track (Table 26, column 5).

With decreasing *Luzula* ground cover, other species become established, particularly in gaps or in situations where the substrate is composed of loose rocky material. These include *Jasione montana*, *Succisa pratensis*, *Angelica sylvestris*, *Rhodiola rosea*, *Geum rivale*, *Calluna vulgaris*, *Primula vulgaris* and *Thymus praecox*. Ferns are poorly represented in the *Luzula* community and are usually of small stature. Of these, *Hymenophyllum wilsonii* is the most common species, found in sheltered areas beneath small terraces. Other ferns occasionally recorded include *Blechnum spicant* and *Asplenium adiantum-nigrum*. Rarer Arctic-Alpine ferns recorded from this community by Praeger (1911), although not relocated in the current survey, include *Asplenium viride*, *Cystopteris fragilis* and *Polystichum lonchitis*.

Bryophytes are diverse, though few contribute significantly to ground cover. They are mainly found growing through the dense, moist vegetation. The common occurrences include *Rhytidiadelphus squarrosus*, *Polytrichum commune*, *Hylocomium splendens* and *Thuidium tamariscinum*. Occasionally, *Racomitrium lanuginosum* is found in peripheral locations abutting dry grasslands of the Achilleo-Festucetum tenuifoliae. The bryophytes are more significant where the wood-rush communities overhang cliffs or grow directly on bare rock/scree. Along seepage zones in gullies on the mountain slopes or on craggy rocks, *Herbertus aduncus* attains dominance.

Phytosociology

CALCIFUGOUS VEGETATION

Betulo-Adenostyletea Br.-Bl. 1948
Adenostyletalia Br.-Bl. 1931
Dryoptero-Calamagrostidion purpureae Nordhagen 1943
Luzula sylvatica-Vaccinium myrtillus **Association** Birks 1973 (Table 26, columns 1–6)

Diagnostic species: *Luzula sylvatica*, *Vaccinium myrtillus*

Associated species: *Anthoxanthum odoratum*, *Agrostis capillaris*, *Festuca ovina*, *Rhytidiadelphus squarrosus*

Table 26
Relevé table for the cliff vegetation

Column	1	2	3	4	6	5	7	8	Synoptic value
Relevé	3		3	3	3	3	3		
	9	6	4	4	9	9	4	6	
	5	0	1	2	7	6	3	2	
Luzula sylvatica-Vaccinium myrtillus Association									
Luzula sylvatica	3	4	4	2	2	4	2	1	V
Vaccinium myrtillus	+	1	+	1	+	2		r	V
Deschampsia flexuosa	3	2	+	1					III
Eriophorum vaginatum	1	1		1					II
Nardetalia species									
Anthoxanthum odoratum	1	1	2	2	+	2	3	1	V
Agrostis capillaris	2	2	3	2	2	2		2	V
Hylocomium splendens	+	+	1	1	+				IV
Oxalis acetosella	+	1	+		+			+	IV
Thuidium tamariscinum	+	+	1			+	+		IV
Rhytidiadelphus squarrosus		1	1	1	1	1	1		IV
Potentilla erecta		1	1	1	2	+			IV
Carex binervis			+	+	1	+			III
Festuca vivipara			2	2		+	3		III
Galium saxatile			+	2	+		+		III
Holcus lanatus			+	1		1	1		III
Festuca ovina			2	2	2	1	2	4	IV
Polytrichum commune		1	+	+		+	+	1	IV
Hypnum jutlandicum				1	1		+	1	III
Herberteto-Polytrichetum alpini									
Saxifraga spathularis						+	2	2	II
Herbertus aduncus							1		I
Rhodiola rosea								1	I
Bryum pseudotriquetrum								1	I
Cladonia portentosa								1	I
Companion species									
Lophocolea bidentata				+	+	+			II
Sphagnum capillifolium				1	2				II
Dicranum scoparium					+	+			II
Aulacomnium palustre				+			+		II
Calluna vulgaris				3	+		+		II
Mnium hornum					+	+	1	+	II
Plagiothecium undulatum					+	+	+		II
Jasione montana						+	+	+	II
Lepidozia reptans						+	+		II
Hymenophyllum wilsonii						+	2	1	II
Frullania tamarisci							+	2	II
Thymus praecox							+	1	II
Plagiochila punctata	+							1	II
Angelica sylvestris							+	+	II
Primula vulgaris							+	+	II
Succisa pratensis							+	+	II
Viola riviniana							+	+	II
Vaccinium oxycoccos	+					2			II
Eriophorum angustifolium		+			1				II
Racomitrium lanuginosum		1	+						II
Eurhynchium praelongum	+					+			II
Rhytidiadelphus loreus	1								I
Deschampsia caespitosa						1			I
Erica cinerea					1				I
Eurhynchium striatum			+						I
Pseudoscleropodium purum			+						I
Sphagnum papillosum				+					I
Polygala serpyllifolia				+					I
Brachypodium sylvaticum						+			I
Sedum anglicum						+			I
Juncus squarrosus					+				I
Blechnum spicant					+				I
Pleurozia purpurea					+				I
Cladonia gracilis					+				I
Agrostis stolonifera							+		I
Carex viridula							+		I
Oxyria digyna							+		I
Carex flacca							+		I
Selaginella selaginoides							+		I
Solidago virgaurea							+		I
Rhizomnium punctatum							+		I
Bryum capillare							+		I
Dicranum majus							+		I
Ranunculus acris							+		I
Euphrasia officinalis							+		I
Hypnum cupressiforme							+		I
Sphagnum subnitens							+		I
Cerastium fontanum							+		I
Scapania gracilis							+		I
Orchis mascula								+	I
Cladonia uncialis								+	I
Rhinanthus minor				r					I
Asplenium adiantum-nigrum						r			I
Plantago lanceolata							r		I
Polypodium vulgare								r	I
Number of species	13	14	19	23	24	26	41	25	

Heritage Council analogue: Exposed siliceous rock—ER1 / Montane heath—HH4

Cliff vegetation dominated by *Luzula sylvatica* has not been formally described from Ireland, although it forms a distinctive community on many damp areas on the mountain ranges which has been noted by a number of authors. Braun-Blanquet and Tüxen (1952) described a *Luzula sylvatica* gesellschaft (extrasilvaticum subassociation of the Blechno-Quercetum) from Carrowkeel in County Sligo, a subassociation that has not gained general acceptance in the phytosociological literature (see White and Doyle 1982). Braun-Blanquet and Tüxen (1952) were undecided as to the syntaxonomic validity of their subassociation, due in part to the proximity of their samples and also the perceived paucity of stands from other areas in Ireland. Braun-Blanquet and Tüxen (1952) noted that the subassociation was generally free of tree species, although possible relicts of the former forest community were occasionally recorded and included *Sorbus aucuparia* and *Corylus avellana*. They considered that it was clearly an upland form of the Blechno-Quercetum, the acidophilous oak wood community. Associated species in their community included *Vaccinium myrtillus*, *Festuca ovina*, *Agrostis capillaris*, *Potentilla erecta*, *Deschampsia flexuosa* and *Oxalis acetosella*. The intrusion of Arrhenatheretalia species was indicative of local forest clearance and pasture establishment. The canopy and shrub layer had disappeared, and was replaced by a luxuriant carpet of *Luzula*. Braun-Blanquet and Tüxen (1952) noted that the ground flora associated with forests was poorly developed, though 'forest mosses' extended into the heath and grassland that surrounded the *Luzula* community in Sligo. It developed in windy, exposed situations, and was only found on cliffs that have an associated band of calcareous rocks. The only other comprehensive description of *Luzula*-rich vegetation from Ireland was from the Wicklow Mountains (Pethybridge and Praeger 1905). That community, which developed over granite rock types, differed in that it was dominated by *Vaccinium myrtillus* with only occasional lawns of *Luzula sylvatica*.

McVean and Ratcliffe (1962), Birks (1973) and Jermy and Crabbe (1978) have described *Luzula sylvatica-Vaccinium myrtillus* lawns from north-west Britain. These descriptions show affinities with the *Luzula sylvatica-Vaccinium myrtillus* U16 tall herb community described from high cliffs in the wet and cold mountainous regions in northern Britain and the sub-alpine zone in the Scottish highlands (Rodwell 1992). The *Luzula sylvatica-Vaccinium myrtillus* association, recorded from northern Scotland, is a treeless association referred to the *Betula*-Adenostyletea, a class that includes the forest communities of high mountain ranges in Central Europe. This class has not been previously described from Ireland. However, the vegetation described from Scotland has many similarities with the *Luzula*-dominated vegetation on Clare Island and Birk's (1973) account from Skye, in particular, records the association occurring from sea level to elevations of 520 metres.

The Clare Island vegetation develops on flat or sloping ledges (plateaus) and vertical cliff faces in areas underlain by calcareous rocks. Soil types range from peaty podzols to peaty clays with a large humic fraction consisting of decomposing *Luzula* herbage. Soil composition is governed by the nature of the exposed rock outcrop, the amount of broken rock fragments in the soils and increasing drainage with changes in slope. The highly drained situations are generally inaccessible to sheep, or at least less severely grazed than much of the surrounding heathland with which the community is associated.

The wood-rush, *Luzula sylvatica*, is the overwhelming dominant in this vegetation (Table 26, columns 1 and 2). Other species recorded include *Vaccinium myrtillus*, *Anthoxanthum odoratum*, *Agrostis capillaris*, *Festuca ovina*, *Deschampsia flexuosa*, *Eriophorum vaginatum* and *Oxalis acetosella*. They are interspersed among the *Luzula* sward, with no individual species ever providing more than 20% cover. Despite its luxuriant appearance, the vegetation is easily eroded, as it is not fully rooted. It is often treacherous underfoot, as the peaty soils contain fragmented rock material, unlike the surrounding podzols on the commonage. Sheep frequently uproot *Luzula* while grazing and pedestrian traffic can create trails, thereby increasing the rate of vegetation loss and subsequent loss of soil integrity. Floristic and physiognomic variation is evident, especially with increases in slope.

Steeper gradients result in the gradual diminution of the dominance enjoyed by *Luzula* and an increase in species diagnostic of the Nardetalia (Table 26, columns 3–6). Soils are patchily distributed in cracks in the cliffs, among rocks and boulders and also trapped by the dense *Luzula* shoots that commonly hang over the cliffs. There is also a greater diversity in plant species, reflecting the influence of thin bands of calcareous rock that outcrop in three areas on Clare Island, and the abundance of gaps within the vegetation. The taller dicotyledonous species, while noticeable, occur infrequently and include *Angelica sylvestris, Jasione montana, Succisa pratensis, Rhodiola rosea* and *Solidago virgaurea*. Bryophytes are more numerous than higher plant species and in some situations can provide a significant amount of cover. Species include *Mnium hornum, Plagiothecium undulatum, Sphagnum capillifolium, Aulacomnium palustre, Dicranum scoparium* and *Frullania tamarisci*. The treacherous substrate and the general inaccessibility are factors in the continued survival of this community.

CALCICOLOUS VEGETATION

Calluno-Ulicetea Br.-Bl. et Tx. 1943
Vaccinio-Genistetalia Schubert 1960
Vaccinio-Callunion Moore in Mhic Daeid 1979
Herberteto-Polytrichetum alpini Mhic Daeid 1979 (Table 26, columns 7–8)

Herberteto-Polytrichetum alpini diagnostic species: *Herbertus aduncus, Saxifraga spathularis*
Associated species: *Festuca ovina, Anthoxanthum odoratum, Hymenophyllum wilsonii, Frullania tamarisci, Luzula sylvatica* (much reduced)

Heritage Council analogue: Exposed calcareous rock—ER2 / Montane heath—HH4

The most calcicolous form of the vegetation in which *Luzula* is found (Table 26, columns 7 and 8) is confined to near vertical cliffs. Base-rich waters irrigate the vegetation growing directly onto cliff faces in earth-laden fractured rock seams. The community is different to the *Luzula sylvatica-Vaccinium myrtillus* Association in that *Luzula* is greatly reduced, never more than 30%. This community is comparable to the Herberteto-Polytrichetum alpini, recorded on steep sheltered slopes from mountains in Kerry and Connemara (Mhic Daeid 1976). Analogous communities have been described that are confined to inaccessible ground in cold wet uplands of north-western Britain (McVean and Ratcliffe 1962; Birks 1973; Prentice and Prentice 1975; Jermy and Crabbe 1978), while under the BNVC scheme the vegetation is transitional between the *Saxifraga aizoides-Alchemilla glabra* (U15) banks and the *Luzula sylvatica-Geum rivale* tall-herb (U17) cliff vegetation (Rodwell 1992). All report the community occurring at elevations from 350m to 800m.

Overhanging *Luzula* provides a suitable substrate in which small herbaceous plants may root. The community is floristically diverse, (column 7 has 41 species). Bryophytes, though not always the most conspicuous component of the community, are the most numerous component of the vegetation (Pl. XXVI A). The vegetation is characterised by the presence of the extensive cushion forming hepatic *Herbertus aduncus*. Other constant occurrences include *Festuca vivipara, Festuca ovina, Holcus lanatus, Saxifraga spathularis* and *Hymenophyllum wilsonii*. Also recorded in this community is the Arctic-Alpine species *Oxyria digyna*. This species appears to be one of a few surviving constant relicts from the glacial times on the northern cliffs at Knockmore (Pl. XXVI B).

ARCTIC-ALPINE VEGETATION

Montio-Cardaminetea Br.-Bl. et Tx. 1943
Montio-Cardaminetalia Pawlowski 1928
Cratoneurion Koch 1926
Saxifragetum aizoidis McVean et Ratcliffe 1962
No relevés

Diagnostic species: *Saxifraga oppositifolia, Alchemilla glabra, Selaginella selaginoides,* (*Saxifraga aizoides*—not recorded from Clare Island)
Notable species: *Alchemilla alpina, Asplenium viride, Oxyria digyna, Polystichum lonchitis, Saussurea alpina, Rhodiola rosea, Silene acaulis*

Heritage Council analogue: Exposed Calcareous Rock—ER2

The 'Arctic-Alpine' community described by Praeger (1911) is associated with the steep cliffs of Knockmore. These communities are fragmentary, representing the surviving relicts of an Arctic–Alpine community that probably flourished on

Pl. XXVIA Moss-dominated vegetation that develops on flushes within the *Luzula* community. The diminutive *Saxifraga rosacea* is seen in the centre of the photograph.

Pl. XXVIB Vegetation along seepage zones in cliff gullies including *Oxyria digyna* in the centre of the photograph.

north-facing cliffs during the last glaciation. Nutrient-enriched waters seeping down the gullies provide the perfect environment for many of the Arctic-Alpine species. Two paths, the '1200ft' and '1000ft' path facilitated access to the cliffs during the original survey.

List of Praeger's (1911) 'Arctic-Alpine' species

Relocated	Not Found
Oxyria digyna *	*Saussurea alpina* *, **
Silene acaulis *	*Polystichum lonchitis* *
Saxifraga oppositifolia *	*Asplenium viride*
Rhodiola rosea *	*Cystopteris fragilis*
Saxifraga spathularis	*Rubus saxatilis*
Saxifraga x polita	*Salix herbacea*
Saxifraga rosacea	

* These are regarded as true 'Arctic-Alpine' species in the context of the Irish flora (Webb 1983). The remaining species are commonly recorded from montane habitats in Ireland, although some may descend to sea level.
** These species were relocated in 1999 (C. Farrell, pers. comm.)

Subsequent erosion of the 1200ft path has restricted access to the vertical gullies from the 1000ft path. Many of the species listed by Praeger have been relocated, though some remain undetected despite repeated efforts during the New Survey of Clare Island to locate them. Some may yet be refound by lucky botanists.

The vegetation is tentatively classified in the Saxifragetum aizoidis (U15 *Saxifraga aizoides-Alchemilla glabra* banks (Rodwell 1992)), although apart from species lists, no relevés were made. The vegetation is a glacial relict found on north-facing cliffs as a fragmentary element of the calcicolous montane vegetation zonation, a finding confirmed from similar vegetation in Ireland and elsewhere (Winder and Moore 1947; McVean and Ratcliffe 1962; Birks *et al.* 1969; 1973; Huntley 1979). White and Doyle (1982) consider that the association is related to the Cochlearion alpinae in Ireland, based on the character species *Rhodiola rosea*, *Cochlearia officinalis* and *Saxifraga oppositifolia*. Other alpine vegetation includes the Polysticho-Asplenietum viridis, defined on the occurrence of the diagnostic species *Polystichum lonchitis* and *Asplenium viride*. These species, characteristic of the wall fern class in Ireland, were located under crevices on steeply dipping cliffs on Clare Island (Praeger 1911).

WOODLAND COMMUNITIES

Forests once dominated much of the landscape in Ireland. Oak woodlands are regarded as the climax vegetation on acid soils in Ireland and Great Britain (McVean and Ratcliffe 1962; Kelly and Moore 1975; Miles 1981; 1986), with birch woodlands generally forming a transition stage in the vegetation succession toward oakwoods in all but the wettest soil situations where true birch wood survives.

Climate changes combined with anthropogenic influence resulted in much of the forested landscape of the western seaboards being overwhelmed by blanket mire development (Moore 1975; Kirby and O'Connell 1982). The arrival of agricultural settlements hastened the demise of the remaining woodlands. As a consequence of the extensive clearance, the only extant primitive woodland scrub on Clare Island is located in a sheltered valley at Lassau, west of Portlea. Further smaller patches of floristically depauperate, scrub woodland are found in unproductive or wet areas in isolated situations elsewhere on the island.

There is some evidence that much of Clare Island was covered in woodland before man initiated its decline (Coxon 1994). Scrub woodland probably occupied most of the sheltered ground in low-lying areas, and extended up some of the steeper slopes. In the absence of radiocarbon dates for the pollen record taken on Clare Island, Coxon (1987; 1994; 2001) estimated the ages of the important phases in the vegetation record by reference to radiocarbon dated sites elsewhere in Ireland.

The earliest sediments recovered were laid down in standing water that was surrounded by open vegetation characterised by Poaceae and Cyperaceae pollen assemblages. Such pollen assemblages are widely known from the end of the Nahanagan Stadial—approximately 10,000 years ago. The landscape was gradually colonised by *Juniperus* as the climate ameliorated. After the local peak of *Juniperus* (9500 years BP), *Betula* and *Corylus* began to dominate the vegetation. These trees migrated, and expanded into the landscape sometime between 10,000 years and 8000 years BP. Increasing organic matter in the steadily accumulating sediment, resulted in the appearance of other tree types such as *Quercus*, *Alnus* and *Ulmus*, while *Pinus* was gradually increasing in importance. Coxon has suggested that the expansion of *Alnus* on Clare Island occurred sometime between 7500 and 6500 years BP. He also noted a local increase in charcoal fragments within the pollen profile, indicating the frequency with which the vegetation was being burnt. This is also evidenced by the charred nature of some tree stumps at the cut bog at Maum townland (Pl. XXVII A).

Gradually, *Pinus* began to dominate the vegetation, and along with *Betula*, seems to have been the main woodland components. Subfossil *Pinus* remnants of the once widespread woodland vegetation protrude through the peat in many places, especially in the eastern half of the island (Pl. XXVII A). Recent archaeological excavations and road improvement schemes have also unearthed evidence of extensive tracts of previously forested land.

Subsequent to the decline in *Pinus*, approximately 4000 years BP, *Betula*, *Quercus* and *Alnus* woodland (and shrub-dominated heath vegetation) expanded. The increasing occurrence of Poaceae and *Plantago lanceolata* within the profile at this stage indicates the development of open ground, and Coxon has tentatively suggested that these phases and the increase in charcoal represented clearance (Coxon 2001).

Pl. XXVII A Fossil stumps of *Pinus* are a common feature of the landscape in Maum townland. It is estimated that the stumps probably date to 4000 years BP (Coxon 1994; 2001).

Pl. XXVII B Woodland in Lassau valley, dominated by birch. Despite the small extent of the woodland, it is possible to distinguish rudimentary zonation. The peripheral vegetation is willow scrub. In wetter areas alongside the small stream, the vegetation is transitional to primitive alder woodland. The central area supports birch scrub with some hazel on the upper valley slopes.

Floristics

The woodland at Lassau (Pl. XXVII B) comprises four distinct vegetation types, each of which displays considerable floristical variability. On the drier fringes of the woodland, the vegetation is characterised by *Salix cinerea*. It is gradually replaced by *Betula pubescens* in the more mature central woodland. The dense canopy dominated by *Betula pubescens* and shrubby *Salix aurita* is 3–5m high. In some stands of the vegetation *Salix cinerea* replaces *S. aurita*. A single specimen of *Quercus petraea* is located in the centre of the woodland. (Another stunted specimen, 0.5m tall, is located above Park on a rocky outcrop). In younger, more open stands and around the perimeter of the woodland, a more heterogeneous canopy develops with *Sorbus aucuparia*, *Crategus monogyna*, *Ilex aquifolium* and *Corylus avellana* as significant components. Occasional alder (*Alnus glutinosa*) trees occur along a stream skirting the woodland at the edge of the adjacent marshy meadow.

The composition of the herbaceous layer is the most diverse, with zonation reflecting the influence of surrounding vegetation types and edaphic features under the canopy. Vegetation gradients have been reported from wet sites to drier sites in some Irish woodlands, with boundaries fluctuating because of soil saturation (Iremonger and Kelly 1988). The occurrence of species assemblages often reflects the surrounding vegetation (Rodwell 1991), such that at the Clare Island site the majority of ericaceous plants in the scrub woodland are concentrated close to the adjoining shrub-dominated heathland. Few species are consistently distributed among the vegetation. Only *Brachypodium sylvaticum*, *Anthoxanthum odoratum*, *Holcus lanatus*, *Oxalis acetosella* and *Rumex acetosa* are frequently recorded. The heathland element found mainly on slopes includes *Calluna vulgaris*, *Vaccinium myrtillus*, *Potentilla erecta* and *Galium saxatile*. In drier conditions *Hyacinthoides non-scripta*, *Ranunculus repens*, *R. ficaria*, *Poa trivialis*, *Primula vulgaris* and *Dryopteris dilatata* are found under the dense canopy.

The development of *Alnus*-dominated vegetation in the woodland is correlated with soils liable to frequent flooding. In this habitat alder, *Alnus glutinosa*, is the only canopy-forming tree, though *Salix* sometimes forms a subcanopy layer. Grasses dominate the field layer with *Anthoxanthum odoratum*, *Glyceria fluitans*, *Holcus lanatus*, *Poa trivialis* and *Agrostis stolonifera* providing the rest of the ground cover. Clumps of *Juncus effusus*, *Iris pseudacorus* and *Osmunda regalis* are conspicuous among the herbaceous vegetation. Smaller herbs include *Ranunculus repens*, *Oxalis acetosella* and *Potentilla erecta*. Occasionally *Epilobium palustre* and *Equisetum palustre* are recorded. This ground vegetation is gradually replaced by *Filipendula ulmaria*, *Lythrum salicaria* and species of the wet meadows as shading from the alder canopy decreases.

The rich bryophyte flora contributes a significant portion of the species list. Most are epiphytic, growing on tree trunks or among debris littered on the ground. Some of the more frequent bryophytes include *Sphagnum palustre*, *S. papillosum*, *Dicranum majus*, *Mnium hornum*, *Ulota crispa* and *Hypnum jutlandicum*. In densely shaded areas, large patches are devoid of vegetation other than bryophytes. Species including *Thuidium tamariscinum* and *Hypnum cupressiforme* account for 35% of the cover in places. The lichen flora is poorer and confined to the upper reaches of the canopy.

Phytosociology

WILLOW SCRUB

Franguletea Doing 1962 em. Westhoff 1968
Salicetalia auritae Doing 1962 em. Westhoff 1968
Salicion cinereae Th. Müller et Görs 1958
(Table 27, column 1)

Salicion cinereae diagnostic species: *Salix cinerea*

Heritage Council analogue: Scrub—WS1

Although it is not widespread, the willow scrub represents the pioneer stage of the woodland colonisation of the lower slopes of the Lassau woodland (Table 26, column 1). Willow scrub is not well defined in floristic terms in Ireland, as it is often present only in fragmentary form and varies considerably in its overall appearance. In Ireland, it was previously classified as the Osmundo-Salicetum atrocinereae, the Salicion cinereae, Salicetalia auritae, in the class Franguletea. Originally recorded from Mayo, Cork and Kerry by Braun-Blanquet and Tüxen (1952), it has also been described from bog edges and lake shores on Lough Ree (Klein 1975) and on drainage channels on Atlantic blanket bog complexes in Mayo, where peat erosion has revealed the mineral soils (Doyle 1990).

Difficulties with this classification are compounded by the presence of species that are characteristic of the Quercetea robori-petraeae, including *Lonicera periclymenum*, *Ilex aquifolium*, *Sorbus aucuparia* and *Pteridium aquilinum*. A recent revision of the Osmundo-Salicetum atrocinereae placed this association in the Alnion glutinosae (Browne 1999) as *Alnus* was nearly always associated with the willow scrub. More recently, the status of willow scrub as pioneer woody vegetation was defined and the ecological features consolidated, under the BNVC scheme, as the *Salix cinerea-Galium palustre* W1 woodland (Rodwell 1991). However, since this vegetation type is often encountered on wet mineral soils, Rodwell felt that it warranted inclusion within the Alnion glutinosae rather than the Salicion cinereae.

Notwithstanding these syntaxonomic difficulties, willow scrub dominated by *Salix cinerea* on Clare Island is spatially associated with Franguletea woodland, with a gradual succession to more closed birch woodland. The main woody component is *Salix cinerea*, although *Betula pubescens* occurs occasionally in the vegetation. Other contributors to the subcanopy vegetation include minor amounts of *Lonicera periclymenum*, *Ilex aquifolium*, *Sorbus aucuparia* and *Pteridium aquilinum*. The field layer varies in its cover and composition and, in general, has a graminoid component, consisting of *Anthoxanthum odoratum*, *Holcus lanatus*, *Brachypodium sylvaticum* and *Agrostis* spp. There is also a rudimentary heathland element in the vegetation with stunted plants of *Calluna vulgaris*, *Vaccinium myrtillus*, *Luzula sylvatica* and

Table 27
Relevé table for the woodland communities on Clare Island

Column	1	2	3	4	5	6	7	Synoptic value
Relevé	298	299	302	300	301	305	303	
Salicion cinereae								
Betula pubescens	1	5	4	3				III
Salix cinerea	3		4	1				III
Salix aurita			2			1		II
Corylo-Fraxinetum								
Corylus avellana				1	4	5		III
Hedera helix	+		1					II
Dryopteris filix-mas		2		1		2		III
Calluna vulgaris	+	+		+				III
Vaccinium myrtillus	+	+		+				III
Ranunculus ficaria			+		2	1		III
Conopodium majus			+		+	+		III
Hyacinthoides non-scripta			+	+	1			III
Polystichum setiferum			1		1			II
Alnion glutinosae								
Alnus glutinosa							3	I
Glyceria fluitans							2	I
Iris pseudacorus							1	I
Constant species								
Anthoxanthum odoratum	3	3	4	4		3	3	V
Holcus lanatus	2	1	1	1		2	2	V
Oxalis acetosella	+	2	1	1	3	2	+	V
Rumex acetosa	+	1	+	+	+	r	+	V
Hypnum cupressiforme		2	2	2	2	1	1	V
Thuidium tamariscinum	2	2	1	2	+	1		V
Mnium hornum	+	2	+	+	+	+		V
Ulota crispa	+	1	+		+	+	+	V
Brachypodium sylvaticum	2	2	1	2	2			IV
Potentilla erecta	2	1	1	2			+	IV
Rubus fruticosus agg.	1		+	1	1	+		IV
Polypodium vulgare		+	+	+	1	1		IV
Luzula sylvatica	+	+		+		+	+	IV
Frullania tamarisci		+		+	+	+	+	IV
Ilex aquifolium	1		2	1	2			III
Pteridium aquilinum	2	+	1				3	III
Juncus effusus			+	1		1	3	III
Viola riviniana	+		+		+	+		III
Lonicera periclymenum		+	+	+		+		III
Hypnum jutlandicum	2	1			+			III
Ranunculus repens					+	1	1	III
Solidago virgaurea		1	+	+				III
Dicranum majus		+	+	+				III
Lophocolea bidentata	+		+	+				III
Companion species								
Poa trivialis						3	2	II
Agrostis stolonifera					2		2	II
Agrostis capillaris			1	2				II
Festuca ovina		1		1				II
Sphagnum palustre	1		1					II
Rhytidiadelphus squarrosus	1			1				II
Sphagnum papillosum		+		2				II
Nardia scalaris		1		+				II
Sorbus aucuparia	1	+						II
Polytrichum commune			1			+		II
Carex binervis	r			+				II
Carex echinata	+			+				II
Dactylorhiza maculata	+					+		II
Galium palustre	+	+						II
Primula vulgaris					+	+		II
Leucobryum glaucum		+	+					II
Digitalis purpurea						r	+	II
Agrostis canina	2							I
Cynosurus cristatus						2		I
Dryopteris dilatata					2			I
Isothecium myosuroides					2			I
Athyrium filix-femina	1							I
Dactylis glomerata			1					I
Rhytidiadelphus triquetrus						1		I
Lysimachia nemorum					+			I
Eurhynchium striatum						+		I
Eurhynchium praelongum							+	I
Hylocomium splendens				+				I
Angelica sylvestris	+							I
Blechnum spicant					+			I
Cirsium palustre						+		I
Epilobium palustre							+	I

(Continued)

Table 27 (*Continued*)

Column	1	2	3	4	5	6	7	Synoptic value
Relevé	298	299	302	300	301	305	303	
Equisetum palustre							+	I
Erica cinerea		+						I
Galium saxatile	+							I
Geranium robertianum					+			I
Juncus bulbosus				+				I
Lythrum salicaria							+	I
Poa annua			+					I
Trifolium pratense	+							I
Viola palustris	+							I
Brachythecium rutabulum							+	I
Neckera complanata		+						I
Plagiothecium undulatum				+				I
Pleurozium schreberi						+		I
Rhytidiadelphus loreus	+							I
Pellia epiphylla						+		I
Scapania gracilis					+			I
Cladonia uncialis		+						I
Calypogeia sp.				+				I
Crataegus monogyna						r		I
Number of species	35	31	34	37	27	33	23	

Blechnum spicant recorded. These heathland species are found in the willow scrub and continue into less mature stands of the woodland community (Table 27, columns 1 and 2).

BIRCH WOODLAND

Franguletea Doing 1962 em. Westhoff 1968
Salicetalia auritae Doing 1962 em 1962 Westhoff 1968
Salicion cinerae Th. Müller et Görs 1958
Salici-Betuletum pubescens Görs 1961
Table 27, columns 2–4

Salici-Betuletum pubescentis diagnostic species: *Betula pubescens, Salix cinerea, Salix aurita*

Heritage Council analogue: Bog Woodland—WN7

Birch (*Betula pubescens*) woodlands are widespread in both upland and lowland areas in Ireland and are often associated with wetland habitats. Birch woodland survives on acid soils and in scrub woodlands in mountain glens, since it is tolerant both of cold climates (Grime *et al.* 1988) and damp soils (Iremonger 1986). The birch community is rather variable in terms of its structural organisation and floristic composition. Classification of Irish birch woodlands has proven somewhat confusing, since none of the vascular species are entirely confined to this community. Unlike their European counterparts, Irish birch woodlands are lower growing but have denser canopies (Iremonger 1990). It is not uncommon to find vegetation complexes among birch woodland (Rodwell 1991) since *Betula pubescens* and *Salix* species are characteristic components of a variety of woodland communities. Environmental gradations in the edaphic conditions and the surrounding topography can also influence the canopy structure and ground flora assemblage.

The present community is best ascribed to the Salici-Betuletum pubescentis, Salicion cinerea, Salicetalia auritae and class Franguletea. Its soils have a shallow humic horizon with large amounts of fragmented rock material in some of the older stands. The freely draining soils are moderately acid (pH 4.5 to 5). Despite the occurrence of birch dominated woodlands around the country (White and Doyle 1982), the Salici-Betuletum pubescentis has not been comprehensively described from Ireland (O'Connell 1981; Hanrahan 1997). Many species diagnostic of the various associations within the alliance have been recorded in or around the Lassau woodland. Elsewhere, O'Connell (1981) recorded a subassociation, the pyroletosum, from Scragh Bog, in County Westmeath, using *Betula pubescens, Salix repens, Pyrola rotundifolia* and several bryophyte species as diagnostic species. The rare *Pyrola rotundifolia* which is restricted in its distribution in Ireland (Scannell and Synnott 1987; Curtis and McGough 1988), was not found on Clare Island.

Under the BNVC classification system, this community resembles the *Dryopteris dilatata-Rubus fruticosus* subcommunity of the *Betula pubescens-Molinia caerulea* W4 woodland (Rodwell 1991), although *Molinia* is not recorded in the current vegetation and is primarily associated with the heathlands above the woodland.

The canopy is variously dominated by *Betula pubescens*. Birch juveniles (column 4) are restricted to areas of bare soil, as the denser canopy prohibits sapling growth (Grime *et al.* 1988), or to a narrow band on the periphery of the woodland that is dominated by *Salix cinerea* (column 1). The older stands consist of a dense, mature birch canopy (columns 2 and 3). The ground flora is less well developed here owing to the low growing, dense nature of the canopy. The most consistent and abundant species providing ground cover are *Brachypodium sylvaticum, Anthoxanthum odoratum* and *Holcus lanatus*. Between them they generally provide up to 50% cover. The other vascular plants in the ground layer are patchily distributed. These include *Luzula sylvatica, Rumex acetosa, Potentilla erecta, Viola riviniana* and *Rubus fruticosus*. There is a well-developed bryophyte flora with several consistently recorded species including *Hypnum cupressiforme, Thuidium tamariscinum, Ulota crispa, Mnium hornum, Dicranum majus* and *Lophocolea bidentata*. Other species that are sometimes recorded include *Hedera helix, Ranunculus ficaria, Hyacinthoides non-scripta, Polystichum setiferum, Asplenium scolopendrium* and *Conopodium majus*. This last group of species is characteristic of the Corylo-Fraxinetum (White and Doyle 1982), and the ground flora shares spatial affinities with this on Clare Island. These species are generally found under gaps in the canopy, which may appear due to natural turnover or storm damage, or may be part of the typical woodland zonation.

HAZEL SCRUB

Querco-Fagetea Br.-Bl. et Vlieger 1937
Fagatalia sylvaticae Pawlowski 1928
Circaeo-Alnenion (Oberd. 1953) Doing 1962
Corylo-Fraxinetum Br.-Bl. et Tx. 1952 em. Kelly et Kirby 1981

(Table 27, columns 5 and 6)

Corylo-Fraxinetum character species:
Polystichum setiferum, Asplenium scolopendrium, Hyacinthoides non-scripta, Conopodium majus

Heritage Council analogue: Scrub—WS1

Hazel scrub (Table 27, columns 5 and 6) is distinguished by the general dominance of hazel, *Corylus avellana*, with both *Betula* and *Salix* species much reduced or absent (Braun-Blanquet and Tüxen 1952, Kelly and Kirby 1982). The evergreen *Ilex aquifolium* has a greater relative abundance here than in the birch woodland. This assemblage is classified as the Corylo-Fraxinetum, although the floristic composition of this community is less diverse than typical ashwood stands that develop over limestone and are widely distributed in Ireland.

While there is no direct equivalent of this community in Great Britain, hazel scrub is perhaps best considered in relation to the *Fraxinus excelsior-Sorbus aucuparia-Mercurialis perennis* W9 woodland that includes the *Corylus avellana-Oxalis acetosella* association described by Birks (1973). This community, described from Skye, on the north-western seaboard of Scotland, has some similarities with the character of the hazel scrub on Clare Island.

The low-growing hazel stands are mainly found on sloping ground, on the perimeter of the woodland at the top of the valley. The soils are slightly less acidic, with a mean pH of 5.4 compared with the mean pH of 4.9 recorded from woodland soils dominated by *Betula*. Ground layer species characteristic of the Corylo-Fraxinetum that occur only in open canopy vegetation include *Ranunculus ficaria, Conopodium majus* and *Polystichum setiferum*. In certain situations these species may dominate the ground layer to such an extent that grasses typically recorded from the other Clare Island woodland types such as *Anthoxanthum odoratum* and *Holcus lanatus* are excluded or reduced. The quantitative composition of the remaining herbaceous and bryophyte layer is variable, a finding that concurs with Rodwell's (1991) synthesis. The common species include *Oxalis acetosella, Brachypodium sylvaticum, Hypnum cupressiforme* and *Thuidium tamariscinum*. Apart from these species there is little difference in either abundance or composition of both the vascular and bryophyte elements of the ground layer of the hazel scrub and birch woodland.

ALDER WOODLAND

Alnetea glutinosae Br.-Bl. et Tx. 1943 em. Th. Müller et Görs 1958
Alnetalia glutinosae Vlieger 1937 em. Th. Müller et Görs 1958
Alnion glutinosae (Malcuit 1929) Meijer Drees 1936 em. Th. Müller et Görs 1958
Osmundo-Salicetum atrocinereae Br.-Bl. et Tx. 1952 (Table 27, column 7)

Osmundo-Salicetum atrocinereae species: *Alnus glutinosa*
Differential species to other woodland classes: *Iris pseudacorus*

Heritage Council analogue: Riparian Woodland—WN5 / Wet Willow-Alder-Ash Woodland—WN6

Alder, *Alnus glutinosa*, woodlands are difficult vegetation types to satisfactorily classify, since there is a plethora of syntaxonomic appellations reflecting the variability of this wetland vegetation and the range of habitats in which alder is found. There are few phytosociological descriptions for the Alnion glutinosae in Ireland (Duff 1930; White 1932; Braun-Blanquet and Tüxen 1952; Klöztli 1970; Kelly and Moore 1975), despite the fact that *Alnus* is frequently recorded from woodland vegetation mosaics (White and Doyle 1982; Browne 1998).

A recent revision of the classification of Irish Alder woodlands, based on an analysis of the literature from both Ireland (Browne 1998) and Great Britain (e.g. Klötzli 1970; Birks 1973; Klein 1975; Wheeler 1980; Rodwell 1991), considered that the majority of Irish alder vegetation should be classified in the Osmundo-Salicetum atrocinereae. Browne (1998) concluded that this association, previously assigned to the Franguletea (Westhoff and Den Held 1969; Wheeler 1980; White and Doyle 1982), was more properly placed within the Alnion glutinosae, of the Alnetalia glutinosae in the Alnetea glutinosae. Under the BNVC scheme, the alder vegetation on Clare Island corresponds with the *Alnus glutinosa-Filipendula ulmaria* subcommunity of *Salix cinerea-Betula pubescens-Phragmites australis* W2 woodland.

The alder, *Alnus glutinosa*, shows a preference for flooded or relatively enriched mineral soils along lowland streams and ditches, although this 'fen' woodland community is more widespread than these limited stands suggest, occurring in many marshy sites on acid substrates in close proximity to birch woodlands at Lassau (Table 27, column 7). This alder community is situated at the base of the Lassau woodland, bordering on an abandoned field system and a small river.

Apart from *Alnus*, the only diagnostic species of the alliance are *Iris pseudacorus* and *Osmunda regalis*. Other constant species include *Glyceria fluitans, Poa trivialis* and *Agrostis stolonifera*. Tall growing *Iris pseudacorus, Lythrum salicaria* and *Filipendula ulmaria* are found here growing in conjunction with moderately tall grasses and rushes. These taller forbs are intimately associated with the adjacent wetland complex of the neglected grassland belonging to the Junco conglomerati-Molinion, described elsewhere in the text. The lush vegetation, with each of these species generally contributing between 15% to 30% ground cover, prevents the development of a rich herb layer. Unlike typical Osmundo-Salicetum atrocinereae communities (Browne 1998), bryophytes do not contribute greatly to the current community. Those present include *Hypnum cupressiforme, Thuidium tamariscinum, Eurhynchium praelongum, Brachythecium rutabulum* and *Ulota crispa*.

Comparative changes in the vegetation of Clare Island from 1911–1996

The vegetation map (fold-out map 2) shows the native communities identified through the vegetation analysis described in the previous section of this paper and also the agricultural land, which was not separated into its constituent communities (although Praeger (1911) characterised several floristic communities). The distribution of the vegetation units illustrates the major changes that occurred in the vegetation since 1911. Direct comparisons between the vegetation units used by Praeger (1911) and those used in the current vegetation map are listed in Table 28. Despite differences in scale and nomenclature of the vegetation units,

Table 28
Comparison of the vegetation divisions included on Praeger's (1911) vegetation map with those utilised in the current survey

Vegetation divisions (1911)		Main divisions and subdivisions (current survey)		
	①	Beach complex		
Saltmarsh		Saltmarsh community		
		Plantain sward	*	*Armeria* community
Plantago sward			*	Pure sward
			*	Grass-rich sward
	①	Intermediate Coastal/Plantain		
	①	*Phragmites* reed swamp		
Natural Grassland		*Calluna* heath	*	Dry heath
			*	Wet heath
		Agrostis-Festuca grassland	*	Coastal subtype
			*	Species-poor
			*	Damaged
Calluna formation			*	Dry grassland
			*	*Calluna* rich
			*	*Ulex* community
			*	*Racomitrium* community
			*	Agricultural land (lazy bed dominated)
	①	Bracken community		
		Meadow vegetation	*	Wet grazing pasture
Farmland			*	Tall marsh vegetation
			*	Hay fields
			*	Crops
Calluna-Eriophorum	∞			
Calluna-Eriophorum-Juncus	∞			
	③	Bog vegetation	*	Blanket bog
			*	*Schoenus*-rich bog vegetation
			*	Miscellaneous deep peat communities
Arctic-Alpine Vegetation		Cliff community		
Woodland		Woodland communities		

① units were not included on the 1911 map.
∞ units were not recognised during the current survey.
③ distribution of bog communities in the current survey does not correspond with any community identified by Praeger (1911), because these were included in the *Calluna* formation.
* newly recognised community/subdivision in current survey.

the original vegetation map provides an impressive baseline from which comparisons with the present vegetation can be made. There are some similarities in the extent and boundaries of the vegetation. This is in part due to the underlying topography but also the influence of historical land boundaries.

The current distribution map was based primarily upon the vegetation descriptions and field notes, and was enhanced through the use of aerial images. Certain limitations in the presentation of the data may cause confusion in relation to the interpretation of the vegetation descriptions. While the relative proportions of some of the main vegetation types are similar, reflecting both topographical and climatic constraints and the pattern of land use, the detail and extent of certain fragmentary communities prohibits their satisfactory display for cartographic purposes. The most obvious differences between the two maps are related to the resolution of individual units and the nomenclature of the various communities. The format selected for the presentation of information was adopted as it better reflects the complex vegetation mosaic that occurs on Clare Island, particularly outside of the agricultural land, and reflects the modern phytosociological methodology employed, which allows distinction between communities on broad floristic criteria, rather than the reliance on dominance, which underlies the vegetation descriptions presented by Praeger (1911).

Beach and saltmarsh vegetation

There is little doubt that the area occupied by these maritime vegetation units has not altered significantly. The saltmarsh community at Kinnacorra, situated behind the distinctive boulder beach in the extreme east of the island, has not expanded outside its original range. Its distribution is connected with the gathering of salt water directly behind the boulder beach, with further extension beyond this range prevented by grazing, which occurs to the edges of the marsh.

The beach community was characterised in the original description of the vegetation, but was not included on the vegetation map. The overall composition of the vegetation is similar to that described in 1911. Despite the constant reworking of the sediment, mediated by wind and wave, the extent of the beach has remained relatively similar.

Plantago sward

The *Plantago* sward first described from Clare Island (Praeger 1903; 1911) is concentrated on the western end of the island and also around the lighthouse at the northernmost tip of the island. The original vegetation description consisted of two sub-communities, the pure plantain sward and a second sub-community, with a greater diversity of acid grassland species. Praeger (1911) mapped both as a single vegetation unit. The vegetation graded into the natural grassland, which occurs as a restricted community along coastal areas on Clare Island not occupied by the plantain vegetation. Some of the natural grassland community from the original survey overlaps with the grass-rich plantain sward recognised in this survey. The land occupied by the *Armeria*-rich community is small, surviving only where fencing prohibits grazing.

The area occupied by the plantain sward has increased since it was previously mapped in 1911. While there is little difference in the extent of the pure sward, the grass-rich community has extended its range, probably due to the removal of scraw turf and consequent heavy grazing of the dwarf vegetation as much as to the maritime influence.

Reed swamp

The expansion of *Phragmites australis* is noteworthy, since at the time of the original survey it only occurred in deep water, forming a fringe around Lough Avullin. It has since fully colonised the lake, forming a dense reed swamp that is quite distinct from the surrounding landscape. A second smaller area of reeds occurs alongside the road from the harbour towards the lighthouse. While it has expanded its range over the past number of years, it is not possible to comment whether it was extant at the time of the original survey.

Calluna formation

The most significant vegetation changes that have occurred on Clare Island have been to the *Calluna* formation. This formation, as mapped by Praeger, occupied the majority of the commonage, later calculated as 49% of the total land area (Doyle and Foss 1986a).

There is considerable evidence that heather-dominated communities are declining in extent across north-western Europe (Thompson *et al.* 1995). Stevenson and Birks (1995) summarise data from a number of sources showing a decline in the relative abundance of *Calluna* pollen over the last 200 years. Hypotheses that have been advanced for the decline include grazing, afforestation, climate changes and atmospheric pollution.

The change to pastoralism and the massive increase in sheep numbers has resulted in this heather-dominated community being replaced by species-poor *Agrostis-Festuca* grassland. This is common in densely stocked areas, especially in the western Irish counties, resulting in the alteration of upland vegetation. This fragmentation and loss of species diversity is well documented, with comparable vegetation replacing much of the heath and moorland communities in many upland areas throughout Ireland (O'Sullivan 1982; MacGowan and Doyle 1996; 1997; McKee *et al.* 1998; Bleasdale 1998), and the British Isles (Ball *et al.* 1982; Ratcliffe and Thompson 1988; Bunce 1989; Thompson *et al.* 1995a; 1995b; 1995c).

Sheep preferentially graze grasses (MacDonald 1990; Thompson *et al.* 1995a), only turning to the woodier heather when the grass resource is depleted. This usually occurs in late summer and early autumn (Grant *et al.* 1976; 1987; 1996a; 1996b). Sheep show a preference for grassy patches with a lower proportion of dead vegetation (Miles 1988). The introduction of hardier sheep breeds has resulted in less selective grazing, so that heather located near patches of grass is more heavily grazed than more distant heather (Grant *et al.* 1985; 1987; MacDonald 1990).

Heather grows from shoot tip meristems and is relatively slow growing in comparison with grasses that may out-compete *Calluna*, owing to the damage caused by excessive sheep numbers. Grasses are quicker to recover from the effects of grazing, as growth is from basal meristems. The last remnants of *Calluna* vegetation are confined to the east of Clare Island and along some gullies in riverbeds. The heather has survived only where sheep have been excluded from the land by means of fences, or in inaccessible areas. Most of the shrub-dominated community is dry heath with an extensive cover of *Calluna vulgaris*, similar in appearance to many heathlands in the British Isles.

Small enclaves of dry heathland vegetation remain in upland areas where access is prevented either through fencing or on inaccessible cliffs. The majority of the extant dry heathland community is confined to the eastern side of Clare Island, where the land is relatively more sheltered and where there are considerably fewer sheep than elsewhere on the commonage. There has also been an attempt, using fencing, at halting sheep damage through overgrazing and poaching, ensuring the survival of the intact heathland vegetation. Attempts at promoting heather regeneration through the reduction of sheep numbers started in the late 1980s (Ballytoohy Beg) has resulted in the rapid establishment of moderately tall heather vegetation. The developing wet heath vegetation is interspersed with *Molinia caerulea*. Photographs from the original survey suggest that much of the wetter commonage, particularly at lower elevations, supported similar wet heath vegetation.

Both *Agrostis capillaris* and *Festuca ovina* are nutritionally important in the hill pastures that are extensively grazed by sheep and are well suited to the cool, damp climate (Grime *et al.* 1988). These grasses account for a considerable proportion of the herbivore diet during the summer months, as sheep tend not to forage in the shrubby heather until the grass resource is depleted in late autumn (Gimingham 1989). However, because of overstocking of sheep on Clare Island and the lack of sufficient grazing pasture, heather is grazed earlier in the season. It does not get a chance to regenerate, thus favouring the spread of more acidic grasses.

Most of the *Agrostis-Festuca* grassland retains remnants of its heathland past, particularly in the centre of the island, where there is an abundance of heather shoots, both dead and alive. On some of the drier slopes on the commonage, particularly Knockmore, the vegetation is comparable with that described from agricultural land. It often grades into coastal grassland, separated on the basis of soil composition, rather than floristic organisation. Within the dry grassland other subcommunities similar to the dry grassland vegetation exist. These are found in areas often underlain by scree material, especially at higher elevations. Vegetation dominated by *Racomitrium lanuginosum* is found at one site on rock debris deposited after field clearance. Isolated patches of *Ulex europaeus* vegetation have developed at a few locations on relicts of drier heath vegetation left on drumlins. *Ulex europaeus* vegetation occurs on raised ground, with extremely shallow soils, generally 20cm or less.

Exploitation of the rough grasslands has led to the dramatic increase of *Nardus stricta*, often co-dominant with *Agrostis* and *Festuca*, or succeeding them as the dominant graminoid species in severely degraded sites. It is a common constituent of grass and heath vegetation on relatively free-draining acidic soils. The rush *Juncus squarrosus* displays similar trends to *Nardus* in that it is indicative of severe grazing pressures, but is most abundant in upland areas of high rainfall. The current management regime in the commonage (year round heavy grazing) on wet, degraded soils favours its continued expansion.

Bracken community

Although listed by Praeger as an extensive component of the flora (Praeger 1911, p. 41), no description of a bracken community is given in the original botany paper, nor was it included on the vegetation map. Bracken is an invasive species and is known to spread rapidly under favourable conditions. The spread of bracken is not confined to abandoned farmland on Clare Island, and it occurs on derelict land in the commonage. This distinctive vegetation type forms dense stands in drier acid grasslands and is often found on abandoned potato drills.

Natural grassland

The natural grassland community described by Praeger (1911) extended from the western end of the island around the northern cliffs to the lighthouse. Other outliers of the grassland community were mapped along a green road on the eastern side of Knocknaveen, at an abandoned farm settlement. Had the agricultural land been divided into constituent vegetation units during Praeger's time, there is little doubt that some of the land lying outside earthen field boundaries along the southern coast could have been included in this group. The majority of Praeger's natural grassland corresponds with the coastal subtype of the Achilleo-Festucetum described in the current survey. It occurs in situations that are less exposed than the plantain sward. This is most clearly seen in the north of the island, where the vegetation grades from the grass-rich plantain sward to intermediate coastal/plantain grassland to dry coastal grassland. Another difference between the present vegetation and that described by Praeger (1911) is related to increased numbers of sheep. A large part of the community outside of the fence along the north-western slopes of Knockmore has been overgrazed, with a consequent increase in the abundance of *Juncus squarrosus* and *Nardus stricta*. This vegetation is also home to the majority of the island's rabbit population.

Farmland

The land given over to agriculture on Clare Island has not changed in extent since Praeger surveyed the island (Praeger 1911). The limit of present day farming is found along the south coast and in the north-eastern corner of the island. For this reason the farmland has been included on the present vegetation map, as there has been little real change to the overall composition of the vegetation.

The majority of land not under tillage was similar in composition to the current acid grassland. With the change to pastoral farming around the turn of the twentieth century, land formerly under tillage reverted to *Agrostis capillaris - Festuca ovina* grasslands, similar in composition to the moderately species-rich dry upland grassland. This land is not subjected to similar grazing pressures as the commonage, as it is usually supports fewer sheep or is used during the summer months.

There is a considerable degree of variability within this vegetation unit, reflecting grazing pressure, soil composition, drainage and management regime. Around margins of the agricultural land, heathy vegetation comparable to that encountered on the commonage is present. In other areas, dwarf heather becomes a major component of the agricultural grassland. This reversion to precursor heath vegetation occurs only where the grazing levels are minimal or absent.

Currently, little arable farming is carried out on Clare Island, with harvests destined primarily for domestic use. Hay is also gathered in a few locations for winter feed. The former extent of arable land is easily recognised by the numerous lazy beds (deserted potato drills). The shallow, nutrient-poor soils relied heavily upon fertiliser inputs (sand and seaweed) to yield any significant returns.

Some agricultural land has been underutilised due to population decreases. The invasion of bracken is quite common in some of the drier lazy beds where grazing animals are no longer kept and on some freely draining slopes on the commonage. Bracken's establishment has been so successful as to warrant its inclusion on the vegetation map. In wetter situations, rushes are a noticeable feature of the vegetation. They are best represented in the marsh vegetation and damp meadows described in this survey and are found within the confines of the agricultural land, usually situated in derelict pasture and along roadside ditches. Some of the derelict sites are indicative of pre-Famine settlements that have been abandoned and now only support a few cattle.

***Calluna-Eriophorum* and *Calluna-Eriophorum-Juncus* vegetation**

In Praeger's account of the moorland, most of the vegetation was categorised in the *Calluna* formation. The only 'true bog' communities, based on current definitions, were confined to the summit of Knockmore. Had Praeger divided the agricultural land, there is no doubt that much of the Maum townland would have had similar vegetation. Both the *Calluna-Eriophorum* and *Calluna-Eriophorum-Juncus* formations were found on the peats that accumulated on top of Knockmore. The peats were deeper than the surrounding *Calluna* dominated slopes. The *Calluna-Eriophorum* vegetation was confined to the summit of Knockmore, while the *Calluna-Eriophorum-Juncus* vegetation occurred on the ridge that runs along the top of Knockmore towards the western end of the island. Presently the two communities, while certainly occurring on deeper peats, are similar in composition to the rough grazings of the commonage. Along the exposed ridge, waterlogged soils and grazing has resulted in the vegetation being replaced with a *Juncus squarrosus*-dominated community. Interestingly, severe weather conditions reported in 1946 led to erosion of the vegetation at the western end of the Knockmore ridge. The climatic erosion of the *Calluna-Eriophorum-Juncus* vegetation revealed peat thickness in excess of three metres.

The limit of present day bog vegetation is more extensive, reflecting the different classification scheme used and the resolution employed in the production of the map. Most bog vegetation is situated in valleys between drumlins and on impervious glacial outwash on the eastern side of the island, in the townlands of Maum and Ballytoohy More. Other outliers occurring on deeper peats are situated in topographical depressions in the commonage. The vegetation is very much fragmented, consisting of a complex mosaic of bog-pools, cutaway hollows, drainage channels and revegetated areas once harvested for turf.

Blanket bog vegetation, dominated by *Molinia caerulea*, occurs mainly on the north-eastern side of the island below the lighthouse. Other smaller patches of blanket bog occur but are too insignificant to be included on the map. In recent times, some of this area occupied by blanket bog has been fenced off in order to allow it recover from the severe poaching.

Vegetation dominated by *Schoenus nigricans* has been mapped separately from Atlantic blanket bog (of which it is a character species), as it is rarely a major component of blanket bog vegetation on Clare Island. It is indicative of flushed areas, where a moderately base-rich vegetation has developed. It is sometimes associated with closely cropped grassland (although the area occupied by this *Schoenus*-rich grassland precludes its inclusion on the vegetation map).

Cliff vegetation

The impressive cliffs on the northern face of Knockmore are home to the surviving vegetation that commonly existed during the last glacial period. The distribution of the 'Arctic-Alpine' vegetation is not given as significant a status as the community described by Praeger (1911). The treacherous nature of this habitat prohibited a comprehensive examination of some of the cliff faces. Despite efforts to locate this unusual element of the Clare Island flora, the frequency of many of the species listed by Praeger (1911) has decreased, and several notable species including *Asplenium viride, Cystopteris fragilis, Salix herbacea* and *Rubus saxatilis* were not relocated during fieldwork. *Polystichum lonchitis* and *Saussurea alpina* were relocated after fieldwork had finished (C. Farrell, pers. comm.). The '1200ft' path described by Praeger (1911), where many of these interesting plants occurred, has eroded as a result of natural causes. Many of the Arctic-Alpine species relocated in the present survey were not abundant or distributed in well-defined communities. These included *Oxyria digyna, Silene acaulis, Saxifraga oppositifolia, S. rosacea* and *S. spathularis*. Most were recorded along seepage zones and gullies on the sheer cliffs and or scattered through the *Luzula*-dominated cliff community of Knockmore. Interestingly, some plants such as *Saxifraga rosacea* and *Oxyria digyna* were recorded in some scree-lined gullies that descended to sea level.

Woodland

Tree-dominated vegetation is no longer extensive on Clare Island. There is only one area supporting scrub woodland vegetation of appreciable size, with a smaller patch of scrub vegetation located behind Lough Avullin. This is despite the fact that much of the land was once afforested. Two other small areas of woodland recorded in the south-eastern section (Praeger 1911) are no longer extant, except possibly as fragments along roadside ditches. Smaller areas of scrub consisting of trees are scattered in ditches and small isolated patches are not included on the map.

Marrs and Hicks (1986) working in the Breckland of East Anglia consider birch a threat to open heathland areas. Birch seedlings are restricted to areas of bare soil, especially where there are gaps in the heath/forest canopy (Grime *et al.* 1988). However, this appears to be only in disturbed habitats and not ancient woodland. The persistence of the woodland is constrained by the physical boundaries of Lassau valley and the suitability of the humic soils. Examination of early photographs from the original survey and aerial photographs from 1973, combined with the present research, suggest that there has been a slight increase in area occupied by the woodland, which now occupies a little under a quarter of the valley at Lassau.

Discussion

It is difficult to interpret the vegetation of Clare Island without recognising the historical influences, both social and agricultural, that have moulded the island's landscape. The first people to arrive on Clare Island encountered a landscape with vegetation dominated by trees in the low-lying areas and shrub-dominated heathland on the slopes. The forests were gradually felled to provide fuel and building materials for shelter for these visiting hunter-farmers. Forest clearance accelerated with the arrival of the first permanent, Neolithic farming communities. The areas that were first settled and cultivated were located along the south coast of Clare Island, where conditions were favourable—sheltered with moderately fertile, moderately drained soils that coincided with areas overlain by glacially derived soils.

Clare Island was in a state of profound flux at the time of the original survey, undergoing significant changes in both social and agricultural structures. Land redistribution by the Congested Districts Board (1894 to 1901) altered many of the original field boundaries, apportioning much of the land equally among farmers. The island's sheep population was approximately 2000 in 1905 and even then was beginning to have an influence upon the vegetation, particularly in the commonage. Praeger commented upon the change between the time of his first visit to Clare Island (in 1903) and his return for the original survey (1909 to 1911)—'The flora here [the moorland] has been disturbed by grazing, by burning, and by turf cutting; and grazing in particular has probably greatly altered the appearance—if not so much the flora—of the far extending moorland formation'. Despite the increase in grazing pressures, the tall heather (up to one metre) remained on the slopes of Knockmore, often obscuring sheep from sight (see Doyle and Foss 1986a; also anecdotal reports) until the middle of this century.

Prior to 1975, sheep farmers were not grant aided and sheep numbers remained fairly static throughout the country—ranging between three and four million (Table 29). An increase seen between 1960 and 1965 was due to an increased national and international demand for lamb. Thereafter, since Ireland's entry into the European Union, there has been a rapid increase in the sheep flock throughout the country. The introduction of headage grants promoted dramatic increases in sheep population, with a near doubling of the national herd between 1980 to 1991 (Table 29). An underlying trend recognised in western counties was that over the same timeframe, sheep farming had become a primary source of income for many (Mitchell 1986; Bleasdale 1998). Environmental and economical concerns at the impacts and spiralling costs of subsidising this mode of agriculture at the end of the twentieth century have coincided with a steady decrease in sheep numbers, with the national total in 2009 at a little over three million.

The data for Clare Island (Table 29) show that sheep numbers had risen from *c.* 2500 in 1965 to a peak of 6200 in 1975, and had steadied at about 5500 during the years that the survey was carried out. Notwithstanding the fact that the latest figures for Clare Island (CSO, due 2013) are not yet available, it is fair to say that in the intervening years since the new survey was carried out that there has been a decrease in the number of sheep on Clare Island, which reflects national trends.

With sheep continuously on the land, the vegetation is given little time to regenerate. Grazing animals tend to affect upland plant communities in a variety of ways: (1) plant defoliation,

Table 29

Total sheep numbers for Ireland (from 1965 to 2009) and Clare Island (1965 to 2000). Data from the Central Statistics Office agricultural census (Central Statistics Office 2013)[1].

Year of Census	Ireland (26 Counties)	Clare Island (District Electoral Division)
1965	5,013,706	2433
1970	4,082,253	4568
1975	3,682,680	6174
1980	3,291,522	5324
1991	8,888,204	5426
2000	5,056,000	5615[1]
2009	3,182,600	

[1] The last census for Clare Island District Electoral Division (for 2010), due to be published in autumn 2011, is not yet available.

(2) nutrient removal and redistribution and (3) mechanical impacts on soils and plant material.

On Clare Island, sheep are left unattended on the commonage for a considerable portion of the year. They range over wide areas in loose flocks on commonage (other than young lambs who tend not to stray too distant), resulting in localised nutrient enrichment. Trampling by sheep is generally not considered widespread in upland communities (Ausden and Treweek 1995), confined mainly to areas near tracks or gates (Thompson *et al.* 1995a). However, poaching has been recognised as a factor in soil erosion in wetter peatland habitats (MacGowan and Doyle 1996). The adverse weather conditions of the winter months compound this situation, with increased water runoff resulting in the development of terraced landscapes and ground dominated by carpets of *Sphagnum* mosses. As a consequence the *Calluna*-dominated vegetation that was once extensive has been replaced, due in large part to excessive numbers of foraging sheep replacing and maintaining it with a mosaic of species-poor dwarf grassland communities.

Praeger's comprehensive account of the flora of Clare Island, together with the vegetation map published in 1911 provide a baseline that has enabled an assessment of the changes that have occurred over the past eighty-five years. Clare Island had a more extensive flora than most areas of the adjacent mainland, of equal dimensions, with its insularity being more than compensated for by the variety of habitats that it afforded (Praeger 1903; 1911). There has been an overall reduction in plant species numbers since the original Clare Island Survey. A total of 384 higher plant species were recorded during the current survey, representing a total reduction of 10% in the diversity of the flora on the island since Praeger's time. This value increases to 25%, if species newly recorded since Praeger's survey are discounted. This represents a significant reduction in the higher plant diversity and highlights fundamental changes that have occurred to the island's vegetation. The reduction in plant species diversity has also been reported from comparable island habitats (Webb and Hodgson 1968; Brodie and Sheehy-Skeffington 1990) and some mainland situations (Doyle and Foss 1986a; Bleasdale and Sheehy-Skeffington 1995) along the west coast of Ireland and is largely attributable to intensive human activities. Jebb (this volume, p. 181) estimates that there has been a significant turnover in the flora in the order of 10%–20% or more per century.

The failure to find some species commonly associated with arable crops during the current survey is not surprising, since tillage as a sustainable method of farming was largely abandoned. The reduction in the extent of arable land resulted in the continuation of tillage only on a small scale, such as in kitchen gardens. The use of modern weed control, coupled with cleaner seed stock has been suggested as an additional cause for the decline of the weedy flora (Doyle and Foss 1986a).

The most dramatic change has occurred in the upland vegetation, where shrub-dominated heathlands were originally widespread and where many of the shade-tolerant species formerly associated with the heather canopy have not been recorded. The *Calluna* formation described by Praeger, encompassing the majority of the commonage vegetation, has been severely altered and almost entirely replaced by severely grazed and poached rough grassland dominated by admixtures of *Agrostis capillaris*, *Festuca ovina* and *Nardus stricta*.

Climate, management and historical land-use, for which there are few quantitative results, have had an impact upon the pattern of the vegetation mosaic on Clare Island. There has been no significant change in the long-term climate patterns in the west of Ireland in general, and Clare Island in particular. The effects of atmospheric pollution that have been variously detailed from the United Kingdom and central Europe (e.g. Bobbink *et al.* 1998) are not experienced on Clare Island, as it is too remote from any sizeable centre of population or industrial area to experience any negative atmospheric effects on the vegetation—Galway, 50km to the south-east has a population of 25,000 (Boyle *et al.* 1993).

While Clare Island has undergone significant changes in the overall vegetation composition and physiognomy (e.g. reduction in height), in particular in upland areas, it has not suffered from any appreciable vegetation degradation or subsequent large-scale erosion. Erosion, either by natural causes, or as a result of loss of vegetation integrity through grazing pressure, is not widespread, unlike other upland and blanket mire habitats from nearby Galway and Mayo (Bleasdale and Sheehy-Skeffington 1992; 1995; MacGowan and Doyle 1996; 1997; Bleasdale 1998; McKee *et al.* 1998). The degradation of the vegetation is primarily associated with the level of stocking. The areas where erosion has occurred on Clare Island have been subjected to the removal of vegetation and to extreme climatic conditions, particularly rainfall. Local accounts described a bog burst in 1946 that had led to some sheet erosion on the flanks of Knockmore. Some of the trails along the steep northern cliffs of Knockmore have eroded through natural subsidence.

Evidence of peat-cutting is widespread, although it is not directly linked to the distribution of human settlements. Cutaways vary in extent, although most are small, as access to the 'extensive' ombrotrophic peats is restricted to just a few families. In past times, turf gathered from small cutaways was adequate for most average households, but the combination of constant peat removal and the opening of associated drainage channels has led to significant modifications of the peatland habitats since the island was first inhabited.

Concluding Remarks

It is worth noting that the habitat descriptions and their portrayal on the accompanying map represent a point in time after four years of survey. In the intervening years, there have been changes in the management and hence distribution of some habitats. Now a full century on, Clare Island is entering another period of change, albeit quite different to that described by Robert Lloyd Praeger during the original Clare Island Survey. The reduction of sheep numbers on Clare Island in the past number of years and their withdrawal from parts of the commonage must be favourably considered. With the reduction in the sheep population and the associated grazing pressure, the commonage is less heavily impacted, particularly during the winter months. It may be some years before the effects of the reliance on headage subsidies are erased and the impacts of other more environmentally beneficial schemes on the upland ecology of Clare Island can be assessed. The work detailed in this paper will hopefully provide a basis for future comparative work where such impacts can be assessed, particularly in light of the growing literature describing habitats similar to those detailed here.

REFERENCES

Adam, P. 1977 On the phytosociological status of *Juncus maritimus* on British saltmarshes. *Vegetatio* **35**, 81–94.

Adam, P. 1978 Geographical variation in British saltmarsh vegetation. *Journal of Ecology* **66**, 339–66.

Ausden, M. and Treweek, J. 1995 Management of grasslands: techniques. In W.J. Sutherland and D.A. Hill (eds), *Managing habitats for conservation*, 205–17. Cambridge. Cambridge University Press.

Averis, A., Averis, B., Birk, J., Horsfield, D., Thompson, D. and Yeo, M. 2004 *An illustrated guide to British upland vegetation*. Peterborough. JNCC.

Bailey, M.L. 1984 *Air quality in Ireland, the present position*. Dublin. An Foras Fórbatha.

Ball, D.F., Dale, J., Sheail, J. and Heal, O.W. 1982 *Vegetation change in upland landscapes*. Bangor. Institute of Terrestrial Ecology.

Barkmann, J.J., Moravec, J. and Rauschert, S. 1986 Code of phytosociological nomenclature. *Vegetatio* **67**, 145–58.

Beckers, A., Brock, T.H. and Klerkx, J. 1976 A vegetation study of some parts of Dooaghtry, Co. Mayo, Republic of Ireland. Unpublished MSc thesis, Laboratory for Geobotany, Catholic University, Nijmegen.

Birks, H.J.B. 1973 *Past and present vegetation of the Isle of Skye. A palaeoecological study*. Cambridge. Cambridge University Press.

Birks, H.J.B., Birks, H.H. and Ratcliffe, D.A. 1969 Mountain plants on Slieve League, Co. Donegal. *Irish Naturalists' Journal* **16**, 203

Birse, E.L. 1980 *Plant communities of Scotland: a preliminary phytocoenonia*. Aberdeen. Macaulay Institute for Soil Research.

Birse, E.L. and Robertson, J.S. 1976 *Plant communities and soils of the Lowland and southern Uplands regions of Scotland*. Aberdeen. Macaulay Institute for Soil Research.

Bleasdale, A. 1998 Overgrazing in the west of Ireland—assessing solutions. In G. O'Leary and F. Gormley (eds), *Towards a conservation strategy for the bogs of Ireland*, 67–78. Dublin. Irish Peatland Conservation Council.

Bleasdale, A. and Conaghan, J. 1995 Flushes and springs in the Connemara hills and uplands, Co. Galway, Ireland. *Bulletin of the British Ecological Society* **26**, 28–35.

Bleasdale, A. and Sheehy-Skeffington, M. 1992 The influence of agricultural practices on plant communities in Connemara. In J. Feehan (ed.), *Environment and Development in Ireland*, 331–6. Dublin. University College Dublin Environment Institute.

Bleasdale, A. and Sheehy-Skeffington, M. 1995 The upland vegetation of north-east Connemara in relation to sheep grazing. In D.W. Jeffrey, M.B. Jones and J.H. McAdam (eds), *Irish grasslands—their biology and management*, 110–24. Dublin. Royal Irish Academy.

Bobbink, R., Hornung, M. and Roelofs, J.G.M. 1998 The effects of air-borne pollutants in natural and semi-natural European vegetation. *Journal of Ecology* **86**, 717–38.

Boyle, G.M., Farrell, E.P. and Cummins, T. 1997 *Monitoring of forest ecosystems in Ireland, FOREM 3 Project Final Report*. Forest Ecosystem Research Group Number 21. Dublin. Department of Environmental resource management, UCD.

Braun-Blanquet, J. and Tüxen, T. 1952 Irische Plflanzengesellschaften. *Veroffentlichungen des Geobotanischen Institutes Rübel in Zurich* **25**, 224–415.

Brock, T., Frigge, P. and van der Ster, H. 1978 A vegetation study of the pools and surrounding wetlands in the Dooaghtry area, Co. Mayo, Republic of Ireland. Unpublished MSc thesis, Laboratory for Geobotany, Catholic University, Nijmegen.

Brodie, J. 1991 Some observations on the flora of Clare Island, western Ireland. *Irish Naturalists' Journal* **23** (9), 376–7.

Brodie, J. and Sheehy-Skeffington, M. 1990 Inishbofin: a resurvey of the flora. *Irish Naturalists' Journal* **23**, 293–8.

Browne, A. 1998 Vegetation–environment interactions in the vicinity of a pharmaceutical plant near Kinsale, Co. Cork. Unpublished PhD thesis, University College Dublin.

Browne, J.F. 1991 The glacial geomorphology of Clare Island, Co. Mayo. Unpublished BA thesis, University of Dublin, Trinity College.

Bunce, R.G.H. 1989 *Heather in England and Wales*. Institute of Terrestrial Ecology, Research publication No. 3. London. Her Majesty's Stationery Office.

Carter, R.W.G. 1988 *Coastal environments: An introduction to the physical, ecological and cultural systems of coastlines*. London. Academic Press.

Caulfield, S. 1978 Neolithic fields: the Irish evidence. In H.C. Bowen and P.J. Fowler (eds), *Early land allotment in the British Isles*, 137–43. British Archeological Reports 48. Oxford.

Caulfield, S. 1983 The Neolithic settlement of N. Connaught. In T. Reeves-Smith and F. Hammond (eds), *Landscape archaeology in Ireland*, 195–215, British Archeological Reports 116. Oxford.

Central Statistics Office 2013 Available at www.cso.ie (last accessed on 18 February 2013).

Cole, G.A.J., Kilroe, J.R., Hallissy, T. and Newell Arber, E.A. 1914 *The geology of Clare Island, Co. Mayo*. Memoirs of the Geological Survey of Ireland. Dublin. HMSO.

Cooper, E., Crawford, I., Malloch, A.J.C. and Rodwell, J. 1992 *Coastal vegetation survey of Northern Ireland*. Lancaster. Unit of Vegetation Science report to the Department of the Environment (Northern Ireland).

Cotton, J. 1975 The National Vegetation Survey of Ireland: Nardo-Callunetea. *Colloques Phyto-sociologiques* **2**, 237–44.

Coxon, P. 1987 A post-glacial pollen diagram from Clare Island, Co. Mayo. *Irish Naturalists' Journal* **22**, 219–23.

Coxon, P. 1994 The glacial geology of Clare Island. In P. Coxon and M. O'Connell (eds), *Clare Island and Inisbofin*. Field Guide No. 17. Dublin. Irish Association for Quaternary Studies.

Coxon, P. 2001 The Quaternary history of Clare Island. In John R. Graham (ed.), *New Survey of Clare Island. Volume 2: geology*, 87–112. Dublin. Royal Irish Academy.

Coxon, P. and O'Connell, M. 1994 *Clare Island and Inishbofin*. Dublin. Irish Association for Quaternary Studies.

Cullen, C. and Gill, P. 1991 *Studying an island: Clare Island, Co. Mayo*. Clare Island Series 1. Clare Island. Centre for Island Studies.

Curran, P.L., O'Toole, M.A. and Kelly, F.G. 1983 Vegetation of terraced hill grazings in north Galway and south Mayo. *Journal of Life Sciences, Royal Dublin Society* **4**, 195–202.

Curtis, T.G.F. and McGough, H.N. 1988 *The Irish Red Data Book 1. Vascular plants*. Dublin. Stationery Office.

Dargie, T.C.D. 1993 *Sand dune vegetation survey of Great Britain: a national inventory Part II: Scotland*. Peterborough. Joint nature Conservancy Committee.

Dierßen, K. 1978 Die wichtigsten Pflanzengesellschaften der Moore NW-Europas. Doctoral thesis, Freiburg.

Doyle, G.J. 1982 Narrative of the excursion of the international society for vegetation science to Ireland, 21–31 July 1980. *Journal of Life Sciences, Royal Dublin Society* **3**, 43–64.

Doyle, G.J. 1982 The vegetation, ecology and productivity of Atlantic blanket bog in Mayo and Galway, western Ireland. *Journal of Life Sciences, Royal Dublin Society* **3**, 147–64.

Doyle, G.J. 1990 Phytosociology of Atlantic blanket bog complexes in north-west Mayo. In G.J. Doyle (ed.), *Ecology and conservation of Irish peatlands*, 75–90. Dublin. Royal Irish Academy.

Doyle, G.J. 1997 Blanket bogs: an interpretation based on Irish blanket bogs. In L. Parkyn, R.E. Stoneman and H.A.P. Ingram (eds), *Conserving peatlands*, 25–34. Oxford. CAB International.

Doyle, G.J. and Foss, P.J. 1986 A resurvey of the Clare Island flora. *Irish Naturalists' Journal* **22**, 85–9.

Doyle, G.J. and Foss, P.J. 1986 *Vaccinium oxycoccus* L. growing in the blanket bog area of West Mayo (H27). *Irish Naturalists' Journal* **22**, 101–4.

Doyle, G.J. and Moore, J.J. 1980 Western blanket bog (*Pleurozio purpureae–Ericetum tetralicis*) in Ireland and Great Britain. *Colloques Phytosociologiques* **7**, 217–23.

Doyle, G.J. and Whelan, S. 1991 Proposal for a new survey of Clare Island: 1991–1995. Unpublished report to Royal Irish Academy, Dublin.

Doyle, G.J., O'Connell, C.A. and Foss, P.J. 1987 The vegetation of peat islands in bog lakes in County Mayo, western Ireland. *Glasra* **10**, 23–5.

Duff, M. 1930 The ecology of the Moss Lane region, Lough Neagh. *Proceedings of the Royal Irish Academy* **39**B, 477–96.

Evans, E.E. 1957 *Irish folkways*. London. Rutledge and Kegan.

Farrell, E.P., Cummins, T., Boyle, G.M., Smillie, G.W. and Collins, J.F. 1993 Intensive monitoring of forest ecosystems. *Irish Forestry* **50**, 70–83.

Foss, P.J. 1986 The distribution, phytosociology, autecology and post glacial history of *Erica erigena* R.Ross in Ireland. Unpublished PhD thesis, University College Dublin.

Foss, P.J. and Doyle, G.J. 1988 Why has *Erica erigena* (the Irish heather) such a markedly disjunct European distribution? *Plants Today* 161–8.

Foss, P.J., Doyle, G.J. and Nelson, E.C. 1987 The distribution of *Erica erigena* R. Ross in Ireland. *Watsonia* **16**, 311–27.

Fossitt, J.A. 2000 *A guide to habitats in Ireland*. Kilkenny. The Heritage Council.

Gaynor, K. 2008 The phytosociology and conservation value of Irish sand dunes. Unpublished PhD thesis, University College Dublin.

Géhu, J.-M. 1975 Essai pour un système de classification phytosociologiques des landes Atlantiques planitiares Françaises. *Colloques Phytosociologiques* **2**, 361–77.

Gillham, M.E. 1953 An ecological account of the vegetation of Grassholm Island, Pembrokeshire. *Journal of Ecology* **41**, 84–99.

Gillham, M.E. 1955 Ecology of the Pembrokeshire Islands III. The effect of gazing on the vegetation. *Journal of Ecology* **43**, 172–206.

Gillham, M.E. 1956a Ecology of the Pembrokeshire Islands IV. Effects of treading and burrowing by birds and mammals. *Journal of Ecology* **44**, 51–82.

Gillham, M.E. 1956b Ecology of the Pembrokeshire Islands V. Manuring by colonial seabirds and mammals, with a note on seed distribution by gulls. *Journal of Ecology* **44**, 429–54.

Gimingham, C.H. 1964 Maritime and submaritime communities. In J.H. Burnett (ed.), *The vegetation of Scotland*, 67–141. Edinburgh. Oliver and Boyd.

Gimingham, C.H. 1989 Heather and heathlands. *Botanical Journal of the Linnean Society* **101**, 263–8.

Goodman, G.T. and Gillham, M.E. 1954 Ecology of the Pembrokeshire Islands II. Skolhom, environment and vegetation. *Journal of Ecology* **42**, 296–327.

Graham, J.R. 1994 Pre-Pleistocene geology of Clare Island. In P. Coxon and M. O'Connell (eds), *Clare Island and Inishbofin*, 8–10. Dublin. Irish Association for Quaternary Studies.

Graham, J.R. (ed.) 2001 *New Survey of Clare Island. Volume 2: Geology*. Dublin. Royal Irish Academy.

Grant, S.A., Lamb, W.I.C., Kerr, C.D. and Bolton, G.R. 1976 The utilisation of blanket bog vegetation by grazing sheep. *Journal of Applied Ecology* **13**, 857–69.

Grant, S.A., Barthram, G.T., Lamb, W.I.C. and Milne, J.A. 1978 Effect of season and level of grazing on the utilisation of heather by sheep. I. Responses of the sward. *Journal of the British Grassland Society* **33**, 289–300.

Grant, S.A., Suckling, D.E., Smith, H.K., Torvell, L., Forbes, T.D.A. and Hodgson, J. 1985 Comparative studies of diet selection by sheep and cattle: the hill grasslands. *Journal of Ecology* **73**, 987–1004.

Grant, S.A., Torvell, L., Smith, H.K., Suckling, D.E., Forbes, T.D.A. and Hodgson, J. 1987 Comparative studies of diet selection by sheep and cattle: blanket bog and heather moor. *Journal of Ecology* **75**, 947–60.

Grant, S.A., Torvell, L., Common, T.G., Sim, E.M. and Small, J.L. 1996 Controlled grazing studies on

Molinia grassland: effects of different seasonal patterns and levels of defoliation on *Molinia* growth and responses of swards to controlled grazing by cattle. *Journal of Applied Ecology* **33**, 1267–80.

Grant, S.A., Torvell, L., Sim, E.M., Small, J.L. and Armstrong, R.H. 1996 Controlled grazing studies on *Nardus* grassland: effects of between-tussock sward height and species of grazer on *Nardus* utilization and floristic composition. *Journal of Applied Ecology* **33**, 1053–64.

Grime, J.P., Hodgson, J.G. and Hunt, R. 1988 *Comparative plant ecology*. London. Unwin Hyman.

Grolle, R. 1976 Verzlichnis der lebermoose Europas und benachbarter gebiete. *Feddes Repertorium* **87**, 171–279.

Hallissy, T. 1914 Clare Island Survey, part 7. Geology. *Proceedings of the Royal Irish Academy* **31**, 1–22.

Hammond, R.F. 1979 *The peatlands of Ireland*. Dublin An Foras Talúntais.

Hanrahan, J. 1997 The effects of grazing on vegetation and soils of an oak woodland in Glendalough, County Wicklow. Unpublished MSc thesis, University College Dublin.

Heil, G.W. and Bruggink, M. 1987 Competition for nutrients between *Calluna vulgaris* (L.) Hull and *Molinia caerulea* (L.) Moench. *Oecologia* **73**, 105–7.

Huntley, B. 1979 The past and present vegetation of the Caenlochan National Nature Reserve, Scotland. I. Present vegetation. *New Phytologist* **83**, 215–83.

Iremonger, S.F. 1986 An ecological account of Irish wetland woods; with particular reference to the principal tree species. Unpublished PhD thesis, University of Dublin, Trinity College.

Iremonger, S.F. 1990 Structural analysis of three Irish wooded wetlands. *Journal of Vegetation Science* **1**, 359–66.

Ivimey-Cook, R.B. and Proctor, M.C.F. 1966 The plant communities of the Burren, Co. Clare. *Proceedings of the Royal Irish Academy* **64**B, 211–301.

Jermy, A.C. and Crabbe, J.A. 1978 *The island of Mull: a survey of its flora and environment*. London. British Museum.

Jordan, C. 1997 Mapping of rainfall chemistry in Ireland 1972–4. *Biology and Environment: Proceedings of the Royal Irish Academy* **97**B, 53–73.

Keane, T. 1986 *Climate, weather and Irish agriculture*. Dublin. Agmet.

Kelly, D.L. and Kirby, E.N. 1982 Irish native woodlands over limestone. *Journal of Life Sciences, Royal Dublin Society* **3**, 181–98.

Kelly, D.L. and Moore, J.J. 1975 Preliminary sketch of the acidophilous oakwoods. In J.-M. Géhu (ed.), *La végétation des forêts caducifoliées acidophiles*, 375–87. Vaduz. Cramer.

Kent, M. and Coker, P. 1992 *Vegetation description and analysis—a practical approach*. Belhaven, London.

Kirby, E.N. and O'Connell, M. 1982 Shannawoneen wood, County Galway, Ireland: the woodland and saxicolous communities and the epiphytic flora. *Journal of Life Sciences, Royal Dublin Society* **4**, 73–96.

Klein, J. 1975 An Irish landscape. A study of natural and semi-natural vegetations in the Lough Ree area of the Shannon basin. PhD thesis, Rijksuniversiteit, Utrecht.

Klotzli, F. 1970 Eichen-, Edellaub- und Bruchwalder der Britischen Inseln. *Schweirerischen Zeitschrift fur Forstwesen* **121**, 329–66.

Lamb, H.H. 1977 *Climate—present, past and future*. Volume 1. London. Methuen.

Lee, J.A. and Caporn, S.J.M. 1998 Ecological effects of atmospheric reactive nitrogen deposition on semi-natural terrestrial ecosystems. *New Phytologist* **139**, 127–34.

Lee, J.A., Caporn S.J.M. and Read, D.J. 1992 Effects of increasing nitrogen deposition and acidification on heathlands. In T. Schneider (ed.), *Acidification research, evaluation and policy applications*, 97–106. Amsterdam. Elsevier Science.

Lewis, S. 1837 *A topographical dictionary of Ireland*.Vol. I. London. Samuel Lewis and Co.

Little, D.J. and Collins, J.F. 1995 Anthropogenic influences on soil development at a site near Pontoon, Co. Mayo. *Irish Journal of Agriculture and Food Research* **34**, 151–63.

Lockhart, N. 1991 The phytosociology and ecology of blanket bog flushes in west Galway and north Mayo. Unpublished PhD thesis, University College Galway.

Lowday, J.E. 1984 The effects of cutting and Asulam on the frond and rhizome characteristics of bracken (*Pteridium aquilinum* (L.) Kuhn). *Aspects of Applied Biology* **5**, 275–82.

Lowday, J.E. and Marss, R.H. 1992 Control of bracken and restoration of heathland. I. Control of bracken. *Journal of Applied Ecology* **29**, 204–11.

MacCárthaigh, C. 1999 Clare Island folklife. In C. MacCárthaigh and K. Whelan (eds), *New survey of Clare Island. Volume 1: History and cultural landscape*, 41–72. Dublin. Royal Irish Academy.

MacDonald, A. 1990 *Heather damage: a guide to the types of damage and their causes*. Peterborough. Nature Conservancy Council.

MacGowan, F. and Doyle, G.J. 1996 The effects of sheep grazing and trampling by tourists on lowland blanket bog in the west of Ireland. In P.S. Giller and A.A. Myers (eds), *Disturbance and recovery in ecological systems*, 20–32. Dublin.Royal Irish Academy.

MacGowan, F. and Doyle, G.J. 1997 Vegetation and soil characteristics of damaged Atlantic blanket bog in the west of Ireland. In J.H. Tallis, R. Meade and P.D. Hulme, *Blanket mire degradation: causes, consequences and challenges. Proceedings of the Mires research Group Conference, University of Manchester 9–11 April 1997*, 54-63. Aberdeen. British Ecological Society and Macaulay Land Use Research Institute.

Malloch, A.J.C. 1971 Vegetation of the maritime cliff-tops of the Lizard and Land's End Peninsulas, West Cornwall. *New Phytologist* **70**, 1155-1197.

Malloch, A.J.C. 1972 Salt-spray deposition on the maritime cliffs of the Lizard peninsula. *Journal of Ecology* **60**, 103–12.

Marrs, R.H. 1987 Studies on the conservation of lowland *Calluna* heaths. I. Control of birch and bracken and its effects on heath vegetation. *Journal of Applied Ecology* **24**, 163–75.

Marrs, R.H. 1993 An assemblage of change in *Calluna* heathlands in Breckland, Eastern England between 1983 and 1991. *Biological Conservation* **65**, 133–9.

Marrs, R.H. and Hicks, M.J. 1986 Study of vegetation change at Lakenheath warren: a re-examination of A.S. Watt's theories of bracken dynamics in relation to succession and vegetation management. *Journal of Applied Ecology* **23**, 1029–46.

Marrs, R.H., Johnson, S.W. and Le Duc, M.G. 1998 Control of bracken and the restoration of heathlands. VI. The response of bracken to 18 years of continued control or 6 years of control followed by recovery. *Journal of Applied Ecology* **35**, 479–90.

Marrs, R.H., Johnson, S.W. and Le Duc, M.G. 1998 Control of bracken and the restoration of heathlands. VII. The response of bracken rhizomes to 18 years of continued control or 6 years of control followed by recovery. *Journal of Applied Ecology* **35**, 748–57.

Marrs, R.H., Johnson, S.W. and Le Duc, M.G. 1998 Control of bracken and the restoration of heathlands. VIII. The regeneration of the heathland community after 18 years of continued bracken control or 6 years of control followed by recovery. *Journal of Applied Ecology* **35**, 857–70.

McCarthy, P.M. 1988 The lichens of Inishbofin, Co. Galway. *Irish Naturalists' Journal* **22**, 403–7.

McFerran, D.H., McAdam, J.H. and Montgomery, W.I. 1994a Seed bank associated with stands of heather moorland. *Irish Naturalists' Journal* **24**, 480–5.

McFerran, D.M., Montgomery, W.I. and McAdam, J.H. 1994b Effects of grazing intensity on heathland vegetation and ground beetle assemblages of the uplands of County Antrim, north-east Ireland. *Biology and Environment: Proceedings of the Royal Irish Academy* **94**B, 41–52.

McKee, A.M., Bleasdale, A.J. and Sheehy-Skeffington, M. 1998 The effects of different grazing pressures on the above-ground biomass of vegetation in the Connemara uplands. In G. O'Leary and F. Gormley (eds), *Towards a conservation strategy for the bogs of Ireland*, 177–88. Dublin. Irish Peatland Conservancy Council,

McVean, D.N. and Ratcliffe, D.A. 1962 *Plant communities of the Scottish Highlands: a study of Scottish mountain, moorland and forest vegetation*. London. Her Majesty's Stationery Office.

Mhic Daeid, C. 1976 A phytosociological and ecological study of the vegetation of peatlands and heaths in the Killarney Valley. Unpublished PhD thesis, University of Dublin, Trinity College.

Miles, J. 1981 Problems in heathland and grasslands dynamics. *Vegetatio* **46**, 61–74.

Miles, J. 1986 What are the effects of trees on soils? In D. Jenkins (ed.), *Trees and wildlife in the Scottish uplands*. 55–62. Huntingdon. Institute of Terrestrial Ecology

Miles, J. 1988 Vegetation and soil change in the uplands. In M.B. Usher and B.A. Thompson (eds), *Ecological change in the uplands*, 57–70. Oxford. Blackwell Scientific Publishers.

Minchin, D. and Minchin, C. 1996 The sea pea *Lathyrus japonicus* (Willd) in Ireland, and an addition to the flora of West Cork (H3) and Wexford (H12). *Irish Naturalists' Journal* **25**, 165–9.

Mitchell, F. 1986 *The Shell guide to reading the Irish landscape*. Dublin. Country House.

Moore, J.J. 1955 The distribution and ecology of *Scheuchzeria palustris* on a raised bog in Co. Offaly. *Irish Naturalists' Journal* **11**, 1–7.

Moore, J.J. 1960 A resurvey of the vegetation of the district lying south of Dublin (1905–1956). *Proceedings of the Royal Irish Academy* **61**B, 1–36.

Moore, J.J. 1968 A classification of the bogs and wet heaths of northern Europe (*Oxycocco–Sphagnetea* Br.-Bl. et Tx. 1943). In R. Tüxen (ed.), *Pflanzensoziologische systematik. Bericht über das internationale symposium in Stozenau/Weser 1964 der internationale vereinigung für vegetationskunde*, 306–20. Den Haag. Junk.

Moore, J.J., Fitzsimons, P., Lambe, E. and White, J. 1970 A comparison and evaluation of some phytosociological techniques. *Vegetatio* **20**, 1–20.

Mueller-Dombois, D. and Ellenberg, H. 1974 *Aims and methods of vegetation ecology*. London. Wiley.

Ní Ghráinne, E. 1993 Palaeoecological studies towards the reconstruction of vegetation and landuse history of Inishbofin, western Ireland. Unpublished PhD thesis, University College Galway.

Ní Lamhna, É. 1982 The vegetation of saltmarshes and sand-dunes at Malahide Island, County Dublin. *Journal of Life Sciences, Royal Dublin Society* **3**, 111–29.

Ó Críodáin, C. 1988 Parvocaricetea in Ireland. Unpublished PhD thesis, University College Dublin.

Ó Críodáin, C. and Doyle, G.J. 1994 An overview of Irish small-sedge vegetation: syntaxonomy and a key to communities belonging to the *Scheuchzerio–Caricetea nigrae* (Nordh. 1936) Tx. 1937. *Biology and Environment: Proceedings of the Royal Irish Academy* **94**B, 127–44.

Ó Críodáin, C. and Doyle, G.J. 1997 *Schoenetum nigricantis*. The *Schoenus* fen and flush vegetation of Ireland. *Biology and Environment: Proceedings of the Royal Irish Academy* **97**B, 203–18.

O'Connell, M. 1977 A palaeoecological and phytosociological study of Scragh Bog, Co. Westmeath. Unpublished PhD thesis, National University of Ireland.

O'Connell, M. 1980 The developmental history of Scragh Bog, Co. Westmeath and the vegetational history of its hinterland. *New Phytologist* **85**, 301–19.

O'Connell, M. 1981 The phytosociology and ecology of Scragh Bog, Co. Westmeath. *New Phytologist* **87**, 139–87.

O'Connell, M. 1990 Origins of lowland blanket bog. In G.J. Doyle (ed.), *Ecology and conservation of Irish peatlands* 49–71. Dublin. Royal Irish Academy.

O'Sullivan, A.M. 1965 A phytosociological survey of Irish lowland meadows and pastures. Unpublished PhD thesis, University College Dublin.

O'Sullivan, A.M. 1976 The phytosociology of the Irish wet grasslands belonging to the order Molinietalia. *Colloques Phytosociologiques* **5**, 259–67.

O'Sullivan, A.M. 1982 The lowland grasslands of Ireland. *Journal of Life Sciences, Royal Dublin Society* **3**, 131–42.

O'Toole, M.A. 1984 *Renovation of peat and hillland pastures*. Dublin. An Foras Talúntais.

Packham, J.R. and Willis, A.J. 1997 *Ecology of dunes, saltmarsh and shingle*. London. Chapman and Hall.

Pakeman, R.J. and Marrs, R.H. 1996 Modelling the effects of climate change on the growth of bracken (*Pteridium aquilinum*) in Britain. *Journal of Applied Ecology* **33**, 561–75.

Pethybridge, G.H. and Praeger, R.L. 1905 The vegetation of the district lying south of Dublin. *Proceedings of the Royal Irish Academy* 25**B**, 124–80.

Phillips, W.E.A. 1965 The geology of Clare Island, County Mayo. Unpublished PhD thesis, University of Dublin, Trinity College.

Phillips, W.E.A. 1973 The pre-Silurian rocks of Clare Island, Co. Mayo, Ireland and the age of metamorphism of the Dalradian in Ireland. *Journal of the Geological Society, London* **129**, 585–606.

Praeger, R.L. 1903 The flora of Clare Island. *The Irish Naturalist* **12**, 277–94.

Praeger, R.L. 1904 The flora of Inishturk. *The Irish Naturalist* **13**, 113–25.

Praeger, R.L. 1905 The flora of the Mullet and Inishkea. *The Irish Naturalist* **14**, 229–44.

Praeger, R.L. 1907 The flora of Achill Island. *The Irish Naturalist* **16**, 265–89.

Praeger, R.L. 1911 Clare Island Survey. Part 10 Phanerogamia and Pteridophyta. *Proceedings of the Royal Irish Academy* **31**, 1–112.

Prentice, H.C. and Prentice, I.C. 1975 The hill vegetation of North Hoy, Orkney. *New Phytologist* **75**, 313–67.

Randall, R.E. 1989 Shingle habitats in the British Isles. *Botanical Journal of the Linnean Society* **101**, 3–18.

Ratcliffe, D.A. and Thompson, D.B.A. 1988 The British uplands: their ecological character and international significance. In M.B. Usher and D.B.A. Thompson, *Ecological change in the uplands*, 9–36. Oxford. Blackwell Scientific Publications.

Rawes, M. 1981 Further results of excluding sheep from high level grasslands in the North Pennines. *Journal of Ecology* **69**, 651–69.

Rodwell, J.S. 1991–2000 *British plant communities*. Vols I–V. Cambridge. Cambridge University Press.

Rodwell, J.S., Dring, J.C., Averis, A.B.G., Proctor, M.C.F., Malloch, A.J.C., Schaminée, J.N.J. and Dargie, T.C.D. 2000 *Review of coverage of the National Vegetation Classification*. JNCC Report no. 302. Peterborough. JNCC.

Rohan, P.K. 1986 *The climate of Ireland*. 2nd edn. Dublin. Irish Stationery Office.

Ruttledge-Fair, R. 1892 *Congested Districts Board: baseline reports of local Inspectors*. London. Her Majesties Stationery Office.

Ryle, T. and Doyle, G.J. 1998 The elusive sea-pea on Clare Island. *Newsletter of the New Survey of Clare Island* 4.

Scannell, M.J.P. and Synnott, D.M. 1987 *Census catalogue of the flora of Ireland*. 2nd edn. Irish Stationery Office, Dublin.

Schoof van Pelt, M.M. 1973 Littorelletea, a study in the vegetation of some amphiphytic communities of western Europe. Dissertation, Catholic University, Nijmegen.

Schouten, M.G.C. and Nooren, M.J. 1977 Coastal vegetation types and soil features in south-east Ireland. *Acta Botanica Neerlandica* **26**, 357–8.

Sheffield, E., Wolf, P.G. and Haufler, C.H. 1989 How big is a bracken plant? *Weed Research* **29**, 455–60.

Smith, A.J.E. 1990 The bryophytes of Achill Island—Musci. *Glasra* (new series) **1**, 27–46.

Stace, C. 1997 *New flora of the British Isles*. 2nd edn. Cambridge. Cambridge University Press.

Stevenson, A.C. and Birks, H.J.B. 1995 Heaths and moorland: long-term ecological changes and interactions with climate and people. In D.B.A. Thompson, A.J. Hester and M.B. Usher (eds), *Heaths and moorland: cultural landscapes*, 224–39. Edinburgh. Scottish Natural Heritage.

Stevenson, A.C. and Thompson, D.B.A 1993 Long-term changes in the extent of heather moorland in upland Britain and Ireland: palaeoecological evidence for the importance of grazing. *The Holocene* **3**, 70–6.

Synge, F.M. 1968 The glaciation of west Mayo. *Irish Geography* **5**, 372–86.

Tallis, J.H. 1973 The terrestrialization of lake basins in north Chesire. *Journal of Ecology* **61**, 537–67.

Tansley, A.G. 1911 *Types of British vegetation*. Cambridge. Cambridge University Press.

Tansley, A.G. 1939 *The British Islands and their vegetation*. Cambridge. Cambridge University Press.

Taylor, J.A. 1986 The bracken problem: a local hazard and global issue. In R.T. Smith and J.A. Taylor (eds), *Bracken: ecology, landuse and control technology*, 21–42. Carnforth. Parthenon Press.

Thompson, D.B.A. and Miles, J. 1995 Heaths and moorland: some conclusions and questions about environmental change. In D.B.A. Thompson, A.J. Hester and M.B. Usher (eds), *Heaths and moorland: cultural landscapes*, 362–85. Edinburgh. Scottish Natural Heritage.

Thompson, D.B.A., MacDonald A.J. and Hudson, P.J. 1995 Upland moors and heaths. In W.J. Sutherland and D.A. Hill (eds), *Managing habitats for conservation*, 292–328. Cambridge. Cambridge University Press.

Thompson, D.B.A., MacDonald, A.J., Marsden, J.H. and Gailbraith, C.A. 1995 Upland heather moorland in Great Britain: a review of international importance, vegetation change and some objectives for nature conservation. *Biological Conservation* **71**, 163–78.

Tutin, T.G., Heywood, V.H., Burgess, N.A., Moore, D.M., Valentine, D.H., Walter, S.M. and Webb, D.A. 1964–1980 *Flora Europaea* Volumes 1–5. Cambridge. Cambridge University Press.

University of Glasgow 1992 1:7500 map of Clare Island. Unpublished map prepared for the Royal Irish Academy.

Usher, M.B. and Thompson, D.B.A. 1993 Variation in upland heathlands of Great BritaIn conservation importance. *Biological Conservation* **66**, 69–81.

Van Groenendael, J.M., Hochstenbach, S.M.H., Van Mansfeld, M.J.M. and Roozen, A.J.M 1979 The influence of the sea and of parent material on wetlands and blanket bog in western Connemara, Ireland. Unpublished MSc thesis, Laboratory for Geobotany, Catholic University, Nijmegen.

Van Groenendael, J.M., Hochstenbach, S.M.H., Van Mansfeld, M.J.M. and Roozen, A.J.M. 1983 Soligenous influences on wetlands and blanket bog in western Connemara, Ireland. *Journal of Life Sciences, Royal Dublin Society* **4**, 103–28.

Van Groenendael, J.M., Hochstenbach, S.M.H., Van Mansfeld, M.J.M. and Roozen, A.J.M. 1983 Plant communities of lakes, wetlands and blanket bogs in western Connemara, Ireland. *Journal of Life Sciences, Royal Dublin Society* **4**, 129–37.

Vullings, W., Collins, J.F. and Smillie, G. 2013. Soils and soil associations on Clare Island. New Survey of Clare Island. Dublin. Royal Irish Academy.

Webb, D.A. 1983 The flora of Ireland in its European context. *Journal of Life Sciences, Royal Dublin Society* **4**, 143–60.

Webb, D.A. and Hodgson, J. 1968 The flora of Inishbofin and Inishshark. *Proceedings of the Botanical Society of the British Isles* **7** (3), 345–63.

Welch, D. 1984 Studies in the grazing of heather moorland in north-east Scotland. I. Site description and patterns of utilisation. *Journal of Applied Ecology* **21**, 179–95.

Welch, D. 1984 Studies in the grazing of heather moorland in north-east Scotland. II. Responses of heather. *Journal of Applied Ecology* **21**, 197–207.

Welch, D. 1986 Studies on the grazing of heather moorland in north-east Scotland. V. Trends in *Nardus stricta* and other unpalatable graminoids. *Journal of Applied Ecology* **23**, 1047–58.

Westhoff, V. and Den Held, A.J. 1969 *Plantengemeenschappen in Nederland*. Zutphen. Thieme.

Westhoff, V. and Van Der Maarel, E. 1973 The Braun-Blanquet approach. In R.H. Whittaker (ed.), *Handbook of vegetation science*, 617–726. The Hague. Junk.

Westhoff, V. and Van Der Maarel, E. 1978 The Braun-Blanquet approach. In R.H. Whittaker (ed.), *Classification of plant communities*, 2nd edn, 287–399. The Hague. Junk.

Wheeler, B.D. 1980 Plant communities of rich fen systems in England and Wales. II. Communities of calcareous mires. *Journal of Ecology* **64**, 405–20.

White, J.M. 1932 The fens of north Armagh. *Proceedings of the Royal Irish Academy* 40**B**: 233–83.

White, J. 1982 The *Plantago* sward, *Plantaginetum Coronopodo maritimi*. *Journal of Life Sciences, Royal Dublin Society* **3**, 105–10.

White, J. and Doyle, G.J. 1982 The vegetation of Ireland: a catalogue raisonné. *Journal of Life Sciences, Royal Dublin Society* **3**, 289–368.

Whitehead, S.J., Caporn, S.J.M. and Press, M.C. 1997 Effects of elevated CO_2, nitrogen and phosphorus on the growth and photosynthesis of two upland perennials: *Calluna vulgaris* and *Pteridium aquilinum*. *New Phytologist* **135**, 201–11.

Whittow, J.B. 1974 *Geology and scenery in Ireland*. Penguin. Middlesex.

Winder, F.G. and Moore, J.J. 1947 Some notes on the rarer plants of the Ben Bulben range. *Irish Naturalists' Journal* **9**, 68–71.

Zuidhoff, A.C., Rodwell, J.S. and Schaminée, J.H.J. 1995 The *Cynosurion cristati* Tx. 1947 of central, southern and western Europe: a tentative overview, based on the analysis of individual relevés. *Annali di Botanica* **53**, 25–47.

APPENDIX 1

Grid references

Location of the 407 relevés made during the current survey. Grid references correspond with the National grid reference scheme.

Relevé	Grid reference Easting	Northing	Townland	Relevé	Grid reference Easting	Northing	Townland
1	69750	288400	Ballytoohy More	40	69025	287400	Ballytoohy More
2	69760	288250	Ballytoohy More	41	69460	287890	Ballytoohy More
3	69875	288125	Ballytoohy More	42	69330	288000	Ballytoohy More
4	69900	288000	Ballytoohy More	43	69825	285600	Commonage
5	70550	286700	Maum	44	69610	285480	Commonage
6	70500	286700	Maum	45	70200	285400	Commonage
7	70625	286300	Maum	46	70200	285300	Commonage
8	70650	286050	Maum	47	69870	285200	Commonage
9	70550	285800	Fawnglass	48	69670	285225	Commonage
10	70600	285550	Fawnglass	49	68600	286775	Commonage
11	70725	285475	Fawnglass	50	68700	286800	Commonage
12	70850	286350	The mill	51	69550	285730	Lecarrow
13	71085	286200	Capnagower	52	69500	285320	Lecarrow
14	71090	285950	Capnagower	53	69300	284300	Kill
15	70250	286450	Maum	54	68625	284250	Kill
16	70040	286750	Ballytoohy Beg	55	68100	285450	Commonage
17	69090	286150	Commonage	56	67725	285550	Commonage
18	69110	285460	Lecarrow	57	67970	289100	Commonage
19	68950	285625	Lecarrow	58	66300	285450	Commonage
20	68100	286550	Commonage	59	66890	286200	Commonage
21	68000	286425	Commonage	60	67090	286255	Commonage
22	65325	285100	Commonage	61	67090	286210	Commonage
23	65400	285100	Commonage	62	66725	286250	Commonage
24	65560	285400	Commonage	63	69875	286940	Ballytoohy Beg
25	65560	285700	Commonage	64	69750	287250	Ballytoohy More
26	66550	285600	Commonage	65	69800	287325	Ballytoohy More
27	66740	285850	Commonage	66	70060	287575	Ballytoohy More
28	66900	284125	Commonage	67	70820	286200	Commonage
29	67000	285250	Commonage	68	70850	286060	Commonage
30	67050	285725	Commonage	69	71000	286330	Commonage
31	67100	286000	Commonage	70	71360	286450	Commonage
32	68700	285475	Commonage	71	71650	286440	Capnagower
33	67960	285800	Commonage	72	71440	286210	Capnagower
34	69900	285875	Lecarrow	73	68940	285960	Park
35	65050	285510	Commonage	74	68840	285800	Lecarrow
36	65125	285300	Commonage	75	68690	285840	Lecarrow
37	69100	287745	Ballytoohy More	76	68515	286300	Commonage
38	69550	287390	Ballytoohy More	77	68750	285510	Commonage
39	69400	287500	Ballytoohy More	78	69550	286300	Commonage

(Continued)

Appendix 1 *(Continued)*

Relevé	Grid reference Easting	Northing	Townland	Relevé	Grid reference Easting	Northing	Townland
79	69350	286035	Lecarrow	123	65725	284850	Commonage
80	69450	285840	Lecarrow	124	65625	284830	Commonage
81	69360	285710	Lecarrow	125	65575	284700	Commonage
82	69220	285550	Lecarrow	126	65410	284700	Commonage
83	70380	286175	Maum	127	65300	284900	Commonage
84	71440	284850	The quay	128	65400	284900	Commonage
85	71350	284875	The quay	129	65545	285170	Commonage
86	71250	284940	The quay	130	65880	285250	Commonage
87	71130	284825	Glen	131	65865	284440	Commonage
88	71025	284950	Glen	132	65440	285700	Commonage
89	70810	284660	Glen	133	65275	285450	Commonage
90	70720	284850	Gorteen	134	65735	284900	Commonage
91	70500	284700	Gorteen	135	65800	284475	Bunnamohaun
92	69610	284350	Gorteen	136	65950	284440	Bunnamohaun
93	69300	284275	Gorteen	137	65945	284460	Bunnamohaun
94	71890	285900	Gorteen	138	65955	284460	Bunnamohaun
95	71940	285925	Capnagower	139	66000	284315	Bunnamohaun
96	71935	285999	Capnagower	140	65950	284240	Bunnamohaun
97	68425	286565	Commonage	141	65810	284240	Bunnamohaun
98	68450	286750	Ballytoohy Beg	142	69050	285620	Lecarrow
99	68785	286710	Ballytoohy Beg	143	69050	285600	Lecarrow
100	69080	286555	Commonage	144	69060	285600	Lecarrow
101	66800	284575	Commonage	145	69050	285510	Lecarrow
102	66770	284585	Commonage	146	70450	285790	Ballytoohy Beg
103	66600	284730	Commonage	147	70460	285750	Ballytoohy Beg
104	66590	284735	Commonage	148	70440	285755	Ballytoohy Beg
105	66600	285000	Commonage	149	70440	285775	Ballytoohy Beg
106	66730	285100	Commonage	150	70360	285800	Ballytoohy Beg
107	66210	285410	Commonage	151	70380	287100	Ballytoohy Beg
108	65800	284790	Commonage	152	70025	287940	Ballytoohy More
109	65850	284400	Commonage	153	70020	287930	Ballytoohy More
110	65565	284510	Commonage	154	69900	287900	Ballytoohy More
111	65575	284500	Commonage	155	70160	287210	Ballytoohy More
112	65600	284500	Commonage	156	71350	285235	The quay
113	65275	284560	Commonage	157	71350	285240	The quay
114	65175	284290	Commonage	158	71350	285240	The quay
115	65350	284290	Commonage	159	71350	285200	The quay
116	65800	284680	Commonage	160	71360	285180	The quay
117	65750	284550	Commonage	161	71325	285260	The quay
118	65780	284545	Commonage	162	68370	285980	Commonage
119	65690	284390	Commonage	163	68200	285950	Commonage
120	65660	284220	Commonage	164	67950	286000	Commonage
121	70500	286840	Ballytoohy Beg	165	67815	286060	Commonage
122	70460	286920	Ballytoohy Beg	166	67800	286000	Commonage

(Continued)

Appendix 1 *(Continued)*

Relevé	Grid reference Easting	Northing	Townland	Relevé	Grid reference Easting	Northing	Townland
167	66995	284195	Commonage	211	68000	286250	Commonage
168	67060	284150	Commonage	212	68810	286150	Park
169	67180	284165	Commonage	213	68360	286250	Commonage
170	67265	284225	Commonage	214	68945	285920	Lecarrow
171	67475	284240	Commonage	215	68540	285850	Commonage
172	68775	284135	Kill	216	68310	285890	Commonage
173	68720	284260	Kill	217	68150	285955	Commonage
174	68515	284185	Kill	218	67850	286275	Commonage
175	68300	284175	Kill	219	68250	286275	Commonage
176	69730	287650	Ballytoohy More	220	68750	286310	Commonage
177	69500	288000	Ballytoohy More	221	68800	286975	Ballytoohy Beg
178	69500	288070	Ballytoohy More	222	68815	287015	Ballytoohy Beg
179	69575	288060	Ballytoohy More	223	68800	287100	Ballytoohy More
180	69620	288140	Ballytoohy More	224	68655	287170	Ballytoohy More
181	69585	288190	Ballytoohy More	225	68750	287200	Ballytoohy More
182	69420	287755	Ballytoohy More	226	68945	287205	Ballytoohy More
183	69770	287725	Ballytoohy More	227	69045	287320	Ballytoohy More
184	66900	284715	Commonage	228	69300	287200	Ballytoohy More
185	66900	285000	Commonage	229	69525	287170	Ballytoohy More
186	67070	285075	Commonage	230	69415	287050	Ballytoohy More
187	67260	285125	Commonage	231	69000	285150	Commonage
188	67625	288200	Commonage	232	69050	285000	Kill
189	67590	285305	Commonage	233	69200	284905	Kill
190	67510	285475	Commonage	234	69255	285160	Commonage
191	67905	285400	Commonage	235	69360	285240	Commonage
192	67720	285645	Commonage	236	69500	284950	Gorteen
193	67600	285700	Commonage	237	69800	285400	Commonage
194	67640	285740	Commonage	238	69900	285445	Commonage
195	67400	285950	Commonage	239	69740	285495	Commonage
196	67100	285505	Commonage	240	69590	285560	Commonage
197	66450	285510	Commonage	241	70150	287550	Ballytoohy More
198	66850	285820	Commonage	242	70100	287725	Ballytoohy More
199	66595	286000	Commonage	243	69990	287800	Ballytoohy More
200	72000	285950	Capnagower	244	69850	287770	Ballytoohy More
201	71390	285350	The quay	245	69700	287850	Ballytoohy More
202	70700	285580	Fawnglass	246	69900	287470	Ballytoohy More
203	70360	285615	Fawnglass	247	70120	287390	Ballytoohy More
204	68925	286390	Commonage	248	70295	286525	Maum
205	69210	286700	Ballytoohy Beg	249	70050	287200	Ballytoohy More
206	69365	286750	Ballytoohy Beg	250	70150	287050	Ballytoohy Beg
207	69300	286880	Ballytoohy Beg	251	69900	287250	Ballytoohy More
208	69150	286995	Ballytoohy Beg	252	70040	286970	Ballytoohy Beg
209	68880	285125	Kill	253	70080	286900	Ballytoohy Beg
210	68050	286525	Commonage	254	69240	285780	Lecarrow

(Continued)

Appendix 1 *(Continued)*

Relevé	Grid reference		Townland
	Easting	Northing	
255	69340	285810	Lecarrow
256	69600	285810	Commonage
257	69745	286010	Commonage
258	69950	286250	Commonage
259	70250	285795	Fawnglass
260	70300	285535	Lecarrow
261	66450	284900	Commonage
262	66350	284985	Commonage
263	66300	285350	Commonage
264	66200	285630	Commonage
265	66200	285810	Commonage
266	65840	285625	Commonage
267	66700	285580	Commonage
268	66540	285565	Lecarrow
269	69750	285800	Commonage
270	69260	285400	Commonage
271	69000	285305	Commonage
272	68850	286065	Park
273	69060	285950	Park
274	68400	285945	Commonage
275	68250	286100	Commonage
276	67850	286060	Commonage
277	68475	285450	Commonage
278	70125	286975	Ballytoohy Beg
279	70140	286950	Ballytoohy Beg
280	70160	286860	Ballytoohy Beg
281	70160	286810	Ballytoohy Beg
282	70135	286710	Ballytoohy Beg
283	70450	286560	Maum
284	70480	286700	The mill
285	70825	286485	The mill
286	70900	286400	The mill
287	70550	286400	Capnagower
288	71100	286350	Capnagower
289	71125	286350	Capnagower
290	71250	286400	Capnagower
291	71150	286100	Capnagower
292	71000	286020	Fawnglass
293	70925	285975	Fawnglass
294	70900	285825	Fawnglass
295	70600	285700	Fawnglass
296	70280	286040	Maum
297	70100	286260	Lecarrow
298	70150	286750	Ballytoohy Beg
299	70175	286750	Ballytoohy Beg
300	70200	286750	Ballytoohy Beg
301	70250	286760	Ballytoohy Beg
302	70250	286750	Ballytoohy Beg
303	70250	286770	Ballytoohy Beg
304	70385	286840	Ballytoohy Beg
305	70160	285250	Maum
306	70110	285525	Maum
307	70260	285500	Fawnglass
308	70335	285390	Fawnglass
309	70335	285275	Glen
310	69350	285400	Lecarrow
311	69400	285420	Lecarrow
312	69425	285450	Lecarrow
313	69450	285475	Lecarrow
314	69425	285495	Lecarrow
315	69500	285380	Lecarrow
316	69560	285300	Lecarrow
317	70540	285890	Fawnglass
318	70200	286080	Maum
319	68260	286700	Commonage
320	68200	286700	Commonage
321	67800	286560	Commonage
322	68800	285550	Lecarrow
323	68750	285600	Lecarrow
324	68750	285600	Lecarrow
325	68750	285600	Lecarrow
326	68750	285600	Lecarrow
327	68750	285600	Lecarrow
328	68750	285600	Lecarrow
329	68750	285600	Lecarrow
330	69000	285610	Lecarrow
331	69010	285610	Lecarrow
332	69050	285600	Lecarrow
333	69050	285550	Lecarrow
334	69050	285525	Lecarrow
335	69060	285525	Lecarrow
336	69055	285525	Lecarrow
337	69050	285525	Lecarrow
338	69045	285330	Lecarrow
339	69020	285540	Lecarrow
340	69010	285575	Lecarrow
341	67160	285275	Commonage
342	67165	285275	Commonage

(Continued)

Appendix 1 *(Continued)*

Relevé	Grid reference		Townland	Relevé	Grid reference		Townland
	Easting	Northing			Easting	Northing	
343	66750	286250	Commonage	375	65560	284930	Commonage
344	66900	286300	Commonage	376	65450	284935	Commonage
345	67000	286315	Commonage	377	65350	285240	Commonage
346	69475	287350	Ballytoohy More	378	65200	284840	Commonage
347	69665	287550	Ballytoohy More	379	71325	286140	Capnagower
348	69350	287340	Ballytoohy More	380	71550	286450	Capnagower
349	69275	287340	Ballytoohy More	381	71600	286450	Capnagower
350	69150	287425	Ballytoohy More	382	71790	286350	Capnagower
351	69010	287500	Ballytoohy More	383	70670	285600	Fawnglass
352	69030	287600	Ballytoohy More	384	70775	285590	Fawnglass
353	69350	287600	Ballytoohy More	386	70975	285610	Commonage
354	69355	287595	Ballytoohy More	387	70970	285640	Lecarrow
355	70670	286300	Maum	388	68755	285850	Lecarrow
356	70700	286295	Maum	389	68750	285870	Lecarrow
357	70730	286250	Maum	390	38800	285840	Lecarrow
358	70740	286250	Maum	391	68875	285680	Commonage
359	70735	286275	Maum	392	68350	285440	Strake
360	70760	286275	Maum	393	68290	285250	Strake
361	70775	286300	Maum	394	67440	284295	Commonage
362	70725	286200	Maum	395	67300	286300	Commonage
363	70650	286210	Maum	396	67225	286325	Commonage
364	70475	286125	Maum	397	66675	286250	Commonage
365	68975	286100	Park	398	67000	286150	Commonage
366	68900	286025	Park	399	66950	286200	Commonage
367	68800	286040	Park	400	69200	284200	Kill
368	68500	286025	Lecarrow	401	70490	284840	Glen
369	68100	285660	Commonage	402	71050	285100	Glen
370	67650	285900	Commonage	403	70375	286170	Maum
371	67450	285775	Commonage	404	70375	286170	Maum
372	67180	285680	Commonage	405	70370	286170	Maum
373	66920	285500	Commonage	406	70360	286170	Maum
374	65645	284920	Commonage	407	70365	286170	Maum

AN ANNOTATED CHECKLIST OF THE VASCULAR FLORA OF CLARE ISLAND

Matthew Jebb

ABSTRACT

A résumé of Praeger's survey work, particularly his analysis of the flora, is given. More recent surveys by Gerry Doyle and Peter Foss in 1986, Juliet Brodie in 1991, Tim Ryle in 2000 and most recently a BSBI field trip in 2007 have added numerous taxa, but have failed to find taxa previously recorded by Praeger. In light of these surveys a brief summary of the biogeographical implications of the changing flora over time is presented.

Introduction

Praeger had an abiding interest in islands and their floras, and a special fondness for Clare Island, which he described so eloquently in his book *Beyond Soundings* (Praeger 1930). This enthralment led him to publish florulas for eleven other Irish islands, and in the case of Lambay and Clare Island, detailed vegetation maps. His analysis of these floras was driven by a curiosity to understand how their floras had arisen, and thus to discover how the flora of the island of Ireland in turn had formed. The value of this body of work is of great consequence today, providing not only a baseline against which to compare the present, but also an insight into biogeographic changes over time. In the case of Clare Island, Praeger was unstinting in the analysis he brought to bear on the flora.

Praeger's analyses

Praeger first studied the flora of Clare Island during the course of a week in July 1903 (Praeger 1903). During this first exploration he discovered the remarkable assemblage of alpines on the Knockmore scarp (*Silene acaulis, Oxyria digyna, Saxifraga rosacea, Sedum rosea* and *Saussurea alpina*). Some of the other more remarkable plants of the island, including *Erica erigena* and *Saxifraga* × *polita* were only found on his later visits as part of the Clare Island Survey work in 1909 and 1911 (Praeger 1911a).

Over half Praeger's contribution on the 'Phanerogamia and Pteridophyta' (1911a, part 10) was given over to the origin of the island's plants, as well as to a lengthy and detailed comparison of the flora of the neighbouring islands, Inishbofin and Inishturk, which he had previously recorded (Praeger 1907; 1911b). As to the origin of the native flora, he divided the species according to the likely routes by which they may have arrived (via water, wind or birds) and noted whether a land-bridge provided a more likely pathway. The results were inconclusive, but no doubt stimulated Praeger to continue his investigations for a more complete analysis of the buoyancy of the seeds of the entire Irish flora (Praeger 1913).

Of the 393[1] taxa he recorded, he regarded 54 as non-native, having been introduced by the activities of man, and he took pains to investigate the possible means by which such plants had reached and were still reaching the island. He purchased seeds from agricultural merchants on Achill Island (from whence the Clare Island population obtained its supplies) as well as examining samples of oats imported as chicken feed and horse fodder and sweepings from hay imported from fields near Roonagh Quay, and he even sieved the mud scraped from the shoes of a local man (Pat Grady) when he returned from a few days in Louisburg. All seeds were examined and identified. His enquiries underlined the prevalence of those species characteristic of roadsides in all these samples and indicated the ease with which they dispersed in the wake of human activities in the west of Ireland.

Praeger pointed out that 80% of the flora at the time of his work comprised 'Universal' plants, i.e. those taxa found in all 40 Irish botanical divisions (vice-counties). He also remarked upon the fact that the prevalence of annuals is often a consequence of their life histories being especially suited for life in cultivated soils. Thus the flora of Howth (a comparable size, if a little smaller than Clare Island) comprised 24% annuals, while that of Clare Island comprised 18%. Praeger was interested in the standing of these plants in order to understand which species were truly native and which had been introduced or were reliant for their continued survival upon 'the hand of man'.

During the survey work, Praeger visited several of the surrounding islands, including the smaller islets—the Bills, Mweelaun and Caher Island. He undertook a more detailed comparison of the floras of Inishbofin and Inishturk than with that of Clare Island. He compiled lists of species that were unique to each of the three islands and those found in common to two or to all three islands. He presented the affinities in the form of a modified Venn diagram (Fig. 1).

Praeger applied Colgan's 'index of floral diversity' (Colgan 1901). This is the ratio of the total of species not common to both areas to the total flora of the two areas combined. This gave figures for Clare Island and Inishbofin of 0.300 ((76+60)/454); for Inishturk and Inishbofin of 0.309 ((39+90)/417); and for Clare Island and Inishturk of 0.283 ((26+93)/420). Praeger interpreted these figures to imply a greater divergence between the floras of Inishturk and Inishbofin, and related this to Inishturk being the worst provided with fresh water and Inishbofin the best.

The differences, Praeger concluded, were entirely explicable in terms of Clare Island's unique possession of species characteristic of the 'Big Hill', the woodland plants found at Portlea, and the island's overall greater level of protection from the Atlantic weather.

Island biogeography

The theory of island biogeography was started in the 1960s by the ecologists MacArthur and Wilson (1967). They found that islands exhibit a remarkable correlation between their area and the number of species they support. Further, they found that islands denuded of species would return to a similar diversity level, but the composition would be unpredictable and often very different on each occasion. The theory proposes that the number of species found on an island is determined by the equilibrium of the counteracting effects of immigration and extinction. Islands that are more distant from a mainland or another island are likely to receive a smaller number of immigrants than islands that are less isolated. Likewise, the rate of extinction is affected by population size, and thus is a direct consequence of habitat or island size, larger populations being less likely to become extinct because of random stochastic effects. Further, since immigration and extinction are continuous processes, the species composition of an island will not be fixed but will change over time.

While numerous studies have failed to reveal any statistically valid observations on extinction

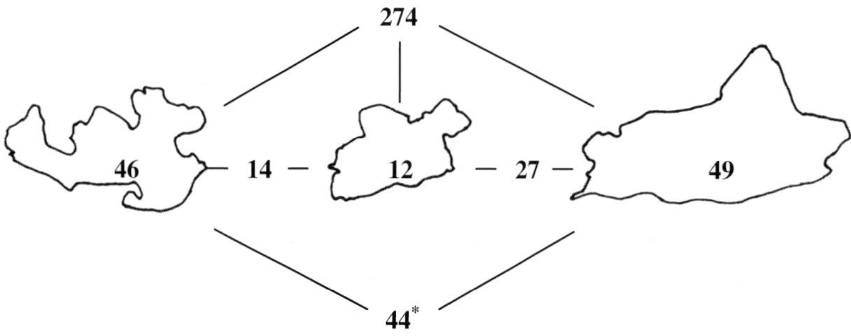

Fig. 1 Diagram of floral affinities between Inishbofin, Inishturk and Clare Island (left to right). Numbers on each island indicate taxa unique to that island; numbers between indicate those in common between each island. The combined flora list for all three islands was 466 (data from Praeger 1911a).

* 43 in 1911a paper due to the omission of *Mentha arvensis*, see note 1, page 185.

or immigration rates, it appears that physical packing of taxa onto an island is a genuine constraint to colonisation. Observations confirm that the relationship of island biotas in similar geographical locations is constant and predictable, and a straight line is formed in a log–log plot of area versus species number. Using Praeger's figures for the three islands, we find that the floras do indeed conform to McArthur and Wilson's predictions (even though it is a minimal dataset) giving a trendline with a good fit (R^2 = 0.9898) (Fig. 2). The slope of this line indicates that for every doubling of area there is a 14% increase in floral diversity. Conversely the halving of land area will reduce the floral diversity by just over 12%.

Any drop in species diversity for one of these islands is therefore indicative of a reduction in the functional size of the island from the floral point of view. The impact of changing land use, particularly agricultural, could therefore be estimated through species diversity.

Recent surveys

Gerry Doyle and Peter Foss visited the island for four days in 1984 (Doyle and Foss 1986). Their visit coincided with a force 7 gale, which curtailed much exploration of the northern and western cliffs. They found 22 taxa that were not listed by Praeger. They provided the first confirmation of five further taxa until then only mapped in the *Atlas of the Distribution of British Plants* (Perring and Walters 1962), i.e. *Avena sativa*, *Carex demissa*, *Carex oederi* (Praeger mentions this species as *C. serotina*, but omitted it from his combined list (1911a) on page 17), *Catapodium marinum* and *Festuca vivipara*. Nine of these 27 taxa were classed as garden escapes. They refound a further 56 species out of a total of 86 species listed by Praeger that had not been recorded since 1930. In addition they reconfirmed a further 251 of the post-1930 records. In total they recorded 301 of the 413 listed by Praeger.

In July 1989, Juliet Brodie visited the island for the specific purpose of locating taxa that the Doyle and Foss survey had been unable to refind. She added four further species—*Atriplex glabriuscula*, *Anagallis minima*, *Agrostis canina* and *Dactylorhiza kerryensis* (Brodie 1991). She also added two other cultivated taxa—*Oxalis articulata* and *Tanacetum vulgare*. She reconfirmed a further 17 taxa recorded by Praeger.

Tim Ryle (2000) conducted a major investigation into the vegetation of Clare Island (see this volume, pp 27–177). He located 384 plant species, 328 of which were included in Praeger's earlier list (1911a). He recorded 22 of the taxa recorded by Doyle and Foss or by Brodie, and added 11 new taxa to the island's flora. Many of the newly recorded species were located in and around the harbour. Of these, most can be classed as introductions that have accompanied freight or pedestrian traffic disembarking from the ferry. However, 79 species listed in the earlier surveys were not relocated during this work.

In 1986 the rare clubmoss *Lycopodiella inundata* was discovered during a National Parks and Wildlife Service survey. In 2007 a second colony of this plant was found during a BSBI field trip that took place on 7 and 8 July 2007. The BSBI field trip, which was jointly led by Gerry Sharkey and Robert Northridge, added a further 33 taxa and confirmed a large proportion of previous records (Table 1). One other significant find was the gametophytic plants of *Trichomanes speciosum* in a fissure on the rockface of Knocknaveen by Robert Northridge. A visit by David McNeill of Belfast in July 1999 yielded a record for *Carex lepidocarpa*, and *Equisetum sylvaticum* was recorded by Sasha Bosbeer of Sylvan Consulting Ecologists in 2000 in the Maum woodland, south of Portlea.

While it is clear that the floral diversity of the island has declined, with 48 taxa that Praeger found not having been located from 1980 to today, nonetheless 71 new taxa have been added to the island list since Praeger's publication. All of these may not be present today, and over the past century a total of 480 species have been recorded from the island at some point. These figures are complicated by the fact that some taxa that Praeger regarded as solely cultivated have since become established in the wild (e.g. *Acer pseudoplatanus*).

The figures suggest a significant turnover in the flora of the order of 10%–20% or more per century.

The flora in relation to that of Inishbofin and Inishturk

Praeger listed fourteen taxa found on both Inishbofin and Inishturk that were absent from Clare Island. Of these, two have since been found by Doyle *et al.* (*Asplenium ruta-muraria*, *Puccinellia maritima*), two by Brodie (*Agrostis canina*, *Anagallis*

Table 1
Number of plants from each survey common to previous surveys and new species

	Praeger 1911	Doyle/Brodie 1980s	Ryle 1990	BSBI 2007
Praeger	403	314	328	266 (282)
Doyle and Foss (1986) Brodie (1991)		26	22	14 (16)
Ryle			11	4 (5)
BSBI and others				35
Totals	**403**	**340**	**361**	**319 (338)**

Note: Figures in brackets represent the final figure when taxa have been confirmed by Donal Synnott, Gerry Sharkey or Matthew Jebb.

Table 2
Numbers of indubitably wild taxa (excludes all planted taxa) from surveys showing total numbers; numbers and percentage in common with Praeger's list and numbers and percentage of novelties

	Total	Praeger	Novelties
Praeger 1911	393		
Doyle and Foss 1986 + Brodie 1991	340	305 (78%)	26 (8%)
Ryle 1996	361	326 (83%)	11 (3%)
BSBI 2007	338	274 (70%)	35 (10%)

minima) and one by Ryle (*Allium babingtonii*). However, the following taxa continue to be absent from Clare Island:

Species still absent from Clare Island
Fumaria capreolata subsp. *babingtonii*
Lobelia dortmanna
Lolium temulentum
Populus tremula
Rubus dumnoniensis
Trifolium campestre
Trifolium medium
Tuberaria guttata
Viola canina

Common plants absent from Clare Island

Praeger was curious to see which common species (those found in all 40 vice-counties and also frequent for maritime species) in West Mayo or in coastal counties were missing from the three island floras. His list is still pretty much the same today (he erroneously placed *Scrophularia nodosa* on the list, since he recorded the species on Clare Island (see 1911a, p. 78); *Alnus glutinosa* has also been recorded from the island, but as an introduced plant only).

Alnus glutinosa[1]
Alopecurus pratensis
Arum maculatum
Barbarea vulgaris
Berula erecta
Bidens cernua
Briza media
Carex disticha
Carex hirta
Carex remota
Crepis paludosa
Cymbalaria muralis
Cytisus scoparius
Epilobium hirsutum
Equisetum sylvaticum
Erophila verna
Galium odoratum
Geum urbanum
Glechoma hederacea
Hippuris vulgaris
Knautia arvensis
Medicago lupulina
Melampyrum pratense
Petasites hybridus
Platanthera bifolia
Potamogeton perfoliatus
Reseda luteola
[*Rosa tomentosa*[2]]

Rubus idaeus
Rumex sanguineus
[*Sagina apetala*[3]]
Salicornia europaea
Salix caprea
Schedonorus giganteus
[*Scrophularia nodosa*[4]]
Silene vulgaris
Smyrnium olusatrum
Sparganium emersum
Suaeda maritima
Typha latifolia
Viburnum opulus

[1] Recorded as an introduced plant today.
[2] In Praeger's day this name included *R. sherardii*, therefore on the strict definition of the name, this is a rare plant, and not a universal.
[3] Recorded by BSBI in 2007.
[4] Erroneously placed on the list by Praeger, who himself recorded it on Clare Island in 1911 (see 1911, p. 78).

Species otherwise rare or absent in Mayo
Lastly, it is of note that a number of taxa found on Clare Island are notable for their rarity in Mayo as a whole (Synnott 1986).
Allium babingtonii
Bromus commutatus
Dactylorhiza kerryensis
Dryopteris affinis subsp. *borreri*
Empetrum nigrum
Geranium sanguineum
Lamium amplexicaule
Mentha spicata
Ophioglossum vulgatum
Orobanche alba
Papaver dubium
Pulicaria dysenterica
Rosa canina
Saxifraga rosacea
× *Schedolium loliaceum*
Sedum telephium
Silene acaulis
Veronica polita

Some observations on the vegetation types existing today compared to Praeger's day
A detailed analysis of the vegetation of Clare Island has been undertaken by Tim Ryle (Ryle 2000) and is presented in this volume also (see Ryle, this volume, pp 27–177). Some comments from a floral recorder, however, are also in order, and are provided here. The commonage area is now characterised by large areas of remarkably short, sheep-grazed sward. The only plants that appear immune to the intense grazing, besides pockets of bracken and gorse, are stands of *Juncus effusus* and *J. squarrosus*. Within *J. effusus* a herb layer of *Oxalis acetosella*, *Dryopteris* spp and other small herbs thrive. In areas of intensive grazing, colonies of *J. squarrosus* form distinct circular patches up to one metre in diameter, standing 5–8cm above the surrounding sward and harbouring plants such as *Potentilla erecta*, which is able to flower here because the sheep avoid grazing these patches. Areas of *Calluna* and *Erica cinerea* are mostly grazed to a 3–5cm high canopy. Judging from several of the photographs taken during the 1909 survey and the census of farm animals then and now, the number of grazing animals is not exceptionally high today, and photographs reveal that a similarly short sward existed then also. Praeger's mapping of a heath vegetation, however, suggests that the southerly slopes of Knockmore were covered by a blanket of taller heather plants that is largely absent there today but persists in a limited way elsewhere.

Across the hilly region between the lighthouse and the eastern flanks of Knockmore, patches of *Juncus squarrosus*, *Nardus stricta*, *Calluna vulgaris*, *Sphagnum* and other mosses all occur in profusion, almost to the exclusion of other plants, and each may be absolutely dominant.

Since Praeger's day, even greater areas of the fields with cultivation ridges have been abandoned to grassland. Depending on the intensity of grazing and drainage, these have developed a number of distinct vegetations. In some the entire field is a close-cropped sward, in others the rush *J. effusus* forms distinct strips in the furrows. In still others a low canopy of *Calluna*, *Erica* and grasses has developed. Bracken thickets often appear to colonise abandoned fields in preference to surrounding hillside. Praeger noted the clumps of *Osmunda regalis* as being 'like boulders' in the fields. These have now gone, and the species is only to be found in drains (notably on the Lighthouse Road as it leads up from the harbour) or beneath rock overhangs.

Turf-cutting has now largely stopped, with only a few tiny areas still being harvested in the far west of the island. In many areas the retaining walls used for holding family turf ('clamps') are still evident, often filled by a dense thicket of *Juncus effusus* where the shelter of the wall and the deposited peat provide a hospitable niche.

Aboriginal vegetation survives, much as Praeger would have seen it, beyond the field fence that skirts the northern cliffs of Knockmore. The *Luzula sylvatica* forms a deep springy layer of rhizomes up to 50cm deep that, combined with the steep topography, makes progress perilous to the walker. The small number of sheep that break through the fence are sufficient to maintain the remanants of the 1200-foot path used by Praeger and his colleagues to access the cliffs. Fencing along the north-west facing cliffs leading to the lighthouse has allowed areas of *Calluna* and *Erica* to form a natural canopy.

The promontory fort at Doon in Bunnamohaun (L 668 842) and the narrow point of land immediately north of Budawanny (L 652 848) appear to have remained ungrazed for innumerable years. This has led to the development of a deep, springy, almost impenetrable thatch largely of *Agrostis* and *Festuca* stems that reach over 30cm deep in places. This develops a polygonal pattern of interlocking stems. Praeger found similar development of ungrazed sward on the sea stacks at Doontraneen near the lighthouse, and at Kinatevdilla (Beetle Head) at the most westerly point of the island (1911a, p. 40).

The *Plantago*-sward highlighted by Praeger on his 1911 map is still extant at the western end of the island. South of Budawanny (L 653 847) an extreme form has developed, forming a dense canopy of small-leaved *Plantago maritima* and *Armeria* some 15–30cm deep over the mineral soil. *Angelica* and *Holcus* are about the only other plants appearing in this association, and even *Plantago coronopus* seems to be excluded. This area is not currently fenced against the ever-present sheep, but for some reason they appear to have left this rather singular vegetation undamaged.

The roadside verges were not mentioned by Praeger as a notable habitat, which they undoubtedly have become today. This is probably on account of the taller growth and flowering state of so much that is otherwise grazed within the fields each side of the road. For example, it is a rarity to see *Dactylorhiza maculata* subsp. *ericetorum* in bloom except on the banks at the side of roads and laneways.

A rather striking habitat is the tarmac roads, which not only support the ubiquitous *Sagina procumbens*, but also *Gnaphalium uliginosum*, which is tolerably common in the median growth of the roads in parts of Lecarrow.

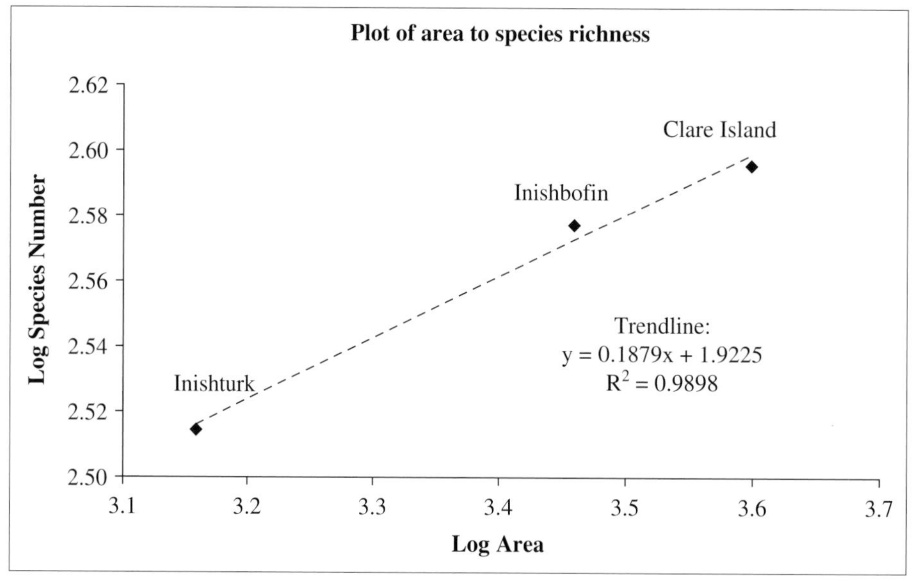

Fig. 2 Log plot of island area to species number from Praeger's 1911 data.

Acknowledgements
Cristina Armstrong, Gerry Sharkey, Donal Synnott and Noeleen Smyth have provided useful advice or assistance with fieldwork on the Clare Island flora.

NOTE

1. He recorded a further 21 taxa he regarded as cultivated but not established on the island.

APPENDIX

A checklist of vascular plants on Clare Island

Synonyms

Only those synonyms, or misapplications of names, when used by Praeger or later authors appear after the current species name.

Explanation of symbols used in the text

P = Praeger (1911a)

SUPERSCRIPTS FOR PRAEGER 1911 RECORDS

** = those species found in cultivated land and dependent on continuance of cultivation for their survival (29 taxa)

* = those species found chiefly on roadsides, banks and pastures, or about houses and gardens, less dependent on man, and could maintain themselves indefinitely in the absence of man (26 taxa);

1 = species seen in one station

2 = species seen in two stations

3 = species seen in three stations;

r = rare;

f, frequent;

c, common;

v.c. = very common;

local = local

D = Doyle and Foss (1986)

B = Brodie (1991)

R = Ryle (2000)

BSBI (BSBI field card 8 July 2007)

McNeill = David McNeill field card July 1999

MJ = Matthew Jebb, field observations 2009–2011 (these are only noted for taxa not recorded by other recent workers)

Others are written out in full;

Brackets indicate that the plant is planted or being cultivated.

Taxa within square brackets - [] - are probable errors.

Annotations in italics indicate a verbatim quote from an historic record.

BM denotes British Natural History Museum.

The taxa are arranged alphabetically under families, nomenclature follows the third edition of Clive Stace's *New Flora of the British Isles* (2010). Locations are only given for taxa with restricted ranges, and then only when this information is available.

LYCOPODIACEAE

Huperzia selago (L.) Bernh. ex Schrank & C. Martius
Synonym: *Lycopodium selago*
P^2; D; R; MJ

Lycopodiella inundata (L.) Holub
NPWS; BSBI
A large and luxuriant colony is present at the edge of a former peat-cutting zone (L 711 863). National Parks and Wildlife Service has a further record from 1986 'located south of Ballytoohy cut away' (L 69 86).

SELAGINELLACEAE

Selaginella selaginoides (L.) P. Beauv.
P^3; D; R; BSBI
Knocknaveen, northern cliffs of Knockmore

EQUISETACEAE

Equisetum arvense L.
P^f; D; R; BSBI

Equisetum fluviatile L.
Synonym: *Equisetum limosum*
P^f; D; R; BSBI

Equisetum fluviatile × arvense = E. × litorale Kühl. ex Rupr.
BSBI
This hybrid is abundant in many of the drains and hedge banks around the harbour and at Portnakilly.

Equisetum palustre L.
P^2; D; R; BSBI
Knockmore, Capnagower

Equisetum sylvaticum L.
Sasha Bosbeer, Sylvan Consulting Ecologists
Woodland SE of Portlea: report for owner of woodland at Ballytoughey More (*sic*) (Sylvan Consulting Ecologists 2000).
This seems an unlikely find, but could have been introduced as part of the reforestation work.

Equisetum telmateia Ehrh.
Synonym: *Equisetum maximum*
P^c; D; R; BSBI

OPHIOGLOSSACEAE

Botrychium lunaria (L.) Sw.
P^1

Ophioglossum vulgatum L.
P^3; R
Praeger recorded this as part of the maritime turf, with plants ½ inch tall and barren.

OSMUNDACEAE

Osmunda regalis L.
P^c; D; R; BSBI
Praeger reported large clumps of *Osmunda* in the fields as resembling large bolders. Today it is a rather scarce plant of drains and rock overhangs.

HYMENOPHYLLACEAE

Hymenophyllum wilsonii Hook.
Synonym: *Hymenophyllum unilaterale*
P^f; D; R; BSBI
Still abundant on the northern cliffs, and a few colonies on the N face of Knocknaveen. Also in Portlea woodlands on sheltered vertical rock faces.

Trichomanes speciosum Willd.
Robert Northridge, BSBI
Knocknaveen, L6985
Gametophyte plants, confirmed by Fred Rumsey (voucher at **BM**).

POLYPODIACEAE

Polypodium vulgare *sensu lato*
P^f; D; R

Polypodium interjectum Shivas
BSBI
Knocknaveen, Maum

Polypodium vulgare L. *sensu stricto*
BSBI
Maum

HYPOLEPIDACEAE

Pteridium aquilinum (L.) Kuhn in Decken
Synonym: *Pteris aquilina*
P^c; D; R; BSBI

ASPLENIACEAE

Asplenium adiantum-nigrum L.
P^f; D; R; BSBI

Asplenium marinum L.
P^f; D; R; BSBI

Asplenium ruta-muraria L.
D; R

Asplenium trichomanes agg.
P^2; D; R

Asplenium trichomanes* subsp. *quadrivalens D.E. Meyer
BSBI

Asplenium viride Hudson
P^1
Knockmore scarp

Asplenium ceterach L.
Synonym: *Ceterach officinarum*
BSBI

Asplenium scolopendrium L.
Synonyms: *Phyllitis scolopendrium*; *Scolopendrium vulgare*
P^r; D; R; BSBI

ATHYRIACEAE

Athyrium filix-femina (L.) Roth
P^c; D; R; BSBI

Cystopteris fragilis (L.) Bernh.
P^1
'Several small colonies on the Croaghmore scarp at 1,200–1,400 feet' (Praeger 1911, p. 31).

DRYOPTERIDACEAE

Dryopteris aemula (Aiton) Kuntze
Synonym: *Lastrea aemula*
P^c; D; R; BSBI

Dryopteris affinis (Lowe) Fraser-Jenkins
R; BSBI
Maum, Capnagower, Knocknaveen, harbour area

Dryopteris borreri Newm.
Synonym: *Dryopteris affinis* subsp. *borreri*
R

Dryopteris dilatata (Hoffm.) A. Gray
Synonym: *Lastrea dilatata*
P^f; B; R; BSBI
Wood SE of Portlea

Dryopteris filix-mas (L.) Schott
Synonym: *Lastrea filix-mas*
P^f; D; BSBI

Polystichum aculeatum (L.) Roth
Synonym: *Aspidium aculeatum*
P^1

Polystichum lonchitis (L.) Roth
Synonym: *Aspidium lonchitis*
P^1
'Knockmore scarp' (Praeger 1911, p. 10)

Polystichum setiferum (Forsskål) T. Moore ex Woynar
Synonym: *Aspidium angulare*
P^1; D; R; BSBI
Knocknaveen

BLECHNACEAE

Blechnum spicant (L.) Roth
P^{vc}; D; R; BSBI
Possibly no longer as common except on the northern cliffs.

PINACEAE

Picea sitchensis (Bong.) Carrière
{D}; R; BSBI

Pinus sylvestris L.
{Sasha Bosbeer}
Portlea woodland: report for owner of woodland at Ballytoughey More (*sic*) (Sylvan Consulting Ecologists 2000).

Pinus contorta Douglas ex Loudon
{D}; R

CUPRESSACEAE

Juniperus communis* subsp. *nana (Hook.) Syme
Synonym: *Juniperus nana*
P^1
The small island mentioned by Praeger was explored without success.

TAXACEAE

Taxus baccata L.
BSBI

NYMPHAEACEAE

Nuphar lutea (L.) Smith in Sibth. & Smith
Synonym: *Nuphar luteum*
P^1; D; R; BSBI
Lough Avullin

Nymphaea alba L.
P^1; D; R; MJ
Lough Merrignagh

RANUNCULACEAE

Anemone nemorosa L.
P^1
Knockmore scarp

Caltha palustris L.
P^1; R

Ficaria verna Huds.
Synonym: *Ranunculus ficaria*
P^3; B; R
Wood NW Portlea.

Ficaria verna* subsp. *fertilis (Lawralrée ex Laegaard) Stace
Synonym: *Ranunculus ficaria* subsp. *ficaria*
BSBI

Ranunculus acris L.
P^r; D; R; BSBI

Ranunculus bulbosus L.
D; R; BSBI
Harbour sands

Ranunculus flammula L.
P^c; D; R; BSBI

Ranunculus hederaceus L.
P^c; D; R; BSBI

Ranunculus repens L.
P^c; D; R; BSBI

BERBERIDACEAE

Berberis vulgaris L.
R
Ryle indicated some doubt as to the identity of the species.

PAPAVERACEAE

Papaver dubium L.
R

Papaver somniferum L.
{P}

FUMARIACEAE

Fumaria bastardii Boreau
Synonym: *Fumaria confusa*
P^{**2}; D; R

Fumaria muralis Sonder ex Koch
BSBI
Harbour area

URTICACEAE

Urtica dioica L.
P^{*f}; D; R; BSBI

Urtica urens L.
P^{**f}; D; R

MYRICACEAE

Myrica gale L.
P^f; D; R; BSBI
At Portlea, bog myrtle forms thickets among the bracken on fairly steep slopes.

FAGACEAE

Castanea sativa Mill.
{BSBI}

Quercus petraea (Mattuschka) Liebl.
Synonym: *Quercus sessiliflora*
P^1; B; R
Wood SE of Portlea

Quercus robur L.
BSBI
Maum and Scalpatruce, L6838 8649, 165m

BETULACEAE

Alnus glutinosa (L.) Gaertner
{P}; D; R; BSBI
Nowhere did I see alder naturalising, and all trees appear to be planted.

Betula pubescens Ehrh.
P^3; D; R; BSBI

CORYLACEAE

Corylus avellana L.
P^1; D; R; BSBI
Knocknaveen, Maum

CHENOPODIACEAE

Atriplex glabriuscula Edmondston
B; McNeill 1999
Near harbour

Atriplex patula L.
P^{**f}; D; R

Atriplex prostrata Boucher ex DC.
Synonym: *Atriplex hastata*
P^f; D; R; BSBI

Beta vulgaris* subsp. *maritima (L.) Arcang.
Synonym: *Beta maritima*
P^f; B; R; MJ
While Praeger claimed this species as frequent, he noted that it was 'widely spread'. Today it is found high up on the western sea cliffs in remarkably small quantities (MJ), and a single plant in the harbour sand in 2009 (MJ). The north-west coast of Mayo is strikingly devoid of this species.

Chenopodium album L.
P^{**r}; D; R; BSBI

Salsola kali* subsp. *kali L.
Synonym: *Salsola kali*
P^1; D
Harbour sand

PORTULACACEAE

Montia fontana L.
P^c; R

CARYOPHYLLACEAE

Cerastium diffusum Pers.
Synonym: *Cerastium tetrandrum*
P^f; D; R

Cerastium fontanum Baumg.
Synonym: *Cerastium triviale*
P^f; D; R; BSBI

Cerastium glomeratum ThuilL.
P^f; D; R; BSBI

Honckenya peploides (L.) Ehrh.
Synonym: *Arenaria peploides*
P^1; D; R; BSBI
Harbour sand

Sagina apetala Ard.
BSBI
Harbour area

Sagina maritima G. Don
P^f; D; R; BSBI

Sagina nodosa (L.) Fenzl
P^1; R; McNeill 1999
Sand dunes at harbour

Sagina procumbens L.
P^c; D; R; BSBI

Sagina subulata (Sw.) C. Presl
P^1; BSBI

Silene acaulis (L.) Jacq.
P^1; D; R; BSBI
Knockmore scarp
Prior to Praeger's record in 1903, this species was known from Benbulben, Co. Sligo, with small outlying populations in counties Donegal and Derry. Clare Island still remains the most southerly population in Ireland.

Silene dioica (L.) Clairv.
Synonym: *Lychnis diurna*
P^2; BSBI
Inaccessible rock ledges

Silene flos-cuculi (L.) Greuter & Burdet
Synonym: *Lychnis flos-cuculi*
P^f; D; R; McNeill 1999

Silene uniflora Roth
Synonym: *Silene maritima*
P^c; B; R; MJ
W sea cliffs

Spergula arvensis L.
P^{**vc}; R; BSBI
Ballytoohy, harbour area

Spergularia marina (L.) Griseb.
Synonym: *Spergularia salina*
P^2; D; R

Spergularia rupicola Lebel ex Le Jolis
Synonym: *Spergularia rupestris*
P^{c}; D; R; BSBI

Stellaria alsine Grimm
Synonym: *Stellaria uliginosa*
P^{f}; D; R; McNeill 1999

Stellaria graminea L.
P^{f}; D; R; BSBI

Stellaria media (L.) Villars
P^{**f}; D; R; BSBI

POLYGONACEAE

Fagopyrum esculentum Moench
BSBI
In a field to the right of the road between the harbour and the Abbey, also in foredunes at harbour.

Fallopia convolvulus (L.) Á. Löve
Synonym: *Polygonum convolvulus*
P^{**r}; BSBI
Ballytoohy

Oxyria digyna (L.) Hill
P^{1}; D; R; MJ
'Croaghmore scarp' (p. 23)

Persicaria amphibia (L.) Gray
Synonym: *Polygonum amphibium*
P^{1}; D; R; BSBI
Portnakilly, Maum

Persicaria hydropiper (L.) Spach
Synonym: *Polygonum hydropiper*
P^{f}; D; R; BSBI
Ruderal

Persicaria lapathifolia (L.) Gray
Synonym: *Polygonum lapathifolium*
P^{**f}; D; R; BSBI

Persicaria maculosa Gray
Synonym: *Polygonum persicaria*
P^{**f}; D; R; BSBI
Ruderal

Polygonum arenastrum Boreau
BSBI

Polygonum aviculare L.
P^{f}; D; R; BSBI

Polygonum oxyspermum* subsp. *raii (Bab.) D. Webb & Chater
Synonym: *Polygonum raii*
P^{1}; MJ
On beach near harbour (2011)

Rumex acetosa L.
P^{f}; D; R; BSBI

Rumex acetosella L.
P^{c}; D; R; BSBI

Rumex conglomeratus Murray
P^{*1}; MJ
Praeger was surprised to find but a single plant, which he considered casual. The species is present today on the margins of the saltmarsh at Kinnacorra, N of the harbour, Capnagower, July 2009.

Rumex crispus L.
P^{c}; D; R; BSBI

Rumex crispus × obtusifolius = R. × pratensis Mert. & Koch.
BSBI

Rumex crispus* subsp. *littoreus (J. Hardy) Akeroyd
BSBI

Rumex obtusifolius L.
P^{c}; D; R; BSBI
Abundant on roadsides

PLUMBAGINACEAE

Armeria maritima (Mill.) Willd.
P^{c}; D; R; BSBI

HYPERICACEAE

Hypericum androsaemum L.
P^{3}; D; R; BSBI
Knocknaveen, Maum, Portnakilly

Hypericum elodes L.
P^{vc}; D; R; BSBI

Hypericum humifusum L.
P[3]; D; R; MJ
On a freshly gravelled gateway near the Community Centre (2011)

Hypericum perforatum L.
BSBI

Hypericum pulchrum L.
P[f]; D; R; BSBI
Praeger cited the presence of the variety *procumbens* on the Knockmore scarp. This dwarf form with prostrate stems occurs in exposed habitats in N and W Scotland and in islands off W Ireland.

Hypericum tetrapterum Fries
P[f]; D; R; BSBI

MALVACEAE

Althaea officinalis L.
D; R
Sand dunes

Malva arborea (L.) Webb & Berthel.
Synonym: *Lavatera arborea*
{P, MJ}
Still growing near Granuaile guesthouse

DROSERACEAE

Drosera anglica Huds.
P[2]; D; R; BSBI
'*... patch of wet bog near Lough Avullin,*' (Praeger 1911a, p. 20) and still present in quaking *Sphagnum* bog around Lough Leinapollbauty.

Drosera intermedia Hayne in Dreves
R; BSBI
The BSBI record notes the identity as unconfirmed.

Drosera rotundifolia L.
P[c]; D; R; BSBI

VIOLACEAE

Viola arvensis Murray
P[*1]; R
Sand dunes at Harbour

[***Viola canina*** L.
Listed by Ryle as refound by Doyle, but neither a Doyle nor Praeger record have been traced.]

Viola palustris L.
P[c]; D; R; BSBI

Viola riviniana Reichb.
P[c]; D; R; BSBI

SALICACEAE

Populus nigra L.
{P}

Populus nigra × deltoides* = P. × *canadensis Moench
{P}
There is no sign of these plants today.

Salix aurita L.
P[f]; D; R; BSBI

Salix caprea L.
BSBI
Ballytoohy

Salix cinerea* subsp. *oleifolia Macreight
Synonym: *Salix cinerea*
P[f]; D; R; BSBI

Salix fragilis L.
BSBI

Salix herbacea L.
P[1]; D; MJ
Only seen at one site above the beginning of the 1200-foot path. The tree is quite submerged in a *Sphagnum* blanket.

Salix pentandra L.
{P; R}

Salix repens L.
P[c]; D; R; BSBI

Salix viminalis L.
{P; D; R; BSBI}

Salix viminalis × caprea × cinerea* = S. × *calodendron Wimm.
{BSBI}

Salix viminalis × cinerea* = S. × *smithiana Willd.
{P; BSBI}
Portnakilly

BRASSICACEAE

Brassica oleracea L.
BSBI

Brassica rapa L.
Synonym: *Brassica rapa* var. *briggsii*
P**c; D; R
Ruderal

Cakile maritima Scop.
D; R; McNeill 1999; MJ 2010
On sand beach near harbour. Praeger noted this as an unusual absence, and with the small area of beach sand it is not surprising that it may be missing in some years.

Capsella bursa-pastoris (L.) Medik.
P**f; D; R; BSBI

Cardamine flexuosa With.
P1; D; R; BSBI

Cardamine hirsuta L.
Pf; B; R
Bank by roadside

Cardamine pratensis L.
Pf; D; R; BSBI

Cochlearia danica L.
P2; D; R; BSBI

Cochlearia officinalis agg.
Pc; D; R; BSBI

Cochlearia officinalis subsp. ***scotica*** (Druce) P.S. Wyse Jackson
Synonym: *Cochlearia groenlandica*
Plocal
West end

Hesperis matronalis L.
{D}; R; McNeill 1999

Lepidium coronopus (L.) Al-Shehbaz
Synonyms: *Coronopus squamatus*; *Senebiera coronopus*
P*c; D; R; BSBI

Lepidium didymum L.
Synonym: *Coronopus didymus*
D; R; BSBI
Cliff S of harbour and on sand dunes

Nasturtium officinale agg.
Synonyms: *Rorippa nasturtium-aquaticum* agg.; *N. officinalis*
Pf; D; R; BSBI

Raphanus raphanistrum* subsp. *raphanistrum L.
Synonym: *Raphanus raphanistrum*
P**3; R

Sinapis alba L.
Synonym: *Brassica alba*
P**c

Sinapis arvensis L.
Synonym: *Brassica sinapis*
P**f; D; R; BSBI

Sisymbrium officinale (L.) Scop.
P*2; D; R

EMPETRACEAE

Empetrum nigrum L.
Pr; D; R; BSBI
Capnagower and area around Creggan Lough

ERICACEAE

Calluna vulgaris (L.) Hull
Pc; D; R; BSBI

Erica cinerea L.
Pc; D; R; BSBI

Erica erigena R. Ross
Synonym: *Erica mediterranea*
P1; D; R; BSBI

Erica tetralix L.
Pc; D; R; BSBI

Rhododendron ponticum L.
R
It is likely that this record is of a cultivated plant as no other records are known.

Vaccinium myrtillus L.
Pf; D; R; BSBI

Vaccinium oxycoccos L.
D; R; DS & MJ
Cutaway bog in Creggan townland

PRIMULACEAE

Anagallis arvensis L.
P**f; R; BSBI
Ruderal

Centunculus minimus L.
Synonym: *Anagallis minima*
B; BSBI
E of Signal tower. Praeger listed this species as present on Inishturk and Inishbofin, but absent from Clare Island.

Anagallis tenella (L.) L.
Pvc; D; R; BSBI
As in Praeger's day this is still a remarkably abundant plant on the island, characterising even the most tightly grazed grassland in the extreme west of the island.

Glaux maritima L.
P2; D; R

Lysimachia nemorum L.
Pc; D; R; BSBI

Primula vulgaris Hudson
Pc; D; R; BSBI

Samolus valerandi L.
Pf; D; R; BSBI

ESCALLONIACEAE

Escallonia macrantha Hook. & Arn.
{D}; R; BSBI

GROSSULARIACEAE

Ribes nigrum L.
{D}; R

Ribes rubrum L.
R

Ribes sanguineum Pursh
{P}; D
Planted at hotel (Praeger, p. 43)

Ribes uva-crispa L.
Synonym: *Ribes grossularia*
{P}; R

CRASSULACEAE

Sedum acre L.
P1; D; R
Sand dunes at harbour

Sedum album
{MJ}
Planted on several graves about the Abbey. The species has the potential to spread.

Sedum anglicum Hudson
Pc; D; R; BSBI

Sedum rosea (L.) Scop.
Synonym: *Sedum rhodiola*
Plocal; D; R; BSBI
The population by Signal Tower mentioned by Praeger has now gone (1911, p. 29).

Sedum telephium L.
{P}

Umbilicus rupestris (Salisb.) Dandy in Riddelsd.
Synonym: *Cotyledon umbilicus*
P1; D; R; McNeill 1999

SAXIFRAGACEAE

Chrysosplenium oppositifolium L.
P2; B; R; BSBI
Wood S of Portlea, Knocknaveen

Saxifraga oppositifolia L.
P1; R; MJ
Along 1200-foot path

Saxifraga rosacea Moench
Synonym: *Saxifraga decipiens*
P1; R; MJ
Throughout northern sea cliffs

Saxifraga spathularis Brot.
Synonym: *Saxifraga umbrosa sensu* Praeger
Pc; D; R; BSBI

Saxifraga spathularis × hirsuta = S. × polita Haw.
Synonym: *Saxifraga geum*
P1; R; MJ
Praeger noted hybrids between what he took to be true *S. hirsuta* (his *S. geum*) and *S. spathularis*. (1911, p. 29). The population he saw is still extant in a west-facing wet gully on the north cliffs near the 1200-foot path.

ROSACEAE

Alchemilla glabra Neyg.
R; MJ
Abundant on the 1200-foot path

Alchemilla xanthochlora Rothm.
Synonym: *Alchemilla vulgaris*
P^{r}; MJ
Knocknaveen

Aphanes arvensis agg.
Synonym: *Alchemilla arvensis*
P^{**r}; R; BSBI
Knocknaveen

Cotoneaster microphyllus Wall. ex Lindl.
D; R; BSBI (as *C. integrifolius*, and as *Lonicera microphyllus*)
On Knocknaveen cliff.

Crataegus monogyna Jacq.
Synonym: *Crataegus oxyacantha*
{P}; D; R; BSBI
In Praeger's day this was recorded as a single 'wild' plant on Inishturk, and as hedging near the harbour of Clare Island (1911, p. 43). Now more widespread.

Filipendula ulmaria (L.) Maxim.
Synonym: *Spiraea ulmaria*
P^{f}; D; R; BSBI

Fragaria vesca L.
P^{1}; D; R
Knocknaveen

Geum rivale L.
P^{1}; R
Croaghmore scarp (p. 23); Knocknaveen.

Malus sylvestris *sensu lato*
Synonym: *Pyrus malus*
P^{1}
'One very old tree … by the Ooghganny stream' (p. 29)

Potentilla anglica Laich.
Synonym: *Potentilla procumbens*
P^{2}; BSBI

Potentilla anserina L.
P^{c}; D; R; BSBI

Potentilla erecta (L.) Räusch.
Synonym: *Potentilla tormentilla*
P^{c}; D; R; BSBI

Potentilla fruticosa L.
{R}
Ryle reports this as self-sowing around its planted sites. All the plants I saw were pale-flowered cultivars.

Comarum palustre L.
Synonym: *Potentilla palustris*
P^{c}; D; R; BSBI

Potentilla sterilis (L.) Garcke
Synonym: *Potentilla fragariastrum*
P^{2}; D

Prunus cerasus L.
{P}; R+

Prunus spinosa L.
P^{local}; D; R; BSBI

Rosa canina L.
P^{1}; D; R; BSBI

Rosa spinosissima L.
Synonym: *Rosa pimpinellifolia*
P^{r}; R

Rubus agg.
P^{f}; D; R; BSBI

Rubus iricus W.M. Rogers
P^{f}; BSBI
BSBI record is qualified with a question mark

Rubus plicatus Weihe & Nees
P^{f}

Rubus polyanthemus Lindeb.
Synonym: *Rubus pulcherrimus*
P^{f}

Rubus saxatilis L.
P^{1}; MJ
Croaghmore scarp (p. 23)

Rubus ulmifolius Schott
P^{f} (as *Rubus rusticanus*); BSBI

Sorbus aucuparia L.
Synonym: *Pyrus aucuparia*
P^2; D; R; BSBI
A single multi-trunked individual still present on Knocknaveen, and also present in the Lassau woodland, probably the two sites that Praeger lists. A further individual grows a few metres to the west of the site.

LEGUMINOSAE

Anthyllis vulneraria L.
P^f; D; R; BSBI

Lathyrus japonicus* subsp. *maritimus (L.) P. Ball
R
Found in the summer of 1993 (Ryle and Doyle 1998). It was located on a shingle beach along the southern coast near Portnakilly. A rare and sporadic plant of western coasts.

Lathyrus linifolius (Reichard) Bässler
Synonym: *Lathyrus macrorrhizus*
P^3; D; R; BSBI

Lathyrus pratensis L.
P^f; D; R; BSBI

Lotus corniculatus L.
P^c; D; R; BSBI

Lotus pedunculatus Cav.
Synonym: *Lotus uliginosus*
P^1; D; R; BSBI

Trifolium dubium Sibth.
P^{*2}; D; R; BSBI

Trifolium pratense L.
P^f; D; R; BSBI

Trifolium repens L.
P^c; D; R; BSBI

Ulex europaeus L.
P^1; D; R; BSBI

Vicia cracca L.
P^c; D; R; BSBI

Vicia hirsuta (L.) Gray
P^{*1}

Vicia sepium L.
P^f; D; R; BSBI

ELAEAGNACEAE

Hippophae rhamnoides L.
{P}; R+
As a hedging plant

HALORAGACEAE

Myriophyllum alterniflorum DC.
P^1; B; R; BSBI
Stream, W end

Gunnera tinctoria (Molina) Mirbel
D; R; BSBI
First recorded by David McClintock in 1968 (McClintock 1969). Alongside most water courses crossed by the southern road; extremely abundant in Fawnglass on all slopes behind the beach and to south of GAA pitch. Extensive colonies on seacliffs near Alnamarnagh. Occasional single plants on the cliffs of the south and east coasts. This plant is currently the target of an eradication programme funded by the Heritage Council, and run by Mayo County Council and the National Botanic Gardens.

LYTHRACEAE

Lythrum portula (L.) D. Webb
Synonym: *Peplis portula*
P^f; R

Lythrum salicaria L.
P^c; D; R; BSBI

ONAGRACEAE

Circaea lutetiana L.
P^{*1}; R
Cottage wall on way to lighthouse

Epilobium brunnescens (Cockayne) Raven & Engelhorn
Synonym: *Epilobium nerterioides*
D; R; BSBI

[***Epilobium ciliatum*** Raf.
BSBI
This species has been marked on the Ballytoohy field card (L6987) of 8 July 2007, but then marked with an x at each end, implying an error. Its presence on the composite list for the BSBI list is therefore open to doubt.]

Epilobium montanum L.
P^{f}; B; R; MJ
Bluff near Knocknaveen

Epilobium obscurum Schreb.
P^{f}; R; BSBI

Epilobium palustre L.
P^{f}; D; R; BSBI

Epilobium parviflorum Schreb.
P^{f}; D; R; BSBI

Fuchsia magellanica Lam.
Synonym: *Fuchsia riccartonii*
{P}; D; R; BSBI

AQUIFOLIACEAE

Ilex aquifolium L.
P^{r}; D; R; BSBI
Knocknaveen

EUPHORBIACEAE

Euphorbia helioscopia L.
P^{**f}; D
Ruderal

Euphorbia peplus L.
P^{**r}; B; R; BSBI
Churchyard

LINACEAE

Linum catharticum L.
P^{f}; D; R; BSBI

Radiola linoides Roth
P^{f}; D; R; BSBI
Remarkably abundant everywhere on tracks or broken turf surfaces

POLYGALACEAE

Polygala serpyllifolia Hose
Synonym: *Polygala depressa*
P^{f}; D; R; BSBI

Polygala vulgaris L.
P^{1}; D; R; MJ
Knockmore scarp; cliff summits near lighthouse

STAPHYLEACEAE

[***Staphylea pinnata*** L.
An Atlas 2000 record, dated 1910, seems highly improbable and most likely an error of transcription.]

ACERACEAE

Acer pseudoplatanus L.
{P}; D; R; BSBI
Portnakilly (2007)

OXALIDACEAE

Oxalis acetosella L.
P^{f}; D; R; BSBI

{***Oxalis articulata*** Savigny
{B}; R
Planted on graves in churchyard}

GERANIACEAE

Geranium dissectum L.
P^{*2}; D; R; BSBI

Geranium molle L.
P^{f}; D; R; BSBI

Geranium robertianum L.
P^{f}; D; R; BSBI

[***Geranium sanguineum*** L.
R
The presence of this species other than as a possible garden escape seems highly improbable.]

ARALIACEAE

Hedera helix L.
P^{f}; D; R; BSBI

APIACEAE

Aegopodium podagraria L.
{P}; D; R

Angelica sylvestris L.
P^{c}; D; R; BSBI
This species is remarkably common on steep sea cliffs at the western end of the island.

Apium graveolens L.
D; R
Saline flat

Apium inundatum (L.) Reichb.f. in Reichb. & Reichb. f.
P^{2}; D; R; BSBI

Apium nodiflorum (L.) Lagasca
P^{3}; D; R; BSBI

Conium maculatum L.
P^{*r}; D; R; BSBI
Granuaile's castle (Tower House)

Conopodium majus (Gouan) Loret in Loret & Barrandon
Synonym: *Conopodium denudatum*
P^{3}; D; R; BSBI
Maum

Crithmum maritimum L.
P^{f}; B; R; MJ
W sea cliffs

Daucus carota L.
P^{c}; D; R; BSBI

Heracleum sphondylium L.
P^{f}; D; R; BSBI

Hydrocotyle vulgaris L.
P^{c}; D; R; BSBI

Oenanthe crocata L.
P^{c}; D; R; BSBI

Sanicula europaea L.
P^{2}; B; R; BSBI
In the wood NW of Portlea

GENTIANACEAE

Centaurium erythraea Rafn
Synonym: *Erythraea centaurium*
P^{f}; D; R; BSBI

Gentianella campestris (L.) Boerner
Synonym: *Gentiana campestris*
P^{f}

SOLANACEAE

Solanum tuberosum L.
{BSBI}
Washed up on beach

CONVOLVULACEAE

Calystegia sepium (L.) R. Br.
P^{*2}; D; R; BSBI

Calystegia silvatica (Kit.) Griseb.
BSBI

Convolvulus arvensis L.
P^{1}
Still present as a weed of crops.

MENYANTHACEAE

Menyanthes trifoliata L.
P^{f}; D; R; BSBI

BORAGINACEAE

Myosotis arvensis (L.) Hill (P^{f}; D; R)
Myosotis discolor Pers.
Synonym: *Myosotis versicolor*
P^{1}

Myosotis laxa Lehm.
Synonym: *Myosotis caespitosa*
P^{c}; D; R; BSBI

Myosotis secunda Al. Murray
Synonym: *Myosotis repens*
P^{c}; R; BSBI

Symphytum officinale L.
P^{*1}; B; R
Ballytoohy; garden (Brodie 1991)
P^{f}; D; R; BSBI

Ajuga reptans L.
P^{1}; R; BSBI
Maum

Galeopsis tetrahit L.
P^{**1}; D; BSBI

LAMIACEAE

[***Lamium amplexicaule*** L.
Listed by Ryle as a Praeger record, but no such record has been traced.]

Lamium confertum Fries
Synonym: *Lamium intermedium*
P^{**2}

Lamium hybridum Villars
P^{**1}; D; R; BSBI

Lamium purpureum L.
P**f; D; R; BSBI

Mentha aquatica L.
Synonym: *Mentha hirsuta*
Pf; D; R; BSBI

Mentha arvensis L.
P*; R
Praeger omitted this species from his flora list, but mentions it as a roadside associate (1911a, p. 47).

Mentha spicata L.
{D}; R
In gardens (Doyle and Foss 1986).

Prunella vulgaris L.
Pc; D; R; BSBI

Stachys arvensis (L.) L.
P**2
Ruderal

Stachys palustris L.
Pf; D; R; BSBI

Stachys sylvatica L.
P2; D; BSBI
Knocknaveen (p. 21), Maum

Teucrium scorodonia L.
Pf; D; R; BSBI

Thymus polytrichus A. Kerner ex Borbás
Synonym: *Thymus serpyllum, T. chamaedrys*
Pc; D; R; BSBI
In Praeger's day there was a more narrow definition of the species and thyme was split into two segregates, a distinction ignored from 1950s onwards. Praeger recorded *T. serpyllum* as common, and *T. chamaedrys* as rare (1911, p. 26).

CALLITRICHACEAE

Callitriche brutia Petagna
Synonym: *Callitriche pedunculata*
P2

Callitriche stagnalis *sensu lato*
Pf; D; R; BSBI

PLANTAGINACEAE

Littorella uniflora (L.) Asch.
Synonym: *Littorella lacustris*
P3; D; R; BSBI

Plantago coronopus L.
Pc; D; R; BSBI

Plantago lanceolata L.
Pc; D; R; BSBI

Plantago major L.
P*r; D; R; BSBI

Plantago maritima L.
Pc; D; R; BSBI

OLEACEAE

Fraxinus excelsior L.
{P1}; D; R; BSBI

Ligustrum ovalifolium Hassk.
BSBI

Ligustrum vulgare L.
Synonym: *Ligustrum* sp.
{P}; D; R

Digitalis purpurea L.
Pf; D; R; BSBI

SCROPHULARIACEAE

Euphrasia arctica Lange ex Rostrup
Synonym: *Euphrasia brevipila*
P; BSBI

[***Euphrasia arctica* subsp. *borealis*** (F. Towns.) Yeo
An Atlas 2000 record, dated 1910, possibly an error]

Euphrasia arctica* × *micrantha* = *Euphrasia* × *difformis
BSBI

Euphrasia micrantha Reichb.
BSBI
Capnagower

Euphrasia nemorosa (Pers.) Wallr.
Synonym: *Euphrasia curta* var. *glabrescens*
P

Euphrasia officinalis agg.
Synonym: *Euphrasia gracilis*
P^c; D; R; BSBI

Euphrasia scottica Wettst.
P

Euphrasia tetraquetra (Bréb.) Arrond.
Synonym: *Euphrasia occidentalis*
P; BSBI

Hebe elliptica × speciosa = H. × franciscana (Eastw.) Souster
Synonym: *Hebe speciosa*
{D}; R; BSBI
Occasional individuals that have self-seeded outside gardens, on cliffs near harbour.

Odontites vernus (Bellardi) Dumort.
Synonym: *Bartsia odontites*
P^f; D; R; BSBI

Pedicularis palustris L.
P^2; B; R; MJ
Bog near Lough Leinapollbauty

Pedicularis sylvatica L.
P^f; D; R
These records are not identified below species level and may all be subsp. *hibernica*.

Pedicularis sylvatica* subsp. *hibernica D. Webb
BSBI

Rhinanthus minor L.
Synonym: *Rhinanthus crista-galli*
P^c; D; R

Rhinanthus minor* subsp. *stenophyllus O. Schwarz
BSBI

Scrophularia nodosa L.
P^1; DS & Gerry Sharkey
By cattle grid in Ballytoohy More (2007).

Veronica agrestis L.
P^{**2}; D

Veronica arvensis L.
P^{*f}; D; R

Veronica beccabunga L.
P^f; D; R; McNeill 1999

Veronica chamaedrys L.
P^f; D; R; BSBI

Veronica officinalis L.
P^3; D

Veronica persica Poiret
Synonym: *Veronica tournefortii*
P^{**2}; D; McNeill 1999

Veronica polita Fries
P^{**2}

Veronica scutellata L.
P^3; BSBI

Veronica serpyllifolia L.
P^f; B; R; BSBI

OROBANCHACEAE

Orobanche alba Stephan ex Willd.
Synonym: *Orobanche rubra*
P^1
Cliffs below lighthouse

LENTIBULARIACEAE

Pinguicula lusitanica L.
P^r; D; R; BSBI

Pinguicula vulgaris L.
P^f; D; R; BSBI

Utricularia intermedia sensu lato
P^1; D
Pools on the cliff near the lighthouse

Utricularia minor L.
P^r; D; R; BSBI

CAMPANULACEAE

Campanula rotundifolia L.
P^1; D; R
Knockmore scarp

Jasione montana L.
P^f; D; R; BSBI

RUBIACEAE

Galium aparine L.
P[c]; D; R; BSBI

Galium palustre L.
P[f]; D; R; BSBI

Galium saxatile L.
P[c]; D; R; BSBI

Galium verum L.
P[f]; D; R; BSBI

CAPRIFOLIACEAE

Lonicera periclymenum L.
P[f]; D; R; BSBI

Sambucus nigra L.
{P}; D; R; BSBI
Planted, mostly about cottages (Praeger, p. 43)

Symphoricarpos albus (L.) S.F. Blake
BSBI

VALERIANACEAE

Valeriana officinalis L.
Synonym: *Valeriana sambucifolia*
P[f]; D; R; BSBI

DIPSACACEAE

Succisa pratensis Moench (*Scabiosa succisa*)
P[c]; D; R; BSBI

ASTERACEAE

Achillea millefolium L.
P[f]; D; R; BSBI

Achillea ptarmica L.
P[f]; D; R
Roadside, Kill; west end

Antennaria dioica (L.) Gaertner
P[2]

Arctium minus (Hill) Bernh.
Synonym: *Arctium newbouldii*
P[1]; D; R; BSBI
Restricted to harbour sand, in Praeger's day (1911, p. 45). Now also at side of roads on south side of island and Ballytoohy. This species has spread in Ireland since Praeger's day.

Artemisia vulgaris L.
P[*f]; D; R; BSBI

Aster tripolium L.
P[f]; D; R

Bellis perennis L.
P[f]; D; R; BSBI

Bidens tripartita L.
P[3]
Roadside ditch near chapel

Carduus nutans L.
Colgan and Scully (1898, p. 193)
A Miss Lawless collected a specimen during a brief visit to the island and sent it to A.G. More. Praeger discovered that A.G. More's original edition of *Cybele Hibernica* in the Royal Irish Academy notes the record in his own hand, indicating that the identity is almost certainly correct (1911a, p. 30). Praeger doubted the record. However, it is possible that the plant, a seed contaminant in Ireland, arrived briefly with imported material.

Centaurea nigra L.
P[f]; D; R; BSBI

Glebionis segetum (L.) Fourr.
Synonym: *Chrysanthemum segetum*
P[**1]; R; BSBI
Ballytoohy

Cirsium arvense (L.) Scop.
Synonym: *Cnicus arvensis*
P[*f]; D; R; BSBI

Cirsium dissectum (L.) Hill
Synonym: *Cnicus pratensis*
P[c]; D; R; BSBI

Cirsium palustre (L.) Scop.
Synonym: *Cnicus palustris*
P[c]; D; R; BSBI

Cirsium vulgare (Savi) Ten.
Synonym: *Cnicus lanceolatus*
P[*c]; D; R; BSBI

Crepis capillaris (L.) Wallr.
Synonym: *Crepis virens*
P[*1]; D; R; BSBI
Ballytoohy

Eupatorium cannabinum L.
P^{2}
Praeger found this growing only on seacliffs. He remarked on it occuring in the same habitat as *Angelica sylvestris*—the steep soil slopes above cliffs at the extreme western end of the island (Praeger 1911a, p. 16).

Filago vulgaris Lam.
Synonym: *Filago germanica*
P^{1}

Gnaphalium sylvaticum L.
P^{2}

Gnaphalium uliginosum L.
P^{c}; B; R; BSBI
Growing on the weathered tarmac surfaces in several places, and on newly laid gravel surfaces.

Hieracium anglicum Fries
P^{2}; MJ
Sea-stacks SE of lighthouse

Hieracium hypochaeroides Gibson
P^{1}

Hypochaeris radicata L.
P^{c}; D; R; BSBI

Lapsana communis L.
P^{*f}; D; R; BSBI

Scorzoneroides autumnalis (L.) Moench
Synonym: *Leontodon autumnalis*
P^{c}; D; R; BSBI

Leontodon saxatilis Lam.
Synonym: *Leontodon hirtus*
P^{2}; D; R; BSBI

Leucanthemum vulgare Lam.
Synonym: *Chrysanthemum leucanthemum*
P^{r}; D; R; BSBI
Portnakilly, harbour area

Matricaria discoidea DC.
P^{*c}; D; R; BSBI

Olearia macrodonta Baker
{R}
There are several large hedges of *Olearia traversii* about the island in the townlands of Glen, Kill and Maum. There is no indication that this or the latter species has escaped as yet.

Pilosella officinarum F. Schultz & Schultz-Bip.
Synonym: *Hieracium pilosella*
P^{f}; D; R; BSBI

Pulicaria dysenterica (L.) Bernh.
P^{1}; R

Saussurea alpina (L.) DC.
P^{1}
Knockmore scarp. Without a thorough survey of the cliffs it is difficult to ascertain the status of this species.

Senecio aquaticus Hill
P^{f}; D; R; BSBI

Senecio jacobaea L.
P^{f}; D; R; BSBI

Senecio sylvaticus L.
P^{r}

Senecio vulgaris L.
P^{**c}; D; R; BSBI

Solidago virgaurea L.
P^{f}; D; R; BSBI

Sonchus arvensis L.
P^{2}; D; R; BSBI
Harbour area

Sonchus asper (L.) Hill
P^{**f}; D; R; BSBI

Sonchus oleraceus L.
P^{f}; D; R; BSBI

Tanacetum vulgare L.
{B}; R
In a garden near Ballytoohy

Taraxacum agg.
Synonym: *Taraxacum officinale* agg.
P^{f}; D; R; BSBI

Tripleurospermum maritimum (L.) Koch
(as *Matricaria inodora*)
P^{c}; D; R; BSBI

Tussilago farfara L.
P^{f}; D; R; BSBI

ALISMATACEAE

Alisma plantago-aquatica L.
Synonym: *Alisma plantago*
P[1]

Baldellia ranunculoides (L.) ParL.
Synonym: *Alisma ranunculoides*
P[1]; R

JUNCAGINACEAE

Triglochin maritima L.
Synonym: *T. maritimum*
P[1]; D; R
Saline flat by GAA pitch

Triglochin palustris L.
Synonym: *Triglochin palustre*
P[f]; D; R; BSBI

POTAMOGETONACEAE

Potamogeton natans L.
P[3]; R; MJ

Potamogeton polygonifolius Pourret
P[c]; D; R; BSBI

Potamogeton pusillus L.
P[2]

LEMNACEAE

Lemna minor L.
P[1]; D; R; MJ
In drains on road towards Abbey

JUNCACEAE

Juncus acutiflorus Ehrh. ex Hoffm.
P[f]; D; R; McNeill 1999
Harbour area

Juncus articulatus L.
Synonym: *Juncus lamprocarpus*
P[f]; D; R; BSBI

Juncus bufonius *senu. lato*
P[c]; D; R

Juncus bufonius *sensu stricta* L.
BSBI

Juncus bulbosus L.
Synonym: *Juncus supinus*
P[f]; D; R; BSBI

Juncus conglomeratus L.
D; R; BSBI
Knocknaveen, Capnagower, SE of the lighthouse

Juncus effusus L.
P[c]; D; R; BSBI

Juncus effusus var. ***spiralis***
BSBI

Juncus gerardii Lois.
P[1]; D; R; McNeill 1999

Juncus maritimus Lam.
P[1]; D; R

Juncus squarrosus L.
P[f]; D; R; BSBI

Juncus subnodulosus Schrank
Synonym: *Juncus obtusiflorus*
P[3]; D; R

Luzula campestris (L.) DC. in Lam. & DC.
P[f]; D; R

Luzula multiflora (Ehrh.) Lej.
Synonym: *Luzula erecta*
P[f]; D; R; BSBI
Harbour area

Luzula multiflora* subsp. *congesta (Thuill.) Arcang. (BSBI)
Maum

Luzula sylvatica (Hudson) Gaudin
Synonym: *Luzula maxima*
P[3]; D; R; BSBI

CYPERACEAE

Bolboschoenus maritimus (L.) Palla in Koch
Synonym: *Scirpus maritimus*
P[1]; D; R

Carex arenaria L.
P[1]; D; R; BSBI
Sand dunes at Harbour

Carex binervis Smith
P[c]; D; R; BSBI
On the ridge of Knockmore (Praeger 1911, p. 22).

Carex binervis* × *hostiana
BSBI

Carex caryophyllea Latour.
Synonym: *Carex praecox*
P^r; R

Carex demissa Horem.
Synonym: *Carex viridula* subsp. *oedocarpa*
D; BSBI
Knockmore, Knocknaveen, Capnagower, harbour area

Carex dioica L.
P^1; D; R; BSBI
Harbour area

Carex distans L.
P^f; D; R

Carex echinata Murray
P^f; D; R; BSBI

Carex extensa Gooden.
P^1
It is surprising that this very characteristic long-bracted species has not been refound, Praeger gave no indication of its single site. Modern investigators have tended to record the eastern rather than the western end of the island.

Carex flacca Schreb.
Synonym: *Carex glauca*
P^c; D; R; BSBI

Carex hostiana DC.
Synonym: *Carex hornschuchiana*
P^2; D; R

Carex lepidocarpa Tausch
Synonym: *C. viridula* subsp. *brachyrrhyncha*
Along eastern or southern coasts, L78, David McNeill recording card 15 July 1999.

Carex leporina L.
Synonym: *Carex ovalis*
P^f; D; R; BSBI

Carex limosa L.
P^1; D; R; BSBI
Lough Leinapollbauty, Lough Merrignagh

Carex nigra (L.) Reichard
Synonym: *Carex vulgaris*
P^c; D; R; BSBI

Carex oederi Retz.
Synonyms: *C. viridula* subsp. *viridula; C. serotina*
P; D
Praeger omitted this from his list, but mentioned it elsewhere in the text (1911, p. 17).

Carex otrubae Podp. (as *Carex vulpina*)
P^2; D; R; BSBI

Carex panicea L.
P^f; D; R; BSBI

Carex paniculata L.
P^3; R; BSBI
Area around Creggan Lough

Carex pilulifera L.
P^r; R

Carex pulicaris L.
P^r; D; R; BSBI
Knocknaveen, SE of lighthouse

Carex rostrata Stokes in With.
Synonym: *Carex ampullacea*
P^f; D; R; BSBI

Carex sylvatica Hudson
P^1; R

Carex viridula *sensu lato* Michx.
Synonyms: *Carex flava; Carex viridula*
P^c; R

Eleocharis multicaulis (Smith) Desv.
P^c; D; R; BSBI

Eleocharis palustris (L.) Roemer & Schultes
P^f; D; R; BSBI

Eriophorum angustifolium Honck.
P^r; D; R; BSBI

Eriophorum vaginatum L.
P^r; D; R; BSBI

Isolepis cernua (M. Vahl) Roemer & Schultes
Synonym: *Scirpus savii*
P^r; D; R

Isolepis fluitans (L.) R. Br.
Synonym: *Scirpus fluitans*
P^f; D; R; BSBI

Isolepis setacea (L.) R. Br.
Synonym: *Scirpus setaceus*
P^2; B; R; BSBI
Knocknaveen, Ballytoohy, harbour area

Rhynchospora alba (L.) M. Vahl
P^1; D; R; BSBI
Harbour area

Schoenoplectus lacustris (L.) Palla
Synonym: *Scirpus lacustris*
P^2; D

Schoenus nigricans L.
P^f; D; R; BSBI

Trichophorum caespitosum (L.) Hartman
Synonym: *Scirpus cespitosus*
P^r; D; R; BSBI
SE of lighthouse

POACEAE

Agrostis canina L.
B; R; BSBI
Harbour area, Capnagower; damp grassland

Agrostis capillaris L.
Synonym: *Agrostis vulgaris*
P^c; D; R; BSBI

Agrostis stolonifera L.
Synonym: *Agrostis alba*
P^1; D; R; BSBI

Agrostis vinealis Schreb.
BSBI
Capnagower

Aira caryophyllea L.
P^f; D; R; BSBI

Aira praecox L.
P^c; D; R; BSBI

Alopecurus geniculatus L.
P^1; D; R; BSBI
Ballytoohy

Ammophila arenaria (L.) Link
D; R; BSBI
Sand dunes

Anthoxanthum odoratum L.
P^f; R; BSBI

Arrhenatherum elatius (L.) P. Beauv. ex J.S. Presl & C. Presl
Synonym: *Arrhenatherum avenaceum*
P^f; D; R; BSBI

Arrhenatherum elatius var. ***bulbosum*** (Willd.) St-Amans
BSBI
Maum, Alnamarnagh, cliff-top grasslands

Avena sativa L.
{D}; R
Grown as a crop plant

Brachypodium sylvaticum (Hudson) P. Beauv.
P^f; D; R; BSBI

Bromus commutatus Schrader
P^{*f}; D; R; BSBI
Near Abbey

Bromus hordeaceus L.
Synonym: *Bromus mollis*
P^{*1}; D; R; BSBI

Bromus racemosus L.
BSBI

[***Calamagrostis epigejos*** (L.) Roth
Ryle cited this as a 1911 record, but there is no mention of it by Praeger.]

Catabrosa aquatica (L.) P. Beauv.
BSBI

Catapodium marinum (L.) C.E. Hubb.
Synonym: *Desmazeria marina*
D; R; BSBI

Cynosurus cristatus L.
P^2; D; R; BSBI
Portnakilly, Capnagower, Knocknaveen

Dactylis glomerata L.
P^f; D; R; BSBI

Danthonia decumbens (L.) DC. in Lam. & DC.
Synonym: *Triodia decumbens*
P^f; D; R; BSBI

Deschampsia caespitosa (L.) P. Beauv.
As: *Deschampsia caespitose* in earlier works
P^1; R; BSBI
SE of Knockmore

Deschampsia flexuosa (L.) Trin.
P^3; D; R; BSBI
SE of Knockmore

Elytrigia juncea* subsp. *boreoatlantica (Simonet & Guin.) N. Hylander
Synonym: *Agropyron junceum*
P^1; D; R; McNeill 1999
Harbour sand

Elytrigia repens (L.) Desv. ex Nevski
Synonym: *Agropyron repens*
P^1; R

Festuca ovina L.
P^f; D; R; BSBI

Festuca rubra L.
P^f; D; R; BSBI

Festuca vivipara (L.) Smith
D; R; BSBI
Knocknaveen, Capnagower, harbour area

Glyceria fluitans (L.) R. Br.
P^c; D; R; BSBI

Holcus lanatus L.
P^{*1}; D; R; BSBI
Now widespread

Koeleria macrantha (Ledeb.) Schultes in Schultes & Schultes f.
Synonym: *Koeleria cristata*
P^f; D; R; BSBI

Lolium perenne L.
{P as a crop}; D; R; BSBI

Molinia caerulea (L.) Moench
P^f; D; R; BSBI

Nardus stricta L.
P^f; D; R; BSBI

Phalaris arundinacea L.
P^1

Phleum pratense L.
BSBI

Phragmites australis (Cav.) Trin. ex Steudel
Synonym: *Phragmites communis*
P^2; D; R; BSBI

Poa annua L.
P^{*f}; D; R; BSBI

Poa humilis Ehrh. ex Hoffm.
BSBI

Poa pratensis *sensu lato*
P^f; D; R

Poa trivialis L.
P^f; D; R; BSBI

Puccinellia maritima (Hudson) ParL.
D; R
Saline flat at harbour.

× *Schedolium loliaceum* (Huds.) Holub
Synonym: *Festuca rottboellioides*
P^2
Sand dunes at harbour.

Vulpia bromoides (L.) Gray
Synonym: *Festuca sciuroides*
P^f; D; R; BSBI

SPARGANIACEAE

Sparganium angustifolium Michaux
Synonym: *Sparganium affine*
P^2; D; R

Sparganium erectum L.
Synonym: *Sparganium ramosum*
P^f; D; R; McNeill 1999

Sparganium natans L.
Synonym: *Sparganium minimum*
P^2; D

LILIACEAE

Allium babingtonii Borrer
Synonym: *Allium ampeloprasum* var. *babingtonii*
R; BSBI
A. ampeloprasum in Ryle

Hyacinthoides non-scripta (L.) Chouard ex Rothm.
Synonym: *Scilla nutans*
P^2; D; R
Probably more widespread than in Praeger's day

Narthecium ossifragum (L.) Hudson
P[f]; D; R; BSBI

IRIDACEAE

Crocosmia pottsii* × *aurea* = *C.* × *crocosmiiflora (Lemoine ex Burb. & Dean) N.E. Br.
D; R; BSBI
While exceptionally abundant in a few sites, such as the mill stream at Maum, this species is still not widespread on the island.

Iris pseudacorus L.
P[f]; D; R; BSBI

PHORMIACEAE

Phormium tenax J.R. & G. Forst.
{D}; R; BSBI

ORCHIDACEAE

Cephalanthera longifolia (L.) Fritsch
Synonym: *Cephalanthera ensifolia*
P[1]
The Portlea *Erica erigena* site is a large area of bracken and bog myrtle thicket, and the refinding of Praeger's record of this and *Listera ovata* would require considerable attention (1911, p. 30).

[***Coeloglossum viride*** (L.) Hartm.
Synonym: *Dactylorhiza viridis*
Listed by Ryle as having been refound by Doyle, but Doyle and Foss (1986) merely report a BSBI atlas record, which dates from 1910 with no further details, which suggests an error has occurred].

Dactylorhiza incarnata* subsp. *incarnata (L.) Soó
Synonym: *Orchis incarnata*
P[f]; D; R

Dactylorhiza incarnata* subsp. *pulchella (Druce) Soó
BSBI
The forgoing subspecies was not recorded by the BSBI visit of 2007 and this taxon was marked with a question mark also, suggesting that the population does not fit a particular variant of *D. incarnata*.

Dactylorhiza kerryensis (Wilm.) P. Hunt & Summerh.
Synonym: *Dactylorhiza majalis* subsp. *occidentalis*
B; R; BSBI
Ditches and water logged areas by GAA pitch near harbour, and Portnakilly.

Dactylorhiza maculata* subsp. *ericetorum (E.F. Linton) P. Hunt & Summerh.
Synonym: *Orchis maculata*
P[c]; D; R; BSBI

Neottia cordata (L.) Rich.
Synonym: *Listera cordata*
P[1]
Praeger records this as 'the only mountain plant … (800 to 1,520 ft), amongst an otherwise continuous *Calluna* canopy' (Praeger 1911a, p. 22).

Neottia ovata (L.) Bluff & Fingerh.
Synonym: *Listera ovata*
P[1]
Portlea *Erica erigena* site.

Orchis mascula (L.) L.
P2; R
Northern cliffs; inaccessible rock ledges.

Platanthera chlorantha (Custer) Reichb. in Moessler
Synonym: *Habenaria chloroleuca*
P3; R; McNeill 1999.

REFERENCES

Brodie, J. 1991 Some observations on the flora of Clare Island, western Ireland. *Irish Naturalists' Journal* **23** (9), 376–7.

Colgan, N. 1901 Notes on Irish Topographical Botany, with some remarks on floral diversity. *Irish Naturalists' Journal* **10**, 232–40.

Colgan, N. and Scully, R.W. 1898 *Contributions towards a Cybele Hibernica*. 2nd edn. Dublin. Edward Ponsonby.

Doyle, G.J. and Foss, P.J. 1986 A resurvey of the Clare Island flora. *Irish Naturalists' Journal* **22**, 85–9.

MacArthur, R.H. and Wilson, E.O. 1967 *The theory of island biogeography*. New Jersey. Princeton University Press.

McClintock, D. 1969 Field meetings, 1968. Mayo and Galway, 14th–20th April. *Proceedings Botanical Society of the British Isles* **7**, 634.

More, D. and Moore, A.G. 1866 *Contributions towards a Cybele Hibernica*. Dublin. Curry.

Perring, F.H. and Walters, S.M. 1962 *Atlas of the distribution of British plants*. London and Edinburgh. Nelson.

Praeger, R.L. 1903 The flora of Clare Island. *The Irish Naturalist* **12**, 277–94.

Praeger, R.L. 1907 The flora of Inishturk. *The Irish Naturalist* **16**, 113–25.

Praeger, R.L. 1911a Clare Island Survey. Part 10 Phanerogamia and Pteridophyta. *Proceedings of the Royal Irish Academy* **31**, 1–112.

Praeger, R.L. 1911b Notes on the flora of Inishbofin. *The Irish Naturalist* **20**, 165–72.

Praeger, R.L. 1913 On the buoyancy of the seeds of some Britannic plants. *Scientific Proceedings of the Royal Dublin Society* **15**, 13–62.

Praeger, R.L. 1930 *Beyond soundings*. Dublin. Talbot Press Ltd.

Ryle, T. and Doyle, G.J. 1998 The elusive sea-pea on Clare Island. *Newsletter of the New Survey of Clare Island* **4**.

Synnott, D.M. 1986 An outline of the flora of Mayo. *Glasra* **9**, 13–117.

Webb, D.A. 1980 The flora of the Aran Islands. *Journal of Life Sciences, Royal Dublin Society* **2**, 51–83.

Webb, D.A. 1982 The flora of Aran: additions and corrections. *Irish Naturalists' Journal* **20**, 45

Webb, D.A. and Hodgson, J. 1968 The flora of Inishbofin and Inishshark. *Proceedings of the Botanical Society of the British Isles* **7** (3), 345–63.

BRYOPHYTES OF CLARE ISLAND

Donal Synnott

ABSTRACT

A history of recording and an annotated list of the hornworts, liverworts and mosses of Clare Island are presented. Specimens from the original survey have been reassessed. One hornwort, one hundred and one liverworts and one hundred and eighty-five mosses are now recorded for the island. Of these, twenty-eight liverworts and forty-three mosses are additions to the original list. The flora is compared to that of the larger, nearby Achill Island.

History of recording

Canon Henry William Lett was on the steering committee for the first Clare Island Survey. He participated in the fieldwork and wrote the accounts for the mosses and liverworts (Lett 1912). These accounts included records for the adjacent mainland about Louisburgh and for Achill Island. Comparison of the bryophyte lists for this greater area with those for Clare Island is a main feature of Lett's account.

Lett spent a total of eight weeks on Achill, being joined at various times by W.H. Pearson, D. McArdle, C.H. Waddell and R.L. Praeger. In 1911 a distinguished team of four British bryologists, J.B. Duncan, D.A. Jones, S.J. Owen and J.C. Wilson, spent a week on Achill concentrating on the Slievemore and Doogort area. They added several species to the Achill flora, notably *Sphenolobopsis pearsonii*, *Leiocolea alpestris*, *Scapania nimbosa* and *Radula lindenbergiana* (Jones 1917). Lett spent two weeks on Clare Island in 1909 and a further week in 1910. The only other bryologist to visit Clare Island until recent decades was David McArdle, who joined Lett on the island for five days in 1909. At the time of the first survey C.H. Waddell was the most able bryologist resident in Ireland, but he did not visit Clare Island, although he contributed many records for the Westport area, spending some eight days in the area, including a day on Croaghpatrick and a day on Achill.

Despite finding a liverwort believed to be new to science (*Adelanthus dugortiensis*, now reduced to synonymy with *A. lindenbergianus*), Lett was less than enthusiastic about the bryophyte diversity of the area. He concluded: 'The omnipresence and abundance of peat, the small amount of land under cultivation, and the frequent occurrence of outcropping rocks would lead a botanist on a first visit to expect a rarer and richer flora of Bryophytes than is actually met with on either the island or the neighbouring mainland' (Lett 1912, p. 4).

Lett used the current printed catalogues for mosses (Dixon 1897) and liverworts (Waddell 1897) to record the bryophytes from Clare Island, Achill and the mainland about Louisburgh. The two volumes are bound together and are in the National Herbarium.

Initials prefixed to the plant names in the catalogue are the initials of the several localities where collections were made, i.e. C for Clare Island, A for Achill, L for Louisburgh, W for Westport and P for Croaghpatrick (Pl. I). Letters on the interleaves, after the number of each species, are the initials of the collectors of the plants, i.e. L for H.W. Lett, McA for D. McArdle, W for C.H. Waddell, P for W.H. Pearson, Pr for R. L. Praeger, M for David Moore, E for J.C. Wilson, J.B. Duncan, S.J. Owen

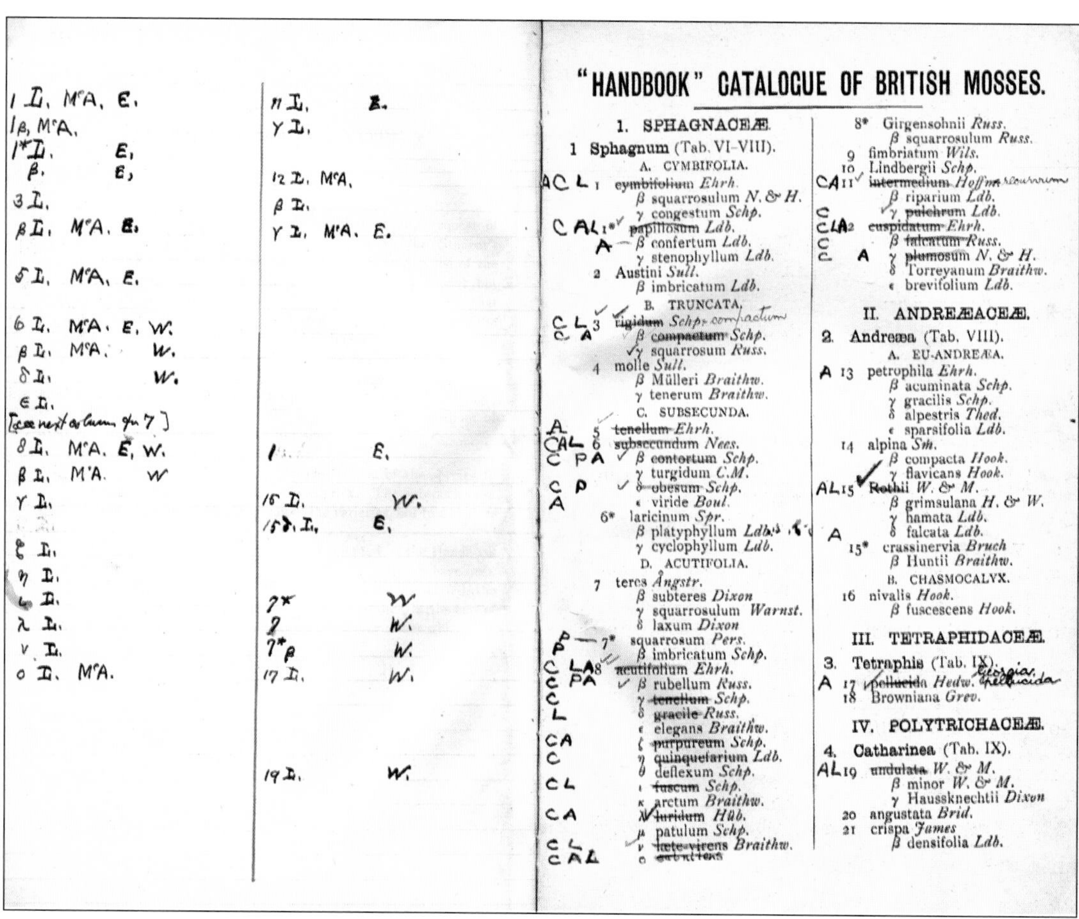

"HANDBOOK" CATALOGUE OF BRITISH MOSSES.

I. SPHAGNACEÆ.

1 Sphagnum (Tab. VI–VIII).

A. CYMBIFOLIA.

1 cymbifolium *Ehrh.*
β squarrosulum *N. & H.*
γ congestum *Schp.*
1* papillosum *Ldb.*
β confertum *Ldb.*
γ stenophyllum *Ldb.*
2 Austini *Sull.*
β imbricatum *Ldb.*

B. TRUNCATA.

3 rigidum *Schp.*
β compactum *Schp.*
γ squarrosum *Russ.*
4 molle *Sull.*
β Mülleri *Braithw.*
γ tenerum *Braithw.*

C. SUBSECUNDA.

5 tenellum *Ehrh.*
6 subsecundum *Nees.*
β contortum *Schp.*
γ turgidum *C.M.*
δ obesum *Schp.*
ε viride *Boul.*
6* laricinum *Spr.*
β platyphyllum *Ldb.*
γ cyclophyllum *Ldb.*

D. ACUTIFOLIA.

7 teres *Ångstr.*
β subteres *Dixon*
γ squarrosulum *Warnst.*
δ laxum *Dixon*
7* squarrosum *Pers.*
β imbricatum *Schp.*
8 acutifolium *Ehrh.*
β rubellum *Russ.*
γ tenellum *Schp.*
δ gracile *Russ.*
ε elegans *Braithw.*
ζ purpureum *Schp.*
η quinquefarium *Ldb.*
θ deflexum *Schp.*
ι fuscum *Schp.*
κ arctum *Braithw.*
λ luridum *Hüb.*
μ patulum *Schp.*
ν laete-virens *Braithw.*
8* Girgensohnii *Russ.*
β squarrosulum *Russ.*
9 fimbriatum *Wils.*
10 Lindbergii *Schp.*
11 intermedium *Hoffm.*
β riparium *Ldb.*
γ pulchrum *Ldb.*
12 cuspidatum *Ehrh.*
β falcatum *Russ.*
γ plumosum *N. & H.*
δ Torreyanum *Braithw.*
ε brevifolium *Ldb.*

II. ANDREÆACEÆ.

2. Andreæa (Tab. VIII).

A. EU-ANDREÆA.

13 petrophila *Ehrh.*
β acuminata *Schp.*
γ gracilis *Schp.*
δ alpestris *Thed.*
ε sparsifolia *Ldb.*
14 alpina *Sm.*
β compacta *Hook.*
γ flavicans *Hook.*
15 Rothii *W. & M.*
β grimsulana *H. & W.*
γ hamata *Ldb.*
δ falcata *Ldb.*
15* crassinervia *Bruch*
β Huntii *Braithw.*

B. CHASMOCALYX.

16 nivalis *Hook.*
β fuscescens *Hook.*

III. TETRAPHIDACEÆ.

3. Tetraphis (Tab. IX).

17 pellucida *Hedw.*
18 Browniana *Grev.*

IV. POLYTRICHACEÆ.

4. Catharinea (Tab. IX).

19 undulata *W. & M.*
β minor *W. & M.*
γ Haussknechtii *Dixon*
20 angustata *Brid.*
21 crispa *James*
β densifolia *Ldb.*

Pl. I Page from H.W. Lett's annotated catalogue for mosses (Dixon 1897) and liverworts (Waddell 1897), recording the bryophytes from Clare Island, Achill and the mainland around Louisburgh. The two volumes are bound together and are in the National Herbarium.

and D.A. Jones. Species from Clare Island are marked with a red tick and those from Achill and the adjoining mainland are marked with a blue tick. A small number of species have a lead pencil tick, and these are not usually supported by herbarium specimens.

This is a useful and interesting document that mostly confirms but sometimes contradicts or casts doubt on the published report (see, for example, the accounts for *Scapania subalpina* and *Sphagnum fuscum* below).

Vouchers for most of the bryophyte records from the original survey are preserved in the herbarium at the National Botanic Gardens, often with just 'Clare Island' and the year of collection (1909 or 1910). Lett's collection shows that Praeger collected some of the specimens from the more difficult high ground on Knockmore, where Lett apparently did not venture (See *Metzgeria leptoneura, Campylopus gracilis, Racomitrium lanuginosum, Neckera crispa*). Lett was seventy-three at the start of the survey in 1909 and may not have ventured far from the precipitous 1200-foot and 1000-foot contour paths described by Praeger (1911). Only remnants of these paths survive, and some of the alpine species, e.g. *Metzgeria leptoneura* and *Neckera crispa*, were not refound, though some of the rarest vascular plants, e.g. *Polystichum lonchitis* and *Saussurea alpina*, were refound by C. Farrell (Ryle 2000, p. 195).

Since the first survey, a number of bryologists, mainly from Britain, have made visits to Achill (Table 1): W.N. Tetley in April 1917; J.B. Duncan, C.V.B. Marquand and others in June and July 1937, following the British Bryological Society meeting in Donegal (Armitage 1938); E.C. Wallace in July 1953; D.A. Ratcliffe in September 1961 (Ratcliffe 1962); E.F. Warburg in August 1962 (Warburg 1963) and August 1965 (Warburg 1966); J.A. Paton in May 1968 and D.M. Synnott in 1982, 1983 and

Table 1
Bryologists' field trip to Achill after the first Clare Island Survey

Researchers	Achill field visit	Reference
W.N. Tetley	April 1917	
J.B. Duncan, C.V.B. Marquand and others	June and July 1937	Armitage 1938
E.C. Wallace	July 1953	
D.A. Ratcliffe	September 1961	Ratcliffe 1962
E.F. Warburg	August 1962	Warburg 1963
E.F. Warburg	August 1965	Warburg 1966
J.A. Paton	May 1968	
D.M. Synnott	1982	
	1983	
	1984	

1984. Following the Donegal meeting of the British Bryological Society thirteen participants spent a week on Achill and, in mostly wet weather, collected at Slievemore, Croaghaun, Keel, Dooagh and Valley. The most interesting find was *Bryum turbinatum*, which remains the only Irish record. Most of these bryologists concentrated on the most celebrated locality on the island, Slievemore near Doogort. Warburg however, made some significant discoveries in previously unworked areas. It was his report (Warburg 1963) of *Cyclodictyon laetevirens, Teleranea europaea, Jubula hutchinsiae, Bryum marratii* and *Bryum calophyllum* and his subsequent discovery in 1965 of *Catoscopium nigritum* that suggested that further exploration might be worthwhile.

While working on the vascular flora of Mayo this author visited Achill in 1982, 1983 and 1984, spending a total of five days on the island and also visiting Achill Beg (Synnott 1986). During this time vascular plants and bryophytes were recorded. While few additions were made to the bryophyte flora it was felt that further fieldwork by a team of experienced bryologists would yield worthwhile results. The British Bryological Society enthusiastically took up the idea of working towards a bryophyte flora of Achill. During the 1987 meeting of the society all the well-known localities were visited and many additional localities, some of which proved to be rich in rare and interesting species. Many new records for the island were made. Reports of the meeting have been published by Rothero (1988) and Synnott (1988). A comprehensive account of the bryophytes of Achill was subsequently published in three parts (Long 1990; Smith 1990; Synnott 1990).

Some interesting additions were made to the bryophyte list for Achill, including *Geocalyx graveolens*, a rare hepatic that had been first found in Ireland in Kerry in 1967. Experienced members of the BBS were convinced that it must surely occur on Clare Island and resolved to visit the island at some time in the future. This author did not manage to find the species for himself in Achill despite being shown it in situ, 'on peaty soil in declivity on damp rocky slope, N. coast of Achill' (Long 1988), an indication of the limitations of at least one bryologist and the value of teamwork in field surveys.

Working at the National Herbarium at Glasnevin provided an opportunity to examine the bryophyte collections of the earlier surveys carried out in Lambay and Clare Island. In general, the Clare Island specimens, mostly collected by Canon H.W. Lett, were correctly identified, whereas a sizeable proportion of the Lambay specimens, all collected by David McArdle, were not correct. All of the Lambay specimens were re-examined and following three days of fieldwork on the island, at the invitation of the late Lord Revelstoke, a bryophyte flora of Lambay was published (Synnott 1990b). It seemed like a natural progression then to take an interest in Clare Island, following the pattern laid down by Praeger *et al.* in the early twentieth century.

Meanwhile a modern listing of the flowering plants of Clare Island was published by Doyle and Foss (1986). Soon afterwards a new survey of Clare Island was proposed (Doyle and Whelan 1991) and adopted by the Royal Irish Academy. This early enthusiasm resulted in visits to the island in 1990 and again in 1992 and 1993, when bryophytes were collected on Knocknaveen, Portlea, Knockmore, the Harbour area, and Loughanaphuca, adding *Adelanthus decipiens, Scapania compacta, Scapania aspera, Pleurozia purpurea, Drepanolejeunea hamatifolia, Lejeunea cavifolia, Cololejeunea minutissima, Anthoceros punctatus, Sphagnum squarrosum, S. fimbriatum, S. tenellum, S. flexuosum, Polytrichastrum alpinum, Atrichum undulatum, Dicranella varia, Dicranodontium uncinatum, Campylopus introflexus, Weissia brachycarpa, Didymodon tophaceus, Cinclidotus fontinaloides, Physcomitrium pyriforme, Pohlia melanodon, Rhizomnium pseudopunctatum, Plagiomnium ellipticum, Philonotis calcarea, Neckera complanata, Thuidium delicatulum, Drepanocladus aduncus, Oxyrrhynchium hians, Rhynchostegiella tenella, Ctenidium molluscum* var. *condensatum* and *Hyocomium armoricum. Hennediella heimii* and *Brachythecium glareosum* were added later.

The British Bryological Society came to Ireland again in 1994, spending a week in County Clare and a second week at Clifden, the latter allowing for a one-day visit to Clare Island. A party of some ten bryologists from Ireland, England, Scotland, Norway and Germany spent a day on the island, and each of three groups concentrated on different areas (Blockeel 1995). The combined effort resulted in the addition of 27 species to the island flora, including the *Geocalyx* predicted earlier.

Dispersal of the party to the various habitats was greatly facilitated by the RIA jeep. Daniel Kelly and Richard Bowen studied the woodland at Portlea. Tim Ryle, who had already spent part of a year on the island recording the vegetation and had made some forays along the sheep tracks on the intimidating slopes of Knockmore, led access to the cliffs on the Atlantic side of Knockmore from the western side, resulting in the discovery of the *Geocalyx* and of *Leiocolea fitzgeraldiae, Scapania scandica, Leptoscyphus cuneifolius* and *Weissia perssonii* by Gordon Rothero. David Long concentrated on the northern flank of the cliffs and added *Anastrepta orcadensis, Leiocolea collaris, Colura calyptrifolia* and *Isothecium myosuroides* var. *brachythecioides*. The main party, which included Jean Paton and Tom Blockeel, spent much of the time on the eastern side of the island about Knocknaveen, adding *Barbilophozia floerkii, Barbilophozia attenuata, Jungermannia atrovirens, Jungermannia pumila, Marsupella funckii, Chiloscyphus pallescens, Plagiochila bifaria, Colura calyptrifolia, Fossombronia pusilla, Blasia pusilla, Pellia endiviifolia* and *Riccia subbifurca* and the mosses *Pleuridium acuminatum, Pseudephemerum nitidum, Gymnostomum calcareum, Pohlia drummondii, Plagiobryum zierii* and *Anomodon viticulosus*.

Blockeel (1995) reporting on the visit, was much more enthusiastic than Lett had been in 1912. He concluded, 'After we boarded the boat for the crossing to the mainland, shower clouds descended over Clare Island. During the blustery crossing, with gannets above our heads, we were able to reflect on our good fortune on another favourable and productive day in such an excellent place'.

Ryle (this volume, pp 27–177) included bryophytes in the more than 300 relevés recorded for his vegetation studies. These have been incorporated into the species lists below. When the records for a particular species have been nine or fewer, the location (townland) and number of each relevé is listed (see Ryle, Appendix 1, this volume, pp 173–177). For more widespread species the total number of relevés alone is given. Ryle's bryophyte records include four species not otherwise recorded—*Cladopodiella fluitans, Sphagnum molle, Plagiomnium elatum* and *Campyliadelphus elodes*. David Holyoak added *Tritomaria exsectiformis* and *Bryum rubens* in 2003.

Comparison of first and second surveys of Clare Island

In the first survey 74 liverworts and 139 mosses were recorded for the island. Eighteen liverworts recorded by Lett were not refound in the recent survey. Of these, four (*Marchantia polymorpha, Cephalozia catenulata, Harpanthus scutatus* and *Metzgeria leptoneura*) are supported by correctly identified specimens. The remainder are not supported by specimens and are considered doubtful.

Of the sixteen mosses not recorded in the latest survey, six (*Dicranella cerviculata, Ulota hutchinsiae, Hedwigia stellata, Neckera crispa, Sciurohypnum populeum* and *Hypnum lacunosum*) are supported by correctly identified specimens, ten were incorrectly identified and have been redetermined (*Encalypta vulgaris, Tortula marginata, Bryoerythrophyllum ferruginascens, Pseudocrossidium revolutum, Enthostodon fascicularis, Bryum uliginosum, Bryum amblyodon, Bryum radiculosum, Brachytheciastrum velutinum* and *Campyliadelphus chrysophyllus*) and four are not supported by voucher specimens and are considered doubtful (*Sphagnum fuscum, Pogonatum nanum, Pogonatum aloides* and *Amblystegium serpens*).

The number of species now recorded is 1 hornwort, 101 liverworts and 185 mosses. One hornwort, 36 liverworts and 50 mosses have been added since the first survey (see Table 2).

The only one of these species sure to have arrived since the first survey is *Campylopus introflexus*, a species of wide distribution in the southern hemisphere, now widespread in Europe and first found in Ireland on Howth Head in Dublin in 1942. All of the other species added to the list in the present survey would have had suitable habitats available at the time of the first survey. The increase in species numbers is likely a function of the greater number of experienced field bryologists to have visited the island in recent decades.

Table 2

List of bryophyte species recorded since the first Clare Island Survey

Hornworts

Anthoceros punctatus

Liverworts

Blasia pusilla
Riccia subbifurca
Pellia endiviifolia
Fossombronia pusilla
(Fossombronia foveolata)
Pleurozia purpurea
Riccardia palmata
Radula aquilegia
Cololejeunea calcarea
Cololejeunea minutissima
Colura calyptrifolia
Drepanolejeunea hamatifolia
Lejeunea cavifolia
Chiloscyphus pallescens
Leptoscyphus cuneifolius
Plagiochila bifaria
Adelanthus decipiens
Cladopodiella fluitans
Odontoschisma sphagni
Odontoschisma denudatum
Barbilophozia floerkii
Barbilophozia attenuata
Anastrepta orcadensis
Tritomaria quinquedentata
Lophozia excisa
Scapania compacta
Scapania scandica
Scapania aspera
Calypogeia fissa
Leiocolea fitzgeraldiae
Leiocolea collaris
Leiocolea turbinata
Jungermannia atrovirens
Jungermannia pumila
Geocalyx graveolens
Marsupella funckii

Mosses

Sphagnum squarrosum
Sphagnum fimbriatum
Sphagnum molle
Sphagnum tenellum
Atrichum undulatum
Polytrichastrum alpinum
Polytrichum strictum
Diphyscium foliosum
(Enthostodon attenuatus)
(Enthostodon obtusus)
Physcomitrium pyriforme
(Racomitrium sudeticum)
Archidium alternifolium
Pleuridium acuminatum
Pseudephemerum nitidum
Ditrichum gracile
Dicranella varia
Dicranodontium uncinatum
Campylopus introflexus
Weissia perssonii
Weissia brachycarpa
Gymnostomum calcareum
(Pseudocrossidium revolutum)
Hennediella heimii
Cinclidotus fontinaloides
Ulota bruchii
Philonotis calcarea
Plagiobryum zierii
(Bryum riparium)
Pohlia cruda
Pohlia drummondii
Pohlia annotina
Pohlia melanodon
Plagiomnium elatum
Plagiomnium ellipticum
Palustriella commutata
(Campylium protensum)
Campyliadelphus elodes
(Hygrohypnum luridum)
Thuidium delicatulum
Rhynchostegiella tenella
Oxyrrhynchium hians
Brachythecium glareosum
(Brachythecium mildeanum)
Calliergonella lindbergii
Ctenidium molluscum var. *condensatum*
Hyocomium armoricum
Neckera complanata
Isothecium myosuroides var. *brachythecioides*
Anomodon viticulosus

Species in parentheses refer to Lett specimens reidentified

Comparison with bryophytes of Achill Island
Clare Island is less than six kilometres from Achill and shares the same extreme Atlantic climate, which is an important influence on the bryophyte floras of the two islands. Notable oceanic bryophyte species on Clare Island include the liverworts *Adelanthus decipiens, Colura calyptrifolia, Lepidozia cupressina* and *Leptoscyphus cuneifolius* and the mosses *Dicranum scottianum*, several species of *Campylopus, Ulota hutchinsiae* and *Hyocomium armoricum*.

Achill is about ten times the area of Clare Island. It has a greater range of habitats, from coastal dunes and machair to high north-facing corries and block scree and more extensive areas of shared habitats.

One hundred and fifty eight liverworts (Long 1990) and 246 mosses (Smith 1990) are recorded for Achill. Of these, 63 liverworts and 94 mosses have not been found on Clare Island. Comparatively few species occur on the smaller island that have not been found on Achill—eight liverworts (*Cephaloziella spinigera, Cephaloziella divaricata, Leiocolea bantriensis, Leiocolea fitzgeraldiae, Marsupella funckii, Cololejeunea calcarea, Reboulia hemisphaerica* and *Riccia subbifurca*) and nineteen mosses (*Sphagnum molle, Pleuridium acuminatum, Dichodontium pellucidum, Weissia perssonii, Weissia brachycarpa, Cinclidotus fontinaloides, Physcomitrium pyriforme, Pohlia cruda, Pohlia drummondii, Pohlia melanodon, Plagiobryum zierii, Anomobryum julaceum* var. *concinnatum, Bryum amblyodon, Plagiomnium ellipticum, Neckera crispa, Anomodon viticulosus, Campyliadelphus elodes, Hygrohypnum luridum* and *Brachythecium glareosum*). Given the size and location of Clare Island and the range of habitats, the total number of bryophytes recorded is close to what might be predicted. However, there are a number of moss species that might be expected but have not been found, including, for example, *Tetraphis pellucida, Pogonatum aloides, Tortula truncata, Pseudocrossidium hornschuchianum, Bryum dichotomum, Bryum argenteum, Climacium dendroides, Amblystegium serpens* and *Pseudotaxiphyllum elegans*.

Further fieldwork will undoubtedly add more species to the total now recorded.

Acknowledgements
Thanks are due to the people of Clare Island for their hospitality and assistance, especially to Ciara Cullen, Beth Moran, Chris O'Grady and Peter Gill; to the Royal Irish Academy for financial and logistical support, especially to Sara Whelan and Roisín Jones; to the National Botanic Gardens for time and facilities to carry out the work, especially to Matthew Jebb and the late Aidan Brady; to the following participants in the British Bryological Society visit to Clare Island on 24 July 1994: J. Blackburn, T. Blockeel, R. Bowen, S. Drangard, T. Homm, D. Kelly, D. Long, K. Long, J. Paton, A. Pedersen, G. Rothero and T. Ryle. I am indebted to Jean Paton, MBE, to Gordon Rothero and to Tom Blockeel for assistance with identifications and to David Holyoak for easing the manuscript into the twenty-first century and for much helpful comment and advice. Tim Ryle, Matthew Jebb, David Cabot and Theresa, Niamh and John were companions on memorable visits.

CATALOGUE

Hornworts, liverworts and mosses of Clare Island
Arrangement and nomenclature follow Hill *et al.* 2008. The names of species used in the original survey report (Lett 1912) are given, as published, in brackets. Additional synonyms are given where the synonymy trail is particularly tortuous or ambiguous. Explanations are given where specimens have been redetermined. BBS refers to the visit of the British Bryological Society in 1994. Ryle's bryophyte records are those from his relevés (see Ryle, this volume, p. 131). Records not attributed are those of the author.

DBN = Voucher specimen in Botanic Gardens, Glasnevin, Dublin

HORNWORTS

***Anthoceros punctatus* L.**
Vertical clay bank on S side of gully, E side of Knocknaveen.

LIVERWORTS

***Blasia pusilla* L.**
Green Road, E of Knocknaveen, BBS.

***Marchantia polymorpha* L.**
There are two references to the species in Lett (1912): 'of Marchantia polymorpha only one colony was seen; it was in a little gully near the sea-shore south of Knocknaveen' (p. 8), and 'Clare I. (A.D. Cotton). Not seen anywhere within the area except in this one locality, near Kill' (p. 15).
A specimen in DBN, labelled 'near Kill, Aug 1909, Praeger', is referrred to subsp. polymorpha, det. D.G. Long 1990.

Preissia quadrata (Scop.) Nees ***(Preissia commutata)***
Lett 1912.
E side of Knockmore, Long, BBS.
N-facing steep dry slope of Knocknaveen. Flush on S side of gully on SE side of Knocknaveen.
E side of island, BBS.

Reboulia hemisphaerica (L.) Raddi **(*Asterella hemisphaerica*)**
Lett 1912.
N-facing, steep, dry slope of Knocknaveen.

Conocephalum conicum (L.) Dumort.
Lett 1912.
Stream at Portlea, 1990.
Knocknaveen. E side of island, BBS.

Riccia subbifurca Warnst. ex Croz.
Green Road, E of Knocknaveen, BBS.

Pellia epiphylla (L.) ***Corda***
Lett 1912.
E side of island, BBS.
Ryle: 31 releves, some may be *Pellia neesiana*.

Pellia neesiana (Gottsche) Limpr. *(P. calycina)*
Lett 1912.
Marshy field below Lassau Wood. With *Juncus effusus* by the Green Road at Knocknaveen. In valley bog below *Lycopodiella inundata* habitat, N of the Wall, 2008, Ryle and Synnott.
E side of island, BBS.

Pellia endiviifolia (Dicks.) Dumort.
E side of island, BBS.

Fossombronia foveolata Lindb.
Clare Island, 1909, Lett, specimen in DBN, fide A.L.K. King (Paton 1964, p. 713).

Fossombronia angulosa (Dicks.) Raddi
Lett 1912: 'Near Maum (Lett)'. (DBN). Near the sea at Loughanaphuca. Cliffs below Knockmore, Long, BBS. W end of Knockmore.

Fossombronia pusilla (L.) Nees
Green Road, E of Knocknaveen, BBS.

Pleurozia purpurea Lindb.
Bog above Lassau wood.
Ryle: 9 releves—Commonage (78, 127, 215, 231, 256, 391, 397); Ballytoohy More (249); Lecarrow (368).
Confined to degraded blanket bog and associated with *Campylopus atrovirens*, *Sphagnum papillosum* and *Sphagnum subnitens* (Ryle, this volume, p. 129).

Metzgeria furcata (L.) Dumort.
Lett MS.
E side of island, BBS.
var. *prolifera*. Lett 1912. Recorded but not confirmed in MS. No voucher specimen in DBN.
Neither *M. temperata* nor *M. fruticulosa* was discovered during the recent survey.

Metzgeria conjugata Lindb.
Lett 1912
DBN: Inland cliff, 4/1909, Praeger.
N-facing cliff, Knocknaveen, with *Plagiochila porelloides*. E side of island, BBS.
Knockmore, BBS.

Metzgeria leptoneura Spruce **(*M. hamata*)**
Lett 1912
DBN: Knockmore, 1910, Praeger.

Aneura pinguis (L.) Dumort.
Lett 1912.
E side of island, BBS.
Ryle: 27 releves.

Riccardia multifida (L.) Gray (*Aneura multifida*)
Lett 1912.
Top of gully on E side of Knockmore. E side of island, BBS.

Riccardia chamedryfolia (With.) Grolle (*Aneura sinuata*)
Lett 1912. Boulder clay sea cliff, Portlea.
Knockmore. E side of island, BBS.

Riccardia palmata (Hedw.) Carruth. (*Aneura palmata*)
E side of Knockmore. Knockmore, BBS.
E side of island, BBS.

Riccardia latifrons (Lindb.) Lindb. (*Aneura latifrons*)
Lett 1912.
Boggy ground N of Knocknaveen, BBS.

Radula complanata (L.) Dumort.
Lett 1912.
Knocknaveen, 1990.
Outcrop N of tower. Crevice in rock by the sea, Loughanaphuca. Tree trunk in Lassau Wood.
Ryle: Lecarrow (268, 339).

Radula lindenbergiana Gottsche ex C. Hartm. (*R. lindbergii*)
Lett 1912
DBN: On vertical rocks at Toormore and near Loughanaphuca, 1909, Lett.
N-facing crag, Knocknaveen, BBS.

Radula aquilegia (Hook. f. & Taylor) Gottsche *et al.*
E side of Knockmore. N-facing crag, Knocknaveen, BBS.
Knockmore, Rothero, BBS.

Radula carringtonii J.B. Jack
Lett 1912.
DBN: 1910, Lett, McArdle and Praeger.
Relatively exposed habitat on N-facing rocks of crag, Knocknaveen, BBS.
Knockmore, Rothero, BBS.

Frullania tamarisci (L.) Dumort.
Lett 1912.
Lassau Wood. With apiculate leaves, on the ground on boggy slope, Knocknaveen. Wall on W side of the harbour.
E side of island, BBS.
Ryle: 65 releves.
var. *cornubica* (Lett). Knockmore, BBS.

Frullania teneriffae (F. Weber) Nees (*F. germana*)
Lett 1912.
Lassau Wood. Rocks between the road and the sea E and W of the church on S side of the island.
Outcrop N of tower. Moraine, Ballytoohy Beg.
Knockmore, Rothero, BBS.
E side of island, BBS.

Frullania microphylla (Gottsche) Pearson
Lett 1912 but not confirmed in Lett MS nor is there a voucher specimen in DBN.
Rocks in field near the sea on E side of the church.

Frullania dilatata (L.) Dumort.
Lett 1912.
Lassau Wood. Roadside rocks, Capnagower.
On *Crataegus*, roadside W of the harbour.
E side of island, BBS.

Cololejeunea calcarea (Lib.) Schiffn.
N-facing crag, Knocknaveen, BBS.
Knockmore, Rothero, BBS.

Cololejeunea minutissima (Sm.) Schiffn.
On *Frullania* on *Betula*, Lassau Wood.

Colura calyptrifolia (Hook.) Dumort.
N-facing crag, Knocknaveen, BBS.
E. side of Knockmore, Long, BBS.

Drepanolejeunea hamatifolia (Hook.) Schiffn.
N-facing crag, Knocknaveen, BBS.
Knockmore, Rothero, BBS.
Top gully on E side of Knockmore.

Harpalejeunea molleri (Steph.) Grolle (*Lejeunea ovata*)
Lett 1912.
Knockmore, conf. Rothero, BBS.

Lejeunea cavifolia (Ehrh.) Lindb. (*L. serpyllifolia* var. *cavifolia*)
Underside of rocks in field near the sea E of the church.

Lejeunea lamacerina (Steph.) Schiffn. (*L. serpyllifolia*)
Lett 1912.
Lassau Wood, 1990.
Knocknaveen. E side of island, BBS.

Lejeunea patens Lindb.
Lett 1912.
E side of Knockmore. Dry stream bed above the road, S side of island.
Dead branch of *Salix aurita* in gully on SE side of Knocknaveen. E side of island, BBS.

Marchesinia mackaii (Hook.) Gray (*Phragmicoma mackaii*)
Lett 1912.
N-facing bluff, Knocknaveen. E side of island, BBS.

Microlejeunea ulicina (Taylor) A. Evans (*Lejeunea ulicina*)
Lett 1912.
On rotted stem, Lassau wood.

Blepharostoma trichophyllum (L.) Dumort.
Lett 1912.
E side of Knockmore with *Tritomaria quinquedentata*. Rock overhang, crag of Knocknaveen and on N-facing slope, Knocknaveen.

Herbertus aduncus (Dicks.) Gray subsp. ***hutchinsiae*** (Gottsche) R.M. Schust. (*Herberta adunca*)
Lett 1912.
E side of Knockmore, and Long, BBS.

(*Bazzania trilobata* (L.) Gray)
Lett 1912, but not recorded in Lett MS and no specimen in DBN.

Bazzania tricrenata (Wahlenb.) Lindb. (*B. deflexa*)
Lett 1912.
DBN: 'Clare Island, April, 1909, Praeger' is one stem.
E side of Knockmore, Long, BBS.

(***Bazzania pearsonii*** Steph.)
Lett 1912 but not listed in Lett MS and no specimen in DBN.

Kurzia pauciflora (Dicks.) Grolle (*Lepidozia setacea*)
Lett 1912.
Listed in Lett MS but there is no specimen in DBN.
Boggy ground N of Knocknaveen, BBS.

(***Kurzia trichoclados*** (Müll. Frib.) Grolle (*Lepidozia trichoclados*))
Listed but not confirmed in Lett MS.
No specimen in DBN.

Lepidozia reptans (L.) Dumort.
Lett 1912.
Lassau Wood, 1990.
Bank of old drain in boggy field near the sea, E of the church, 1990. Knockmore; E side of island, BBS.
Ryle: 11 releves.
Cliff, peat islands.

Lepidozia cupressina (Sw.) Lindenb. (*L. pinnata* (Hook.) Dum.)
Lett 1912.
E side of Knockmore, 1990.
Turfy ledges of knoll, E of Knocknaveen, BBS.

Lophocolea bidentata (L.) Dumort. (*L. cuspidata; L. lateralis*)
Lett 1912.
Loughanaphuca. Marshy field below Lassau Wood. Top of gully on E side of Knocknaveen. E side of island, BBS.
Ryle: 24 releves.

Lophocolea heterophylla (Schrad) Dumort.
Lett 1912 (attributed to D. McArdle but not confirmed in Lett MS).
No specimen in DBN.

(*Lophocolea fragrans* (Moris & De Not.) Gottsche *et al.* (*Lophocolea spicata*)
Lett 1912. Recorded but not confirmed in Lett MS.
No specimen in DBN.

Chiloscyphus polyanthos (L.) Dumort.
Lett 1912.
L. Merrigagh. Stone in Owenmore R. at bridge.

Chiloscyphus pallescens (Ehrh. ex Hoffm.) Dumort.
E side of island, BBS.

Leptoscyphus cuneifolius (Hook.) Mitt.
Knockmore, Rothero, BBS.

Plagiochila porelloides (Torr ex Nees) Lindenb. (*Plagiochila asplenioides*)
Lett 1912.
Lassau Wood. Knocknaveen. Track to Knocknaveen above the harbour. E side of island, BBS.
Ryle: Glen (309).

Plagiochila spinulosa (Dicks.) Dumort.
Lett 1912.
Knocknaveen. Top gully on E side of Knockmore.
Ryle: Commonage (163, 398), Capnagower (382).

Plagiochila bifaria (Sw.) Lindenb. (*P. killarniensis* Pearson)
N-facing crag, Knocknaveen, BBS.

Plagiochila punctata (Taylor) Taylor
Lett 1912. E side of Knockmore.
Ryle: Commonage (92, 109, 395), Ballytoohy More (352), Fawnglass (384).

(***P. exigua*** Taylor)
Lett 1912 (Recorded but not confirmed in MS. No specimen in DBN).

Adelanthus decipiens (Hook.) Mitt.
Top gully on E side of Knockmore.

Cephalozia bicuspidata (L.) Dumort. (including var. *lammersiana* (Hueb.) Schuster)
Lett 1912. By stream, Portlea, 1990. E side of island, BBS.

Cephalozia catenulata (Huebener) Lindb.
Lett 1912.
There are five DBN vouchers: 'With *C. leucantha* and *C. multiflora* (fide W.H. Pearson), Clare Island, July 1910, Lett/Praeger'

Cephalozia leucantha Spruce
Lett 1912
DBN: Found growing in cushions of *Campylopus brevipilus*, Lett. On *Sphagnum*, Ballytoohy, Aug. 1909, Lett.
Turfy ledges of knoll E of Knocknaveen, BBS.
E side of island, BBS.

Cephalozia lunulifolia (Dumort.) Dumort.
Lett 1912.
Fen at Loughanaphuca, 1990.

Cephalozia connivens (Dicks.) Lindb.
Lett 1912.
DBN: Found growing in cushions of *Campylopus brevipilus*, Lett. Lighthouse marsh, June 1909, Praeger.

Cladopodiella fluitans (Nees) H. Buch
Ryle: Ballytoohy Beg (250)

Cladopodiella francisci (Hook.) H. Buch. ex Jorg. (*Cephalozia francisci*)
Lett 1912.
Lett's DBN specimen is a mixture of hepatics. *C. francisci* was not relocated among them.

Odontoschisma sphagni (Dicks.) Dumort.
E side of island, BBS.
Ryle: 15 releves.

Odontoschisma denudatum (Mart.) Dumort.
E side of island, BBS.

Cephaloziella spinigera (Lindb.) Warnst. (*Cephalozia striatula*)
Lett 1912
DBN: Maum, Aug 1909, growing on *Campylopus brevipilus*, Lett.

Cephaloziella divaricata (Sm.) Schiffn. (*C. starkei* (Funck) Schiffn.)
Lett 1912.
Track W of Knocknaveen and N of church, 1990.

(***Cephaloziella stellulifera*** (Spruce) Schiffn. (*Cephalozia stellulifera*))
Lett 1912. Recorded but not confirmed in Lett MS.
No specimen in DBN.

Gymnocolea inflata (Huds.) Dumort. var. ***inflata***
Lett 1912.
Bare patches on Green Road at Knocknaveen.

Barbilophozia floerkii (F. Weber & D. Mohr) Loeske
Turfy ledges of knoll E of Knocknaveen, BBS.

Barbilophozia attenuata (Mart.) Loeske
Rock outcrop on knoll, E of Knocknaveen, BBS.

Anastrepta orcadensis (Hook.) Schiffn.
Lett 1912.
E. side of Knockmore, Long, BBS.

Tritomaria quinquedentata (Huds.) H. Buch
E. side of Knockmore, 1990.
N-facing crag, Knocknaveen, BBS.

Lophozia ventricosa (Dicks.) Dumort.
(*Jungermannia ventricosa*)
Lett 1912.
E side of Knockmore, 1990.
Face of peat cutting on E side of Knockmore, 1990.
Track W. of Knocknaveen and N of church, 1990.
Woodland floor and base of *Osmunda* stems, Portlea. E side of island, BBS.
Knockmore, BBS.

Lophozia excisa (Dicks.) Dumort.
E side of island, BBS.

Lophozia incisa (Schrad.) Dumort.
Lett 1912. L. Merrigagh. E side of island, BBS.

Diplophyllum albicans (L.) Dumort.
Lett 1912.
Knockmore. Ballytoohy Beg. E side of island, BBS.
Ryle: Commonage (133, 193, 210, 398); Ballytoohy More (155, 182, 242); Lecarrow 389).

Scapania compacta (A. Roth) Dumort.
Track on W side of Knocknaveen. Track above the harbour on E side of Knocknaveen.

Scapania scandica (Arnell & H. Buch) Macvicar
Green Road, E. of Knocknaveen, BBS.
Knockmore, Rothero, BBS.

(***Scapania curta*** (Mart.) Dumort.)
Lett 1912.
The Clare Island specimen in DBN, 'June 1909, Praeger' is *Scapania gracilis* fide J.A. Paton.

Scapania umbrosa (Schrad.) Dumort.
Lett 1912.
E side of island, BBS.

Scapania nemorea (L.) Grolle (*S. nemorosa*)
Lett 1912.
Lassau Wood.

Scapania irrigua (Nees) Nees
Lett 1912.
Marshy ground between *Juncus effusus* tussocks N of Knocknaveen. Track above the harbour.
E side of island, BBS.
Ryle: Commonage (104); Fawnglass (203).

Scapania undulata (L.) Dumort. (*S. dentata*)
Lett 1912.
Flushes and ditches at Portlea. Gully on E side of Knocknaveen. E side of island, BBS.
Ryle: Commonage (131); Lecarrow (323, peat islands).

(***Scapania subalpina*** (Nees ex Lindenb.) Dumort.)
Lett 1912 'Clare I.... Rare'.
Listed but not confirmed in Lett MS.
No specimen in DBN.

Scapania aspera Bernet & M. Bernet
Knockmore.

Scapania gracilis Lindb. (*S. resupinata*)
Lett 1912.
Moorland between road and Portlea.
Lassau Wood. Common about Knocknaveen.
E side of island, BBS.
Knockmore, BBS.
Ryle: 48 releves.

Mylia taylorii (Hook.) Gray
Lett 1912.
E. side of Knockmore, 1990.
Turfy ledges of knoll, E. of Knocknaveen, BBS.
Ryle: Commonage (188, 204, 269, 276); Lecarrow (325, 329); Gorteen (236); The quay (86).

Mylia anomala (Hook.) Gray
Lett 1912: 'Clare I. (Lett). This is a new record for West Mayo. Very rare.'
Boggy ground N. of Knocknaveen, BBS.
Ryle: Commonage (267, 398); Maum (360, 364).

Calypogeia fissa (L.) Raddi (*Kantia sprengellii*)
E side of island, BBS.

Calypogeia muellerana (Schiffn.) Müll. Frib.
(*Kantia trichomanis*)
Lett 1912.
Lassau Wood, 1990. E. side of Knockmore, with *Cephalozia bicuspidata*, 1990.
E side of island, BBS.

Calypogeia arguta Nees & Mont.
(*Kantia arguta*)
Lett 1912.
E side of island, BBS.
Knockmore, BBS.

Leiocolea bantriensis (Hook.) Jorg. (*Jungermannia bantriensis*)
Lett 1912.

Leiocolea fitzgeraldiae Paton & A.R. Perry
On irrigated ledges of more or less basic sea cliffs, 300m alt., L68, 1994, G.P. Rothero, BBS.

Leiocolea collaris (Nees) Schljakov (*L. alpestris; L. muelleri*)
Cliffs E. side of Knockmore,
Long, BBS.

Leiocolea turbinata (Raddi) H. Buch
N-facing crag, Knocknaveen, BBS.
Cliffs E side of Knockmore,
Long, BBS.

Jungermannia atrovirens Dumort. (*Solenostoma triste*)
Dry stream bed above the road, S side of island, 1990.
E side of island, BBS.

Jungermannia pumila With.
E side of island, BBS.

Leiocolea turbinata (Raddi) H. Buch
N-facing crag, Knocknaveen, BBS.
Cliffs E side of Knockmore, Long, BBS.

Nardia scalaris Gray (*Alicularia scalaris*)
Lett 1912.
E side of island, BBS.
Ryle: Ballytooohy More (230, 251); Commonage (163); Lecarrow (387).

(***Solenostoma sphaerocarpum*** (Hook) Steph. (*Aplozia sphaerocarpa*))
Lett 1912.
Listed but not confirmed in Lett MS.
No DBN specimen. Two Lett specimens from Achill are *Nardia scalaris* fide D.G. Long 1988.

Solenostoma gracillimum (Sm.) R.M. Schust.

(***Solenostoma crenulatum*** (Smith) Mitt.)
Lett 1912: Treated as a separate species.

(Aplozia crenulata; A. gracillima)
Lett 1912.
Stream at Portlea, 1990.
Bare peat in bog between road and Portlea, 1990.
E side of island, BBS.

(Solenostoma obovatum (Nees) C. Massal. (*Southbya obovata*))
Lett 1912.
Not listed for Clare Island in Lett MS but for 'other parts of Ireland'.

Geocalyx graveolens (Schrad.) Nees
Peaty hollow on N-facing cliff, Knockmore, Rothero, BBS.

(***Harpanthus scutatus*** (Web. & Mohr) Spruce)
Lett 1912.
Not recorded for Clare Island in Lett MS.
No specimen in DBN.

Saccogyna viticulosa (L.) Dumort.
Lett 1912.
Lassau Wood. Rocks in field near the sea, E of the church. Roadside W of the harbour. Gully on E side of Knocknaveen. E side of island, BBS.

Marsupella emarginata (Ehrh.) Dumort. var. ***emarginata***
Lett 1912.
E side of island, BBS.

Marsupella funckii (F. Weber & D. Mohr) Dumort.
Stony track N side of Knocknaveen, Paton, BBS.

MOSSES

Sphagnum papillosum Lindb.
Lett 1912
DBN: 1910, Praeger.
Bog near Portlea. Lough Merrigagh. E side of island, BBS.
Ryle: 52 relevés.

Sphagnum palustre L. (*S. cymbifolium*)
Lett 1912.
Bog near Portlea. E side of island, BBS.
Ryle: 37 relevés.

Sphagnum squarrosum Crome
Boggy ditch on NW side of Lassau Wood.

Sphagnum fimbriatum Wilson
Knockmore.

Sphagnum capillifolium (Ehrh.) Hedw.
(*S. acutifolium*)
Lett 1912.
Lakes NW of Knocknaveen (*S. rubellum*).
Bog N of wall, Capnagower. E side of island, BBS.
Ryle: 45 relevés.

(***Sphagnum fuscum***)
Lett 1912.
Marked but not confirmed in Lett MS.
No specimen located

Sphagnum subnitens Russow & Warnst.
Lett 1912.
DBN: 1909 & 1910, Lett.
Bog by the road N of Portlea. Knocknaveen.
E side of island, BBS.
Ryle: 116 relevés.

Sphagnum molle Sull.
Ryle: Lecarrow (368, 390); The mill (286)

Sphagnum compactum Lam. & DC. (*S. rigidum* var. *compactum*)
Listed but not confirmed in Lett MS.
Knockmore. Rocky heath N of wall, Capnagower.
Ryle: Commonage (191, 242, 239); Ballytoohy More (241, 242, 350); Maum (356); Lecarrow (327); The mill (285).

Sphagnum inundatum Russow (*S. subsecundum pro parte*)
Lett 1912
DBN: Toormore Lough 1909. Lough Merrigagh.
Ryle: (as *S. auriculatum* agg.) Lecarrow (143, 338); Fawnglass (295); Park (366); Maum (248); Commonage (219); Ballytoohy More (241).

Sphagnum denticulatum Brid. (*S. subsecundum* pro parte)
Lett 1912
DBN: 1910, Praeger; 1911, Lett.
Bog between the road and Lassau Wood. Lough Merrigagh. Knocknaveen. E side of island, BBS.

Sphagnum tenellum (Brid.) Pers. ex Brid.
(*S. molluscum*)
Dried out pool between road and Lassau Wood, with *S. cuspidatum*.
Ryle: Lecarrow (142, 388, 389); Commonage (77, 369); Ballytoohy More (176, 241).

Sphagnum cuspidatum Ehrh. ex Hoffm.
Listed but not confirmed in Lett MS. Bog pool dried out in early June 1990, NW of Portlea.
Lough Merrigagh. Knocknaveen. E side of island, BBS.
Ryle: 34 relevés.

Sphagnum fallax (H. Klinggr.) H. Klinggr.
(*S. intermedium?*)
Moorland/blanket bog above Lassau Wood.
Lough Merrigagh. E side of island, BBS.

Sphagnum recurvum Beauv. (*S. intermedium?*)
Lett 1912.
DBN: 1909 & 1910, Lett.
Lassau Wood. Sloping bog,
SW side of Knockmore.

Sphagnum flexuosum Dozy & Molk.
(*S. intermedium?*)
Lough Merrigagh.

Atrichum undulatum (Hedw.) P. Beauv. var. ***undulatum*** (*Catharinea undulata*)
Knocknaveen. Road verge, N of wall, Capnagower.
Ryle: Leecarrow (323); The mill (286).

(***Pogonatum nanum*** (Hedw.) P. Beauv.
(*Polytrichum subrotundum*))
Lett 1912. Pencil tick in MS. No voucher specimen in DBN.

(***Pogonatum aloides*** (Hedw.) P. Beauv.
(*Polytrichum aloides*))
Lett 1912.
Doubtfully confirmed in Lett MS.
No voucher specimen in DBN.

Pogonatum urnigerum (Hedw.) P. Beauv.
(*Polytrichum urnigerum*)
Lett 1912.
DBN, 1909, Lett.
E side of island, BBS.

Polytrichastrum alpinum (Hedw.) G.L. Sm.
Summit of Knockmore.
Knockmore, BBS.

Polytrichastrum formosum (Hedw.) G.L. Sm.
(*Polytrichum attenuatum*)
Listed but not confirmed in Lett MS.
By stream near Portlea. Roadside bank, N of wall, Capnagower.

Polytrichum commune Hedw.
Listed but not confirmed in Lett MS.
E side of island, BBS.
Ryle: 74 relevés.

Polytrichum piliferum Hedw.
Lett 1912.
DBN: 1909, Lett.
E side of island, BBS.

Polytrichum juniperinum Hedw.
Lett 1912.
Capnagower. E side of island, BBS.
Ryle: 22 relevés.

Polytrichum strictum Menzies ex Brid.
Near the summit of Knockmore.
N of Knocknaveen.
Ryle: Commonage (55).

Diphyscium foliosum (Hedw.) D. Mohr
Outcrop E side of Knocknaveen, BBS.

(***Encalypta vulgaris*** Hedw. (*Leersia extinctoria*))
Lett's specimen in DBN is *Barbula unguiculata*.

Funaria hygrometrica Hedw.
Listed but not confirmed in Lett MS.
E side of Knockmore.
Cemetery.

Enthostodon attenuatus (Dicks.) Bryhn
(*Funaria attenuata*)
Lett, DBN, as *E. fascicularis*, no locality.
Rock in stream at Portlea. Knockmore. Top of gully on E side of Knockmore. Gully on E. side of Knocknaveen. E. side of island, BBS.
With *Preissia quadrata* on N-facing, steep, dry slope of Knocknaveen.

(***Enthostodon fascicularis*** (Hedw.) Müll. Hal. *Funaria fascicularis*))
Lett specimens in DBN are *E. obtusus* and *E. attenuatus*.

Enthostodon obtusus (Hedw.) Lindb.
(*Funaria obtusa*)
Lett, unlocalised specimen in DBN (as *E. fascicularis*).
Knockmore. Edge of runnel with *Potamogeton polygonifolius* and *Hypericum elodes*, on W side of road, N of Wall, Capnagower. E side of island, BBS.

Physcomitrium pyriforme (Hedw.) Bruch & Schimp.
In extensive sheets on sides of recently cleaned, deep drains between the road and Portlea.

Schistidium maritimum (Sm. ex R. Scott) Bruch & Schimp. (*Grimmia maritima*)
Lett 1912.
DBN: West side, 1909, Lett.
Above Loughanaphuca.

Schistidium apocarpum (Hedw.) Bruch & Schimp. var. ***apocarpum*** (*Grimmia apocarpa*)
Lett 1912
DBN: 1909 and 1910, Lett.
Old gate-pier, Portlea, 1990. Owenmore R. at bridge. Above Loughanaphuca. Stone in flush from old well. E side of island, BBS.

Grimmia pulvinata (Hedw.) Sm.
Lett 1912.
Old gate-pier, Portlea.

Grimmia trichophylla Grev.
Listed but not confirmed in Lett MS.
Large roadside rock E of church.

Racomitrium aciculare (Hedw.) Brid. (*Grimmia acicularis*)
Lett 1912.
DBN: 1909, Lett.
Stone in stream at Portlea. Gully on E side of Knocknaveen. E side of island, BBS.

(***Racomitrium aquaticum*** (Schrad.) Brid. (*Grimmia aquatica*))
The DBN, 1910, Lett, specimen is *Racomitrium sudeticum* (det. D.S. after Blockeel 1991).

Racomitrium fasciculare (Hedw.) Brid. (*Grimmia fascicularis*)
Listed but not confirmed in Lett MS.
E side of island, BBS.

Racomitrium heterostichum (Hedw.) Brid. (*Grimmia heterosticha*)
Listed but not confirmed in Lett MS.
E side of island, BBS.

Racomitrium sudeticum (Funck) Bruch & Schimp.
Unlocalised, Lett specimen in DBN, labelled *R. aquaticum*, is *R. sudeticum* (det. D.S. after Blockeel 1991).

Racomitrium lanuginosum (Hedw.) Brid. (*Grimmia hypnoides*)
Lett 1912.
DBN: cliff, April 1909, Praeger.
Boggy slope near summit of Knockmore.
E side of island, BBS.
Ryle: 50 relevés.

Racomitrium ericoides (Brid.) Brid. (*Grimmia canescens*)
Listed (*R. canescens*) but not confirmed in Lett MS.
Trench, NW of Knocknaveen. Path W of Knocknaveen. Bare patches on Green Road at Knocknaveen. E side of island, BBS.
Knockmore, BBS.

Ptychomitrium polyphyllum (Dicks. ex Sw.) Bruch & Schimp. (*Glyphomitrium polyphyllum*)
Lett 1912.
Rocks and walls about Portlea, frequent.
Roadside E of church. Lane above the harbour.
E side of island, BBS.

Blindia acuta (Hedw.) Bruch & Schimp.
Lett 1912.
E side of Knockmore at 1200ft. Knocknaveen, BBS.

Archidium alternifolium (Hedw.) Mitt.
Green Road between old school and Knocknaveen, BBS.

Fissidens viridulus (Sw. ex anon.) Wahlenb.
Lett 1912.
Bank of Owenmore at bridge, with immature *Trichostomum brachydontium* and *Pohlia* sp.

Fissidens bryoides Hedw.
Noted but not confirmed for Clare Island in Lett MS.
E side of island, BBS.

Fissidens osmundoides Hedw.
Lett 1912.
Knockmore cliffs. Bog N of wall, Capnagower.
E side of island, BBS.

Fissidens taxifolius Hedw.
Lett 1912.
Bank of Owenmore, at bridge, with *Funaria attenuata*. Dry stream, S side of island. E side of island, BBS.
Ryle: Commonage (372).

Fissidens dubius P. Beauv. (*F. decipiens*. *F. cristatus*)
Noted but not confirmed in Lett MS.
Knocknaveen. E side of island, BBS.

Fissidens adianthoides Hedw. (*F. adiantoides*)
Noted but not confirmed in Lett MS.
Bog near the road N of Portlea.
Ryle: Commonage (105, 112, 118, 169, 171); Strake (117) 'found in a runnel in marshy grassland, ... The vegetation is characterised by the presence of *Fissidens adianthoides*, *Callitriche stagnalis*, *Potamogeton polygonifolius* and *Aneura pinguis*,'.

Pleuridium acuminatum Lindb.
Green Road E of Knocknaveen, BBS.
Roadside, Knocknagower.

Pseudephemerum nitidum (Hedw.) Loeske
E side of island, BBS.

Ditrichum gracile (Mitt.) Kuntze (*D. flexicaule*)
E side of Knockmore.

Ceratodon purpureus (Hedw.) Brid.
Lett 1912.
DBN: Knockmore, 1910, Praeger.
Roadside bank at Portlea. Top of gully on E. side of Knockmore. E. side of island, BBS.

Amphidium mougeotii (Schimp.) Schimp. (*Zygodon mougeotii*)
Lett 1912.
Knocknaveen, 1990.
Dry stream above the road on S side of island.
Knockmore.

Dichodontium pellucidum (Hedw.) Schimp.
Lett 1912.
Roadside ditch, E of church. Owenmore R. at bridge. Road verge, N of wall, Capnagower. E side of island, BBS.

Dichodontium palustre (Dicks.) M. Stech
(*Anisothecium squarrosum*)
Lett 1912.
DBN: 1909, Lett.
E side of island, BBS.

Dicranella varia (Hedw.) Schimp.
Clay seacliffs at Portlea with *Erica erigena*. Loughanaphuca, with *Cratoneuron filicinum*.

Dicranella cerviculata (Hedw.) Schimp.
Lett 1912.
DBN: Knockmore, 1910, Praeger, conf. H.N. Dixon.

Dicranella heteromalla (Hedw.) Schimp.
Lett 1912.
Knockmore. E side of island, BBS.
Ryle: Commonage (97, 168).

Dicranum bonjeanii De Not.
Lett 1912.
Lakes NW of Knocknaveen. Edge of runnel with *Potamogeton polygonifolius* and *Hypericum elodes*, N of wall, Capnagower.

Dicranum scoparium Hedw.
Lett 1912.
DBN: Knockmore summit & Ballytoohy, 1909, Lett.
S slope of Knockmore summit.
L. Merrigagh. Lassau Wood, 1990.
E side of island, BBS.
Ryle: 57 relevés.

Dicranum majus Sm.
Lett 1912.
Lassau Wood. Knockmore.
Ryle: 15 relevés.

Dicranum scottianum Turner ex R. Scott (*D. scottii*)
Lett 1912.
Rock outcrop on knoll on eastern end of Knocknaveen, BBS.
Knockmore. Sloping bog, N of tower at W end of the island.

Dicranodontium uncinatum (Harv.) A. Jaeger
Knockmore cliffs, E side of island, at 1200 ft.

Campylopus gracilis (Mitt) A. Jaeger
(*C. schwarzii*)
Lett 1912.
DBN: Knockmore, June 1909, Praeger.

Campylopus fragilis (Brid.) Bruch & Schimp.
Lett 1912.
DBN: on lighthouse, 1909, Lett.
Side of stream at Portlea. E of Knockmore. Rocks above the shore, E of church, 1990. E side of island, BBS.
Ryle: Commonage (67, 216, 377).

Campylopus pyriformis (Shultz) Brid. (incl. var. *azoricus* M.F.V.Corley)
Lett 1912.
Lassau Wood. E of Knockmore. Rocky heath N of wall, Capnagower.
Ryle: 10 relevés.
A Lett specimen 'Clare Island, 1910' is referred to var. *azoricus* by Corley.

Campylopus flexuosus (Hedw.) Brid.
(*C. paradoxus*)
Lett 1912.
DBN: Maum, 1909, Lett.
Lassau Wood, with *Saccogyna*.
E side of island, BBS.
Ryle: 27 relevés.

Campylopus atrovirens De Not.
Lett 1912.
Boggy slopes of Knocknaveen. E side of island, BBS.

Campylopus introflexus (Hedw.) Brid.
Turf-cutter's track, E side of Knockmore. E side of island, BBS.
Ryle: 12 relevés.

Campylopus brevipilus Bruch & Schimp.
Lett 1912.
DBN: Loughnaphuca and Maum, 1909, Lett.
Frequent on boggy ground at Portlea. Boggy slopes of Knocknaveen.

Leucobryum glaucum (Hedw.) Angstr.
Lett 1912.
DBN: May 1909, Lett.
E side of island, BBS.
Ryle 57 relevés.

Eucladium verticillatum (With.) Bruch & Schimp.
(*Mollia verticillata*)
Noted (attr. D. McArdle) but not confirmed in Lett MS.
Bridge at Owenmore. Knocknaveen crag, BBS.

Weissia controversa Hedw. var. ***controversa***
(*Mollia viridula*)
Noted but not confirmed in Lett MS.
Roadside bank, W of harbour. On bank of laneway, E side of Knocknaveen. E side of island, BBS.

Weissia perssonii Kindb.
Knockmore, BBS.

Weissia brachycarpa (Nees & Hornsch.) Jur.
(*W. microstoma*)
Knockmore.

Tortella tortuosa (Hedw.) Limpr. (*Mollia tortuosa*)
Lett 1912.
DBN: Knockmore, 1910, Praeger.
Knocknaveen. Cliffs on E side of Knockmore.
E side of island, BBS.

(Tortella fragilis (Drumm.) Limpr. (*Mollia fragilis*))
This species does not occur in Ireland. Lett's specimen in DBN is *Tortella tortuosa*.

(***Tortella nitida*** (Lindb.) Broth. (*Mollia nitida*))
The specimen in DBN (1909, Lett) is not *T. nitida* and may be *T. inclinata* or *flavovirens* fide A.J.E. Smith?

(***Tortella inclinata*** (R.Hedw.) Limpr.)
The specimen in DBN (1909, Lett) named *T. nitida* by Lett, may be *T. inclinata* or *flavovirens* fide A.J.E. Smith.

Tortella flavovirens (Bruch) Broth. var. ***flavovirens***
Boulder clay sea cliff, Portlea.

Trichostomum brachydontium Bruch. (*Mollia litoralis; M. brachydontia*)
Lett 1912.
DBN: 1909, Lett; 1910, Praeger.
Roadside wall N of Portlea; boulder clay sea cliff at Portlea. Roadside rock E of church. Bank of Owenmore R. at bridge. E and S side of Knocknaveen. Top of gully on E side of Knockmore. Knockmore, BBS.
E side of island, BBS.

Trichostomum crispulum Bruch. (*Mollia crispula*)
Noted but not confirmed in Lett MS.
Cliffs, E side of Knockmore. Rocks near the sea, Loughanaphuca. Roadside verge, Capnagower.
E side of island, BBS.

Hymenostelium recurvirostrum Hedw. (*Barbula curvirostris*)
Lett 1912.
Knockmore.

Gymnostomum calcareum Nees & Hornsch.
Base rich crag of Knocknaveen, BBS.

Gymnostomum aeruginosum Sm. (*Mollia aeruginosa*)
Lett 1912.
Top of gully on E side of Knockmore.
E side of island, BBS.

Pseudocrossidium revolutum (Brid.) R.H. Zander
(*Barbula revoluta*)
Lett, unlocalised, sandy specimen in DBN, labelled *Barbula tophacea*, is mostly *Pseudocrossidium revolutum* with scattered stems of *Barbula fallax*.

Bryoerythrophyllum recurvirostrum (Hedw.) P.C. Chen
Lett 1912.
Track by Lough Merrigagh.

(***Bryoerythrophyllum ferruginascens*** (Stirt.) Giacom. (*Barbula rubella*))
Lett specimen (?Knockmore), 1909, in DBN is *Ceratodon purpureus* (det. D.S., 1987).

Barbula convoluta Hedw.
Lett 1912.
Roadside, W of harbour, with *Bryum bicolor*. Road verge, Capnagower. E side of island, BBS.

Barbula unguiculata Hedw.
Lett 1912.
E side of island, BBS.
Roadside verge, N of wall, Capnagower.

Didymodon rigidulus Hedw. (*Barbula rigidula*)
Listed but not confirmed in Lett MS.
Knockmore.

Didymodon insulanus (De Not.) M.O. Hill (*Barbula cylindrica*)
Listed but not confirmed in Lett MS. Lane at Portlea. Sandy bluff near harbour. Road verge, Capnagower. E side of island, BBS.

Didymodon tophaceus (Brid.) Lisa (*Barbula tophacea*)
Lett's specimen in DBN is *Barbula revoluta* (det D.S.).
Boulder clay sea cliff at Portlea. Stream on S side of island. E side of island, BBS.

Didymodon fallax (Hedw.) R.H. Zander (*Barbula fallax*)
Lett 1912.
Path in cemetery at church. Above Loughanaphuca. Dune at harbour. Cliff, Knockmore. Mud by bog, N of wall, Capnagower.

Anoectangium aestivum (Hedw.) Mitt. (*Pleurozygodon aestivus*)
Lett 1912.
Scattered stems with *Fissidens osmundoides*, 1200ft on E side of Knockmore.

Gyroweisia tenuis (Hedw.) Schimp. (*Weisia tenuis*)
Listed but not confirmed in Lett MS.
Rock outcrop in blanket bog/heath, high on the E slope of Knockmore.

(***Tortula marginata*** (Bruch & Schimp.) Spruce)
DBN specimen, 1909, Lett, is *Tortula muralis* (det A.J.E.Smith, 1988).

Tortula muralis Hedw.
Lett 1912.
Old gate pier, Portlea. Stone at dune near harbour. E side of island, BBS.

Hennediella heimii (Hedw.) R.H. Zander
Below the road at back of sand dune, harbour.

Syntrichia ruralis var. ***ruraliformis*** (Besch.) Delogne (*Tortula ruralis* var. *arenicola*)
Listed but not confirmed in Lett MS.
Dune near the harbour.
Ryle: The quay (84, 160); Kill (400).

Cinclidotus fontinaloides (Hedw.) P. Beauv.
Owenmore, at bridge.

Zygodon viridissimus (Dicks.) Brid. var. ***viridissimus***
Lett 1912
DBN: 1909, Lett.
E side of island, BBS.

Ulota crispa (Hedw.) Brid. var. ***crispa*** (*Weissia ulophylla*)
Lett 1912.
DBN: Ballytoohy, 1909, Lett.
Lassau Wood. On *Salix aurita* in gully on E side of Knocknaveen.
Ryle: Ballytoohy More (1153); Commonage 321.

Ulota bruchii Hornsch. ex Brid. (*U. crispa* var. *norvegica* (Groenvall) Smith & Hill)
Lassau Wood.

Ulota hutchinsiae (Sm.) Hammar. (*Weissia americana*)
Lett 1912.
DBN: unlocalised, May 1909, Lett. Lett's collection consists of ten or twelve robust tufts. The species was not rediscovered.

Ulota phyllantha Brid. (*Weissia phyllantha*)
Lett 1912.
DBN: Loughanaphuca, Aug 1909, Lett.
Rocks N of Portlea, 1990. Trees, Lassau Wood. Above Loughanaphuca. On *Salix aurita* in gully and on stone wall of lane E side of Knocknaveen.

Hedwigia stellata Hedenäs (*H. albicans*)
Lett 1912.
DBN: unlocalised, Lett, 1909.
Not rediscovered.

Philonotis fontana (Hedw.) Brid.
Lett 1912.
Track N of Portlea, 1990.
L. Merrigagh. Flush at old well. Roadside, Capnagower. E side of island, BBS.

Philonotis calcarea (Bruch & Schimp.) Schimp.
E. side of Knockmore, with *Drepanocladus revolvens*.

Breutelia chrysocoma (Hedw.) Lindb.
(*B. arcuata*)
Lett 1912.
DBN: Cliff, April 1909, Praeger; July 1910, Lett. Near summit of Knockmore, with *Fissidens cristatus*. Boggy slope below Knocknaveen.
Ryle: 11 relevés.

Plagiobryum zierii (Hedw.) Lindb.
Base rich crag, Knocknaveen, BBS.

Anomobryum concinnatum (Spruce) Lindb.
Lett 1912.
DBN: Knockmore, 1910, Praeger; Clare Island, 1910, Lett.
E slope of Knockmore.

Bryum pallens Sw. ex anon.
Lett 1912.
DBN: May 1909, Lett.
Knocknaveen, with *Trichostomum brachydontium*.
E side of island, BBS.

Bryum capillare Hedw.
Lett 1912.
DBN: May 1909 and W end, Aug 1909.
Rocks above the shore, E of the church.
E side of island, BBS.
Ryle: Commonage (70, 192, 264, 320, 343).

Bryum archangelicum Bruch & Schimp.
(*B. inclinatum*).
Lett 1912.
DBN: Clare Island, 2 June 1909, Lett.

Bryum pseudotriquetrum (Hedw.) P. Gaertn. *et al.* (*B. ventricosum*)
Lett 1912.
Bog, W of wall near Portlea. L. Merrigagh. Road verge, Capnagower. E side of island, BBS.
Ryle: 18 relevés.

Bryum radiculosum Brid.
Lett 1912.
Lett specimen in DBN, unlocalised, May 1907.

Bryum riparium I.Hagen.
Lett unlocalised specimen in DBN, labelled '*B. uliginosum*', May 1909, is *B. riparium* (conf. H.L.K. Whitehouse).
The specimen was overlooked or disregarded for Lett's 1912 paper.

Bryum alpinum Huds. ex With
Lett 1912.
E side of island, BBS. Roadside, Capnagower.

Pohlia cruda (Hedw.) Lindb.
Hard packed sand, dune, with *Trichostomum brachydontium*.

Pohlia nutans (Hedw.) Lindb.
Listed but not confirmed in Lett MS.
E side of island, BBS.
Ryle: Bunnamohaun (137, 141); Commonage (377).

Pohlia drummondii (Müll.Hal.) A.L. Andrews
Green Road to Knocknaveen, BBS.

Pohlia annotina (Hedw.) Lindb.
Bare patches on Green Road below Knocknaveen.

Pohlia melanodon (Brid.) A.J. Shaw (*P. carnea*)
Bank of Owenmore R. at bridge.

Pohlia wahlenbergii (Web. & Mohr) Andrews
(*P. albicans*)
Lett 1912.
Laneway at Portlea. Flush at Knocknaveen. E side of island, BBS.

Mnium hornum Hedw.
Lett 1912: 'The most frequent [moss] … found everywhere on the island, from sea level to the top of Croaghmore [Knockmore], the discoloured but flourishing foliage of which, deeply tinged with a tawny hue, gives evidence of the harmless effect of the salt on its tissues'.
In acid turf, common.
Ryle: 61 relevés.

Rhizomnium punctatum (Hedw.) T.J. Kop. (*Mnium punctatum*)
Lett 1912.
DBN: May 1909 and July 1910. Knockmore, 1910, Praeger—as *Mnium pseudopunctatum*.
Knockmore, BBS.
Roadside ditch on S side of island. E side of Knocknaveen. E side of island, BBS.
Ryle: Lecarrow (335); Commonage (343).

Rhizomnium pseudopunctatum (Bruch & Schimp.) T.J. Kop. (*Mnium subglobosum*)
Praeger specimens in DBN are *Rhizomnium punctatum*.
Flush on E side of Knockmore, 1990.

Plagiomnium elatum (Bruch and Schimp.) T.J. Kop.
Ryle: Park (365).

Plagiomnium ellipticum (Brid.) T.J. Kop.
L. Merrigagh, 1990.

Plagiomnium undulatum (Hedw.) T.J. Kop. (*Mnium undulatum*)
Lett 1912.
DBN: Cliff, May 1909, Lett.
Roadside at sand dune and W of harbour. E side of island, BBS.

Aulacomnium palustre (Hedw.) Schwagr.
Lett 1912.
Knocknaveen area, BBS.
Ryle: 47 relevés.

Hookeria lucens (Hedw.) Sm. (*Pterygophyllum lucens*)
Lett 1912.
Lassau Wood with *Atrichum undulatum*.
Roadside ditch, W of harbour, 1990.

Palustriella commutata agg. (*Amblystegium glaucum/falcatum*)
Ryle: Commonage (166).

Palustriella commutata (Hedw.) Ochyra var. ***commutata*** (*Amblystegium glaucum*)
Lett 1912.
E side of island, BBS.

Palustriella falcata (Brid.) Hedenäs (*Amblystegium falcatum*)
Lett 1912
DBN: Loughanaphuca, 1909, Lett.
Stream, E side of Knockmore, with *Carex paniculata*. Above Loughanaphuca. Flush on SE side of Knocknaveen.

Cratoneuron filicinum (Hedw.) Spruce var. ***filicinum*** (*Amblystegium filicinum*)
Lett 1912.
DBN: Near Loughanaphuca, Aug 1909, Lett.
Bank at Portlea. Knocknaveen. L. Merrigagh. Above Loughanaphuca. Road verge, Capnagower. E side of island, BBS.
Ryle: Commonage (112); Lecarrow (335).

Campylium stellatum (Hedw.) J. Lange & C. Jens. var. ***stellatum*** (*Amblystegium stellatum*)
Lett 1912.
DBN: Knockmore, 1909, Praeger—leaf with one long faint nerve. Boggy ground at side of stream, N of the wall towards Portlea. Flush on S side of gully on SE side of Knocknaveen. E side of island, BBS.
Ryle: 9 relevés.

Campylium protensum (Brid.) Kindb.
Unlocalised Lett specimen in DBN (as *C. chrysophyllum*).
Boulder clay sea cliff at Portlea.

(*Campyliadelphus chrysophyllus* (Brid.) R.S. Chopra (*Amblystegium chrysophyllum*))
Lett DBN specimen is *C. protensum*.

Campyliadelphus elodes (Lindb.) Kanda
Ryle: Commonage (112, 370).

(*Amblystegium serpens* (Hedw.) Schimp.)
Listed but not confirmed in Lett MS.
No specimen in DBN.

Drepanocladus aduncus (Hedw.) Warnst.
L. Merrigagh (cf. *D. polycarpus*—pronounced, incrassate, angular cells).

Hygrohypnum luridum (Hedw.) Jenn.
As *Calliergonella lindbergii*, May 1909, Lett (DBN).

Warnstorfia fluitans (Hedw.) Loeske var. ***fluitans***
(*Amblystegium fluitans*)
Lett 1912.
In boggy fields with *Juncus effusus* on the E side of the church below the road.
Ryle: Ballytoohy More (228); Commonage (237).

Sarmenthypnum exannulatum (Schimp.) Hedenäs
(*Amblystegium exannulatum*)
Lett 1912.
DBN: W end of island, 1909, Lett.
Bog, N of wall, Capnagower. L. Merrigagh and other lakes NW of Knocknaveen.
Ryle: Ballytoohy Beg (147).

Sarmenthypnum sarmentosum (Wahlenb.) Tuom. & T.J. Kop.
Lett 1912.
DBN: Marsh near lighthouse, June 1909, Praeger.
Flush at Portlea.

Scorpidium revolvens (Sw. ex anon.) Rubers
(*Amblystegium revolvens, Drepanocladus revolvens*)
Lett 1912.
Drains and runnels by roadside, N side of the wall towards Portlea. Flush on S side of gully on SE side of Knocknaveen. E side of island, BBS.

Scorpidium scorpioides (Hedw.) Limpr.
(*Amblystegium scorpioides*)
Lett 1912.
DBN: Near Loughnaphuca, 1909, Lett.
Lighthouse marsh, 1909, Praeger.
In a flush at Portlea. L. Merrigagh, with *Apium inundatum*.
Ryle: 16 relevés.

Thuidium tamariscinum (Hedw.) Schimp.
(*T. tamariscifolium*)
Lett 1912.
DBN: Ballytoohy, 1909, Lett.
Lassau Wood. E side of island, BBS.
Ryle: 83 relevés.

Thuidium delicatulum (Hedw.) Schimp.
Boggy field sloping to sea, E of church. E side of island, BBS.

Pseudoscleropodium purum (Hedw.) M. Fleisch.
(*Hypnum purum*)
Lett 1912.
E side of island, BBS.
Ryle: 27 relevés.

Eurhynchium striatum (Hedw.) Schimp. (*Hypnum striatum*)
Lett 1912.
Roadside, N of church. Wall, W of harbour, with *Hypnum cupressiforme*. E side of island, BBS.
Ryle: Ballytoohy More (348, 350, 351); Commonage (341, 372); Capnagower (13); Maum (305); Park (365).

Platyhypnidium riparioides (Hedw.) Dixon
(*Hypnum rusciforme*)
Lett 1912.
DBN: May 1909, Lett; W end, June 1909, Praeger.
Owenmore R. at bridge.

(*Platyhypnidium lusitanicum* (Schimp.) Ochyra & Bednarek-Ochyra (*Hypnum rusciforme* var. *atlanticum; Rhynchostegium lusitanicum*)
Lett specimen, 1909, in DBN is *R. riparioides* (det A.J.E. Smith)

Rhynchostegiella tenella (Dicks.) Limpr. var. ***tenella*** (*Hypnum tenellum*)
Knocknaveen. On stones at end of dune, harbour.

Oxyrrhynchium hians (Hedw.) Loeske (*Hypnum swartzii*)
Wall, W. of harbour. E. side of island, BBS.

Kindbergia praelonga (Hedw.) Ochyra (*Hypnum praelongum; Eurhynchium praelongum*)
Lett 1912.
Lighthouse, 1909, HWL (as *Brachythecium velutinum*).
Dune near the harbour.
Ryle: 19 relevés.

Sciuro-hypnum populeum (Hedw.) Ignatov & Huttunen (*Hypnum viride; Brachythecium populeum*)
Lett 1912 (DBN: Clare Island, May 1909, Lett)

Sciuro-hypnum plumosum (Hedw.) Ignatov & Huttunen (*Hypnum pseudoplumosum; Brachythecium plumosum*)
Listed but not confirmed in Lett MS.
No specimen in DBN.
Stone at side of lane, Portlea. E. side of island, BBS.

Brachythecium albicans (Hedw.) Schimp. (*Hypnum albicans*)
All three Lett specimens in DBN from near the harbour are *Brachythecium mildeanum*. Dune near the harbour.

Brachythecium glareosum (Bruch ex Spruce) Schimp.
Gravelly moraine, Ballytoohy Beg.

Brachythecium mildeanum (Schimp.) Schimp.
DBN: Near harbour, May 1909, Lett (as *B. albicans*).
Sandy bluff facing the sea, harbour.

Brachythecium rutabulum (Hedw.) Schimp. (*Hypnum rutabulum*)
Lett 1912.
Edge of roadside at dune near the harbour.
Ryle: Ballytoohy Beg (278, 279).

Brachythecium rivulare Schimp. (*Hypnum rivulare*)
Listed but not confirmed in Lett MS.
no specimen in DBN.
Flush on E side of Knockmore, with *Aulacomnium palustre* and *Barbula fallax*, Roadside ditch, W of harbour. Marshy field below Lassau Wood.
E side of island, BBS.

(***Brachytheciastrum velutinum*** (Hedw.) Ignatov & Huttunen (*Hypnum velutinum; Brachythecium velutinum*))
Lett specimen in DBN (lighthouse, Clare Island), is *Eurhynchium praelongum*.

Homalothecium sericeum (Hedw.) Schimp. (*Hypnum sericeum*)
Lett 1912.
DBN: near Loughnaphuca, 1909, Lett.
Rock W of church, 1990.
E side of island, BBS.

Homalothecium lutescens (Hedw.) H. Rob. (*Hypnum lutescens*)
Lett 1912.
Knocknaveen.

Calliergonella cuspidata (Hedw.) Loeske. (*Acrocladium cuspidatum*)
Lett 1912.
DBN: near Loughanaphuca, Aug. 1909, Lett.
L. Merrigagh. E side of island, BBS.
Ryle 17 relevés.

Calliergonella lindbergii (Mitt.) Hedenäs
Not recorded in Lett 1912.
A Lett specimen in DBN, 'May 1909' is *Hygrohypnum luridum*.
Roadside, S side of island. Green Road, Knocknaveen, BBS.
Roadside N of wall, Capnagower. E side of island, BBS.

Hypnum cupressiforme Hedw. var. ***cupressiforme*** (*Stereodon cupressiforme*)
Lett 1912.
DBN: near Loughanaphuca, Aug 1909, Lett.
Wall, W of harbour. Boulder at Portlea. E side of island, BBS.
Ryle: 18 relevés.

Hypnum cupressiforme var. ***lacunosum*** Brid. (*Stereodon cupressiforme* var. *tectorum*)
Lett 1912.
DBN: May 1909, Lett.

Hypnum cupressiforme var. ***resupinatum*** (Taylor) Schimp. (*Stereodon resupinatus*)
Lett 1912.
Wall W of harbour. Bridge at Owenmore R. Stone wall of abandoned house, Portlea. E side of island, BBS.

Hypnum andoi **A.J.E. Sm.** (*Stereodon cupressiforme* var.; *H. mammillatum*)
DBN: Ballytoohy, 1909, Lett.
Lassau Wood.

Hypnum jutlandicum Holmen & E. Warncke (*Stereodon cupressiforme* var. *ericetorum*)
Lett 1912.
With *Polytrichum juniperinum* specimen in DBN, Lett.
E. side of island, BBS.
Ryle: 118 relevés.

Ctenidium molluscum (Hedw.) Mitt. var. ***molluscum***
Lett 1912 (DBN: cliff, Apr. 1909, Praeger). Boggy roadside, N of the wall. Boulder clay cliffs by the sea with *Calliergon cuspidatum*, Portlea. S-facing slope, top of Knocknaveen. E side of island, BBS.

Ctenidium molluscum var. ***condensatum*** (Schimp.) Broth.
Cliffs on E. side of Knockmore, with *Brachythecium plumosum*, *Tortella tortuosa* and *Isothecium myurum*. N-facing bluff at Knocknaveen.

Hyocomium armoricum (Brid.) Wijk & Marg.
Lassau Wood.

Pleurozium schreberi (Willd ex Brid.) Mitt.
(*Hylocomium parietinum*)
Lett 1912.
Portlea. Knocknaveen; E side of island, BBS.
Ryle: 16 relevés.

Rhytidiadelphus triquetrus (Hedw.) Warnst.
(*Hylocomium triquetrum*)
Lett 1912.
E side of island, BBS.
Ryle: 40 relevés.

Rhytidiadelphus squarrosus (Hedw.) Warnst.
(*Hylocomium squarrosum*)
Cliff, 1909, Lett and 1909, Praeger.
E side of island, BBS.
Ryle: 139 relevés.

Rhytidiadelphus loreus (Hedw.) Warnst.
(*Hylocomium loreum*)
On cliff, 1909, Lett (DBN, as *R. triquetrus*).
Wall top, E of church / W of harbour.
Knocknaveen. Portlea. E side of island, BBS.
Ryle: 17 relevés.

Hylocomium splendens (Hedw.) Schimp.
(*Hylocomium proliferum*)
Lett 1912.
E side of island, BBS.
Ryle: 47 relevés.

(***Plagiothecium denticulatum*** auct.)
Lett 1912.
Prior to the revision by Greene (1957) *P. denticulatum*, *P. succulentum* and *P. nemorale* were often confused. Earlier records for these species are unreliable (Holyoak 2003).

Plagiothecium succulentum (Wilson) Lindb.
DBN: Clare Island, May 1909, Lett.
Lassau wood. E side of island, BBS.
S-facing slope, top of Knocknaveen.

Plagiothecium nemorale (Mitt.) A. Jaeger
DBN: Clare Island, May 1909, Lett.
Rock crevice near the sea, Loughanaphuca.

Plagiothecium undulatum (Hedw.) Schimp.
Lett 1912.
Lassau Wood. Common on Knocknaveen. E side of island, BBS.

Neckera crispa Hedw.
Lett 1912.
DBN: cliffs, Knockmore, 1910, Praeger.
Not rediscovered.

Neckera complanata (Hedw.) Huebener
Knocknaveen, 1990.
E side of island, BBS.

Thamnobryum alopecurum (Hedw.) Gangulee
(*Porotrichum alopecurum*)
Lett 1912.
DBN: 1909, Lett.
Cliff, Knocknaveen. E side of island, BBS.

Isothecium myosuroides Brid. var. ***myosuroides***
Lett 1912.
On roots of trees at Lassau Wood. Field sloping to sea, E of church. Roadside,
S side of island. Rocks N of tower. E side of island, BBS.
Ryle: 12 relevés.

Isothecium myosuroides var. ***brachythecioides*** (Dixon) Braithw.
E slopes of Knockmore, D.G. Long, BBS.

Isothecium alopecuroides (Lam. ex Dubois) Isov.
(*I. viviparum*, *I. myurum*)
Lett 1912.
Top of Knockmore, 1909, Lett.
Base-rich crag of Knocknaveen. Gully SE side of Knocknaveen. E. side of island, BBS.

Anomodon viticulosus (Hedw.) Hook. & Taylor
Base-rich crag, Knocknaveen, BBS.

REFERENCES

Armitage, E. 1938 The British Bryological Society. *Journal of Botany* **76**, 171–4.

Blockeel, T.L. 1991 The *Racomitrium heterostichum* group in the British Isles. *Bulletin of the British Bryological Society* **58**, 29–35.

Blockeel, T.L. 1995 Summer field meeting, 1994, second week, Clifden. *Bulletin of the British Bryological Society* **65**, 12–18.

Dixon, H.N. 1897 *Handbook. Catalogue of British Mosses.* Eastbourne and London.

Doyle, G.J. and Foss, P.J. 1986 A resurvey of the Clare Island flora. *Irish Naturalists' Journal* **22**, 85–9.

Doyle, G.J. and Whelan, S. 1991 Proposal for a new survey of Clare Island: 1991–1995. Unpublished report to Royal Irish Academy, Dublin.

Hill, M.O., Blackstock, T.H., Long, D.G. and Rothero, G.P. 2008 *A checklist and census catalogue of British and Irish bryophytes.* The British Bryological Society.

Holyoak, D.T. 2003 *The distribution of bryophytes in Ireland.* Dinas Powys. Broadleaf Books.

Jones, D.A. 1917 Muscineae of Achill Island. *Journal of Botany* **55**, 240–46

Lett, H.W. 1912 Clare Island Survey. Parts 11–12. Musci and Hepaticae. *Proceedings of the Royal Irish Academy* **31**, (11–12), 1–18.

Long, D.G. 1988 New vice-county records and amendments to the Census Catalogue. Hepaticae. *Bulletin of the British Bryological Society* **52**, 30–3.

Long, D.G. 1990 The bryophytes of Achill Island—Hepaticae. *Glasra* (new series) **1**, 47–54.

Paton, J.A. 1999 *The liverwort flora of the British Isles.* Colchester. Harley Books.

Praeger, R.L. 1911 Clare Island Survey. Part 10 Phanerogamia and Pteridophyta. *Proceedings of the Royal Irish Academy* **31**, 1–112.

Ratcliffe, D.A. 1962 The habitat of *Adelanthus unciformis* (Tayl. Mitt. and *Jamesoniella carringtonii* (Balf.) Spr. in Ireland. *Irish Naturalists' Journal* **14**, 38–40.

Rothero, G.P. 1988 The summer meeting, 1987, Co. Mayo. Second week at Westport: 12–18 August. *Bulletin of the British Bryological Society* **51**, 12–15.

Ryle, T.J. 2000 Vegetation/environment interactions on Clare Island. Co. Mayo. Unpublished PhD thesis, National University of Ireland, Galway.

Smith, A.J.E. 1990 The bryophytes of Achill Island—Musci. *Glasra* (new series) **1**, 27–46.

Smith, A.J.E. 1990 *The moss flora of Britain and Ireland.* 2nd edn. Cambridge. Cambridge University Press.

Smith, A.J.E. 2004 The moss flora of Britain and Ireland. Cambridge. Cambridge University Press.

Synnott, D.M. 1986 An outline of the flora of Mayo. *Glasra* **9**, 13–117.

Synnott, D.M. 1988 The summer meeting, 1987, County Mayo. First week in Achill Island: 5–12 August. *Bulletin of the British Bryological Society* **51**, 7–10.

Synnott, D.M. 1990a The bryophytes of Achill Island—a preliminary note. *Glasra* (new series) **1**, 21–6.

Synnott, D.S. 1990b The bryophytes of Lambay Island. *Glasra* (new series) **1**, 65–81.

Waddell, C.H. 1897 *Moss Exchange Club Catalogue of British Hepaticae.* London.

Warburg, E.F. 1963 Notes on the bryophytes of Achill Island. *Irish Naturalists' Journal* **14**, 139–45.

Warburg, E.F. 1966 New vice-county records and amendments—Musci. *Transactions of the British Bryological Society* **5**, 199.

LICHEN FLORA OF CLARE ISLAND

M.R.D. Seaward and D.H.S. Richardson

ABSTRACT

To date, 355 lichenized fungi, 12 lichenicolous fungi and 6 non-lichenized fungi have been recorded from Clare Island, of which 24 recorded in the first survey of the island (1909–1911) were not refound in the 1990 and 1991 surveys (four of the original records are doubtful and two are improbable). The island's lichens are greatly influenced by oceanic effects and, despite the lack of mature woodland, such trees and scrub as survive, together with timberwork and the varied geology and associated soils, provide substrata for a highly diverse flora. The sensitivity of lichens to natural and human disturbances makes them ideal environmental monitors; therefore, periodic re-evaluation of Clare Island's lichen flora is recommended in view of increased pressures, particularly tourism. Significant changes to the island's lichen flora since the surveys of the 1990s may have already taken place, so a re-evaluation should be undertaken in the near future.

Introduction

Clare Island is the largest island in Clew Bay, Co. Mayo, being approximately 16km^2 (about 8km long × 4km wide). Two small mountains, Knockmore (464m) and Knockaveen (222m), dominate the island. There are steep cliffs on the northern and western sides of the island and a series of coves (Pl. I), a harbour and a sandy beach, on the southern and eastern sides respectively. The geology of the island is complex (see Graham 2001), the varied siliceous and calcareous outcrops and the derived soils providing substrata for a rich saxicolous and terricolous lichen flora. The lichen flora is highly significant in terms of biogeographical patterns due to the island's extreme westerly position, with, for example, *Arthonia atlantica* and *Degelia ligulata* (Pl. II) representing members of the Macaronesian element.

There is a wide variety of habitats suitable for lichen colonisation, from the upland hilly areas to the ridges, cliffs and coves lining the shore. Unfortunately, due to the lack of mature woodland, there are few habitats for epiphytic lichens. However, there are some trees (e.g. *Betula*, *Corylus*, *Salix* and *Sorbus*) and shrubs (e.g. *Ilex* and *Sambucus*) on the island, with a few dense thickets, often heavily encrusted with mosses, which support a relatively diverse lichen flora (Pl. III). Timberwork also provides an important substratum for both lignicolous species and some epiphytes whose preferred habitat is rare or absent on the island.

Factors influencing the lichen flora

An overriding factor affecting the lichen flora is the maritime effect, which generates a distinctive coastal zonation (Pl. IV A and IV B) and influences the whole island, ombrogenously depositing chemicals that influence lichen distribution. This is particularly marked in habitats low on nutrients, such as peat bogs and thin mineral soils over siliceous rocks. Locally, nutrient levels are raised at sites frequented by birds or sheep, resulting in lichen assemblages that are characteristic of eutrophicated substrata. Sheep populations, which have increased due to European Community policies, have enhanced eutrophication, soil erosion

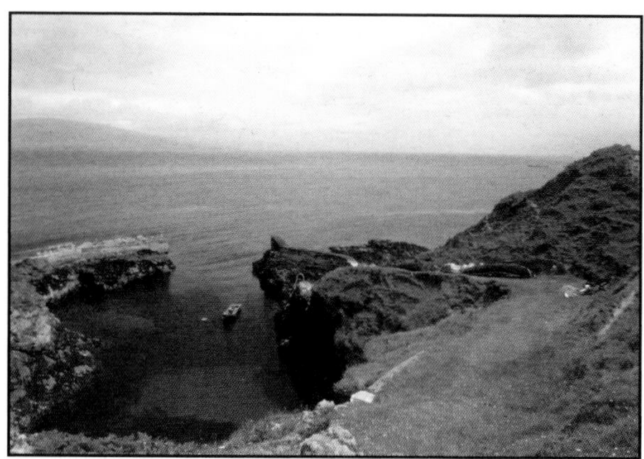

Pl. I Steep cliffs surrounding typical small cove providing varied habitats for lichens.

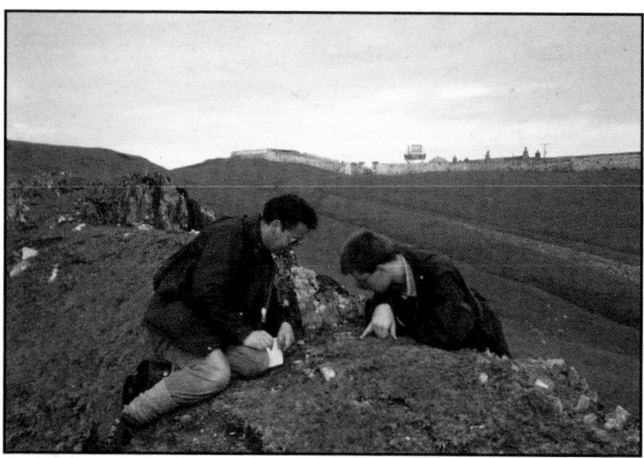

Pl. II Rocky outcrops in short turf near lighthouse colonised by the rare lichen *Degelia ligulata*.

Pl. III Dense thicket of *Corylus* etc. providing habitats for mosses and the locally rare lichens *Degelia atlantica* and *D. plumbea*.

and overgrazing, especially in the upland areas. Human influence is also evident; in the early 1990s there were only about 150 inhabitants, but in the nineteenth century, prior to the potato famine, the farming and fishing population was ten times this figure. There are many man-made substrata, particularly in the harbour area (Pl. V), for lichen species known to favour such colonisation sites.

Pl. IV Conspicuous rocky shore (littoral) lichen zonation: A. orange zone dominated by *Xanthoria parietina* and *Caloplaca* spp; B. grey zone dominated by *Ramalina* spp, *Tephromela atra* and *Lecidella asema*.

History of lichenology on Clare Island

Our knowledge of the island's lichen flora goes back almost 100 years: visits by Matilda C. Knowles (1864–1933) in 1909 and 1910 and Annie Lorrain Smith (1854–1937) and William West (1848–1914) in 1910 provided the material for the first Clare Island Survey (Smith 1911); supplementary material, now housed in the National Botanic Garden, Glasnevin (DBN), was collected by J. Adams in 1909 and 1910, A.D. Cotton in 1911, H.W. Lett in 1909 and 1910 and R.L. Praeger in 1909. According to Mitchell (1995), this first survey revitalised

Pl. V Harbour area providing varied man-made substrata and habitats for lichens.

lichenological fieldwork in Ireland, as well as adding taxa to the Irish lichen flora (cf. Knowles 1929). Included among the fourteen new Irish taxa in Smith (1911) are eight species (*Bacidia egenula, Buellia coniops, Cladonia chlorophaea, Collema bachmanianum, Collema tenax* var. *ceranoides, Peltigera didactyla, Ramalina cuspidata* and *R. subfarinacea*) from Clare Island itself (Mitchell 1993). These early workers also surveyed the nearby mainland, and the resulting data were embodied within the published account by Smith (1911). Her paper provides a good impression of County Mayo's lichen flora in the early part of the twentieth century and lists 289 taxa (approximately 243 taxa now, due to re-identification and modern taxonomic interpretation—about half of those currently known from the county—cf. Seaward 1994), but only 154 taxa (approximately 134 taxa by modern taxonomic interpretation) relate specifically to Clare Island.

Many of the lichens collected during this early work are now in the herbarium of the National Botanic Garden, Glasnevin, Dublin (DBN) and The Natural History Museum, London (BM). Numerous records of the former are included in Knowles (1929) and Porter (1948) and most of the latter are listed in Smith (1918; 1926). The Clare Island material in both herbaria has been consulted by one of us (MRDS), and in some cases redetermination has been necessary; furthermore, extra species not included in Smith (1911) have been found in the material. However, a few of the Clare Island Survey records are doubtful in the absence of supporting herbarium material, and some additional records not included in Smith (1911) are to be found in Smith (1918; 1926), Knowles (1929), Porter (1948) and Coppins (1983). The result of this reappraisal is that approximately 150 taxa are confirmed as having been recorded during the first Clare Island Survey; of these, 24 (*Amandinea coniops, Bacidia egenula, Dactylospora saxatilis, Lecanora carpinea, Lobaria pulmonaria, Melanelia exasperata, Micarea ternaria, Parmotrema reticulatum, Peltigera didactyla, P. horizontalis, Pertusaria pertusa, Placopsis gelida, Platismatia glauca, Porpidia speirea, Ramalina calicaris, Tylothallia biformigera, Usnea florida* and *Verrucaria margacea*, plus the doubtful records of *Buellia spuria, Collema nigrescens, Lecanora albella* and *Xanthoria candelaria* and the improbable records of *Cladonia phyllophora* and *Peltigera malacea*) were not refound in the 1990 and 1991 surveys (see below). However, only a few of them are likely to have become extinct in view of the limited habitat modification that occurred on Clare Island between 1911 and 1991. A further determined search is clearly needed to establish the current status of these taxa.

In July 1990, David H.S. Richardson, Mark R.D. Seaward and Howard F. Fox visited the island in order to make a preliminary assessment of the changes to the lichen flora since the initial studies undertaken during period 1909 to 1911. In all, 170 taxa were recorded (specimens to support many of these records are to be found in the personal herbaria of M.R.D. Seaward and H.F. Fox). In July 1991, one of us (DHSR) was the local organizer for a visit to Clare Island by the British Lichen Society (including MRDS and Howard Fox). As a consequence, 168 taxa additional to those recorded in 1990 were noted; specimens to support many of these are to be found in the herbaria of CABI, Egham (IMI) and the Botanical Museum, Berlin (B), as well as in private herbaria (including those of M.R.D. Seaward and H.F. Fox). Despite detailed searches, herbarium material (and supporting documentation) transported to The Natural History Museum, London (BM) for further study has not been found. This unfortunate loss was the main reason why an account of the 1991 meeting was never published. Howard Fox made a further visit to Clare Island in March 1996, but other than the inclusion of some lichenicolous fungi in Fox (1996), details of any other lichens recorded are apparently unavailable.

Summary and concluding remarks

To date, 355 lichenized, 12 lichenicolous and 6 non-lichenized fungi listed below have been recorded from Clare Island, of which 18 taxa were not refound on surveys in the 1990s. A few are doubtful in the absence of supporting herbarium material but it is unlikely that many have become extinct as there have been no major natural resource developments and the island is not downwind of any nearby sources of toxic air pollutants. Further research to rediscover the unconfirmed lichen taxa and make new discoveries is clearly warranted, especially since lichens are such remarkable indicators of a wide range of environmental disturbances. It is thus important to establish a good baseline dataset for future monitoring programmes. Several factors are potentially damaging to the island's lichen flora, including intensive sheep grazing, soil erosion, the increased use of agrochemicals, tourism and global climate changes. In addition, recent developments in salmon aquaculture off the more sheltered shoreline of Clare Island could affect the seashore lichen communities that include extensive beds of *Lichina pygmaea* (Pl. VI).

With increased tourist accessibility, it is clearly a challenge to develop management plans for islands such as Clare Island to conserve their biodiverse lichen flora. Once considered to be reasonably remote, such islands are now facing demands from population expansion and tourism (cf. Lambay Island—Seaward and Richardson 2000). Surveys since the 1990s have shown that such changes can lead to a deterioration of the lichen flora, which in the case of Clare Island is clearly worth preserving to the greatest possible extent, not only because of the island's situation of the extreme western fringe the British Isles, but also because there has been such a long documented history of lichenology there.

Pl. VI Locally abundant *Lichina pygmaea* part of lower littoral zone that also includes *Verrucaria* spp (on rocks) and *Collemopsidium foveolatum* (on barnacles).

Pl. VII Locally rare *Lobaria virens* on mossy ultrabasic rocky outcrop.

CATALOGUE

Checklist
Nomenclature is mainly according to Coppins (2002). Where information on the frequency and ecology is provided, the lichen is still extant.

S = recorded in Smith (1911)
* = recorded only in 1990 and/or 1991 and/or 1996
[LF] = lichenicolous fungus
[F] = non-lichenized fungus
(DBN) = Clare Island Survey herbarium material in the National Botanic Garden, Glasnevin, but not included in Smith (1911)

Abrothallus parmeliarum (Sommerf.) Arnold
[LF]
S (as *Buellia parmeliarum*).
Infrequent on thalli of *Parmelia omphalodes* and *P. saxatilis*.

* ***Acarospora fuscata*** (Schrad.) Th.Fr
Common on exposed siliceous boulders.

Acarospora smaragdula (Wahlenb.) A.Massal.
S (as *Lecanora smaragdula*).
Occasional on siliceous rocks and walls.

* ***Acrocordia salweyi*** (Leight. ex Nyl.) A.L.Sm.
Occasional on mortar of harbour wall; rare on vertical calcareous cliff and coastal acid rocks.

* ***Agonimia gelatinosa*** (Ach.) Brand & Diederich
Occasional on peaty soil.

Agonimia tristicula (Nyl.) Zahlbr.
Rare on mosses over calcareous substrata.

Amandinea coniops (Wahlenb. ex Ach.) M.Choisy ex Scheid. & H.Mayrhofer
In Knowles 1929 (record of *B. coniopta* in **S** interpreted as *B. coniops*; see also Smith 1926)—not refound.

Amandinea punctata (Hoffm.) Coppins & Scheid.
S (as *Buellia myriocarpa*; also incorrectly as *B. verruculosa*).
Locally frequent on rocks and bark.

* ***Amygdalaria consentiens*** (Nyl.) Hertel, Brodo & Mas.Inoue
Uncommon on exposed siliceous rocks.

* ***Amygdalaria pelobotryon*** (Wahlenb.) Norman
Infrequent on damp siliceous rocks.

Anaptychia runcinata (With.) J.R.Laundon
S (as *Physcia aquila*).
Frequent on supralittoral rocks.

* ***Anisomeridium ranunculosporum*** (Coppins & P.James) Coppins
Uncommon on smooth bark.

* ***Arthonia atlantica*** P.James
Occasional on sheltered siliceous rocks and walls.

* ***Arthonia cinnabarina*** (DC.) Wallr.
Local on bark of *Corylus*.

* ***Arthonia phaeobaea*** (Norman) Norman
Infrequent on siliceous maritime rocks.

* ***Arthonia punctiformis*** Ach.
[F]
Locally frequent on smooth bark.

Arthonia radiata (Pers.) Ach.
S
Occasional on smooth bark of trees and bushes.

* ***Arthopyrenia analepta*** (Ach.) A.Massal.
[F]
Rare on smooth bark.

Arthopyrenia cinereopruinosa (Schaer.) A.Massal.
[F]
S
Uncommon on (mainly smooth) bark.

* ***Arthopyrenia nitescens*** (Salwey) Mudd
Uncommon on *Corylus* bark.

* ***Arthopyrenia salicis*** A.Massal.
Uncommon on *Corylus* bark.

* ***Arthothelium norvegicum*** Coppins & Tønsberg
Rare on *Calluna* stems.

* ***Aspicilia caesiocinerea*** (Nyl. ex Malbr.) Arnold
Locally frequent on (mainly maritime) acid rocks.

Aspicilia calcarea (L.) Mudd
S (as *Lecanora calcarea*).
Locally frequent on walls near harbour.

* ***Aspicilia cinerea*** (L.) Körb.
Locally frequent on wall-tops near harbour.

* ***Aspicilia contorta*** (Hoffm.) Kremp.
Occasional on walls near harbour and on calcareous vertical cliff.

* ***Aspicilia leprosescens*** (Sandst.) Hav.
Locally frequent on supralittoral rocks.

* ***Bacidia arnoldiana*** Körb.
Rare on shaded bark.

* ***Bacidia delicata*** (Larbal. ex Leight.) Coppins
Rare on shaded bark.

Bacidia egenula (Nyl.) Arnold
S—not refound.

* ***Bacidia scopulicola*** (Nyl.) A.L.Sm.
Locally frequent on supralittoral rocks; rare on glaciated rocks.

* ***Baeomyces rufus*** (Huds.) Rebent.
Frequent on acid soils, often spreading over stones, particularly earth banks forming field boundaries.

* ***Biatoropsis usnearum*** Räsänen
[LF]
Rare on thallus of *Usnea flammea*.

* ***Bilimbia lobulata*** (Sommerf.) Hafellner & Coppins
Uncommon on (? basic) soil.

* ***Bilimbia sabuletorum*** (Schreb.) Arnold
Occasional on mosses over calcareous substrata.

* ***Botryolepraria lesdanii*** (Hue) Canals *et al.*
Rare in calcareous rock crevices.

* ***Buellia aethelea*** (Ach.) Th.Fr.
Locally frequent on coastal rocks.

* ***Buellia griseovirens*** (Turner & Borrer ex Sm.) Almb.
Occasional on smooth bark of trees and shrubs.

* ***Buellia ocellata*** (Flot.) Körb.
Record of *B. verruculosa* in **S** = *Amandinea punctata*.
Occasional on exposed (mainly supralittoral) siliceous rocks.

Buellia stellulata (Taylor) Mudd
S (also incorrectly as *B. spuria*).
Infrequent on maritime rocks.

Bunodophoron melanocarpus (Sw.) Wedin
Occasional on shaded mossy rocks and trees.

Caloplaca ceracea J.R.Laundon
S (as *Lecanora caesiorufa*).
Rare on acid rocks.

Caloplaca citrina (Hoffm.) Th.Fr.
S (as *Lecanora citrina*).
Locally frequent on calcareous substrata in harbour area.

Caloplaca crenularia (With.) J.R.Laundon
S (as *Lecanora ferruginea* var. *festiva*).
Locally frequent on acid rocks.

Caloplaca flavescens (Huds.) J.R.Laundon
S (as *Lecanora callopisma*).
Locally frequent on harbour walls; rare inland (on amphibolite).

* ***Caloplaca flavovirescens*** (Wulfen) Dalla Torre & Sarnth.
Occasional on calcareous and acid substrata, mainly in the harbour area.

* ***Caloplaca holocarpa*** (Hoffm.) A.E.Wade
Locally frequent on concrete in harbour area.

Caloplaca marina (Wedd.) Zahlbr. ex Du Rietz
S (as *Lecanora lobulata*).
Common on coastal rocks.

* ***Caloplaca microthallina*** (Wedd.) Zahlbr.
Locally frequent on coastal acid rocks.

Caloplaca saxicola (Hoffm.) Nordin
S (as *Lecanora murorum*).
Occasional on calcareous substrata in harbour area.

* ***Caloplaca thallincola*** (Wedd.) Du Rietz
Frequent on coastal rocks.

* ***Caloplaca verruculifera*** (Vain.) Zahlbr.
Locally frequent on maritime rocks.

* ***Candelariella aurella*** (Hoffm.) Zahlbr.
Occasional on calcareous substrata in harbour area, rare on calcareous cliffs.

* ***Candelariella vitellina*** (Hoffm.) Müll.Arg.
Locally frequent on boulder tops (bird perches).

* ***Catapyrenium cinereum*** (Pers.) Körb.
Locally frequent on soil and mossy schistose rocks near lighthouse, rare elsewhere.

Catapyrenium squamulosum (Ach.) Breuss
Local on earthen banks and crumbling wall mortar.

Catillaria chalybeia (Borrer) A.Massal.
S (as *Biatorina chalybeia*).
Common on sheltered rocks.

Catillaria lenticularis (Ach.) Th.Fr.
S (as *Biatorina lenticularis*).
Frequent on mortar.

* ***Cecidonia xenophana*** (Korb.) Triebel & Rambold
[LF]
Rare on thallus of *Porpidia tuberculosa*.

* ***Cercidospora epipolytropa*** (Mudd) Arnold
[LF]
Rare on thallus of *Lecanora polytropa*.

* ***Cetraria aculeata*** (Schreb.) Fr.
Locally frequent on acid peaty soils.

* ***Cetraria muricata*** (Ach.) Eckfeldt
Infrequent on thin mineral soils over acid rocks.

Cladonia arbuscula (Wallr.) Flot.
S (as *C. sylvatica*).
Occasional on acid soils, mainly in upland areas.

* ***Cladonia bellidiflora*** (Ach.) Schaer.
(DBN)
Uncommon on acid soil in upland area.

* ***Cladonia cervicornis*** (Ach.) Flot. var. ***cervicornis***
Locally frequent on thin soils over acid rocks.

* ***Cladonia cervicornis*** subsp. ***verticillata*** (Hoffm.) Ahti
Rare on thin acid soil.

Cladonia chlorophaea (Florke ex Sommerf.) Spreng.
S (as *C. pyxidata* var. *chlorophaea*).
Frequent in wooded areas; occasional on fence posts.

* ***Cladonia ciliata*** Stirt. var. ***ciliata***
Uncommon on acid soils.

* ***Cladonia ciliata*** var. ***tenuis*** (Flörke) Ahti
Occasional on peat and acid soils.

* ***Cladonia coccifera*** (L.) Willd. s.lat.
Frequent on peat, often overlying boulders.

* ***Cladonia coniocraea*** (Flörke) Spreng.
Uncommon on peat and in wooded areas.

Cladonia fimbriata (L.) Fr.
S
Occasional on acid and peaty soils.

* ***Cladonia firma*** (Nyl.) Nyl.
Rare on coastal (? slightly basic) soil.

* ***Cladonia floerkeana*** (Fr.) Flörke
Locally frequent on peaty acid soils.

Cladonia foliacea (Huds.) Willd.
S (as *C. alcicornis*).
Infrequent on acid soils.

Cladonia furcata (Huds.) Schrad.
S
Common on acid and peaty soils.

Cladonia gracilis (L.) Willd.
S
Infrequent on acid and peaty soils.

* ***Cladonia humilis*** (With.) J.R.Laundon
Occasional on dry acid soils.

Cladonia macilenta Hoffm.
S
Frequent on peat and rotting wood.

* ***Cladonia ochrochlora*** Flörke
Occasional on rotting wood and peat.

* ***Cladonia pocillum*** (Ach.) Grognot
Rare on calcareous cliff.

* ***Cladonia polydactyla*** (Flörke) Spreng.
Locally frequent on peaty soil and rotting wood.

* ***Cladonia portentosa*** (Dufour) Coem.
Locally common on peaty soils.

Cladonia pyxidata (L.) Hoffm.
S
Infrequent on dry acid soils.

Cladonia ramulosa (With.) J.R.Laundon
S (as *C. pityrea*).
Infrequent on acid soils.

* ***Cladonia rangiferina*** (L.) F.H.Wigg.
Rare on peat.

Cladonia rangiformis Hoffm.
S (as *C. pungens*).
Common on soil (usually basic), particularly on field banks near south coast.

* ***Cladonia squamosa*** Hoffm. var. ***squamosa***
Infrequent on mossy substrata and rotting wood.

* ***Cladonia squamosa*** var. ***subsquamosa*** (Nyl. ex Leight.) Vain.
Rare on rotting wood.

* ***Cladonia strepsilis*** (Ach.) Grognot
Occasional on peaty soil.

* ***Cladonia subcervicornis*** (Vain.) Kernst.
Locally frequent in crevices of exposed siliceous rocks.

* ***Cladonia subulata*** (L.) F.H.Wigg.
Uncommon on dry acid soils.

Cladonia uncialis (L.) F.H.Wigg. subsp. ***biuncialis*** (Hoffm.) M.Choisy
S
Locally frequent on peaty soils.

* ***Clauzadea monticola*** (Ach.) Hafellner & Bellem.
Uncommon on walls near harbour; rare on (? basic) rock elsewhere.

* ***Clauzadeana macula*** (Taylor) Coppins & Rambold
Rare on exposed siliceous rocks.

* ***Cliostomum griffithii*** (Sm.) Coppins
Occasional on bark and wood, particularly fence posts.

Cliostomum tenerum (Nyl.) Coppins & S.Ekman
Occasional in crevices of maritime acid rocks, rare elsewhere.

* ***Coccotrema citrinescens*** P.James & Coppins
Uncommon on sheltered acid rocks.

* ***Collema auriforme*** (With.) Coppins & J.R.Laundon
Uncommon on calcareous substrata.

Collema bachmanianum (Fink) Degel.
S (incorrectly as *C. crispum*—Mitchell 1993).
Rare on (? slightly basic) soil near lighthouse.

Collema crispum (Huds.) F.H.Wigg.
S (as *C. cheilum*).
Occasional on mortar.

Collema cristatum (L.) F.H.Wigg.
S (as *C. granuliferum*).
Rare on mortar.

Collema flaccidum (Ach.) Ach.
S (incorrectly as *C. furvum*—cf. Degelius 1954).
Occasional on soil and damp mossy siliceous rocks.

* ***Collema fuscovirens*** (With.) J.R.Laundon
Occasional over mosses on walls near harbour.

* ***Collema tenax*** (Sw.) Ach. var. ***tenax***
Occasional on walls near harbour and mossy soil near lighthouse.

Collema tenax var. ***ceranoides*** (Borrer) Degel.
S (as *C. crispum* var. *ceranoides*).
Rare on sandy (? basic) soil.

Collemopsidium foveolatum (A.L.Sm.) F.Mohr.
[= *Pyrenocollema halodytes* (Nyl.) R.C.Harris].
S (as *Arthopyrenia litoralis*).
Frequent on littoral rocks and seashore shells.

* ***Cystocoleus ebeneus*** (Dillwyn) Thwaites
Infrequent on humid siliceous rock faces.

* ***Dactylospora parellaria*** (Nyl.) Arnold
[LF]
Infrequent on thalli of *Ochrolechia parella*.

Dactylospora saxatilis (Schaer.) Hafellner
[LF]
S (as *Buellia saxatilis*)—not refound.
[If correct, this would imply the presence of *Pertusaria amarescens* (not listed in Smith (1911), the normal host for this species (B.J.Coppins, pers. comm.)]

* ***Degelia atlantica*** (Degel.) P.M.Jørg. & P.James
Occasional on humid mossy trees and shrubs.

Degelia ligulata P.M.Jørg. & P.James (Pl. II)
Locally frequent on soil in short turf rocky outcrops near lighthouse.

Degelia plumbea (Lightf.) P.M.Jørg. & P.James
S (incorrectly as *Pannaria rubiginosa*).
Occasional on humid mossy trees and rocks.

Dermatocarpon miniatum (L.) W.Mann
S
Uncommon on rocks by stream.

Diploicia canescens (Dicks.) A.Massal.
S (as *Buellia canescens*).
Rare on basic rock outcrops.

* ***Diploschistes scruposus*** (Schreb.) Norman
Uncommon on siliceous rocks.

* ***Diplotomma alboatrum*** (Hoffm.) Flot.
Rare on basic rock outcrops.

* ***Diplotomma chlorophaeum*** (Hepp ex Leight.) Kr.P.Singh & S.R.Singh
Uncommon on maritime rocks.

* ***Enterographa crassa*** (DC.) Fée
Uncommon on bark.

* ***Enterographa zonata*** (Körb.) Källsten
Rare on shaded siliceous rock.

Evernia prunastri (L.) Ach.
Occasional on trees, bushes and fence posts.

Flavoparmelia caperata (L.) Hale
S (as *Parmelia caperata*).
Locally frequent on stone walls, occasional on bark.

Fuscidea cyathoides (Ach.) V.Wirth & Vězda
S (as *Lecidea rivulosa*).
Common on acid rock.

* ***Fuscidea kochiana*** (Hepp.) V.Wirth & Vězda
Occasional on acid rock.

* ***Fuscidea lightfootii*** (Sm.) Coppins & P.James
Occasional on *Calluna* stems and twigs of bushes.

Fuscidea lygaea (Ach.) V.Wirth & Vězda
Occasional on acid rock.

Fuscopannaria leucophaea (Vahl) P.M.Jørg.
S (as *Pannularia microphylla*).
Uncommon on till above maritime rocks, and on slate; rare on soil between rocks of vertical calcareous cliff near lighthouse.

Graphina anguina auct.europ.
[= *Graphis britannica* Staiger]
S
Occasional on smooth bark.

Graphis elegans (Borrer ex Sm.) Ach.
S
Locally frequent on bark of bushes and *Calluna* stems.

* ***Graphis scripta*** (L.) Ach.
Occasional on bark of trees and bushes.

Gyalecta jenensis (Batsch) Zahlbr.
S (as *G. cupularis*).
Occasional on acid and damp basic rocks.

Haematomma ochroleucum (Neck.) J.R.Laundon var. ***ochroleucum***
Uncommon on acid rock outcrop near lighthouse.

* ***Haematomma ochroleucum*** var. ***porphyrium*** (Pers.) J.R.Laundon
Rare on acid rock.

* ***Halecania ralfsii*** (Salwey) M.Mayrhofer
Infrequent on siliceous maritime rocks.

Halecania spodomela (Nyl.) M.Mayrhofer
Occasional on siliceous maritime rocks.

Herteliana gagei (Sm.) J.R.Laundon
S (as *Lecidea taylorii*)
Infrequent on damp siliceous rocks.

* ***Homostegia piggotii*** (Berk. & Broome) P.Karst
[LF]
Rare on thallus of *Parmelia saxatilis*.

Hymenelia prevostii (Duby) Kremp.
Rare on calcareous substratum.

Hyperphyscia adglutinata (Flörke) Mayrh. & Poelt
Rare on *Sambucus*.

Hypogymnia physodes (L.) Nyl.
S (as *Parmelia physodes*).
Occasional on bushes, trees and *Calluna* stems; infrequent on acid rock.

* ***Hypogymnia tubulosa*** (Schaer.) Hav.
(DBN)
Uncommon on bushes and trees.

* ***Hypotrachyna laevigata*** (Sm.) Hale
Occasional on bark and *Calluna* stems, rare on acid rock.

* ***Hypotrachyna revoluta*** (Flörke) Hale
Occasional on bark, rare on rocks and walls.

Ionaspis lacustris (With.) Lutzoni
S (as *Lecanora lacustris*).
Locally common in rocky streams.

* ***Koerberiella wimmeriana*** (Körb.) Stein
Rare on wet (? basic) rock.

Lauderlindsaya borreri (Tul.) J.C.David & D.Hawksw.
[LF]
S (as fertile *Normandina pulchella*) –not refound.

* ***Lecania aipospila*** (Wahlenb.) Th.Fr.
Infrequent on acid maritime rocks.

* ***Lecania cyrtella*** (Ach.) Th.Fr.
Uncommon on *Fraxinus* and *Sambucus*.

Lecania erysibe (Ach.) Mudd
S (s.lat.—as *Lecanora erysibe*).
Infrequent on calcareous substrata near harbour.

* ***Lecania hutchinsiae*** (Nyl.) A.L.Sm.
Infrequent on slate and siliceous rocks.

Lecania naegelii (Hepp) Diederich & Van den Boom
S (as *Bilimbia naegelii*).
Rare on shrubs.

Lecania turicensis (Hepp) Müll. Arg.
In Smith 1918 (as *L. albariella*).
Infrequent on upland rocks.

* ***Lecanora actophila*** Wedd.
Locally frequent on maritime acid rocks.

* ***Lecanora aitema*** (Ach.) Hepp
Rare on wood.

Lecanora albescens (Hoffm.) Branth & Rostr.
S (as *L. galactina*).
Locally frequent on mortar in harbour area.

Lecanora campestris (Schaer.) Hue
S (as *L. subfusca* var. *campestris*).
Common on walls in harbour area, less frequent on basic substrata elsewhere.

Lecanora carpinea (L.) Vain.
S (as *L. angulosa*)—not refound.

Lecanora chlarotera Nyl.
S (including *L. rugosa* var. *chlarona*).
Locally frequent on trees, bushes and *Calluna* stems.

* ***Lecanora confusa*** Almb.
Occasional on bark and fence posts

* ***Lecanora crenulata*** Hook.
Infrequent on calcareous substrata near harbour.

Lecanora dispersa (L.) Sommerf.
S (as *L. galactina* subsp. *dispersa*).
Locally frequent on walls, occasional on boulders.

* ***Lecanora expallens*** Ach.
Locally frequent on bark, *Calluna* stems and fence posts.

* ***Lecanora fugiens*** Nyl.
Occasional on acid maritime rocks.

Lecanora gangaleoides Nyl.
S (but *L. coilocarpa* record in **S** = *L. campestris*).
Frequent on siliceous rock.

* ***Lecanora helicopis*** (Wahlenb.) Ach.
Locally frequent on acid maritime rocks.

Lecanora intricata (Ach.) Ach.
Uncommon on siliceous rocks, rare on conglomerate.

* ***Lecanora jamesii*** J.R.Laundon
Uncommon on smooth bark.

* ***Lecanora poliophaea*** (Wahlenb.) Ach.
Locally frequent on acid maritime rocks.

Lecanora polytropa (Hoffm.) Rabenh.
S
Locally frequent on siliceous boulders.

* ***Lecanora rupicola*** (L.) Zahlbr.
Occasional on siliceous rock.

* ***Lecanora soralifera*** (Suza) Räsänen
Occasional on siliceous rock.

* ***Lecanora sulphurea*** (Hoffm.) Ach.
Occasional on siliceous boulders.

* ***Lecanora diducens*** Nyl.
Occasional on acid coastal rocks.

Lecanora fuscoatra (L.) Ach.
As forma *meiosporiza* in Smith (1926).
Locally frequent on siliceous rocks.

* ***Lecanora lactea*** Flörke ex Schaer.
Uncommon on siliceous rocks.

Lecanora lapicida (Ach.) Ach.
S (incorrect as *L. confluens*).
Occasional on siliceous rocks.

Lecanora lithophila (Ach.) Ach.
S
Locally frequent on exposed siliceous rocks.

* ***Lecanora phaeops*** Nyl.
Infrequent in damp crevices of upland and coastal siliceous rocks.

Lecidella anomaloides (A.Massal.) Hertel & H.Kilias
S (as *Lecidea goniophila*).
Uncommon on shaded siliceous rocks.

Lecidella asema (Nyl.) Knoph & Hertel
S (as *Lecidea latypea*).
Locally common on supralittoral rocks, infrequent elsewhere.

Lecidella elaeochroma (Ach.) M.Choisy forma ***elaeochroma***
S (as *Lecidea parasema*).
Locally common on twigs.

* ***Lecidella elaeochroma*** forma ***soralifera*** (Erichsen) D.Hawksw.
Frequent on twigs, rare on wood.

* ***Lecidella meiococca*** (Nyl.) Leuckert & Hertel
Occasional on maritime rocks.

Lecidella scabra (Taylor) Hertel & Leuckert
In Knowles 1929 (as *Lecidea scabra*).
Locally frequent on siliceous rocks and walls.

* ***Lecidella stigmatea*** (Ach.) Hertel & Leuckert
Locally frequent on calcareous substrata, particularly mortar.

* ***Lempholemma polyanthes*** (Bernh.) Malme
Rare on mosses over calcareous substrata.

* ***Lepraria incana*** (L.) Ach. s.lat.
Common on shaded rock.

* ***Lepraria lobificans*** Nyl.
Rare on shaded bark.

* ***Leptogium britannicum*** P.M.Jørg. & P.James
Locally frequent on field banks near south coast.

Leptogium cyanescens (Rabenh.) Körb.
S (as *L. tremelloides*).
Infrequent on mossy trees and rocks.

Leptogium gelatinosum (With.) J.R.Laundon
S (as *L. scotinum*).
Uncommon on mosses over calcareous substrata.

* ***Leptogium plicatile*** (Ach.) Leight.
Uncommon on calcareous substrata.

* ***Leptogium teretiusculum*** (Wallr.) Arnold
Rare on (? slightly basic) soils.

Lichenomphalia alpina (Britzelm.) Redhead *et al.*
[= *Omphalina luteovitellina* (Pilát & Nannf.) M.Lange]
Rare on peat.

* ***Lichenomphalia hudsoniana*** (H.S.Jenn.) Redhead *et al.*
Occasional on peat.

Lichenomphalia umbellifera (L:Fr.) Redhead *et al.*
[= *Omphalina ericetorum* (Pers.:Fr.) M.Lange ex H.E.Bigelow].
Occasional on damp rotting wood.

* ***Lichenomphalia velutina*** (Quél.) Readhead *et al.*
Rare on peat.

Lichina confinis (Müller) C.Agardh
S
Locally frequent on littoral rock.

Lichina pygmaea (Lightf.) C.Agardh (Pl. VI, p.235)
S
Locally frequent on littoral rock below *L. confinis*.

Lobaria pulmonaria (L.) Hoffm.
S—not refound.

Lobaria virens (With.) J.R.Laundon (Pl. VII, p. 236)
S (as *Ricasolia laetevirens*).
Rare on ultramafic (ultrabasic) rock.

* ***Megalaria pulverea*** (Borrer) Hafellner & E.Schreiner
Infrequent on mossy rocks; rare on mossy *Calluna* stems.

Megalospora tuberculosa (Fée) Sipman
Uncommon on shaded bark.

Melanelia exasperata (de Not.) Essl.
S (as *Parmelia exasperata*)—not refound.

Melanelia fuliginosa (Fr. ex Duby) Essl.
S (as *Parmelia fuliginosa*).
Common on siliceous rock outcrops.

* ***Melanelia fuliginosa*** subsp. ***glabratula*** (Lamy) Coppins
Frequent on twigs and fence posts.

* ***Melanelia subaurifera*** (Nyl.) Essl.
Occasional on branches and twigs.

* ***Micarea denigrata*** (Fr.) Hedl.
Occasional on wood.

Micarea lignaria (Ach.) Hedl. var. ***lignaria***
S
Common on peat and peaty soil; occasional on fence posts.

* ***Micarea lignaria*** var. ***endoleuca*** (Leight.) Coppins
Rare on mosses.

* ***Micarea prasina*** Fr. s.lat.
Occasional on fence posts.

Micarea ternaria (Nyl.) Vězda
Old (tentative?) record in Coppins (1983)—not refound.

* ***Moelleropsis nebulosa*** (Hoffm.) Gyeln.
Rare on well-drained earth banks.

* ***Mycomicrothelia confusa*** D.Hawksw.
[F]
Rare on *Corylus*.

* ***Mycoporum antecellens*** (Nyl.) R.C.Harris
[F]
Uncommon on smooth bark.

* ***Neofuscelia pulla*** (Ach.) EssL.
Record in **S** = *N. verruculifera*.
Occasional on coastal siliceous rocks.

Neofuscelia verruculifera (Nyl.) Essl.
S (incorrectly as *Parmelia prolixa*).
Occasional on well-lit rocks.

Nephroma laevigatum Ach.
(DBN)
In Knowles 1929 (as *N. lusitanicum* forma *panniforme*).
Locally frequent on field banks near south coast.

Normandina pulchella (Borrer) Nyl.
S
Locally frequent on liverworts over trees.

* ***Ochrolechia androgyna*** (Hoffm.) Arnold
Occasional on *Calluna* and peat.

Ochrolechia parella (L.) A.Massal.
S (as *Lecanora parella*).
Frequent on siliceous rocks.

Opegrapha atra Pers.
S
Locally frequent on bark.

Opegrapha calcarea Turner ex Sm.
S (also as *O. confluens*).
Locally frequent on walls; rare on (? basic) rock.

* ***Opegrapha cesareensis*** Nyl.
Infrequent on coastal siliceous rocks, rare in sheltered crevices of acid rock near lighthouse.

* ***Opegrapha gyrocarpa*** Flot.
Locally frequent on shaded siliceous walls and rocks.

* ***Opegrapha lithyrga*** Ach.
Rare on shaded siliceous rocks.

* ***Opegrapha multipuncta*** Coppins & P.James
Rare on branches in damp scrubland.

* ***Opegrapha niveoatra*** (Borrer) J.R.Laundon
Rare on bark and wood.

* ***Opegrapha saxigena*** Taylor
Uncommon on (mainly shaded) siliceous rocks.

* ***Opegrapha vulgata*** (Ach.) Nyl.
Locally frequent on bark.

* ***Pannaria rubiginosa*** (Ach.) Bory
Record in **S** is incorrect = *Degelia plumbea*.
Uncommon on *Corylus* and mossy wood.

Parmelia omphalodes (L.) Ach.
S
Frequent on siliceous rock, often in coastal areas.

Parmelia saxatilis (L.) Ach.
S
Common on siliceous walls; occasional on fence posts.

Parmelia sulcata Taylor
S
Common on walls and trees.

Parmotrema crinitum (Ach.) M.Choisy
(DBN)
In Knowles 1929 (as *Parmelia proboscidia* [sic]).
Occasional on peat and boulders.

Parmotrema perlatum (Huds.) M.Choisy
S (as *Parmelia perlata*).
Frequent on rock.

Parmotrema reticulatum (Taylor) M.Choisy
S (as *Parmelia perlata* var. *ciliata*)—not refound.

Peltigera didactyla (With.) J.R.Laundon
S (as *P. spuria*)—not refound.

Peltigera horizontalis (Huds.) Baumg.
S—not refound.

Peltigera hymenina (Ach.) Delise ex Duby
P. polydactyla record in **S** = *P. rufescens*.
Occasional on soil.

Peltigera membranacea (Ach.) Nyl.
S (as *P. canina*).
Occasional on soil.

* ***Peltigera neckeri*** Hepp ex Müll.Arg.
Rare on mossy soil.

* ***Peltigera praetextata*** (Flörke ex Sommerf.) Zopf
In Smith 1918 (as *P. rufescens* var. *praetextata*).
Locally frequent on field banks near south coast.

Peltigera rufescens (Weiss) Humb.
S
Occasional on soil.

* ***Pertusaria albescens*** (Huds.) M.Choisy & Werner
Rare on conglomerate.

* ***Pertusaria amara*** (Ach.) Nyl.
Rare on bark and rock.

* ***Pertusaria aspergilla*** (Ach.) J.R.Laundon
Occasional on siliceous rocks.

* ***Pertusaria chiodectonoides*** Bagl. ex A.Massal.
Occasional on supralittoral rock.

Pertusaria corallina (L.) Arnold
In Porter 1948 (as *P. dealbata* forma *corallina*).
Common on siliceous rock.

* ***Pertusaria excludens*** Nyl.
Occasional on exposed siliceous rock.

* ***Pertusaria flavicans*** Lamy
S (as *P. wulfenii* var. *rupicola*—cf. Knowles 1929).
Locally frequent on exposed coastal siliceous rocks.

* ***Pertusaria hymenea*** (Ach.) Schaer.
Rare on shaded bark.

* ***Pertusaria lactea*** (L.) Arnold
Rare on sheltered rock.

* ***Pertusaria lactescens*** Mudd
Rare on siliceous rock.

* ***Pertusaria leioplaca*** DC.
Occasional on shaded smooth bark.

Pertusaria pertusa (Weigel) Tuck.
S (as *P. communis*)—not refound.

Pertusaria pseudocorallina (Lilj.) Arnold
S (as *P. concreta*, including forma *westringii*).
Common on exposed siliceous rocks.

* ***Peterjamesia circumscripta*** (Taylor) D.Hawksw.
Infrequent on sheltered siliceous coastal rocks.

* ***Phaeophyscia orbicularis*** (Neck.) Moberg
Occasional on twigs.

* ***Phaeophyscia sciastra*** (Ach.) Moberg
Rare on nutrient-enriched rock.

Phaeospora parasitica (Lonnr.) Arnold
[LF]
Rare on *Rhizocarpon reductum*.

* ***Phlyctis argena*** (Spreng.) Flot.
Rare on bark.

* ***Physcia adscendens*** (Fr.) H.Olivier
(DBN)
Occasional on twigs.

* ***Physcia aipolia*** (Ehrh. ex Humb.) Fürnr.
Rare on *Crataegus*.

* ***Physcia caesia*** (Hoffm.) Fürnr.
Locally frequent on exposed rocks, especially nutrient enriched.

* ***Physcia leptalea*** (Ach.) DC.
Rare on bark.

Physcia tenella (Scop.) DC.
S (as *P. stellaris* var. *tenella*).
Common on rocks, especially bird perches.

* ***Placidiopsis custnani*** (A.Massal.) Körb.
Rare on (? basic) soil.

Placopsis gelida (L.) Linds.
In Smith 1918 (as *Lecanora gelida*)—not refound.

* ***Placynthiella icmalea*** (Ach.) Coppins & P.James
Common on rotting wood and peat.

* ***Placynthiella uliginosa*** (Schrad.) Coppins & P.James
Common on peat, occasional on fence posts.

* ***Placynthium nigrum*** (Huds.) Gray
Locally frequent on basic rock and mortar.

Platismatia glauca (L.) W.L.Culb. & C.F.Culb.
S (as *Platysma glauca*)—not refound.

Polyblastia cupularis A.Massal.
S (as *P. intercedens*).
Rare on slate by stream.

* ***Polychidium muscicola*** (Sw.) Gray
Occasional on mossy siliceous rocks.

Porina aenea (Wallr.) Zahlbr.
S (as *P. carpinea*).
Occasional on smooth bark.

* ***Porina borreri*** (Trevis.) D.Hawksw. & P.James
Rare on bark.

Porina chlorotica (Ach.) Müll.Arg.
S
Locally frequent on shady rock.

Porina curnowii A.L.Sm.
Infrequent on coastal siliceous rocks.

Porpidia cinereoatra (Ach.) Hertel & Knoph
S (as *Lecidea albocaerulescens*).
Locally frequent on exposed acid rocks; rare on slightly basic rock.

Porpidia crustulata (Ach.) Hertel & Knoph
S (as *Lecidea crustulata*).
Frequent on acid rocks, rare on slightly basic rock.

* ***Porpidia hydrophila*** (Fr.) Hertel & A.J.Schwab
Occasional on stones in streams.

Porpidia macrocarpa (DC.) Hertel & A.J.Schwab
S (as *Lecidea contigua*).
Common on siliceous rocks and walls.

Porpidia platycarpoides (Bagl.) Hertel
In Porter 1948 (as *Lecidea percontigua*).
Occasional on siliceous rocks.

Porpidia speirea (Ach.) Kremp.
S (as *Lecidea cinerascens*)—not refound.

Porpidia tuberculosa (Sm.) Hertel & Knoph
S (as *Lecidea sorediza*).
Common on siliceous rocks and walls.

Protoblastenia rupestris (Scop.) J.Steiner
S (incorrectly as *Lecanora irrubata* subsp. *calva*—see Smith 1918).
Locally frequent on calcareous substrata.

* ***Protopannaria pezizoides*** (Weber) P.M.Jørg. & S.Ekman
S (incorrectly as *Lecanora hypnorum*).
(DBN)
Occasional on soil near lighthouse.

* ***Psora lurida*** (Ach.) DC.
Occasional on thin soils over calcareous rock.

Psoroma hypnorum (Vahl) Gray
Record of *Lecanora hypnorum* in S = *Protopannaria pezizoides*.
Occasional on acid soil banks.

* ***Punctelia reddenda*** (Stirt.) Krog
Occasional on sheltered mossy rocks and trees.

* ***Punctelia subrudecta*** (Nyl.) Krog s.lat.
Occasional on bark; rare on acidic rock on drystone wall.

Pycnothelia papillaria Dufour
S
Locally frequent on peat.

* ***Pyrenula chlorospila*** Arnold
Locally frequent on shaded smooth bark.

* ***Pyrenula macrospora*** (Degel.) Coppins & P. James
Locally frequent on shaded smooth bark.

* ***Pyrrhospora quernea*** (Dicks.) Körb.
Rare on bark.

* ***Racodium rupestre*** Pers.
Occasional on vertical siliceous rock faces.

Ramalina calicaris (L.) Fr.
S—not refound.

* ***Ramalina canariensis*** J.Steiner
Rare on bark.

Ramalina cuspidata (Ach.) Nyl.
S (including *R. curnowii*)
Locally frequent on siliceous (mainly coastal) rocks.

Ramalina farinacea (L.) Ach.
S
Common on trees; frequent on fence posts.

* ***Ramalina lacera*** (With.) J.R.Laundon
Rare on bark.

Ramalina siliquosa (Huds.) A.L.Sm.
S (as *R. scopulorum*).
Common on rocks; rare on lignum.

Ramalina subfarinacea (Nyl. ex Crombie) Nyl.
S
Frequent on (mainly coastal) rock; rare on lignum.

Rhizocarpon geographicum (L.) DC.
S
Frequent on siliceous rocks.

Rhizocarpon hochstetteri (Körb.) Vain.
S (as *Buellia colludens* and ? *B. atroalba*).
Occasional on siliceous rocks.

* ***Rhizocarpon infernulum*** (Nyl.) Lynge f. ***sylvaticum*** Fryday
Rare on slate pebbles.

* ***Rhizocarpon lavatum*** (Fr.) Hazsl.
Infrequent on siliceous rocks in streams.

Rhizocarpon oederi (Weber) Körb.
S
Occasional on siliceous rocks.

Rhizocarpon petraeum (Wulfen) A.Massal.
S
Locally frequent on acid rocks.

Rhizocarpon reductum Th.Fr.
S (as *R. confervoides*).
Common on siliceous rocks.

Rhizocarpon richardii (Nyl.) Zahlbr.
(DBN)
Occasional on acidic rocks, rare on slightly basic rock.

* ***Rinodina atrocinerea*** (Hook.) Körb.
Occasional on coastal and ultrabasic rocks.

* ***Rinodina confragosa*** (Ach.) Körb.
Infrequent on siliceous coastal rocks.

Rinodina gennarii Bagl.
S (incorrectly as *Lecanora exigua*).
Infrequent on calcareous substrata.

Rinodina luridescens (Anzi) Arnold
S (as *Buellia coniopta*; cf. *Amandinea coniops* interpretation in Knowles 1929).
Locally frequent on supralittoral siliceous rocks, rare elsewhere.

Rinodina sophodes (Ach.) A.Massal.
S (as *Lecanora sophodes* var. *laevigata*).
Occasional on twigs.

Sarcogyne regularis Körb.
S (as *Lecanora pruinosa*).
Infrequent on calcareous substrata.

* ***Schaereria fuscocinerea*** (Nyl.) Clauzade & Cl.Roux
Locally frequent on exposed siliceous rocks.

Scoliciosporum umbrinum (Ach.) Arnold
Knowles 1929 (as *Bacidia umbrina*).
Infrequent on (mainly siliceous) rocks and walls.

Sclerococcum sphaerale (Ach.) Fr.
[LF]
Infrequent on thalli of *Pertusaria corallina*.

* ***Solenopsora holophaea*** (Mont.) Samp.
Locally frequent on coastal rocks and soil.

* ***Solenopsora vulturiensis*** A.Massal.
Locally frequent on acid rocks and thin sheltered coastal soils.

* ***Sphaerophorus fragilis*** (L.) Pers.
Occasional on acid rocks and thin soils; rare on peat.

Sphaerophorus globosus (Huds.) Vain.
S
Locally frequent on (often mossy) rocks; occasional on peaty ledges and soils.

Sticta fuliginosa (Hoffm.) Ach.
S (as *Stictina fuliginosa*).
Occasional on sheltered mossy rocks and trees.

* ***Sticta sylvatica*** (Huds.) Ach.
Occasional on sheltered mossy rocks and trees.

Tephromela atra (Huds.) Hafellner ex Kalb
S (as *Lecanora atra*).
Frequent on siliceous (particularly coastal) rocks and walls.

* ***Thelenella muscorum*** (Fr.) Vain.
Uncommon on mossy bark of *Corylus*.

* ***Thelidium pyrenophorum*** (Ach.) Mudd
Rare on (? basic) rock.

* ***Thrombium epigaeum*** (Pers.) Wallr.
Rare on (? acid) earth bank.

Tomasellia gelatinosa (Chevall.) Zahlbr.
[F]
S (as *Melanotheca gelatinosa*).
Occasional on bark.

Toninia aromatica (Sm.) A.Massal.
S (as *Bilimbia aromatica*).
Occasional on mortar and coastal rocks.

Toninia mesoidea (Nyl.) Zahlbr.
S (as *Bilimbia mesoidea*).
Infrequent on coastal (? basic) rocks.

* ***Toninia sedifolia*** (Scop.) Timdal
Infrequent on calcareous soil.

* ***Trapelia coarctata*** (Sm.) M.Choisy
Common on siliceous rocks and walls.

* ***Trapelia glebulosa*** (Sm.) J.R.Laundon
Frequent on siliceous rock.

* ***Trapelia placodioides*** Coppins & P.James
Common on siliceous rock.

* ***Trapeliopsis granulosa*** (Hoffm.) Lumbsch
Common on peat and lignum.

* ***Trapeliopsis pseudogranulosa*** Coppins & P.James
Locally frequent on peat hags and rotting wood.

* ***Trapeliopsis wallrothii*** (Flörke ex Spreng.) Hertel & Gotth.Schneid.
Occasional on coastal thin soils and earth banks.

* ***Tremolecia atrata*** (Ach.) Hertel
Occasional on exposed siliceous rocks and walls.

Tylothallia biformigera (Leight.) P.James & Kilias
S (as *Biatorina biformigera*)—not refound.

* ***Usnea cornuta*** Körb.
Locally frequent on bark and fence posts; uncommon on siliceous rock.

Usnea flammea Stirt.
S (as *U. ceratina* and *U. plicata*—see Porter 1948).
Locally frequent on wind-swept rocks, wood, shrubs and *Calluna*.

* ***Usnea flavocardia*** Räsänen
Uncommon on bark and mossy rocks.

Usnea florida (L.) F.H.Wigg.
S—not refound.

* ***Usnea fragilescens*** Hav. ex Lynge
Occasional on bark and fence posts; rare on rocks.

* ***Usnea hirta*** (L.) F.H.Wigg.
Occasional on wood, bark and *Calluna*.

* ***Usnea subfloridana*** Stirt.
Locally frequent on bark and fence posts; rare on *Calluna* and mossy rocks.

Usnea wasmuthii Räsänen
Uncommon on bark and mossy rocks.

Verrucaria aethiobola Wahlenb.
S (also as *V. laevata*)
Infrequent on siliceous rocks and slate by streams.

* ***Verrucaria aquatilis*** Mudd
Occasional on siliceous rocks and slate in and by streams.

* ***Verrucaria baldensis*** A.Massal.
Infrequent on calcareous substrata.

* ***Verrucaria bryoctona*** (Th.Fr.) A.Orange
Uncommon on mossy calcareous soils.

* ***Verrucaria funckii*** (Spreng.) Zahlbr.
Occasional on siliceous rocks in streams.

* ***Verrucaria fuscella*** (Turner) Winch
Uncommon on basic stonework.

* ***Verrucaria fusconigrescens*** Nyl.
Locally frequent on exposed siliceous supralittoral rocks, rare elsewhere.

* ***Verrucaria hochstetteri*** Fr.
Occasional on mortar near harbour.

Verrucaria hydrela Ach.
S (as *V. submersa*).
Occasional on rocks in streams.

Verrucaria margacea (Wahlenb.) Wahlenb.
S—not refound.

Verrucaria maura Wahlenb.
S
Common on upper littoral rocks.

Verrucaria mucosa Wahlenb.
S
Frequent on mid-littoral rocks.

Verrucaria muralis Ach.
S
Frequent on walls near harbour.

Verrucaria nigrescens Pers.
S
Common on walls near harbour, infrequent on siliceous rocks elsewhere.

Verrucaria prominula Nyl.
S (as var. *viridans*)
Locally frequent on shaded siliceous rocks near coast.

Verrucaria striatula Wahlenb.
S (as *V. microspora*)—not refound.

Verrucaria viridula (Schrad.) Ach.
S (also as *V. mauroides*).
Occasional on walls near harbour.

* ***Vezdaea leprosa*** (P.James) Vĕzda
Infrequent on peaty banks near coast.

Weddellomyces peripherica (Taylor) Alstrup & D.Hawksw.
[LF]
Rare on thallus of *Pertusaria corallina.*

* ***Xanthoparmelia conspersa*** (Ehrh. ex Ach.) Hale
Locally frequent on exposed siliceous rocks.

Xanthoria parietina (L.) Th.Fr.
S
Common on coastal rocks and walls; locally frequent on bird perches.

* ***Zevadia peroccidentalis*** J.C.David & D.Hawksw.
[LF]
Rare on thallus of *Usnea flammea* (see David & Hawksworth 1995).

Doubtful or tentative records in the absence of supporting herbarium material

Buellia spuria (Schaer.) Anzi
In Smith 1911.

Caloplaca aurantia (Pers.) Hellb.
As *Lecanora callopisma* in Smith 1911.

Cladonia phyllophora Hoffm.
As *C. degenerans* in Smith 1911.
cf. *C. ramulosa* (With.) J.R.Laundon.

Collema nigrescens (Huds.) DC.
In Smith 1911.
cf. *C. subflaccidum* Degel.

Lecanora albella (Pers.) Ach.
In Smith 1911.

Lecidea matildiae H.Magn. = *L. confluentula* Müll. Arg.?
Provenance unknown.

Lepraria umbricola Tønsberg
Details of 1991 collection lacking.

Lobaria scrobiculata (Scop.) DC.
Details of 1991 collection lacking.

Miriquidica leucophaea (Rabenh.) Hertel & Rambold
Details of 1991 collection lacking.

Peltigera malacea (Ach.) Funck
Record in Porter (1948) not this taxon.

Xanthoria candelaria (L.) Th.Fr.
As *Physcia lychnea* in Smith 1911—probably another taxon.
cf. *Candelaria concolor.*

Postscript

Although there has been no further lichenological work undertaken on Clare Island since we submitted this contribution for publication more than five years ago, a considerable improvement has been made to our knowledge of the Irish lichen flora as a result of the LichenIreland initiative. This is reflected in the latest *Census Catalogue* (Seaward 2010) which shows, for example, that West Mayo (which includes Clare Island) currently supports 698 lichen taxa, representing a 41% increase in the number recorded since the previous edition (Seaward 1994, see above). It should also be noted that recent nomenclatural changes and taxonomic reinterpretations have not been incorporated into our chapter.

Acknowledgements

The authors gratefully acknowledge financial assistance from Trinity College Dublin, University College Dublin and the Royal Irish Academy to support the 1990 and 1991 field excursions. We also appreciate the help of the following for providing records, identifications, useful comments on the draft typescript and access to herbarium material: M.A. Allen, B.J. Coppins, J.C. David, H.F. Fox, B.P. Hilton, P.W. James, S. LaGreca, M. Newman, M. Simms and H.J.M. Sipman.

REFERENCES

Coppins, B.J. 1983 A taxonomic study of the lichen genus *Micarea* in Europe. *Bulletin of the British Museum (Natural History)* **11**, 18–214.

Coppins, B.J. 2002 *Checklist of lichens of Great Britain and Ireland*. London British Lichen Society.

David, J.C. and Hawksworth, D.L. 1995 *Zevadia*: a new lichenicolous hyphomycete from western Ireland. *Bibliotheca Lichenologica* **58**, 63–71.

Degelius, G. 1954 The lichen genus *Collema* in Europe. *Symbolae Botanicae Upsalienses* **13** (2), 1–499.

Fox, H.F. 1996 Catalogue of Irish lichenicolous fungi. Unpublished MSC. thesis, University College Dublin.

Graham, J.R. (ed.) 2001 *New Survey of Clare Island. Volume 2: geology*. Dublin. Royal Irish Academy.

Knowles, M.C. 1929 The lichens of Ireland. *Proceedings of the Royal Irish Academy* **38**, 179–434.

Mitchell, M.E. 1993 *First records of Irish lichens 1696–1990*. Galway. Officina Typographica.

Mitchell, M.E. 1995 150 years of Irish lichenology: a concise survey. *Glasra* **2**, 139–55.

Porter, L. 1948 The lichens of Ireland (supplement). *Proceedings of the Royal Irish Academy* **51**, 347–86.

Seaward, M.R.D. 1994 Vice-county distribution of Irish lichens. *Proceedings of the Royal Irish Academy* **94**B, 177–94.

Seaward, M.R.D. 2010 *Census Catalogue of Irish Lichens*. 3rd edition. Holywood, Belfast. National Museums of Northern Ireland.

Seaward, M.R.D. and Richardson, D.H.S. 2000 Lichens of Lambay Island. *Glasra* **4**, 1–6.

Smith, A.L. 1911 Clare Island survey. Part 14 *Lichenes*. *Proceedings of the Royal Irish Academy* **31**, 1–14.

Smith, A.L. 1918 *A monograph of the British lichens*. Part 1. 2nd edn. London. British Museum (Natural History).

Smith, A.L. 1926 *A monograph of the British lichens*. Part 2. 2nd edn. London. British Museum (Natural History).

GRASSLAND FUNGI

David Mitchel

ABSTRACT

Waxcaps, one of the groups of grassland fungi that are now recognised as excellent indicators of unfertilised grassland, can be found in a range of grassland types from dunes to uplands, from lowlands to gardens or churchyards. There are currently over 9800 records of the grassland fungi target species for the whole of Ireland. The results of a survey of grassland fungi on Clare Island are discussed and compared with the adjacent mainland sites in County Mayo, and comparisons are made with surveys of west Cork and west Mayo.

Background

Waxcaps (the genus *Hygrocybe*) have been described as the orchids of the fungi world (Marren 1998). They are often startling in colour, from reds, oranges and yellows to whites and browns. They can smell of honey or cedar wood or, less pleasantly, smell oily or nitrous. They are usually found in grasslands in northern Europe, although they can also be found in woods. They are one of the groups of grassland fungi that are now recognised as excellent indicators of unfertilised grassland or 'waxcap grasslands' (Arnolds 1980). Waxcap grasslands can be rich in other grassland fungi and usually include the *Entolomaceae* (pink-spored gill fungi), the clavarioids (fairy clubs), Geoglossaceae or earth tongues and species from the smaller genera of *Camarophyllopsis, Dermoloma* and *Porpoloma*. Photographs of most of the key species are available at www.nifg.org.uk (NIFG 2012a).

Waxcap grassland can be found in a range of grassland types from dunes to uplands, and from lowlands to gardens or churchyards. Indeed, the last refuges of these species are often gardens and churchyards, isolated areas that have been spared the addition of fertilisers and now give us a glimpse of what our natural grasslands once looked like. Many species are on national Red Lists across Europe, and *Hygrocybe calyptriformis* was on the list of fungal species proposed for inclusion in the Berne Convention in 2003 (Dahlberg and Croneborg 2003): this did not progress for various political reasons, unconnected with the need to protect fungi.

Grassland fungi provide nine of the fifteen fungal species in Northern Ireland's list of species of conservation concern. These are the waxcaps, *Hygrocybe calyptriformis, H. lacmus* and *H. ovina*, the earth tongues, *Geoglossum atropurpureum, Microglossum olivaceum* and *Trichoglossum walteri* along with *Clavaria zollingeri, Entoloma bloxamii* and *Porpoloma metapodium* (see Habitas website (Habitas 2012)).

These species are sensitive to the application of artificial fertilisers, and it is for this reason that they are such a good indicator of 'natural' grasslands. It was estimated in Northern Ireland that the cumulative surplus of phosphorus in the soil was 500,000t (Bailey 1994), meaning that most of the lowland rural Northern Ireland landscape is eutrophicated. There have been various attempts to discover how long it might take before sites recover after intensive fertilisation. Studies in England looking at the improvement in the soil fungal:bacterial biomass ratio due to the cessation of fertiliser application found no improvement after six years (Bardgett and McAlister 1999). Three sites in the Netherlands that had been intensively managed for agriculture but were now managed for nature conservation had only three or fewer species of *Hygrocybe* after twenty years (Arnolds 1994), but the lack of suitable surrounding habitat may have influenced this very slow recovery. Experimental plots also in the Netherlands showed that species of *Hygrocybe* could colonise the plots in a much shorter time period if they were low on phosphorus (Arnolds 1994). Hence recovery is probably more related to the nutrient status of the soils rather than the age of the site, with factors like suitable surrounding habitat also playing a role.

There is now greater interest in managing grasslands sustainably without high fertiliser input. Naturally sustainable grasslands have soils dominated by fungal pathways of decomposition rather than bacterial processes and have a high microbial biomass (Bardgett and McAlister 1999). Given their visual prominence in autumn, waxcaps are an indicator group for 'natural' grasslands that offer a means of rapid site assessment. Their presence indicates a wider nature conservation value beyond mycology. It was noticeable in one study on Fair Head, Co. Antrim, that there was a coincidence between waxcap distribution and the fields most favoured by choughs feeding on leatherjackets (Agri-environment Monitoring Unit 2004).

Waxcap grasslands, however, are often not particularly botanically rich, which can mean that they are missed when designating sites for nature conservation. Statistical studies in Sweden have shown that there is a low congruence between the diversity of *Hygrocybe* spp and diversity of higher plant species (Öster 2008), indicating that reliance on the latter when designating protected sites could lead to sites of high mycological value being missed.

The great unknown, however, is just what these species are actually doing in the soil. One study (Griffith *et al.* 2002) points to some possible answers based on the use of stable isotope analysis. Stable isotopes of carbon (^{13}C) and nitrogen (^{15}N) occur naturally, and it has been shown that the enrichment patterns for these isotopes is quite different in ectomycorrhizal and saprophytic fungi. Waxcaps, however, appear different to normal saprophytic fungi as they are more depleted in ^{13}C and more enriched in ^{15}N. Clavarioids and Geoglossaceae are even more extreme in this trend, but *Entoloma* species are more typical of saprophytic fungi. This may mean that *Hygrocybe* spp, clavarioids and Geoglossaceae could be deep humic decayers rather than normal surface litter decayers adapted to nitrogen-poor conditions.

Assessing site quality from fungal data

Interest has been growing in grassland fungi in Ireland since their first recognition by Feehan and McHugh (1992), as it has been recognised that this unique community is seriously threatened across Europe.

Various systems have been proposed to rank grassland sites for their fungal conservation value. Rald (1985) in Denmark proposed a system based on the number of species of *Hygrocybe*, Nitare (1988) and Jordal (1997) looked at systems in Sweden and in Norway respectively, and the British Mycological Society instigated a survey giving the surveyed sites a CHEG score (Clavariaceae, *Hygrocybe*, *Entoloma* and Geoglossaceae; see Rotheroe (1996)). Rotheroe then proposed a system that included a weighted score for rarer species that are restricted to species-rich sites (Rotheroe 1999). This was further developed by McHugh *et al.* (2002) when we proposed a weighted scoring system for Ireland. One of the main drivers for this was due to the lack of mycological recording in Ireland: we wanted to highlight sites for further visits that had species thought to be rarer or more valuable indicator species. Weighting species is controversial as in reality the data are not available to weight them with confidence (Griffith *et al.* 2006), but the point was to use this in conjunction with standard CHEG scores and highlight sites of possible interest (McHugh *et al.* 2001).

Most of the scoring systems above base their score on species and do not include varieties in the calculation (Rald 1985; Nitare 1988; Boertmann 1995; Vesterholt *et al.* 1999; McHugh *et al.* 2001). However, some surveys have counted varieties (Rotheroe 1999 and Newton *et al.* 2002) so it is very important to be clear about the basis of the system used when comparing data across regions. For this purpose, the definition of species used in all the Irish surveys follows the *Checklist of the Basidiomycetes of the British Isles* (Legon and Henrici 2005) and Spooner's key for Geoglossaceae (Spooner 1998), with three exceptions to remain consistent with the continental surveys.

Hygrocybe pratensis var. *pallida* is the only variety included in the scoring following Vesterholt 1999. Although the *Checklist of the Basidiomycetes of the British Isles* (Legon and Henrici 2005) did list *Hygrocybe conicoides* as a species rather than *Hygrocybe conica* var. *conicoides*, Boertmann's book and his recent interpretation of *Hygrocybe* in *Funga Nordica* (Knudsen and Vesterholt 2008) both still list it as a variety, so it is not counted separately in this study.

Hygrocybe marchii is considered a synonym of *H. coccinea* following *Funga Nordica*. Despite this, any good database can take these differing definitions into account, and a Microsoft Access database is in use for scoring and ranking grassland sites in Ireland.

These systems primarily look at the genus *Hygrocybe* when ranking sites. Inevitably there will be sites that are particularly good for the other target groups, and this is where the value of the CHEG scores is obvious. Some studies (Griffith *et al.* 2006) have added the different CHEG scores together, but this has to be viewed with caution. Entolomataceae are particularly difficult to identify, and to be honest, even very good mycologists may not successfully identify every *Entoloma* to species level. Hence the Entolomataceae are not well recorded and often only partially recorded. Added to this, there are many more species of *Entoloma* than in the other groups so adding CHEG scores together can just end up highlighting sites visited by mycologists who can identify *Entoloma* species.

Table 1 shows the total numbers of CHEG and related species occurring in grasslands in the British Isles according to the *Checklist of the Basidiomycetes of Britain and Ireland* (Legon and Henrici 2005) and (Ridge 1997).

Table 1

Numbers of grassland CHEG and related species occurring in the British Isles

Group	Total grassland species
Clavariaceae	24
Hygrocybe	51
Entolomataceae	99
Geoglossaceae	12
Dermoloma	4
Camarophyllopsis	5
Porpoloma	1

Grassland fungal fruiting patterns

The target species of grassland fungi have been recorded fruiting in Ireland from May to January. The mild Irish maritime climate means that fruiting can be very late. The existing database of grassland fungi in Ireland held by the Northern Ireland Fungus Group gives 27 February as the latest record for a species of *Hygrocybe* (*H. pratensis* at Crawfordsburn House, Co. Down) and shows that as many as eleven species were recorded on Fair Head, Co. Antrim, on 28 December 2001. The earliest record is from 25 May (*H. psittacina* var. *psittacina* from Murlough National Nature Reserve, Co. Down).

These are unusual, however. In addition, the main fruiting periods of the different groups tend to be at slightly different times, which further adds to the complication of scoring sites based on scores combined across all the groups. As the Scottish survey showed (Newton *et al.* 2002), Entolomataceae tend to fruit early in the season and Geoglossaceae right at the end of the season. This means that if you score sites based on all the groups, in reality you need multiple visits at different time periods for the scores to be meaningful.

There are currently over 9800 records of the grassland fungi target species for the whole of Ireland. These records are analysed against week number in Table 2, showing the week numbers of the maximum observations for each species group. Figs 1–4 show the individual phenology charts.

So the peak time for Entolomataceae in Ireland is about three weeks before the peak for *Hygrocybe* and four weeks before the peak for Geoglossaceae,

Table 2

Species group and the week numbers of the maximum observations

Group	Week	2008[1]
C	43	19 October
H	44	26 October
E	41	05 October
G	45	02 November

[1] The date of the start of this week in 2008

which shows that multiple site visits are required before a good overall assessment can be made for any grassland fungi site.

What these figures mask is inter-year variation. As fruiting body production is so dependent on weather conditions, especially rainfall and temperature, the week with the maximum number of observations can considerably differ from year to year. In 2006, 2007 and 2008, I received grants from the Heritage Council to survey grassland fungi in County Clare, west Cork and west Mayo respectively (Mitchel 2006; 2007; 2008). These surveys were all done at the same time (last week of October and first week of November), so the results are comparable (Table 3).

In 2007, warm dry weather in April in Ireland was followed by extremely wet weather in May, June and July (Met Éireann). Waxcaps were recorded fruiting, often in large numbers, in June and July in Wales. This was followed by a very dry September and a very mild October with some rainfall. The actual survey period of 28 November 2007 to 11 November 2007 was marked by mild and very dry weather, with rain on only one morning in the whole two-week

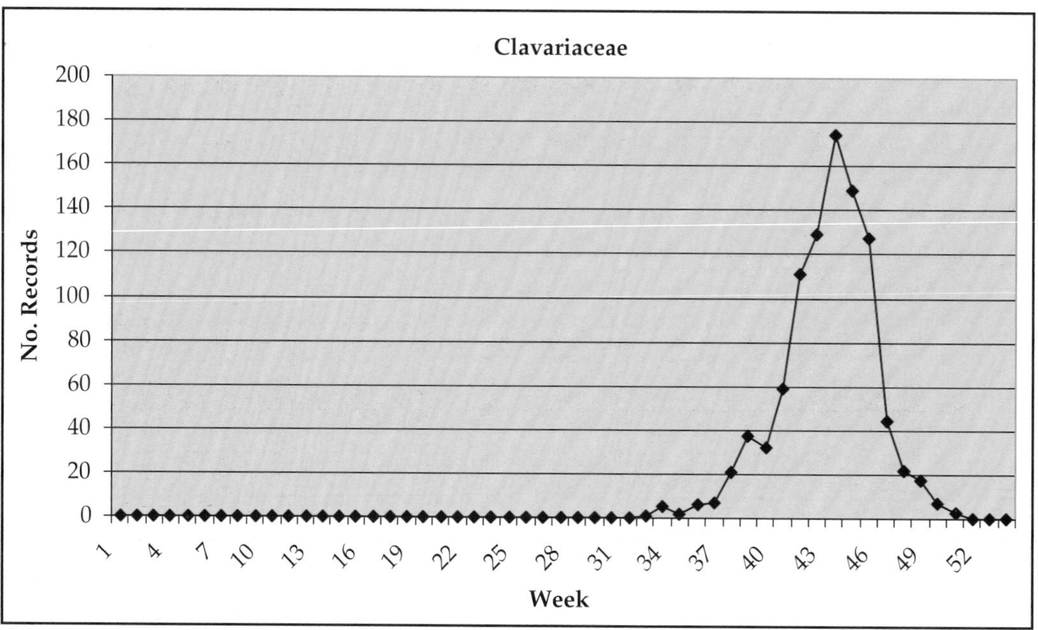

Fig. 1 Phenology chart for Clavariaceae.

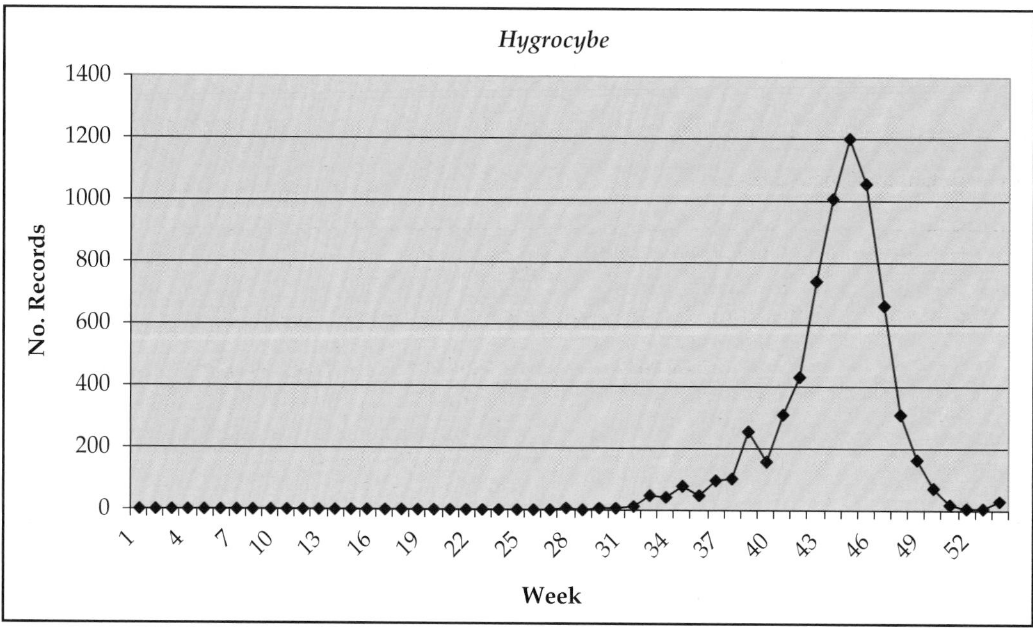

Fig. 2 Phenology chart for *Hygrocybe*.

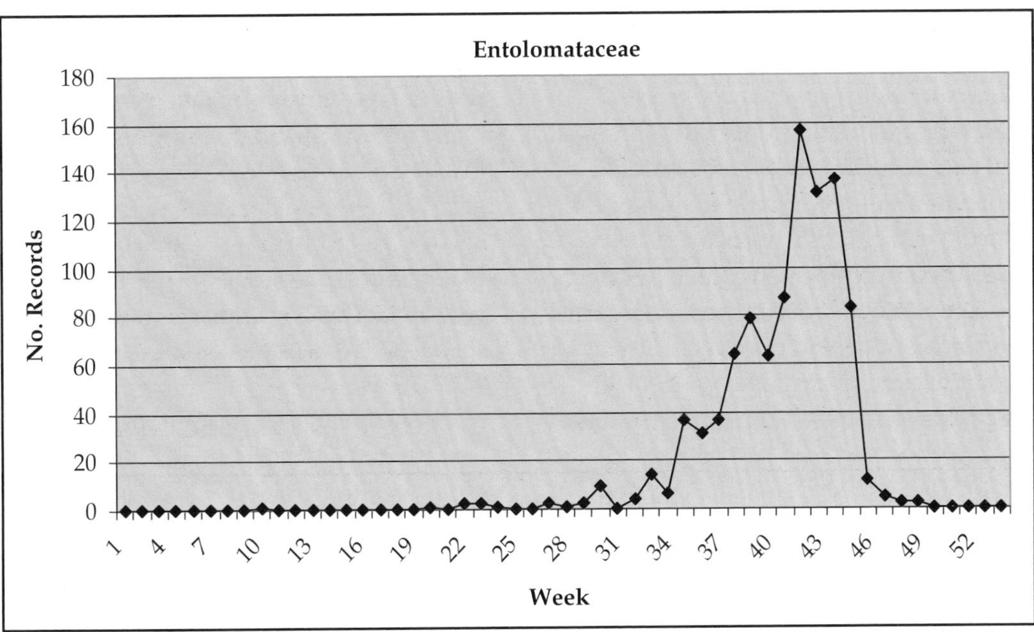

Fig. 3 Phenology chart for Entolomataceae.

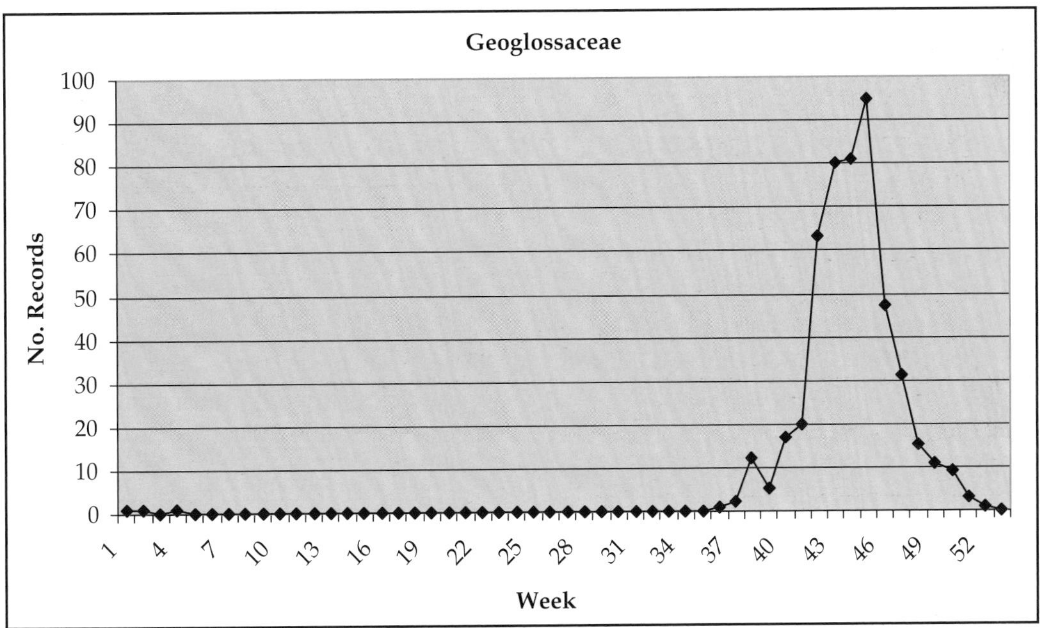

Fig. 4 Phenology chart for Geoglossaceae.

Table 3

Number of species found in the three Heritage Council grassland fungi surveys

	Clare 2006	West Cork 2007	West Mayo 2008	All Ireland to date
Hygrocybe	23	29	25	40
Clavariaceae	10	10	8	16
Entolomataceae	12	20	7	66
Geoglossaceae	5	3	8	11
Other grassland target species[1]	2	2	1	6
Total species[2]	155	206	177	

[1] *Camarophyllopsis*, *Dermoloma* and *Porpoloma*
[2] This includes non-target species

period. It was notable that the ground on many sites was often very dry but at least fruiting was generally reasonable. It was also notable that one site, Saint Matthew's Church of Ireland in Baltimore, was visited twice (31 October 2007 and 10 November 2007), and while the first visit only six species of waxcap were recorded, twelve species were recorded on the second visit. The impression was that the fruiting season was very late (e.g. an abundance of *Entoloma* records and a lack of Geoglossaceae) and that fruiting was improving as time progressed.

The year 2008 was significantly wetter and cooler than 2007, but the mean temperatures were still above average. The weather during the two weeks of survey was marked by a particularly cold first week, with snow even falling on 28 October 2008. The second week was milder, but fruiting did not seem to be affected by these temperatures. Rainfall was significant and there were three days in particular of extremely heavy rain that caused localised flooding. The impression here was that it was the tail end of the season. These two examples illustrate one of the difficulties of mycological surveying, as flexibility in survey dates is desirable but not often achievable.

History of mycological grassland recording on Clare Island and in west Mayo and Ireland

Before the original Clare Island Survey, only one fungus, *Ustilago longissima* (current name = *Ustilago filiformis* (Schrank) Rostr.), had been recorded from the island. Sir Henry Hawley and Carleton Rea were largely responsible for the fungal surveys, although Carleton Rea forayed solely on the mainland (Rea and Hawley 1912). In volume 1 of the *New Survey of Clare Island*, Collins (1999, p. 35) states that Rea and Praeger visited the island in November 1910, but according to Rea, he and Praeger did not visit the island on this date. The only three visits to the island were on 20–27 August 1909, 3–10 October 1910 and 27 April – 4 May 1911 (Rea and Hawley 1912). This is important because, as described above, there is a strong possibility that they missed the main grassland fungi season, which is normally late October to early November.

From these visits, fifteen species of *Hygrocybe* were recorded, although one was *H. obrussea*, which is a *nomen confusum*. From the description, it could have been *H. citrinovirens*, although this is speculation. Twelve Clavariaceae, eight *Entoloma* and two Geoglossaceae species were found, although again *Geoglossum glabrum* is now also considered a *nomen confusum*. The species are listed in Table 4. The notable thing here is the number of Clavariaceae, which makes Clare Island the best site for Clavariaceae in Ireland by a considerable distance. Two other species recorded as *Clavaria* by Hawley were *C. cinerea* and *C. cristata*. These are now both in the genus *Clavulina* and are not target species, being typically woodland species.

Despite being early in the season, only eight *Entoloma* species were recorded. The significant species of *Hygrocybe* recorded were *H. calyptriformis* and *H. ovina*.

On the mainland, even though most of the sites were described as woodland, 18 species of *Hygrocybe* were recorded and 8 Clavariaceae, 21 *Entoloma* and 2 Geoglossaceae were found (Rea 1912). These records are listed in Appendix 2 (pp 279–318). For the mainland sites there is no information from which to judge if the waxcaps were found in the woods, in grassland or in lawns around the houses. The best mainland sites for *Hygrocybe* were Old Deer-Park Wood at Mount Browne (9 species), Knockranny Wood (9) and Westport Park (8).

Since the original survey, mycological recording has been very sparse on Clare Island and indeed in west Mayo. The next significant survey was in 1992 when a Dutch mycologist, Reitze ten Cate, visited the island. He described Clare Island as a paradise for waxcaps (ten Cate 1993). His list uses an older taxonomy, which is reviewed in Tables 5 and 6. Translating species into the concepts of David Boertmann (Boertmann 1995), he recorded fourteen, possibly fifteen species (Table 5).

All these records marked Clare Island as an interesting island for grassland fungi. The numbers of *Hygrocybe* were good (eighteen in total) but the numbers of Clavariaceae indicated that there were possibilities for finding additional species.

Grassland fungi recording in the rest of Mayo was very limited following the original Clare Island Survey. Roland McHugh was the most active, finding good sites on Achill Island (Keem and Keel machair, which were listed as the eighth and twentieth best sites in Ireland in McHugh *et al.* 2001) (see Table 7 for a list of his visits).

Since then, McHugh has visited the area on a number of occasions (pers. comm.) and added the site of Murrevagh machair at Mulranny to the list with sixteen species of *Hygrocybe*. The mycology of this site has generated local interest, and an information board containing photographs taken by myself in earlier surveys was put up as part of a Heritage Council grant.

Table 4
Grassland fungi target species recorded in original Clare Island Survey

Recorded name	Current name	Type
Clavaria acuta	*Clavaria acuta*	C
Clavaria vermicularis	*Clavaria fragilis*	C
Clavaria fumosa	*Clavaria fumosa*	C
Clavaria straminea	*Clavaria straminea*	C
Clavaria amethystina	*Clavaria zollingeri*	C
Clavaria muscoides	*Clavulinopsis corniculata*	C
Clavaria fusiformis	*Clavulinopsis fusiformis*	C
Clavaria dissipabilis	*Clavulinopsis helvola*	C
Clavaria persimilis	*Clavulinopsis laeticolor*	C
Clavaria luteoalba	*Clavulinopsis luteoalba*	C
Clavaria umbrinella	*Clavulinopsis umbrinella*	C
Clavaria kunzei	*Ramariopsis kunzei*	C
Leptonia asprella	*Entoloma asprellum*	E
Eccilia griseorubella	*Nomen dubium*	E
Nolanea pascua	*Entoloma conferendum*	E
Entoloma jubatum	*Entoloma porphyrophaeum*	E
Entoloma prunuloides	*Entoloma prunuloides*	E
Leptonia lampropus	*Entoloma sodale*	E
Microglossum atropurpureum	*Geoglossum atropurpureum*	G
Geoglossum ophioglossoides	*Geoglossum glabrum = nomen confusum*	
Hygrophorus calyptriformis	*Hygrocybe calyptriformis*	H
Hygrophorus ceraceus	*Hygrocybe ceracea*	H
Hygrophorus chlorophanus	*Hygrocybe chlorophana*	H
Hygrophorus coccineus	*Hygrocybe coccinea*	H
Hygrophorus conicus	*Hygrocybe conica*	H
Hygrophorus fornicatus	*Hygrocybe fornicata*	H
Hygrophorus unguinosus	*Hygrocybe irrigata*	H
Hygrophorus laetus	*Hygrocybe laeta*	H
Hygrophorus miniatus	*Hygrocybe miniata*	H
Hygrophorus obrusseus	*Nomen confusum = H. citrinovirens?*	
Hygrophorus ovinus	*Hygrocybe ovina*	H
Hygrophorus pratensis	*Hygrocybe pratensis* var. *pratensis*	H
Hygrophorus psittacinus	*Hygrocybe psittacina* var. *psittacina*	H
Hygrophorus puniceus	*Hygrocybe punicea*	H
Hygrophorus virgineus	*Hygrocybe virginea* var. *virginea*	H

C = Clavariaceae; E = *Entoloma;* G = Geoglossaceae; H = *Hygrocybe*

In the rest of the British Isles, grassland fungi started to gain a higher profile when the British Mycological Society announced the start of a British Isles survey (Rotheroe *et al.* 1996). This included Ireland. While it instigated many site surveys, it was never co-ordinated enough to generate overall summary reports. However, by stimulating interest and by using a system that assessed and ranked sites for their mycological conservation interest, more detailed surveys were instigated and funding was obtained for these. Countrywide surveys were run in Scotland (Newton *et al.* 2002), Wales (Griffith *et al.* 2006) and Northern Ireland (NIFG 2012b). Surveys were more localised in England (Evans 2004).

In Ireland, the first paper on grassland fungi was based on the Curragh (Feehan and McHugh

1992), and after this Roland McHugh produced records almost single-handedly in the Republic of Ireland, with the occasional visit from the Northern Ireland Fungus Group. In 2002, a number of us summarised the state of knowledge about grassland fungi in the whole of Ireland (McHugh *et al.* 2001) and produced a site ranking for waxcap sites for Ireland for the first time. In 2006, I was given a grant to conduct a two-week waxcap survey in County Clare, followed by west Cork in 2007 and west Mayo in 2008. Running surveys with a consistent methodology is now allowing a better picture to be built up of the conservation value of grassland fungi in Ireland (see Tables 7, 8 and 9).

All this effort is now highlighting just how good the British Isles are for grassland fungi. In Europe, Slovakia's best site has only 22 species (Adamcík and Kautmanová 2005), the Netherlands' has 25 (Arnolds 1995) and Denmark and Norway have one site with more than 28 species (Griffith *et al.* 2006). The British Isles have more numerous and better sites than any other known area in Europe and probably the world (waxcap grasslands seem to be largely a European phenomenon, with *Hygrocybe* spp being more common in woodlands in areas like Australia, the Caribbean and the USA) (Griffith *et al.* 2006). In this regard, waxcap grasslands should be a conservation priority and certainly should be highlighted in biodiversity strategies.

Table 5
Species of *Hygrocybe* recorded by ten Cate (1993)

Hygrocybe ceracea
Hygrocybe chlorophana
Hygrocybe coccinea
Hygrocybe conica
Hygrocybe fornicata
Hygrocybe glutinipes
Hygrocybe laeta
Hygrocybe pratensis
Hygrocybe psittacina
Hygrocybe punicea
Hygrocybe reidii
Hygrocybe splendidissima
Hygrocybe virginea var. *fuscescens*
Hygrocybe virginea var. *ochraceopallida*
Hygrocybe virginea var. *virginea*
Hygrocybe vitellina

Methodology

The visit to Clare Island was very short, consisting of a single three day visit (1–3 November) in perfect weather by David Mitchel (DM), Hubert Fuller (HF) and Jolanda Mitchel-Smit (JM). Kieran Connolly (KC) arrived on 2 November.

On 1 November, DM, HF and JM covered the slopes of Knockmore from the church up to the saddle between Knockmore and Knocknaveen and then followed the contour around the southern slopes of Knockmore to the signal tower at Tonadowhy. We returned to

Table 7
Summary results of R. McHugh's visits to Clare Island

Date	C	H	E	G
18–19 September 1992	2	13	4	1
24–26 September 1993	1	11	2	1
30 September – 2 October 1994	2	15	2	1

C = Clavariaceae; E = *Entoloma*; G = Geoglossaceae; H = *Hygrocybe*

Table 6
Other names recorded by ten Cate (1993)

Type	Species	Comment
H	*Camarophyllopsis niveus*	Now *H. virginea*, which he also recorded
H	*Camarophyllopsis phaeophylla*	No other records for Ireland or Britain
H	*Hygrocybe citrina*	*Nomen dubium*, possibly *H. glutinipes*, which he also recorded
H	*Hygrocybe marchii*	There is doubt about this as a valid species, and *Funga Nordica* now relegates it to a synonym of *H. coccinea*
H	*Hygrocybe nigrescens*	Now regarded as a variety of *H. conica*, which he also recorded
H	*Hygrocybe obrussea*	*Nomen dubium*, possibly *H. citrinovirens* or *H. quieta*

Table 8
Grassland fungi records from McHugh, Fuller and Anderson

Species	Type	Date	Recorder
Clavaria acuta	C	18 September 1992	R. McHugh
Clavulinopsis helvola	C	18 September 1992	R. McHugh
Entoloma caesiocinctum	E	18 September 1992	R. McHugh
Geoglossum cookeanum	G	18 September 1992	R. McHugh
Hygrocybe ceracea	H	18 September 1992	R. McHugh
Hygrocybe chlorophana	H	18 September 1992	R. McHugh
Hygrocybe conica	H	18 September 1992	R. McHugh
Hygrocybe persistens var. *konradii*	H	18 September 1992	R. McHugh
Hygrocybe persistens var. *persistens*	H	18 September 1992	R. McHugh
Hygrocybe psittacina var. *psittacina*	H	18 September 1992	R. McHugh
Hygrocybe punicea	H	18 September 1992	R. McHugh
Clavulinopsis helvola	C	19 September 1992	R. McHugh
Entoloma dichroum	E	19 September 1992	R. McHugh
Entoloma poliopus var. *discolor*	E	19 September 1992	R. McHugh
Entoloma triste	E	19 September 1992	R. McHugh
Hygrocybe cantharellus	H	19 September 1992	R. McHugh
Hygrocybe chlorophana	H	19 September 1992	R. McHugh
Hygrocybe conica	H	19 September 1992	R. McHugh
Hygrocybe constrictospora	H	19 September 1992	R. McHugh
Hygrocybe marchii	H	19 September 1992	R. McHugh
Hygrocybe pratensis var. *pratensis*	H	19 September 1992	R. McHugh
Hygrocybe psittacina var. *psittacina*	H	19 September 1992	R. McHugh
Hygrocybe punicea	H	19 September 1992	R. McHugh
Hygrocybe russocoriacea	H	19 September 1992	R. McHugh
Hygrocybe virginea var. *virginea*	H	19 September 1992	R. McHugh
Hygrocybe fornicata var. *fornicata*	H	20 September 1992	R. McHugh
Hygrocybe virginea var. *fuscescens*	H	20 September 1992	R. McHugh
Entoloma conferendum var. *conferendum*	E	24 September 1993	R. McHugh
Hygrocybe laeta var. *laeta*	H	24 September 1993	R. McHugh
Hygrocybe quieta	H	24 September 1993	R. McHugh
Geoglossum cookeanum	G	25 September 1993	R. McHugh
Hygrocybe ceracea	H	25 September 1993	R. McHugh
Hygrocybe chlorophana	H	25 September 1993	R. McHugh
Hygrocybe conica	H	25 September 1993	R. McHugh
Hygrocybe conica var. *conicopalustris*	H	25 September 1993	R. McHugh
Hygrocybe marchii	H	25 September 1993	R. McHugh
Hygrocybe persistens var. *persistens*	H	25 September 1993	R. McHugh
Hygrocybe pratensis var. *pratensis*	H	25 September 1993	R. McHugh
Hygrocybe russocoriacea	H	25 September 1993	R. McHugh
Hygrocybe virginea var. *virginea*	H	25 September 1993	R. McHugh
Clavulinopsis corniculata	C	26 September 1993	R. McHugh
Entoloma hispidulum	E	26 September 1993	R. McHugh
Hygrocybe coccinea	H	26 September 1993	R. McHugh
Hygrocybe reidii	H	26 September 1993	R. McHugh

(Continued)

Table 8 (*Continued*)

Species	Type	Date	Recorder
Entoloma fernandae	E	30 September 1994	R. McHugh
Hygrocybe ceracea	H	30 September 1994	R. McHugh
Hygrocybe insipida	H	30 September 1994	R. McHugh
Hygrocybe marchii	H	30 September 1994	R. McHugh
Hygrocybe miniata	H	30 September 1994	R. McHugh
Hygrocybe persistens var. *persistens*	H	30 September 1994	R. McHugh
Hygrocybe psittacina var. *psittacina*	H	30 September 1994	R. McHugh
Hygrocybe punicea	H	30 September 1994	R. McHugh
Hygrocybe virginea var. *virginea*	H	30 September 1994	R. McHugh
Clavulinopsis corniculata	C	01 October 1994	R. McHugh
Clavulinopsis helvola	C	01 October 1994	R. McHugh
Hygrocybe ceracea	H	01 October 1994	R. McHugh
Hygrocybe chlorophana	H	01 October 1994	R. McHugh
Hygrocybe conica	H	01 October 1994	R. McHugh
Hygrocybe glutinipes var. *glutinipes*	H	01 October 1994	R. McHugh
Hygrocybe laeta var. *laeta*	H	01 October 1994	R. McHugh
Hygrocybe persistens var. *persistens*	H	01 October 1994	R. McHugh
Hygrocybe pratensis var. *pratensis*	H	01 October 1994	R. McHugh
Hygrocybe psittacina var. *perplexa*	H	01 October 1994	R. McHugh
Hygrocybe psittacina var. *psittacina*	H	01 October 1994	R. McHugh
Hygrocybe punicea	H	01 October 1994	R. McHugh
Hygrocybe russocoriacea	H	01 October 1994	R. McHugh
Hygrocybe virginea var. *fuscescens*	H	01 October 1994	R. McHugh
Hygrocybe virginea var. *virginea*	H	01 October 1994	R. McHugh
Hygrocybe vitellina	H	01 October 1994	R. McHugh
Entoloma dichroum	E	02 October 1994	R. McHugh
Geoglossum cookeanum	G	02 October 1994	R. McHugh
Hygrocybe miniata	H	02 October 1994	R. McHugh
Hygrocybe calyptriformis	H	October 1998	H. Fuller
Clavulinopsis fusiformis	C	13 October 2002	R. Anderson
Hygrocybe cantharellus	H	13 October 2002	R. Anderson
Hygrocybe coccinea	H	13 October 2002	R. Anderson
Hygrocybe laeta	H	13 October 2002	R. Anderson
Hygrocybe psittacina	H	13 October 2002	R. Anderson
Hygrocybe quieta	H	13 October 2002	R. Anderson
Hygrocybe russocoriacea	H	13 October 2002	R. Anderson
Clavulinopsis fusiformis	C	14 October 2002	R. Anderson
Hygrocybe chlorophana	H	14 October 2002	R. Anderson

C = Clavariaceae; E = *Entoloma*; G = Geoglossaceae; H = *Hygrocybe*

the harbour via the coast road. On 2 November, DM, HF and JM searched the eastern slopes of Knocknaveen following the Green Road and then searched out to the lighthouse at Lecknacurra. On the third day, HF and KC visited the woodland at Ballytoohy More. Most of the areas searched, therefore, were the commonages, with the tilled land on either side of the coastal road only viewed from over the fence. Large areas were not searched, including the higher slopes of Knockmore, partly due to time constraints and also because preparatory examination of the relief, soil maps and aerial photographs had prioritised areas for survey.

Table 9
Grassland target species found on Clare Island November 2008

Species	Type
Clavulinopsis corniculata	C
Clavulinopsis fusiformis	C
Clavulinopsis laeticolor	C
Entoloma conferendum	E
Entoloma papillatum	E
Geoglossum atropurpureum	G
Geoglossum fallax	G
Geoglossum glutinosum	G
Microglossum olivaceum	G
Trichoglossum walteri	G
Hygrocybe aurantiosplendens	H
Hygrocybe calyptriformis	H
Hygrocybe cantharellus	H
Hygrocybe ceracea	H
Hygrocybe chlorophana	H
Hygrocybe coccinea	H
Hygrocybe conica var. *conica*	H
Hygrocybe fornicata	H
Hygrocybe insipida	H
Hygrocybe irrigata	H
Hygrocybe laeta var. *flava*	H
Hygrocybe laeta var. *laeta*	H
Hygrocybe miniata	H
Hygrocybe mucronella	H
Hygrocybe nitrata	H
Hygrocybe persistens var. *persistens*	H
Hygrocybe pratensis var. *pratensis*	H
Hygrocybe psittacina var. *psittacina*	H
Hygrocybe punicea	H
Hygrocybe quieta	H
Hygrocybe reidii	H
Hygrocybe russocoriacea	H
Hygrocybe splendidissima	H
Hygrocybe virginea var. *ochraceopallida*	H
Hygrocybe virginea var. *virginea*	H

C = Clavariaceae; E = *Entoloma*; G = Geoglossaceae; H = *Hygrocybe*

Identification was based on the available literature (Appendix 1, p. 278). Specimens were kept of the important finds and were sent to the National Botanic Gardens, Glasnevin. All records made have been entered into the biological recording database 'Recorder 6' and have been passed to the FRDBI (Fungus Recording Database of the British Isles) managed by the British Mycological Society and also to the National Biodiversity Data Centre in Waterford.

Results

Clare Island was found to be an exceptional site for waxcaps, with fruiting abundant over the whole island. Much of the commonage on Knockmore and Knocknaveen was exceptionally interesting, with fruiting occurring high on Knockmore. The main areas of interest were sites with short turf where drainage was better. Once soil conditions became too boggy, the number of records dropped. For this reason, areas of abandoned lazy beds and some of the steep thin soils on Knockmore were particularly good. The lower enclosed fields also appeared good when looking over the fences, and these are definitely worth further investigation. The very short coastal turf around the lighthouse and the coastal cliffs yielded fewer records. Table 9 records the list of species identified. Pls I–IV show some of the best areas for grassland fungi.

This gives a CHEG score for the island from this visit of C: 3, H: 23, E: 2, G: 5. The most important records were as follows:

Hygrocybe calyptriformis (Berk. & Broome) Fayod **Pl. V**

This unmistakable pink waxcap is rare across Europe but the British Isles is undoubtedly its stronghold. It is one of Northern Ireland's Priority Species (see Habitas website) and was proposed as one of the 33 species of fungus to be added to the Berne List. Often found in churchyards and lawns, the three sites it occurred at were all upland grassland (Knockmore on Clare Island, Keem on Achill and the Deserted Village on Slievemore on Achill).

Hygrocybe laeta var. ***flava*** Boertm. **Pl. VI**

First record for Republic of Ireland. Two records from Northern Ireland. *Hygrocybe laeta* var. *laeta*

Pl. I Commonage between Knocknaveen and Knockmore.

Pl. II The steep slopes overlooking Loughanaphuca.

Pl. III The steep west facing slopes of Knockmore were particularly rich.

Pl. IV The commonage around the Green Road on Knocknaveen.

Pl. V *Hygrocybe calyptriformis*.

Pl.VI *Hygrocybe laeta* var. *flava*.

was common on the west Mayo survey, but this is the bright yellow variety. Found on Knockmore, Clare Island, on 1 November 2008 (L677854) and Dooghill, Bellacragher Bay on 3 November 2008 (L821986).

Hygrocybe nitrata (Pers.) Wünsche
One of the rarer waxcaps, this was only found on Knockmore on Clare Island on 1 November 2008 (L674854)

Geoglossum atropurpureum (Batsch) Pers.
A rare earth tongue that is one of Northern Ireland's Priority Species (see Habitas website). One of the fungi proposed for the Berne List. Found five times on this survey, which is in itself notable. Recorded at Knockmore on Clare Island on 1 November 2008 at L677854, at Portacloy on 3 November 2008 (F839442), Inishturk on 5 November 2008 (L598752), Cloghmore on Achill on 7 November 2008 (L707937) and the Deserted Village on Achill on 7 November 2008 (F637073).

Microglossum olivaceum (Pers.) Gillet **Pl. VII**
An unmistakable earth tongue that is a Northern Ireland Priority Species (see Habitas website). Found on Knockmore, Clare Island, on 1 November 2008 at L677854 and Knocknaveen on Clare Island on 2 November 2008 at L698858.

Trichoglossum walteri (Berk.) E.J. Durand
Another Priority Species in Northern Ireland. This earth tongue is hardly distinguishable in the field from other earth tongues but is recognised microscopically by its jet black setae (like needles) and seven-septate spores. There are scattered records from Northern Ireland but there are no records for it from the Republic of Ireland in the FRDBI. Found between Ballytoohy and the lighthouse on Clare Island on 2 November 2008 (L698858) and at St Finian's Well, Keel, on Achill on 26 October 2008 (F658031).

Pl. VII *Microglossum olivaceum*.

Non-grassland target species recorded on Clare Island are shown below

Agaricus arvensis
Agaricus urinascens
Arrhenia latispora
Bovista plumbea
Clitocybe dealbata
Collybia butyracea f. *butyracea*
Cordyceps militaris
Cystoderma amianthinum
Dermoloma cuneifolium var. *cuneifolium*
Lepista nuda
Lepista panaeola
Leptosphaeria acuta
Lycoperdon nigrescens
Mycena epipterygia var. *epipterygia*
Omphalina ericetorum
Panaeolus acuminatus
Phragmidium violaceum
Psilocybe coprophila
Psilocybe semilanceata
Rhopographus filicinus
Steccherinum ochraceum
Stropharia pseudocyanea
Stropharia semiglobata
Tremella mesenterica
Tricholomopsis rutilans
Trochila ilicina

Of these, the notable records are as follows:

Arrhenia latispora (J. Favre) Bon & Courtec. **Pl. VIII**

This is an *Arrhenia* with well-developed gills, a short eccentric stipe and clamps on the hyphae. It is similar to *Arrhenia acerosa* but has broader spores. Found among mosses at the western end of Clare Island near to the signal tower on 1 November 2008 at L653852. There are no published records for this species for Ireland or database records in the FRDBI.

Stropharia pseudocyanea (Desm.) Morgan

This attractive *Stropharia* with a mixture of yellow and blue colours in the cap is not rare but is a good find on waxcap grasslands. It was found on the steep west-facing slopes of Knockmore.

Pl. VIII *Arrhenia latispora.*

National and international significance of Clare Island's mycological record

When the 2008 grassland fungi list is compared to the original survey, the only *Hygrocybe* not recorded in 2009 was the very rare *H. ovina*. This is a very distinctive dark-grey waxcap with reddening gills. This is only known from eight sites in Ireland. There is a possible record of *H. citrinovirens* (the *H. obrusseus* record of Rea (1912)) but this is usually recorded earlier in the season than most other waxcaps, and November would be late for it.

McHugh also recorded *H. constrictospora, H. glutinipes* and *H. vitellina* on the island in the early 1990s, so this would give a total modern score for Clare Island of 26 species of *Hygrocybe* (the 1910 finds are not counted). No new Clavariaceae or Entolomataceae were recorded, but six species of Geoglossaceae (including three notables) makes this a significant site for earth tongues. The overall CHEG score for the island is now C: 5 (12); H: 26 (27); E: 8 (14); G: 6 (6). The figures in parentheses include the original survey results.

The curious item of note is that the original survey found Clare Island to be an exceptional site for Clavariaceae, but in recent years only six species have been recorded. This is hard to explain, given that *Hygrocybe* are so well recorded compared to 1910.

In terms of the 2008 west Mayo waxcap survey (Table 10; Fig. 5, p. 271), Clare Island was easily the best site found. This survey found nine sites with ten or more species of *Hygrocybe*, as did the 2007 west Cork survey (there were seven such sites in the 2006 Clare Island survey). The other good sites were Keem Bay on Achill (17), the island of Inishturk (15), Tawnamartola on the slopes of Buckoogh (14), Portacloy near Benwee Head (14), the Deserted Village on Slievemore on Achill (13), St Finian's Well at Keel on Achill (12), Murrevagh machair at Mulranny (11) and Erriff on Maumtrasna (10). Of these, the Deserted Village is worth a mention because it is probably a much more significant site. It was surveyed in the middle of a storm and we had to leave the site before we got hypothermia. The abundance of fruiting was staggering, and it is highly probable that more species were present.

Table 10
Sites ranked by number of species of *Hygrocybe* in west Mayo[1]

Rank	Site	Grid reference	10k grid reference	*Hygrocybe*	Clavaria	*Entoloma*	Geoglossaceae	Irish score
1	Clare Island	L685855	L68	23	3	2	5	53
2	Keem Bay	F560043	F50	17	3	2	1	31
3	Inishturk	L604745	L67	15	3	2	3	30
4	Portacloy	F842440	F84	14	1	1	3	26
4	Tawnamartola	L978992	L99	14	1	2	1	23
6	Deserted Village, Slievemore, Achill	F637073	F60	13	1	1	2	28
7	St Finian's Well, Keel	F658031	F60	12	2	3	2	25
8	Mulranny machair	L840960	L89	11	1	1	1	13
9	Erriff, Maumtrasna	L977696	L96	10	1	2	1	17
10	Cloghmore	L707937	L79	9	1	1	2	13
10	Keel Machair	F645047	F60	9	1	1	1	12
10	Windy Gap	G137014	G10	9	2	1	2	18
13	Deel River Valley	G015085	G00	8	1	1	2	13
14	Ashleam Bay	L688963	L69	7	1	1	1	9
14	Doontrusk	L960970	L99	7	1	0	3	12
14	Glendavoolagh	G013070	G00	7	0	0	1	14
14	Rinnaglana Head	F793435	F74	7	1	1	0	10
18	Bunnahowen RC Church	F759286	F72	6	1	0	0	6
19	Dooghill, Bellacragher Bay	L821986	L89	5	0	0	0	5

[1] Only sites with 5 or more species of *Hygrocybe* are shown.

Once the historical records are added to the 2008 records, Clare Island really begins to stand out nationally and internationally. Table 11 shows that Clare Island is actually one of the best sites in Ireland in terms of waxcaps, second only to the Curragh, which has received markedly more visits. Additionally, the records for the Curragh may need to be reinterpreted against the modern species concepts for *Hygrocybe*. If the other CHEG groups are looked at, Clare Island is the best site in the country for Clavariaceae (see Table 12), second best for Geoglossaceae behind Binevenagh in County Londonderry (see Table 13) and fourth best for Entolomataceae (see Table 14).

If looked at in terms of the British Isles, Clare Island is the joint eleventh best site for *Hygrocybe* (Griffith *et al.* 2006). All of these statistics point to grassland fungi being the most important nature conservation feature on the island, and a statutory nature conservation designation should really be considered on this basis. According to the UK guidelines for the selection of biological Sites of Special Scientific Interest (SSSIs), any site with more than eighteen species of *Hygrocybe* from multiple visits or twelve in a single visit should be considered for SSSI status (Genney *et al.* 2009). European reviews estimate any site with over 22 species is of international importance (Vesterholt *et al.* 1999). Clare Island easily satisfies all these criteria.

Discussion

Why is Clare Island such a good site? Griffith *et al.* 2006) consider the main factors affecting waxcap grasslands to be the addition of synthetic fertilisers, an inappropriate grazing or cutting regime and physical disturbance, such as ploughing. Trampling has been also cited as a possible factor (Griffith *et al.* 2004). The work done on the original Clare Island Survey and the New Survey of Clare Island offers a unique opportunity to look at the waxcaps holistically. To this end, the detailed history of land management and the recent soil survey (Vullings *et al.* 2013) can be related back to these factors and the soil type assessed to see if it is a factor in its own right.

Table 11
Top Irish grassland sites as of 15 October 2009

Rank	Site	County	No. of species	Irish score	No. of visits	Heritage Council survey
1	The Curragh	Kildare	32	73	23	
2	Clare Island	West Mayo	26 (27)	75	8	A
3	Slievenacloy ASSI	Antrim	25	48	14	
4	Crossmurrin NNR	Fermanagh	23	51	7	
5	Binevenagh NNR	Londonderry	22	64	10	
5	Ballyprior	Laois	22	56	5	
7	Kebble NNR	Antrim	22	47	6	
8	Achill Island: Keem Bay	West Mayo	20	47	4	A
8	Monawilkin ASSI	Fermanagh	20	46	6	
10	Aghadachor	West Donegal	19	41	2	
11	Barnett's Park	Antrim	18	46	25	
11	Longmore townland, 1.5km NW of The Sheddings	Antrim	18	38	1	A
11	Hillsborough Parish Church	Down	18	34	7	
11	Dursey Island	West Cork	18	34	3	B
11	Mount Stewart Estate	Down	18	33	10	
16	Murrevagh Maghera	West Mayo	17	34	4	A
16	Bantry House	West Cork	17	32	1	B
16	Ballynacarriga	West Cork	17	29	1	B
19	Agnew's Hill	Antrim	16	38	3	
19	Black Head	Clare	16	30	2	C
19	Silent Valley, Mourne Mountains	Down	16	30	6	
22	Slemish Mountain	Antrim	15	33	2	
22	Inishturk	West Mayo	15	32	1	A
22	John McSparran Memorial Hill Farm	Antrim	15	32	3	
22	Clandeboye Estate	Down	15	31	7	
22	Murlough NNR	Down	15	31	15	
22	Great Heath of Maryborough	Laois	15	31	1	
22	Knockninny ASSI	Fermanagh	15	29	3	
22	East Torr townland, near Torr Head	Antrim	15	28	1	
22	Drum Manor Forest Park	Tyrone	15	28	7	

Sites marked A, B and C have been surveyed in the three recent surveys funded by the Heritage Council. The figures in brackets for Clare Island include the original survey records.

Table 12
Waxcap grassland sites ranked by number of species of Clavariaceae

Rank	Site	County	No. of species	No. of visits	Heritage Council survey
1	Clare Island	West Mayo	5 (12)	8	A
2	Binevenagh NNR	Londonderry	8	9	
2	The Curragh	Kildare	8	23	
4	Bantry House	West Cork	7	1	B
4	Belclare and Prospect House Woods	West Mayo	7	1	
4	Crom Castle Estate	Fermanagh	7	2	
7	Castle Archdale Country Park	Fermanagh	6	3	
7	Castle Coole	Fermanagh	6	5	
7	Dursey Island	West Cork	6	3	B
7	John McSparran Memorial Hill Farm	Antrim	6	3	
7	Murlough NNR	Down	6	15	
7	Slievenacloy ASSI	Antrim	6	12	

Table 13
Waxcap grassland sites ranked by number of species of Geoglossaceae

Rank	Site	County	No. of species	No. of visits
1	Binevenagh NNR	Londonderry	7	10
2	Clare Island	West Mayo	6	8
3	Antrim town cemetery	Northern Ireland	5	1
3	Slemish Mountain	Antrim	5	2
3	The Curragh	Kildare	5	23
6	1st Presbyterian Rectory lawn, Crumlin	Antrim	4	1
6	Boora Lake	Offaly	4	1
6	Castle Archdale Country Park	Fermanagh	4	3
6	Cloghy Dunes	Down	4	8
6	Cuilcagh Gap	Fermanagh	4	1
6	Cuilcagh Mountain ASSI	Fermanagh	4	2
6	Murlough NNR	Down	4	15
6	Silent Valley, Mourne Mountains	Down	4	6
6	White Park Bay	Antrim	4	2

History of land use on Clare Island

Prior to the purchase of Clare Island by the Congested Districts Board in 1895, the island was farmed in the classic rundale fashion. It was marked by a coastal strip of communal tillage that was regularly redistributed among the residents. Potatoes, oats, wheat, barley and rye were grown, and Clare Island was known for its fertile soils. Lazy beds dominated much of the tillage land (Collins 1999). The lazy beds were intensively managed, with soil shovelled from the trenches onto the beds each year, and the beds were often split at the end of each year with one year's bed being the next year's furrow. The advantages of the lazy beds were to improve drainage, reduce wind and frost damage, raise soil pH by adding dug-out subsoil and increase the depth of soil for crops, especially on steep slopes (Vullings 2000).

Since 1925 there has been a steady decline in the acreage of crops farmed on Clare Island. According to the Central Office of Statistics, in 1925, there were 5.25 acres of wheat, 115.25 acres

Table 14

Waxcap grassland sites ranked by number of species of Entolomataceae

Rank	Site	County	No. of species	No. of visits	Heritage Council survey
1	The Curragh	Kildare	32	23	
2	Binevenagh NNR	Londonderry	17	9	
2	Guinness estate	Wicklow	17	6	
4	Clare Island	West Mayo	8 (14)	8	A
5	Aghadachor	West Donegal	14	2	
6	Mulranny machair	West Mayo	13	4	
7	Knockranny	West Mayo	12	1	
7	Muckross Demesne	Kerry	12	4	
9	Bunduff Strand	Sligo	10	1	
9	Tollymore Forest Park	Down	10	7	

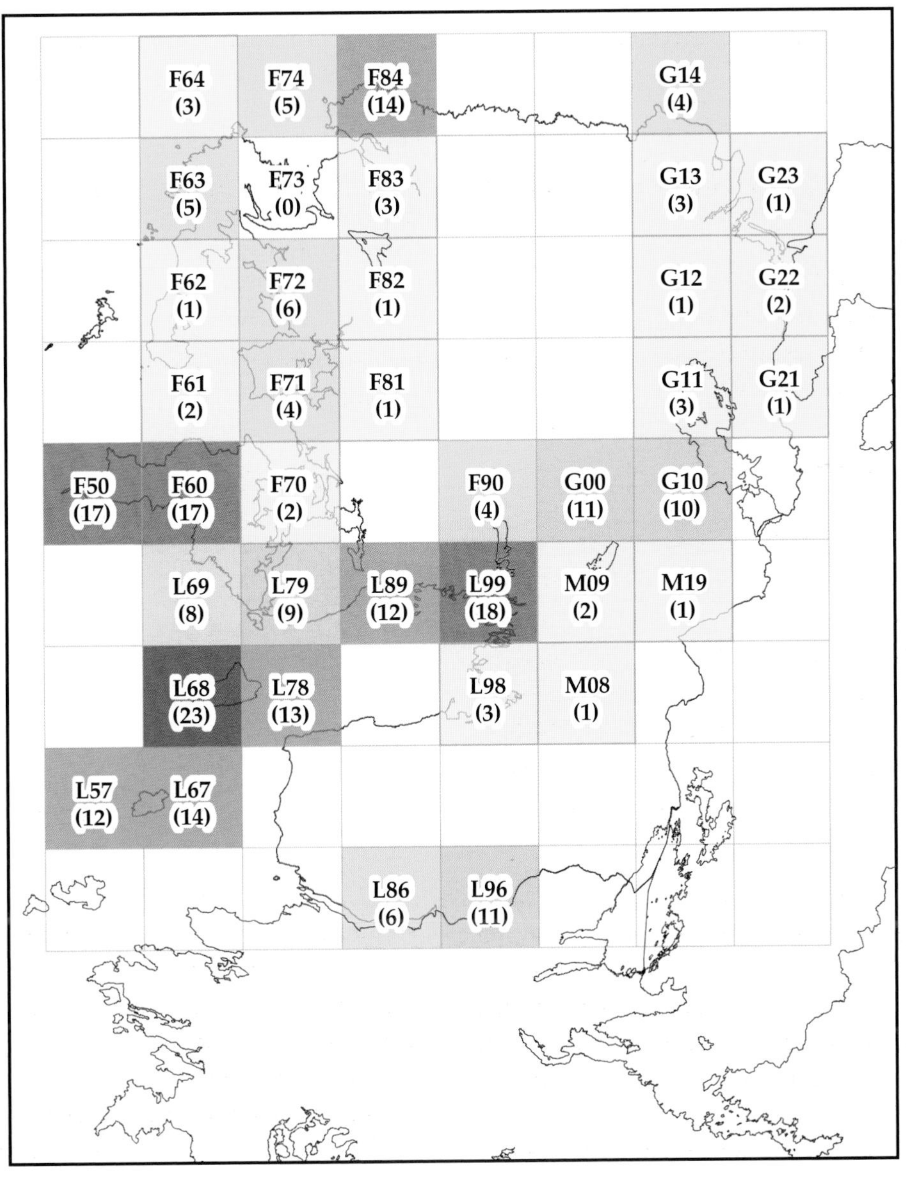

Fig. 5 10km squares surveyed with number of species of *Hygrocybe* recorded.

of oats, 1 acre of barley, 10.75 acres of rye and 124 acres of potatoes (Mac Cárthaigh 1999). In 2000, there was a mere five acres of oats grown (CSO 2012), so the conversion of the land under tillage into pasture and the grassing over and 'fixation' of the lazy beds has been ongoing for many years.

The farm buildings were built at the base of the slope where the tillage ceased and the commonage began. Sheep, cattle and horses grazed the uplands, although in the most heavily populated times lazy beds were created in the higher land, but these rarely lasted for long. This agricultural pattern can still be seen today, with the wall built by the Congested Districts Board dividing the tillage from the commonage (Collins 1999).

Even in 1897, the board inspector was worried about overgrazing on Clare Island. This was in a year, according to the Central Statistics Office, when there were 3246 sheep and 479 cattle of varying ages. In 2000, there were 52 cattle and 5615 sheep, showing the pressure that the commonage is under.

The addition of fertilisers

The Congested Districts Board noted in 1892 that the only fertilisers used were farm manure, shellfish and seaweed (Collins 1999). Fig. 6 shows that for the whole of Ireland, the use of synthetic fertilisers has increased by over fifteen times, when the consumption figures are compared to the 1961 levels (International Fertilizer Industry Association 2012). In the UK, the increase is thought to be as large as 60 times if compared to the 1940 figures (Griffith 1992). It is possible that fertiliser use was not as large on Clare Island, as the almost ubiquitous fruiting of *Hygrocybe* spp even on the old tillage land shows that levels are not deleterious.

Aerial pollution is thought to be another source of artificially high nitrogen levels in the soil (Arnolds 1988), but this is unlikely to have been significant given Clare Island's geographical position.

In England, St Dunstan's Farm is the best site in Sussex, with a CHEG score of C: 8; H: 30; E: 21; G: 1. A number of the richest fields have all been ploughed and even fertilised, although the most recent ploughing was in 1985. Then part of the Big Field was treated with grassland establishment fertiliser (5–10–5 NPK) and limed at three tons per acre (Russell 2005). This was not repeated so it seems that if fertilisers are not added over a number of years that grassland fungi can

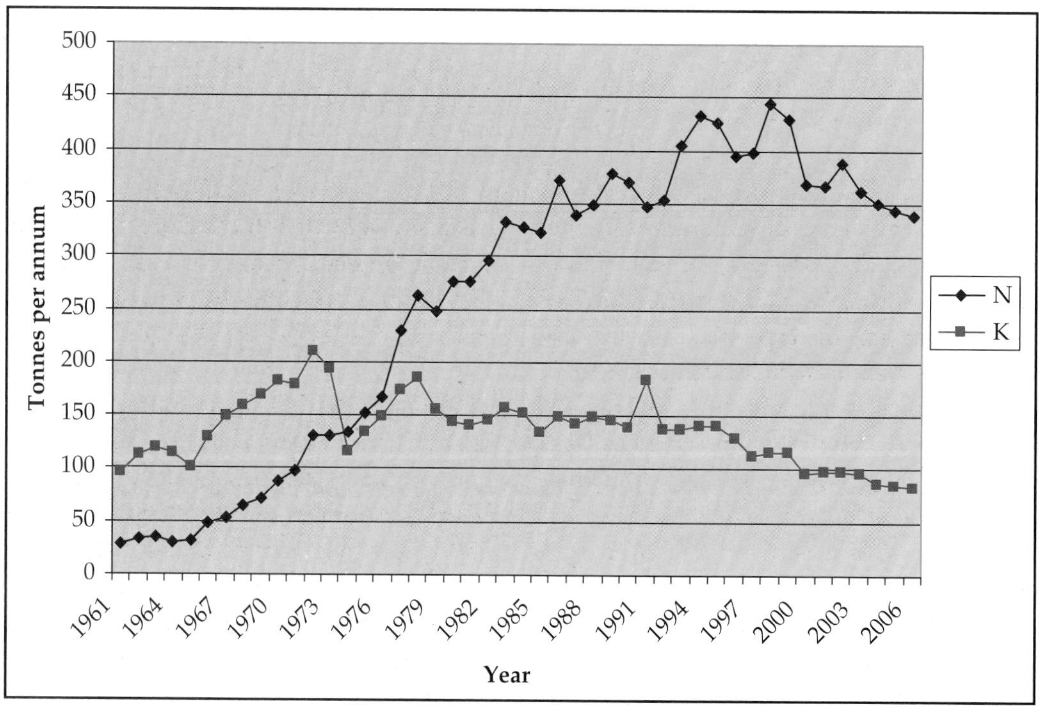

Fig. 6 Annual fertiliser consumption in Ireland (International Fertilizer Industry Association 2012).

recover in a relatively short time scale. However, if significant amounts of fertilisers are added over a number of years, recovery is very slow and can take decades (Arnolds 1994; Bardgett and McAlister 1999; Griffith *et al.* 2002).

Physical disturbance

The area under lazy beds was extremely disturbed land. The beds were being continually worked and rebuilt, and it is hard to imagine this being an area conducive for grassland fungi. However, as the beds are now disused and have become grassed over, it was noted that they were a particularly good place to search for waxcaps. This is likely because the lazy beds are well drained and fruiting is often noted on areas with better drainage (Pl. IX). Similar observations have been made throughout mainland County Mayo.

It has been noted that many waxcap sites are characterised by past disturbance (Griffith *et al.* 2006). Reservoir embankments (e.g. Silent Valley, Co. Down, and the Llanishen and Lisvane Reservoirs in Cardiff, Wales) can be very good for waxcaps, but on sites like this the disturbance was so long ago that it is irrelevant and the improved drainage of the embankments is actually the key factor.

Near Furnace in Mayo it was noted that waxcaps were found along side the road that traversed a bog (Pl. X). The bog itself is inhospitable for waxcaps, but on the grassy roadsides with good drainage waxcaps fruited well.

The key aspects to these sites is the time elapsed since the disturbance and the time taken for for waxcaps to colonise sites after the disturbance ceases. The presence of surrounding natural grassland with waxcaps is likely to play an important role in recolonisation and on sites like Clare Island and Inishturk, the neighbouring commonage is likely to have played an important role. It was noted in my own garden in Bangor, Co. Down, when concrete slabs that had covered the garden for over twenty years were lifted and grass resown, that *H. conica* appeared within two years and *Geoglossum fallax* and *Clavaria acuta* within three years, but this was in an area surrounded by (small) gardens with species like *H. calyptriformis* and *Entoloma porphyrophaeum*.

Pl. IX Lazy beds on Inishturk. Lazy beds are often important sites for waxcaps.

Pl. X Doontrusk near Furnace Lough. Often the sides of roads can be good for waxcaps as the hardcore on top of which the road is built becomes grassed over and remains well drained. The surrounding bog is of little interest for waxcaps.

The example of St Dunstan's Farm above also seems to show that if there are surrounding fields rich in grassland fungi, that, as long as the disturbance does not continue, waxcap grasslands can recover, possibly even quite quickly.

Inappropriate grazing or cutting regime

Grassland fungi *appear* not to be adversely affected by overgrazing. Fruiting is often good on extremely short turf and reduced in rank grassland. The latter, however, maybe because the fungi do not receive the light triggers required to initiate fruiting or perhaps for microclimatic reasons while actually, underground, the mycelia are unaffected. Studies in the US on prairie grasslands found many more species using DNA soil sampling than by collecting fruiting bodies (Griffith and Roderick 2008). In Europe, if grasslands become rank and a thick thatch develops, nutrients may build up in the soil as dead grass rots and this would not be beneficial, especially if nitrogen is being added as fertiliser. On some overgrazed areas, urine from grazing animals like sheep has been shown to negatively affect fungal mycelia (Rooney and Clipson 2009), but on such a dispersed grazing system as Clare Island this is unlikely to have anything but a very localised effect around supplementary feeding areas.

Trampling

Trampling has been reported as a factor affecting fruiting patterns (Griffith *et al.* 2006), but this was on an experimental plot where trampling was concentrated and is unlikely to be an issue on Clare Island. Localised poaching around gates, favoured sheltering spots or supplementary feeding sites will have a negative affect.

Soil type or drainage?

Now that the Environmental Protection Agency have made the National Soils Database available over the internet (https://maps.epa.ie), an analysis of the three recent waxcap surveys funded by the Heritage Council against soil type is now possible. Table 15 shows the different soil types mapped against number of grassland fungi records.

From this, the preference for better-drained mineral soils compared to the wetter gleys or blanket peats is marked. However, national soil datasets are relatively broad scale and do not take the local complexities of soils into account, and this is the scale at which fungal mycelia operate. For instance, the slopes of Ben Creggan and the Devil's Mother in west Mayo would appear to be possible sites from these broad categories, but in reality the sward contained significant amounts of *Sphagnum* spp, leading to a wetter surface layer and no grassland fungi. Fruiting in these sites was restricted to the well-drained stream banks (pers. obs.). However, with these limitations in mind, such maps can help target possible new sites.

The soils of Clare Island are notably very acid and form an extremely intricate pattern that reflects the complex topology. Rankers (lithomorphic soils) dominate on the steep to very steep ground and gleys on the sloping to moderately steep slopes, while peats are dominant on the flat to gentle sloping areas (Vullings 2000).

The best areas for grassland fungi found on 1–3 November 2008 were in Vulling's landform units 4.2, 4.3, 4.5, 6.1 and 6.2, although fruiting was scattered in many other landform units.

Table 15

National Soil Database soil categories and number of grassland fungi records from the Co. Clare (2006), west Cork (2007) and west Mayo (2008) surveys

IFS SOIL type	Description	No. records
AminSRPT	Podzols – Peaty	423
AminDW	Acid Brown Earths – Brown Podzolics	131
BminSW	Renzinas / Lithosols, Basic	128
AminSW	Lithosols / Regosols, Acidic	103
MarSands	Beach sands and gravels	86
AminPD	Surface Water Gleys, Ground Water Gleys, Acidic	85
Made	Man-made soils	43
AeoUND	Aeolian undifferentiated	41
AminPDPT	Peaty Gleys, Acidic	24
BktPt	Blanket peat	23
BminDW	Grey Brown Podzolic Brown Earths, Basic	15
AminSP	Shallow Surface or Ground Water Gleys, Acidic	14
BminPD	Surface Water Gleys, Ground Water Gleys, Basic	6
BminSP	Shallow Surface or Ground Water Gleys, Basic	4
AlluvMIN	Mineral Alluvium	4

The lazy bed areas were only quickly surveyed and were shown to be good, but lack of time prevented a more detailed search. Landform unit 4.2 was mapped as dominated by peaty podzolic ranker, 4.3 mainly as shallow brown earths and 4.5 as mainly peaty podzolic rankers. 6.1 and 6.2 have very intricate soil patterns with the steeper slopes being better drained, and on the less steep slopes of 6.2, where gleys are more important, waxcap fruiting was more common on the old lazy beds and earth banks.

It is not possible to exactly match the fungal fruiting locations to soil type from a desk post-survey due to the resolution of the data. In reality this has to be done in the field. However, the theme emerging is that drainage is the critical factor. On the better drained soils, grassland fungi fruit prolifically and in less well drained soils it is the areas modified by man by the construction of lazy beds or earth banks marking field boundaries that favour fruiting.

This pattern is common. On Inishturk, on the mountain commonage, in between the rock outcrops, peat dominates and conditions become too wet for waxcaps. However, where the turf reeks (locally called turf clamps—rock bases upon which turf was stacked for drying) were built, the soils at the base, being built up, were slightly better draining, and this is where waxcaps were found (Pl. XI).

Slievenacloy in the Belfast Hills is now the best site in Northern Ireland for waxcaps (Table 11). The fields above the Stoneyford River are very wet and fruiting is often exclusively found on earth banks marking the field boundaries.

Pl. XI Inishturk, with turf clamps marked.

Grassland fungi seem to be tolerant of low acid soils, although once they become extremely acid most species disappear and *H. laeta* begins to dominate. A walk up from the beach at Keem Bay on Achill Island was very interesting. The sandy soils beside the beach were carpeted in *Geoglossum cookeanum*. The steeper slopes above the beach harboured a greater diversity of species, and the field below the old coastguard station was particularly rich. However, the higher you walked up this field, the more the diversity dropped off until it was only *H. laeta* fruiting. The field merged into bog as pH dropped. These are only observations and not backed up by soil pH measurements, and it would be very interesting to investigate these areas further.

Conclusion

This grassland fungi survey of Clare Island was extremely short, being just one weekend. The island is an exceptional site and is the second best grassland fungi site known in Ireland to date. It easily qualifies for consideration as a Site of Special Scientific Interest using criteria applied in the UK (Genney *et al.* 2009) and as it has more than 22 species of *Hygrocybe*, it is of international importance (Vesterholt *et al.* 1999). Grassland fungi are thus probably the most important nature conservation feature on Clare Island, and parts of the island merit consideration for statutory site protection.

The detailed knowledge of the land-use history, soils and vegetation of the island mean that this is an excellent site for further research. Full surveys over longer time periods from mid-September to mid-November and over different years would mean more complete species lists, especially of the Entolomataceae, as their peak fruiting period was missed. In addition, the fungal hotspots should be identified, as detailed mapping can help derive appropriate management prescriptions for the key areas. There are a number of questions that could be looked at:

- Why is there such a difference in the number of species of Clavariaceae recorded in recent years compared to the original survey?
- What are the differences in fruiting patterns between the commonage and the old tillage fields? What can this tell us about the recolonisation potential for grassland fungi?
- What can a detailed analysis of fruiting patterns against soil type and soil pH show?

The Fungal Conservation Forum's very attractive leaflet for landowners on grassland fungi (Fungal Conservation Forum 2001) contains the following management guidelines for grassland fungi:

- Keep your grassland well grazed or mown so that the turf is short. Remove clippings wherever possible. Regular cutting does not appear to damage the fungi below ground, but if you want to see what you have, cut less in autumn to allow fruiting.
- Maintain existing field drainage systems where appropriate.
- Avoid fertilisers if possible, as they damage grassland fungi.
- Avoid the use of fungicides or use them sparingly, as they may inadvertently kill useful fungi or fungi you never intended to control.
- Avoid using moss killers since these fungi may form intimate relationships with mosses and may even depend on them.
- Avoid lime or apply it with caution since it may damage fungi.

Acknowledgements

Thanks must go to the Heritage Council, who funded the waxcap surveys of County Clare (2006), west Cork (2007) and west Mayo (2008). As the visit to Clare Island was extra and voluntary, thanks are extended to Hubert Fuller, Kieran Connolly and Jolanda Mitchel for giving their time as well and to Roy Anderson and Graham Wilson for giving their time on the aborted first attempt to get to the island.

REFERENCES

Adamcík, S. and Kautmanová, I. 2005 *Hygrocybe* species as indicators of natural value of grasslands in Slovakia. *Catathelasma* **6**, 24–34.

Agri-environment Monitoring Unit 2004 Monitoring of the chough option in the Antrim Coast, Glens and Rathlin Environmentally Sensitive Area 1998–2002. Belfast. Queen's University Belfast.

Arnolds, E. 1980 De oecologie en Sociologie van Wasplaten (*Hygrophorus* subgenus *Hygrocybe sensu lato*). *Natura* **77**, 17–44.

Arnolds, E. 1988 The changing macromycete flora in the Netherlands. *Transactions of the British Mycological Society* **90**, 391–406.

Arnolds, E. 1994 Paddestoelen en graslandbeheer. In T. Kuyper (ed.), *Paddestoelen en natuurbeheer: wat kan de beheerder?* 74–89. Hoogwoud, Holland. Wetenshappelijke Mededeling KNNV.

Arnolds, E. 1995 Conservation and management of natural populations of edible fungi. *Canadian Journal of Botany* **73**, S987–S998.

Bailey, J.S. 1994 Nutrient balance: the key to solving the phosphate problem. *Topics, Journal of the Milk Marketing Board for Northern Ireland* **16–17**.

Bardgett, R.D. and McAlister, E. 1999 The measurement of soil fungal: bacterial biomass ratios as an indicator of ecosystem self-regulation in temperate meadow grasslands. *Biology and Fertility of Soils* **29**, 282–90.

Boertmann, D. 1995 *The genus Hygrocybe*. Copenhagen. The Danish Mycological society.

Central Statistics Office 2012 Available at www.cso.ie (last accessed on 19 July 2012).

Collins, T. 1999 The Clare Island Survey of 1909–11: participants, papers and progress. *New Survey of Clare Island. Volume 1: history and cultural landscape*, 1–40. Dublin. Royal Irish Academy.

Dahlberg, A. and Croneborg, H. 2003 *33 threatened fungi in Europe: complementary and revised information on candidates for listing in Appendix 1 of the Bern Conventtion, European Council for the Conservation of fungi*. Uppsala. Swedish Environmental Protection Agency and European Council for Conservation of Fungi.

Evans, S. 2004 *Waxcap-grasslands—an assessment of English sites*. English Nature Research Reports. Report no 555. Peterborough. English Nature.

Feehan, J. and McHugh, R. 1992 The Curragh of Kildare as a *Hygrocybe* grassland. *Irish Naturalists' Journal* **24**, 13–17.

Fungal Conservation Forum 2001 *Managing your land with fungi in mind*. Available at http://www.plantlife.org.uk/uploads/documents/management-guide-Managing-land-Fungi.pdf (last accessed on 19 July 2012).

Genney, D.R., Hale, A.D., Woods, R.G., and Wright, M.W. 2009 Chapter 20 Grassland fungi. In *Guidelines for selection of biological SSSIs Rationale Operational approach and criteria: detailed guidelines for habitats and species groups*. Peterborough. JNCC.

Griffith, G.W. and Roderick, K. (eds) 2008 *Saprotrophic Basidiomycetes in grasslands: distribution and function*. London. Elsevier Ltd.

Griffith, G.W., Easton, G.L. and Jones, A.W. 2002 Ecology and diversity of waxcap (*Hygrocybe* spp.) fungi. *Botanical Journal of Scotland* **54**, 7–22.

Griffith, G.W., Bratton, J.L. and Easton, G.L. 2004 Charismatic megafungi: the conservation of waxcap grasslands. *British Wildlife* **15**, 31–43.

Griffith, G.W., Holden, L., Mitchel, D., Evans, D.E., Aron, C., Evans, S. and Graham, A. 2006 Mycological survey of selected semi-natural grasslands in Wales. Unpublished report for the Countryside Council for Wales.

Habitas 2012 Northern Ireland Priority Species List. Available at http://www.habitas.org.uk/priority/ (last accessed 19 July 2012).

International Fertilizer Industry Association 2012 Available at http://www.fertilizer.org/HomePage/STATISTICS (last accessed 23 July 2012).

Jordal, J.B. 1997 *Sopp i naturbeitemarker i Norge. En kunnskapsstatus over utbredelse, okologi, indikatorverdi og trusler i et europeisk perspektiv*. Direktoratet for naturforvaltning, Trondheim.

Knudsen H. and Vesterholt, J. 2008 *Funga Nordica*. Copenhagen. Nordsvamp.

Legon, N.W. and Henrici, A. 2005 *Checklist of the British and Irish Basidiomycota*. London. Royal Botanic Gardens Kew.

MacCárthaigh, C. 1999 Clare Island folklife. In C. MacCárthaigh and K. Whelan (eds), *New survey of Clare Island. Volume 1: history and cultural landscape*, 41–72. Dublin. Royal Irish Academy. pp

MacCárthaigh, C. and Whelan, K. (eds) 1999 *New survey of Clare Island. Volume 1: history and cultural landscape*. Dublin. Royal Irish Academy.

Marren, P. 1998 Fungal flowers: the waxcaps and their world. *British Wildlife* **9**, 164–72.

McHugh, R., Mitchel, D., Wright, M. and Anderson, R. 2001 The fungi of Irish grasslands and their value for nature conservation. *Biology and Environment: Proceedings of the Royal Irish Academy* **101B**, 225–42.

Met Éireann 2012 Monthly values for Belmullet. Available at http://www.met.ie/climate/monthly-data.asp?Num=76 (last accessed 23 July 2012).

Mitchel, D. 2006 Survey of the grassland fungi of County Clare. Unpublished report for the Heritage Council.

Mitchel, D. 2007 Survey of the grassland fungi of the vice county of west Cork. Unpublished report for the Heritage Council.

Mitchel, D. 2008 Survey of the grassland fungi of the vice county of west Mayo. Unpublished report for the Heritage Council.

Newton, A.C., Davy, L.M., Holden, E., Silverside, A., Watling, R. and Ward, S.D. 2002 Status, distribution and definition of mycologically important grasslands in Scotland. *Biological Conservation* **111** (1), 11–23.

NIFG 2012a Available at www.nifg.org.uk/waxcaps.htm (last accessed on 19 July 2012).

NIFG 2012b Available at http://www.nifg.org.uk/home.htm (last accessed on 19 July 2012).

Nitare, J. 1988 Jordtungor, en svampgrupp pa tillbakagang i naturliga fodermarker. *Svensk Botanisk Tidskrift* **82** (5), 485–89.

Öster, M. 2008 Low congruence between the diversity of waxcaps (*Hygrocybe* spp.) fungi and vascular plants in semi-natural grasslands. *Basic and Applied Ecology* **9**, 514–22.

Rald, E. 1985 Vokshatte som indikatorarter for mykologisk vaerdifulde overdrevslokaliteter. *Svampe* **11**, 1–9.

Rea, C. and Hawley, H.C. 1912 Clare Island Survey. Part 13 Fungi. *Proceedings of the Royal Irish Academy* **31**, 1–13.

Ridge, I. 1997 *Simplified key to Geoglossum*. North West Fungus Group.

Rooney, D.C. and Clipson, N.J.W. 2009 Synthetic sheep urine alters fungal community structure in an upland grassland soil. *Fungal Ecology* **2**, 36–43.

Rothero, G.P. 1988 The summer meeting, 1987, Co. Mayo. Second week at Westport: 12–18 August. *Bulletin of the British Bryological Society* **51**, 12–15.

Rotheroe, M. 1999 *Mycological survey of selected semi-natural grasslands in Carmarthenshire*. Countryside Council for Wales.

Rotheroe, M., Newton, A., Evans, S. and Feehan, J. 1996 Waxcap-grassland survey. *Mycologist* **10**, 23–5.

Russell, P. 2005 Grassland fungi and the management history of St. Dunstan's Farm. *Field Mycology* **6**, 85–91.

Spooner, B. 1998 Keys to the British Geoglossaceae (draft). Unpublished.

ten Cate, R.S. 1993 Clare Island Survey en paddestoelen in het westen van Ierland. *In-Nuachta* **IX**, 14–20.

Vesterholt, J., Boertmann, D. and Tranberg, H. 1999 1998—et usaedvanlig godt ar for overdrevssvampe. *Svampe* **40**, 36–44.

Vullings, L.A.E. 2000 Soil variability and pedogenetic trends, Clare Island, Co. Mayo. Unpublished report, Department of Crop Science, Horticulture and Forestry, University College Dublin.

Vullings, W., Collins, J.F. and Smillie, G. 2013 Soils and soil associations on Clare Island. New Survey of Clare Island. Dublin. Royal Irish Academy.

APPENDIX 1

Identification literature

The literature used to identify the grassland target groups were as follows:

Camaropyllopsis
Bas, C., Kuyper, T.H.W., Noordeloos, M.E. and Vellinga, E.C. (eds) 1990 *Flora Agaracina Neerlandica vol. 2*. Leiden. A.A. Balkema.

Hygrocybe
Boertmann, D. 1995 *The Genus Hygrocybe*. Fungi of Northern Europe vol. I. Copenhagen. Danish Mycological Society.

Lavariaceae
Henrici, A. 1997 Keys to British Lavariaceae. Privately circulated.

Entolomaceae
Noordeloos, M.E. 1992 *Entoloma, s.l.* Fungi Europaei vol. 5. Alassio, Italy. Libreria Mykoflora.

Vesterholt, J. 2002 *Contribution to the knowledge of species of Entoloma subgenus Leptonia*. Fungi non delineati vol. 21. Alassio, Italy. Edizioni Candusso

Geoglossaceae
Spooner, B. 1998 Keys to the British Geoglossaceae (draft). Privately circulated.

Dermoloma and Porpoloma
Watling, R. and Turnbull, E. 1998 *Cantharellaceae, Gomphaceae and Amyloid and Xeruloid members of the Tricholomataceae*. British Fungus Flora vol. 8. Edinburgh. Royal Botanic Gardens.

APPENDIX 2

Original Clare Island Survey records of fungi

This list of records is taken from Rea (1912). The Original Name column is the name given by Rea and Hawley, while the Current Name column is the modern name for this species according to the British Mycological Society's GBCHKLST database dated September 2009.

Locality	Original name	Current name
Achill Island	*Armillaria mellea*	*Armillaria mellea*
Achill Island	*Boletus bovinus*	*Suillus bovinus*
Achill Island	*Boletus luteus*	*Suillus luteus*
Achill Island	*Cenangium abietis*	*Cenangium ferruginosum*
Achill Island	*Chlorosplenium aeruginosum*	*Chlorociboria aeruginascens*
Achill Island	*Clavaria contorta*	*Macrotyphula fistulosa* var. *contorta*
Achill Island	*Clavaria cristata*	*Clavulina coralloides*
Achill Island	*Clavaria dissipabilis*	*Clavulinopsis helvola*
Achill Island	*Clavaria fistulosa*	*Macrotyphula fistulosa* var. *fistulosa*
Achill Island	*Coleosporium euphrasiae*	*Coleosporium tussilaginis*
Achill Island	*Collybia conigena*	*Collybia conigena*
Achill Island	*Collybia tenacella*	*Strobilurus tenacellus*
Achill Island	*Coprinus micaceus*	*Coprinellus truncorum*
Achill Island	*Coprinus niveus*	*Coprinopsis nivea*
Achill Island	*Corticium laeve*	*Cylindrobasidium laeve*
Achill Island	*Cortinarius caninus*	*Cortinarius caninus*
Achill Island	*Cortinarius cinnamomeus*	*Cortinarius cinnamomeus*
Achill Island	*Cortinarius decipiens*	*Cortinarius decipiens* var. *decipiens*
Achill Island	*Cortinarius decolorans*	*Cortinarius xanthocephalus*
Achill Island	*Cortinarius elatior*	*Cortinarius livido-ochraceus*
Achill Island	*Cortinarius leucopus*	*Cortinarius leucopus*
Achill Island	*Cortinarius miltinus*	*Cortinarius purpureus*
Achill Island	*Cortinarius paleaceus*	*Cortinarius flexipes* var. *flabellus*
Achill Island	*Cytospora salicis*	*Cytospora salicis*
Achill Island	*Diaporthe salicella*	*Cryptodiaporthe salicella*
Achill Island	*Diaporthe wibbei*	*Cryptodiaporthe aubertii*
Achill Island	*Diatrype stigma*	*Diatrype stigma*
Achill Island	*Ditiola nuda*	*Dacrymyces capitatus*
Achill Island	*Entoloma sericeum*	*Entoloma sericeum* var. *sericeum*
Achill Island	*Flammula scamba*	*Pholiota scamba*
Achill Island	*Fomes annosus*	*Heterobasidion annosum*
Achill Island	*Gloniopsis muelleri*	*Gloniopsis praelonga*
Achill Island	*Gnomonia cerastis*	*Gnomonia cerastis*
Achill Island	*Helminthosporium rhopaloides*	*Helminthosporium rhopaloides*
Achill Island	*Helotium citrinum*	*Bisporella citrina*
Achill Island	*Hyaloscypha hyalina*	*Hyaloscypha hyalina*
Achill Island	*Hydnum alutaceum*	*Hyphodontia alutacea*
Achill Island	*Hymenoscyphus dumorum*	*Lachnum dumorum*
Achill Island	*Hypholoma capnoides*	*Hypholoma capnoides*
Achill Island	*Hypholoma epixanthum*	*Hypholoma epixanthum*

(*Continued*)

Appendix 2 *(Continued)*

Locality	Original name	Current name
Achill Island	*Hypholoma fasciculare*	*Hypholoma fasciculare* var. *fasciculare*
Achill Island	*Irpex obliquus*	*Schizopora paradoxa*
Achill Island	*Laccaria laccata*	*Laccaria laccata*
Achill Island	*Laccaria laccata* var. *amethystea*	*Laccaria amethystina*
Achill Island	*Lachnum niveum*	*Lachnum niveum*
Achill Island	*Lactarius cimicarius*	*Lactarius subumbonatus*
Achill Island	*Lactarius deliciosus*	*Lactarius deliciosus*
Achill Island	*Lactarius pubescens*	*Lactarius pubescens*
Achill Island	*Lactarius quietus*	*Lactarius quietus*
Achill Island	*Lactarius rufus*	*Lactarius rufus*
Achill Island	*Lactarius subdulcis*	*Lactarius subdulcis*
Achill Island	*Lactarius theiogalus*	*Lactarius hepaticus*
Achill Island	*Lactarius turpis*	*Lactarius turpis*
Achill Island	*Lophodermium pinastri*	*Lophodermium pinastri*
Achill Island	*Lycoperdon spadiceum*	*Lycoperdon lividum*
Achill Island	*Mollisia lignicola*	*Mollisia ligni*
Achill Island	*Mycena ammoniaca*	*Mycena leptocephala*
Achill Island	*Mycena elegans*	*Mycena aurantiomarginata*
Achill Island	*Mycena filopes*	*Mycena vitilis*
Achill Island	*Mycena galericulata*	*Mycena galericulata*
Achill Island	*Mycena galopus*	*Mycena galopus* var. *galopus*
Achill Island	*Mycena pura*	*Mycena pura*
Achill Island	*Mycosphaerella ascophylli*	*Mycosphaerella ascophylli*
Achill Island	*Naucoria escharioides*	*Naucoria escharioides*
Achill Island	*Nolanea pascua*	*Entoloma pascuum*
Achill Island	*Nyctalis parasitica*	*Asterophora parasitica*
Achill Island	*Omphalia umbellifera*	*Lichenomphalia umbellifera*
Achill Island	*Panaeolus campanulatus*	*Panaeolus papilionaceus* var. *papilionaceus*
Achill Island	*Patellaria atrata*	*Lecanidion atratum*
Achill Island	*Paxillus involutus*	*Paxillus involutus*
Achill Island	*Peniophora quercina*	*Peniophora quercina*
Achill Island	*Peniophora velutina*	*Phanerochaete velutina*
Achill Island	*Pluteus cervinus*	*Pluteus cervinus*
Achill Island	*Poria mollusca*	*Trechispora mollusca*
Achill Island	*Poria sanguinolenta*	*Physisporinus sanguinolentus*
Achill Island	*Poria terrestris*	*Byssocorticium terrestre*
Achill Island	*Psalliota arvensis*	*Agaricus arvensis*
Achill Island	*Psathyra fibrillosa*	*Psathyrella friesii*
Achill Island	*Psathyrella gracilis*	*Psathyrella corrugis*
Achill Island	*Pseudopeziza petiolaris*	*Pyrenopeziza petiolaris*
Achill Island	*Psilocybe ericaea*	*Hypholoma ericaeum*
Achill Island	*Psilocybe semilanceata*	*Psilocybe semilanceata*
Achill Island	*Psilocybe uda*	*Hypholoma udum*
Achill Island	*Puccinia umbilici*	*Puccinia umbilici*
Achill Island	*Rhizopogon luteolus*	*Rhizopogon ochraceorubens*

(Continued)

Appendix 2 *(Continued)*

Locality	Original name	Current name
Achill Island	*Rhytisma acerinum*	*Rhytisma acerinum*
Achill Island	*Russula adusta*	*Russula adusta*
Achill Island	*Russula caerulea*	*Russula caerulea*
Achill Island	*Russula depallens*	*Russula exalbicans*
Achill Island	*Russula drimeia*	*Russula sardonia*
Achill Island	*Russula fragilis*	*Russula silvestris*
Achill Island	*Russula fragilis* var. *violascens*	*Russula fragilis* var. *fragilis*
Achill Island	*Russula ochroleuca*	*Russula ochroleuca*
Achill Island	*Russula puellaris*	*Russula puellaris*
Achill Island	*Septoria scabiosicola*	*Septoria scabiosicola*
Achill Island	*Stereum rugosum*	*Stereum rugosum*
Achill Island	*Stropharia aeruginosa*	*Stropharia aeruginosa*
Achill Island	*Stropharia semiglobata*	*Stropharia semiglobata*
Achill Island	*Tricholoma cerinum*	*Calocybe chrysenteron*
Achill Island	*Tricholoma flavobrunneum*	*Tricholoma fulvum*
Achill Island	*Tricholoma rutilans*	*Tricholomopsis rutilans*
Achill Island	*Tubaria furfuracea*	*Tubaria furfuracea* var. *furfuracea*
Achill Island	*Typhula grevillei*	*Typhula setipes*
Achill Island	*Xylaria hypoxylon*	*Xylaria hypoxylon*
Belclare and Prospect House Woods	*Amanita muscaria*	*Amanita muscaria* var. *muscaria*
Belclare and Prospect House Woods	*Amanita rubescens*	*Amanita rubescens* var. *rubescens*
Belclare and Prospect House Woods	*Androsaceus rotula*	*Marasmius rotula*
Belclare and Prospect House Woods	*Armillaria mellea*	*Armillaria mellea*
Belclare and Prospect House Woods	*Ascobolus furfuraceus*	*Ascobolus stercorarius*
Belclare and Prospect House Woods	*Bispora monilioides*	*Bispora antennata*
Belclare and Prospect House Woods	*Boletus bovinus*	*Suillus bovinus*
Belclare and Prospect House Woods	*Boletus chrysenteron*	*Boletus chrysenteron*
Belclare and Prospect House Woods	*Boletus edulis*	*Boletus edulis*
Belclare and Prospect House Woods	*Boletus elegans*	*Suillus grevillei*
Belclare and Prospect House Woods	*Boletus luteus*	*Suillus luteus*
Belclare and Prospect House Woods	*Boletus scaber*	*Leccinum scabrum*
Belclare and Prospect House Woods	*Boletus scaber* var. *niveus*	*Boletus niveus*
Belclare and Prospect House Woods	*Bovista nigrescens*	*Bovista nigrescens*
Belclare and Prospect House Woods	*Cantharellus aurantiacus*	*Hygrophoropsis aurantiaca*
Belclare and Prospect House Woods	*Cantharellus cibarius*	*Cantharellus cibarius*
Belclare and Prospect House Woods	*Chlorosplenium aeruginosum*	*Chlorociboria aeruginascens*
Belclare and Prospect House Woods	*Clavaria acuta*	*Clavaria acuta*
Belclare and Prospect House Woods	*Clavaria cinerea*	*Clavulina cinerea f. cinerea*
Belclare and Prospect House Woods	*Clavaria dissipabilis*	*Clavulinopsis helvola*
Belclare and Prospect House Woods	*Clavaria fistulosa*	*Macrotyphula fistulosa* var. *fistulosa*
Belclare and Prospect House Woods	*Clavaria fumosa*	*Clavaria fumosa*
Belclare and Prospect House Woods	*Clavaria fusiformis*	*Clavulinopsis fusiformis*
Belclare and Prospect House Woods	*Clavaria kunzei*	*Ramariopsis kunzei*
Belclare and Prospect House Woods	*Clavaria muscoides*	*Clavulinopsis corniculata*
Belclare and Prospect House Woods	*Clavaria rugosa*	*Clavulina rugosa*

(Continued)

Appendix 2 *(Continued)*

Locality	Original name	Current name
Belclare and Prospect House Woods	*Clavaria vermicularis*	*Clavaria fragilis*
Belclare and Prospect House Woods	*Claviceps purpurea*	*Claviceps purpurea* var. *purpurea*
Belclare and Prospect House Woods	*Clitocybe fragrans*	*Clitocybe fragrans*
Belclare and Prospect House Woods	*Clitopilus undatus*	*Entoloma undatum*
Belclare and Prospect House Woods	*Collybia aquosa*	*Collybia aquosa*
Belclare and Prospect House Woods	*Coprinus atramentarius*	*Coprinopsis atramentaria*
Belclare and Prospect House Woods	*Coprinus friesii*	*Coprinopsis friesii*
Belclare and Prospect House Woods	*Coprinus micaceus*	*Coprinellus truncorum*
Belclare and Prospect House Woods	*Coprinus niveus*	*Coprinopsis nivea*
Belclare and Prospect House Woods	*Coprinus plicatilis*	*Parasola plicatilis*
Belclare and Prospect House Woods	*Coprinus radians*	*Coprinellus radians*
Belclare and Prospect House Woods	*Corticium arachnoideum*	*Athelia arachnoidea*
Belclare and Prospect House Woods	*Cortinarius acutus*	*Cortinarius acutus*
Belclare and Prospect House Woods	*Cortinarius alboviolaceus*	*Cortinarius alboviolaceus*
Belclare and Prospect House Woods	*Cortinarius anomalus*	*Cortinarius anomalus*
Belclare and Prospect House Woods	*Cortinarius biformis*	*Cortinarius tabacinus*
Belclare and Prospect House Woods	*Cortinarius bolaris*	*Cortinarius bolaris*
Belclare and Prospect House Woods	*Cortinarius bovinus*	*Cortinarius bulbosus*
Belclare and Prospect House Woods	*Cortinarius castaneus*	*Cortinarius castaneus*
Belclare and Prospect House Woods	*Cortinarius collinitus*	*Cortinarius collinitus*
Belclare and Prospect House Woods	*Cortinarius elatior*	*Cortinarius livido-ochraceus*
Belclare and Prospect House Woods	*Cortinarius erythrinus*	*Cortinarius vernus*
Belclare and Prospect House Woods	*Cortinarius fasciatus*	*Cortinarius fulvescens*
Belclare and Prospect House Woods	*Cortinarius hinnuleus*	*Cortinarius hinnuleus*
Belclare and Prospect House Woods	*Cortinarius iliopodius*	*Cortinarius parvannulatus*
Belclare and Prospect House Woods	*Cortinarius lepidopus*	*Cortinarius anomalus*
Belclare and Prospect House Woods	*Cortinarius leucopus*	*Cortinarius leucopus*
Belclare and Prospect House Woods	*Cortinarius paleaceus*	*Cortinarius flexipes* var. *flabellus*
Belclare and Prospect House Woods	*Cortinarius rigens*	*Cortinarius acetosus*
Belclare and Prospect House Woods	*Cortinarius tabularis*	*Cortinarius tabularis*
Belclare and Prospect House Woods	*Cortinarius torvus*	*Cortinarius torvus*
Belclare and Prospect House Woods	*Cortinarius varius*	*Cortinarius variiformis*
Belclare and Prospect House Woods	*Coryne urnalis*	*Ascocoryne cylichnium*
Belclare and Prospect House Woods	*Crepidotus calolepis*	*Crepidotus calolepis*
Belclare and Prospect House Woods	*Dacrymyces stillatus*	*Dacrymyces stillatus*
Belclare and Prospect House Woods	*Diaporthe exasperans*	*Diaporthe eres*
Belclare and Prospect House Woods	*Diaporthe tulasnei*	*Diaporthe tulasnei*
Belclare and Prospect House Woods	*Diatrypella favacea*	*Diatrypella favacea*
Belclare and Prospect House Woods	*Entoloma jubatum*	*Entoloma porphyrophaeum*
Belclare and Prospect House Woods	*Entoloma nidorosum*	*Entoloma politum*
Belclare and Prospect House Woods	*Entoloma prunuloides*	*Entoloma prunuloides*
Belclare and Prospect House Woods	*Entoloma sericeum*	*Entoloma sericeum* var. *sericeum*
Belclare and Prospect House Woods	*Exidia albida*	*Exidia albida*
Belclare and Prospect House Woods	*Exoascus turgidus*	*Taphrina betulina*
Belclare and Prospect House Woods	*Fomes annosus*	*Heterobasidion annosum*

(Continued)

Appendix 2 *(Continued)*

Locality	Original name	Current name
Belclare and Prospect House Woods	*Fomes connatus*	*Oxyporus populinus*
Belclare and Prospect House Woods	*Galera hypnorum*	*Galerina hypnorum*
Belclare and Prospect House Woods	*Galera spartea*	*Conocybe rickeniana*
Belclare and Prospect House Woods	*Grandinia granulosa*	*Dichostereum granulosum*
Belclare and Prospect House Woods	*Hebeloma fastibile*	*Hebeloma mesophaeum* var. *crassipes*
Belclare and Prospect House Woods	*Helotium herbarum*	*Calycina herbarum*
Belclare and Prospect House Woods	*Helvella helvelloides*	*Helvella pezizoides*
Belclare and Prospect House Woods	*Heterosphaeria patella*	*Heterosphaeria patella*
Belclare and Prospect House Woods	*Humaria granulata*	*Coprobia granulata*
Belclare and Prospect House Woods	*Hydnum repandum*	*Hydnum repandum*
Belclare and Prospect House Woods	*Hydnum sordidum*	*Leucogyrophana pinastri*
Belclare and Prospect House Woods	*Hygrophorus ceraceus*	*Hygrocybe ceracea*
Belclare and Prospect House Woods	*Hygrophorus chlorophanus*	*Hygrocybe chlorophana*
Belclare and Prospect House Woods	*Hygrophorus distans*	*Hygrophorus distans*
Belclare and Prospect House Woods	*Hygrophorus miniatus*	*Hygrocybe miniata*
Belclare and Prospect House Woods	*Hygrophorus niveus*	*Hygrocybe virginea* var. *virginea*
Belclare and Prospect House Woods	*Hygrophorus obrusseus*	*Hygrocybe obrussea*
Belclare and Prospect House Woods	*Hygrophorus ovinus*	*Hygrocybe ovina*
Belclare and Prospect House Woods	*Hygrophorus psittacinus*	*Hygrocybe psittacina* var. *psittacina*
Belclare and Prospect House Woods	*Hygrophorus puniceus*	*Hygrocybe punicea*
Belclare and Prospect House Woods	*Hygrophorus virgineus*	*Hygrocybe virginea* var. *virginea*
Belclare and Prospect House Woods	*Hygrophorus virgineus* var. *roseipes*	*Hygrocybe virginea* var. *virginea*
Belclare and Prospect House Woods	*Hypholoma candolleanum*	*Psathyrella candolleana*
Belclare and Prospect House Woods	*Hypholoma epixanthum*	*Hypholoma epixanthum*
Belclare and Prospect House Woods	*Hypholoma fasciculare*	*Hypholoma fasciculare* var. *fasciculare*
Belclare and Prospect House Woods	*Hypholoma sublateritium*	*Hypholoma lateritium*
Belclare and Prospect House Woods	*Hypoxylon multiforme*	*Hypoxylon multiforme*
Belclare and Prospect House Woods	*Hysterium angustatum*	*Hysterium angustatum*
Belclare and Prospect House Woods	*Hysterographium fraxini*	*Hysterographium fraxini*
Belclare and Prospect House Woods	*Inocybe destricta*	*Inocybe adaequata*
Belclare and Prospect House Woods	*Inocybe geophylla*	*Inocybe geophylla* var. *geophylla*
Belclare and Prospect House Woods	*Inocybe obscura*	*Inocybe cincinnata* var. *major*
Belclare and Prospect House Woods	*Inocybe rimosa*	*Inocybe rimosa*
Belclare and Prospect House Woods	*Irpex obliquus*	*Schizopora paradoxa*
Belclare and Prospect House Woods	*Laccaria laccata*	*Laccaria laccata*
Belclare and Prospect House Woods	*Lachnum ciliare*	*Incrucipulum ciliare*
Belclare and Prospect House Woods	*Lachnum niveum*	*Lachnum niveum*
Belclare and Prospect House Woods	*Lachnum virgineum*	*Lachnum virgineum*
Belclare and Prospect House Woods	*Lactarius blennius*	*Lactarius blennius*
Belclare and Prospect House Woods	*Lactarius chrysorrheus*	*Lactarius chrysorrheus*
Belclare and Prospect House Woods	*Lactarius glyciosmus*	*Lactarius glyciosmus*
Belclare and Prospect House Woods	*Lactarius mitissimus*	*Lactarius aurantiacus*
Belclare and Prospect House Woods	*Lactarius obnubilus*	*Lactarius obscuratus*
Belclare and Prospect House Woods	*Lactarius piperatus*	*Lactarius piperatus*
Belclare and Prospect House Woods	*Lactarius pubescens*	*Lactarius pubescens*

(Continued)

Appendix 2 *(Continued)*

Locality	Original name	Current name
Belclare and Prospect House Woods	*Lactarius pyrogalus*	*Lactarius pyrogalus*
Belclare and Prospect House Woods	*Lactarius quietus*	*Lactarius quietus*
Belclare and Prospect House Woods	*Lactarius serifluus*	*Lactarius subumbonatus*
Belclare and Prospect House Woods	*Lactarius subdulcis*	*Lactarius subdulcis*
Belclare and Prospect House Woods	*Lactarius torminosus*	*Lactarius torminosus*
Belclare and Prospect House Woods	*Lactarius turpis*	*Lactarius turpis*
Belclare and Prospect House Woods	*Lactarius vellereus*	*Lactarius vellereus*
Belclare and Prospect House Woods	*Lactarius vietus*	*Lactarius vietus*
Belclare and Prospect House Woods	*Leotia lubrica*	*Leotia lubrica*
Belclare and Prospect House Woods	*Lepiota acutesquamosa*	*Lepiota aspera*
Belclare and Prospect House Woods	*Lepiota gracilenta*	*Macrolepiota mastoidea*
Belclare and Prospect House Woods	*Lepiota granulosa*	*Cystoderma granulosum*
Belclare and Prospect House Woods	*Leptonia lampropus*	*Entoloma sodale*
Belclare and Prospect House Woods	*Leptonia sericella*	*Entoloma sericellum*
Belclare and Prospect House Woods	*Lycoperdon depressum*	*Vascellum pratense*
Belclare and Prospect House Woods	*Lycoperdon pyriforme*	*Lycoperdon pyriforme*
Belclare and Prospect House Woods	*Marasmius candidus*	*Marasmiellus vaillantii*
Belclare and Prospect House Woods	*Marasmius oreades*	*Marasmius oreades*
Belclare and Prospect House Woods	*Marasmius ramealis*	*Marasmiellus ramealis*
Belclare and Prospect House Woods	*Melampsoridium betulinum*	*Melampsoridium betulinum*
Belclare and Prospect House Woods	*Mollisia cinerea*	*Mollisia cinerea*
Belclare and Prospect House Woods	*Mycena ammoniaca*	*Mycena leptocephala*
Belclare and Prospect House Woods	*Mycena atrocyanea*	*Mycena atrocyanea*
Belclare and Prospect House Woods	*Mycena capillaris*	*Mycena capillaris*
Belclare and Prospect House Woods	*Mycena corticola*	*Mycena hiemalis*
Belclare and Prospect House Woods	*Mycena discopus*	*Mycena discopus*
Belclare and Prospect House Woods	*Mycena filopes*	*Mycena vitilis*
Belclare and Prospect House Woods	*Mycena galericulata*	*Mycena galericulata*
Belclare and Prospect House Woods	*Mycena galopus*	*Mycena galopus* var. *galopus*
Belclare and Prospect House Woods	*Mycena hiemalis*	*Mycena olida*
Belclare and Prospect House Woods	*Mycena leucogala*	*Mycena galopus* var. *nigra*
Belclare and Prospect House Woods	*Mycena metata*	*Mycena leptocephala*
Belclare and Prospect House Woods	*Mycena polygramma*	*Mycena polygramma*
Belclare and Prospect House Woods	*Mycena pterigena*	*Mycena pterigena*
Belclare and Prospect House Woods	*Mycena pura*	*Mycena pura*
Belclare and Prospect House Woods	*Mycena rorida*	*Mycena rorida*
Belclare and Prospect House Woods	*Mycena rugosa*	*Mycena galericulata*
Belclare and Prospect House Woods	*Mycena tenerrima*	*Mycena adscendens*
Belclare and Prospect House Woods	*Mycena vitilis*	*Mycena filopes*
Belclare and Prospect House Woods	*Naucoria semiorbicularis*	*Agrocybe pediades*
Belclare and Prospect House Woods	*Naucoria tabacina*	*Agrocybe tabacina*
Belclare and Prospect House Woods	*Naucoria temulenta*	*Agrocybe pediades*
Belclare and Prospect House Woods	*Nolanea exilis*	*Entoloma exile*
Belclare and Prospect House Woods	*Nolanea rufocarnea*	*Entoloma rufocarneum*
Belclare and Prospect House Woods	*Nyctalis parasitica*	*Asterophora parasitica*

(Continued)

Appendix 2 *(Continued)*

Locality	Original name	Current name
Belclare and Prospect House Woods	*Omphalia fibula*	*Rickenella fibula*
Belclare and Prospect House Woods	*Omphalia integrella*	*Delicatula integrella*
Belclare and Prospect House Woods	*Omphalia umbellifera*	*Lichenomphalia umbellifera*
Belclare and Prospect House Woods	*Otidea leporina*	*Otidea leporina*
Belclare and Prospect House Woods	*Panaeolus campanulatus*	*Panaeolus papilionaceus* var. *papilionaceus*
Belclare and Prospect House Woods	*Penicillium candidum*	*Penicillium candidum*
Belclare and Prospect House Woods	*Peniophora quercina*	*Peniophora quercina*
Belclare and Prospect House Woods	*Peronospora calotheca*	*Peronospora calotheca*
Belclare and Prospect House Woods	*Peziza succosa*	*Peziza succosa*
Belclare and Prospect House Woods	*Phlebia vaga*	*Phlebiella sulphurea*
Belclare and Prospect House Woods	*Pholiota marginata*	*Galerina marginata*
Belclare and Prospect House Woods	*Pholiota ombrophila*	*Agrocybe ombrophila*
Belclare and Prospect House Woods	*Pholiota squarrosa*	*Pholiota squarrosa*
Belclare and Prospect House Woods	*Phoma nebulosa*	*Phoma nebulosa*
Belclare and Prospect House Woods	*Phragmidium violaceum*	*Phragmidium violaceum*
Belclare and Prospect House Woods	*Phyllosticta ajugae*	*Phyllosticta ajugae*
Belclare and Prospect House Woods	*Phyllosticta primulicola*	*Phyllosticta primulicola*
Belclare and Prospect House Woods	*Phytophthora infestans*	*Phytophthora infestans*
Belclare and Prospect House Woods	*Pilobolus crystallinus*	*Pilobolus crystallinus* var. *crystallinus*
Belclare and Prospect House Woods	*Pistillaria puberula*	*Typhula quisquiliaris*
Belclare and Prospect House Woods	*Polyporus betulinus*	*Piptoporus betulinus*
Belclare and Prospect House Woods	*Polyporus squamosus*	*Polyporus squamosus*
Belclare and Prospect House Woods	*Polystictus versicolor*	*Trametes versicolor*
Belclare and Prospect House Woods	*Poria medulla-panis*	*Perenniporia medulla-panis*
Belclare and Prospect House Woods	*Poria mollusca*	*Trechispora mollusca*
Belclare and Prospect House Woods	*Poria mucida*	*Ceriporiopsis mucida*
Belclare and Prospect House Woods	*Poria sanguinolenta*	*Physisporinus sanguinolentus*
Belclare and Prospect House Woods	*Poria terrestris*	*Byssocorticium terrestre*
Belclare and Prospect House Woods	*Psalliota campestris*	*Agaricus campestris* var. *campestris*
Belclare and Prospect House Woods	*Psathyra corrugis*	*Psathyrella corrugis*
Belclare and Prospect House Woods	*Psathyra fatua*	*Psathyrella fatua*
Belclare and Prospect House Woods	*Psathyra fibrillosa*	*Psathyrella friesii*
Belclare and Prospect House Woods	*Psathyrella atomata*	*Psathyrella atomata*
Belclare and Prospect House Woods	*Psathyrella gracilis*	*Psathyrella corrugis*
Belclare and Prospect House Woods	*Psilocybe ericaea*	*Hypholoma ericaeum*
Belclare and Prospect House Woods	*Psilocybe semilanceata*	*Psilocybe semilanceata*
Belclare and Prospect House Woods	*Puccinia bunii*	*Puccinia bulbocastani*
Belclare and Prospect House Woods	*Puccinia primulae*	*Puccinia primulae*
Belclare and Prospect House Woods	*Puccinia pringsheimiana*	*Puccinia caricina* var. *pringsheimiana*
Belclare and Prospect House Woods	*Puccinia pruni*	*Tranzschelia pruni-spinosae*
Belclare and Prospect House Woods	*Puccinia suaveolens*	*Puccinia punctiformis*
Belclare and Prospect House Woods	*Radulum orbiculare*	*Basidioradulum radula*
Belclare and Prospect House Woods	*Radulum quercinum*	*Ascodichaena rugosa*
Belclare and Prospect House Woods	*Ramularia ajugae*	*Ramularia ajugae*
Belclare and Prospect House Woods	*Rhopographus filicinus*	*Rhopographus filicinus*

(Continued)

Appendix 2 *(Continued)*

Locality	Original name	Current name
Belclare and Prospect House Woods	*Rhytisma acerinum*	*Rhytisma acerinum*
Belclare and Prospect House Woods	*Russula cutifracta*	*Russula cyanoxantha*
Belclare and Prospect House Woods	*Russula cyanoxantha*	*Russula cyanoxantha*
Belclare and Prospect House Woods	*Russula delica*	*Russula delica*
Belclare and Prospect House Woods	*Russula drimeia*	*Russula sardonia*
Belclare and Prospect House Woods	*Russula emetica*	*Russula silvestris*
Belclare and Prospect House Woods	*Russula fellea*	*Russula fellea*
Belclare and Prospect House Woods	*Russula fragilis*	*Russula silvestris*
Belclare and Prospect House Woods	*Russula fragilis* var. *violascens*	*Russula fragilis* var. *fragilis*
Belclare and Prospect House Woods	*Russula integra*	*Russula polychroma*
Belclare and Prospect House Woods	*Russula lepida*	*Russula rosea*
Belclare and Prospect House Woods	*Russula lutea*	*Russula acetolens*
Belclare and Prospect House Woods	*Russula nigricans*	*Russula nigricans*
Belclare and Prospect House Woods	*Russula ochroleuca*	*Russula ochroleuca*
Belclare and Prospect House Woods	*Russula rubra*	*Russula atropurpurea*
Belclare and Prospect House Woods	*Septoria scabiosicola*	*Septoria scabiosicola*
Belclare and Prospect House Woods	*Septoria violae*	*Septoria violae-palustris*
Belclare and Prospect House Woods	*Stereum rugosum*	*Stereum rugosum*
Belclare and Prospect House Woods	*Stereum sanguinolentum*	*Stereum sanguinolentum*
Belclare and Prospect House Woods	*Stropharia aeruginosa*	*Stropharia aeruginosa*
Belclare and Prospect House Woods	*Stropharia semiglobata*	*Stropharia semiglobata*
Belclare and Prospect House Woods	*Stropharia stercoraria*	*Stropharia semiglobata*
Belclare and Prospect House Woods	*Tichothecium rimosicola*	*Phaeospora rimosicola*
Belclare and Prospect House Woods	*Tremella mesenterica*	*Tremella mesenterica*
Belclare and Prospect House Woods	*Tricholoma album*	*Tricholoma stiparophyllum*
Belclare and Prospect House Woods	*Tricholoma cuneifolium*	*Dermoloma cuneifolium*
Belclare and Prospect House Woods	*Tricholoma macrorhizum*	*Leucopaxillus macrocephalus*
Belclare and Prospect House Woods	*Tricholoma panaeolum*	*Lepista panaeola*
Belclare and Prospect House Woods	*Tricholoma rutilans*	*Tricholomopsis rutilans*
Belclare and Prospect House Woods	*Tricholoma ustale*	*Tricholoma ustale*
Belclare and Prospect House Woods	*Tubaria crobula*	*Psilocybe crobula*
Belclare and Prospect House Woods	*Tubaria furfuracea*	*Tubaria furfuracea* var. *furfuracea*
Belclare and Prospect House Woods	*Tubaria inquilina*	*Psilocybe inquilinus*
Belclare and Prospect House Woods	*Typhula erythropus*	*Typhula erythropus*
Belclare and Prospect House Woods	*Uromyces rumicis*	*Uromyces rumicis*
Belclare and Prospect House Woods	*Ustilago hydropiperis*	*Sphacelotheca hydropiperis*
Belclare and Prospect House Woods	*Ustulina vulgaris*	*Kretzschmaria deusta*
Belclare and Prospect House Woods	*Xylaria hypoxylon*	*Xylaria hypoxylon*
Brackloon Wood	*Aegerita candida*	*Bulbillomyces farinosus*
Brackloon Wood	*Amanita mappa*	*Amanita citrina* var. *citrina*
Brackloon Wood	*Amanita muscaria*	*Amanita muscaria* var. *muscaria*
Brackloon Wood	*Amanita rubescens*	*Amanita rubescens* var. *rubescens*
Brackloon Wood	*Androsaceus rotula*	*Marasmius rotula*
Brackloon Wood	*Armillaria mellea*	*Armillaria mellea*
Brackloon Wood	*Ascobolus furfuraceus*	*Ascobolus stercorarius*

(Continued)

Appendix 2 *(Continued)*

Locality	Original name	Current name
Brackloon Wood	*Boletus badius*	*Boletus badius*
Brackloon Wood	*Boletus bovinus*	*Suillus bovinus*
Brackloon Wood	*Boletus edulis*	*Boletus edulis*
Brackloon Wood	*Boletus elegans*	*Suillus grevillei*
Brackloon Wood	*Boletus scaber*	*Leccinum scabrum*
Brackloon Wood	*Boletus scaber* var. *niveus*	*Boletus niveus*
Brackloon Wood	*Boletus subtomentosus*	*Boletus subtomentosus*
Brackloon Wood	*Botrytis cinerea*	*Botrytis cinerea*
Brackloon Wood	*Bulgaria polymorpha*	*Bulgaria inquinans*
Brackloon Wood	*Calocera stricta*	*Calocera stricta*
Brackloon Wood	*Cantharellus cibarius*	*Cantharellus cibarius*
Brackloon Wood	*Cantharellus muscigenus*	*Arrhenia spathulata*
Brackloon Wood	*Cantharellus tubaeformis*	*Cantharellus tubaeformis*
Brackloon Wood	*Chlorosplenium aeruginosum*	*Chlorociboria aeruginascens*
Brackloon Wood	*Claudopus variabilis*	*Crepidotus variabilis*
Brackloon Wood	*Clavaria cristata*	*Clavulina coralloides*
Brackloon Wood	*Clavaria kunzei*	*Ramariopsis kunzei*
Brackloon Wood	*Clavaria vermicularis*	*Clavaria fragilis*
Brackloon Wood	*Clitocybe ditopa*	*Clitocybe ditopa*
Brackloon Wood	*Clitocybe infundibuliformis*	*Lepista flaccida*
Brackloon Wood	*Clitocybe nebularis*	*Clitocybe nebularis*
Brackloon Wood	*Coccomyces coronatus*	*Coccomyces coronatus*
Brackloon Wood	*Collybia dryophila*	*Collybia dryophila*
Brackloon Wood	*Collybia inolens*	*Tephrocybe inolens*
Brackloon Wood	*Collybia maculata*	*Collybia maculata*
Brackloon Wood	*Collybia platyphylla*	*Megacollybia platyphylla*
Brackloon Wood	*Coprinus cinereus*	*Coprinopsis cinerea*
Brackloon Wood	*Coprinus micaceus*	*Coprinellus truncorum*
Brackloon Wood	*Coprinus niveus*	*Coprinopsis nivea*
Brackloon Wood	*Coprinus plicatilis*	*Parasola plicatilis*
Brackloon Wood	*Coprinus radiatus*	*Coprinopsis pseudoradiata*
Brackloon Wood	*Corticium lacteum*	*Phanerochaete tuberculata*
Brackloon Wood	*Cortinarius acutus*	*Cortinarius acutus*
Brackloon Wood	*Cortinarius alboviolaceus*	*Cortinarius alboviolaceus*
Brackloon Wood	*Cortinarius bolaris*	*Cortinarius bolaris*
Brackloon Wood	*Cortinarius brunneus*	*Cortinarius disjungendus*
Brackloon Wood	*Cortinarius callisteus*	*Cortinarius limonius*
Brackloon Wood	*Cortinarius camurus*	*Cortinarius valgus*
Brackloon Wood	*Cortinarius caninus*	*Cortinarius caninus*
Brackloon Wood	*Cortinarius castaneus*	*Cortinarius castaneus*
Brackloon Wood	*Cortinarius collinitus*	*Cortinarius collinitus*
Brackloon Wood	*Cortinarius cotoneus*	*Cortinarius cotoneus*
Brackloon Wood	*Cortinarius decipiens*	*Cortinarius decipiens* var. *decipiens*
Brackloon Wood	*Cortinarius dolabratus*	*Cortinarius dolabratus*
Brackloon Wood	*Cortinarius elatior*	*Cortinarius livido-ochraceus*

(Continued)

Appendix 2 *(Continued)*

Locality	Original name	Current name
Brackloon Wood	*Cortinarius erythrinus*	*Cortinarius vernus*
Brackloon Wood	*Cortinarius glaucopus*	*Cortinarius amoenolens*
Brackloon Wood	*Cortinarius helvelloides*	*Cortinarius helvelloides*
Brackloon Wood	*Cortinarius hemitrichus*	*Cortinarius hemitrichus*
Brackloon Wood	*Cortinarius hinnuleus*	*Cortinarius hinnuleus*
Brackloon Wood	*Cortinarius impennis*	*Cortinarius impennis*
Brackloon Wood	*Cortinarius largus*	*Cortinarius patibilis* var. *scoticus*
Brackloon Wood	*Cortinarius leucopus*	*Cortinarius leucopus*
Brackloon Wood	*Cortinarius myrtillinus*	*Cortinarius anomalus*
Brackloon Wood	*Cortinarius obtusus*	*Cortinarius obtusus*
Brackloon Wood	*Cortinarius ochroleucus*	*Cortinarius ochroleucus*
Brackloon Wood	*Cortinarius paleaceus*	*Cortinarius flexipes* var. *flabellus*
Brackloon Wood	*Cortinarius penicillatus*	*Cortinarius penicillatus*
Brackloon Wood	*Cortinarius purpurascens*	*Cortinarius purpurascens*
Brackloon Wood	*Cortinarius rigidus*	*Cortinarius umbrinolens*
Brackloon Wood	*Cortinarius torvus*	*Cortinarius torvus*
Brackloon Wood	*Coryne sarcoides*	*Ascocoryne sarcoides*
Brackloon Wood	*Cyathus striatus*	*Cyathus striatus*
Brackloon Wood	*Cylindrium flavovirens*	*Cylindrium flavovirens*
Brackloon Wood	*Cyphella muscigena*	*Cyphellostereum laeve*
Brackloon Wood	*Diatrype disciformis*	*Diatrype disciformis*
Brackloon Wood	*Diatrype stigma*	*Diatrype stigma*
Brackloon Wood	*Diatrypella quercina*	*Diatrypella quercina*
Brackloon Wood	*Eccilia griseorubella*	*Entoloma griseorubellum*
Brackloon Wood	*Entoloma nidorosum*	*Entoloma politum*
Brackloon Wood	*Entoloma rhodopolium*	*Entoloma sericatum*
Brackloon Wood	*Eurotium herbariorum*	*Eurotium herbariorum*
Brackloon Wood	*Exidia albida*	*Exidia albida*
Brackloon Wood	*Fistulina hepatica*	*Fistulina hepatica*
Brackloon Wood	*Flammula tricholoma*	*Ripartites tricholoma*
Brackloon Wood	*Galera hypnorum*	*Galerina hypnorum*
Brackloon Wood	*Galera hypnorum* var. *sphagnorum*	*Galerina sphagnorum*
Brackloon Wood	*Galera rubiginosa*	*Galerina vittiformis*
Brackloon Wood	*Galera spartea*	*Conocybe rickeniana*
Brackloon Wood	*Galera tenera*	*Conocybe tenera*
Brackloon Wood	*Geoglossum microsporum*	*Mitrula microspora*
Brackloon Wood	*Grandinia mucida*	*Grandinia mucida*
Brackloon Wood	*Hebeloma crustuliniforme*	*Hebeloma crustuliniforme*
Brackloon Wood	*Hebeloma fastibile*	*Hebeloma mesophaeum* var. *crassipes*
Brackloon Wood	*Helotium aureum*	*Helotium aureum*
Brackloon Wood	*Helotium citrinum*	*Bisporella citrina*
Brackloon Wood	*Helotium herbarum*	*Calycina herbarum*
Brackloon Wood	*Helotium virgultorum*	*Hymenoscyphus calyculus*
Brackloon Wood	*Helotium virgultorum* var. *fructigenum*	*Hymenoscyphus fructigenus*
Brackloon Wood	*Hyaloscypha hyalina*	*Hyaloscypha hyalina*

(Continued)

Appendix 2 *(Continued)*

Locality	Original name	Current name
Brackloon Wood	*Hydnum repandum*	*Hydnum repandum*
Brackloon Wood	*Hydnum rufescens*	*Hydnum repandum*
Brackloon Wood	*Hygrophorus ceraceus*	*Hygrocybe ceracea*
Brackloon Wood	*Hygrophorus chlorophanus*	*Hygrocybe chlorophana*
Brackloon Wood	*Hygrophorus coccineus*	*Hygrocybe coccinea*
Brackloon Wood	*Hygrophorus miniatus*	*Hygrocybe miniata*
Brackloon Wood	*Hygrophorus niveus*	*Hygrocybe virginea* var. *virginea*
Brackloon Wood	*Hygrophorus psittacinus*	*Hygrocybe psittacina* var. *psittacina*
Brackloon Wood	*Hygrophorus virgineus*	*Hygrocybe virginea* var. *virginea*
Brackloon Wood	*Hypholoma capnoides*	*Hypholoma capnoides*
Brackloon Wood	*Hypholoma fasciculare*	*Hypholoma fasciculare* var. *fasciculare*
Brackloon Wood	*Hypholoma hydrophila*	*Psathyrella piluliformis*
Brackloon Wood	*Hypholoma sublateritium*	*Hypholoma lateritium*
Brackloon Wood	*Hypholoma velutinum*	*Lacrymaria lacrymabunda*
Brackloon Wood	*Hysterium angustatum*	*Hysterium angustatum*
Brackloon Wood	*Inocybe geophylla*	*Inocybe geophylla* var. *geophylla*
Brackloon Wood	*Inocybe geophylla* var. *violacea*	*Inocybe geophylla* var. *lilacina*
Brackloon Wood	*Inocybe petiginosa*	*Inocybe petiginosa*
Brackloon Wood	*Inocybe rimosa*	*Inocybe rimosa*
Brackloon Wood	*Irpex obliquus*	*Schizopora paradoxa*
Brackloon Wood	*Isaria arachnophila*	*Isaria arachnophila*
Brackloon Wood	*Laccaria laccata* var. *amethystea*	*Laccaria amethystina*
Brackloon Wood	*Lachnum ciliare*	*Incrucipulum ciliare*
Brackloon Wood	*Lachnum virgineum*	*Lachnum virgineum*
Brackloon Wood	*Lactarius aurantiacus*	*Lactarius aurantiacus*
Brackloon Wood	*Lactarius blennius*	*Lactarius blennius*
Brackloon Wood	*Lactarius chrysorrheus*	*Lactarius chrysorrheus*
Brackloon Wood	*Lactarius cimicarius*	*Lactarius subumbonatus*
Brackloon Wood	*Lactarius deliciosus*	*Lactarius deliciosus*
Brackloon Wood	*Lactarius glyciosmus*	*Lactarius glyciosmus*
Brackloon Wood	*Lactarius hysginus*	*Lactarius hysginus*
Brackloon Wood	*Lactarius insulsus*	*Lactarius zonarius*
Brackloon Wood	*Lactarius mitissimus*	*Lactarius aurantiacus*
Brackloon Wood	*Lactarius piperatus*	*Lactarius piperatus*
Brackloon Wood	*Lactarius pubescens*	*Lactarius pubescens*
Brackloon Wood	*Lactarius quietus*	*Lactarius quietus*
Brackloon Wood	*Lactarius serifluus*	*Lactarius subumbonatus*
Brackloon Wood	*Lactarius subdulcis*	*Lactarius subdulcis*
Brackloon Wood	*Lactarius torminosus*	*Lactarius torminosus*
Brackloon Wood	*Lactarius turpis*	*Lactarius turpis*
Brackloon Wood	*Lactarius uvidus*	*Lactarius uvidus*
Brackloon Wood	*Lactarius vellereus*	*Lactarius vellereus*
Brackloon Wood	*Lactarius vietus*	*Lactarius vietus*
Brackloon Wood	*Lasiosphaeria ovina*	*Lasiosphaeria ovina*
Brackloon Wood	*Leotia lubrica*	*Leotia lubrica*

(Continued)

Appendix 2 *(Continued)*

Locality	Original name	Current name
Brackloon Wood	*Lepiota amianthina*	*Cystoderma amianthinum*
Brackloon Wood	*Leptonia sericella*	*Entoloma sericellum*
Brackloon Wood	*Leptonia serrulata*	*Entoloma serrulatum*
Brackloon Wood	*Lycoperdon perlatum*	*Lycoperdon perlatum*
Brackloon Wood	*Lycoperdon pyriforme*	*Lycoperdon pyriforme*
Brackloon Wood	*Lycoperdon umbrinum*	*Lycoperdon molle*
Brackloon Wood	*Macropodia macropus*	*Helvella macropus*
Brackloon Wood	*Marasmius erythropus*	*Collybia erythropus*
Brackloon Wood	*Marasmius ramealis*	*Marasmiellus ramealis*
Brackloon Wood	*Melampsora hypericorum*	*Melampsora hypericorum*
Brackloon Wood	*Mollisia cinerea*	*Mollisia cinerea*
Brackloon Wood	*Mycena ammoniaca*	*Mycena leptocephala*
Brackloon Wood	*Mycena corticola*	*Mycena hiemalis*
Brackloon Wood	*Mycena discopus*	*Mycena discopus*
Brackloon Wood	*Mycena epipterygia*	*Mycena epipterygia*
Brackloon Wood	*Mycena filopes*	*Mycena vitilis*
Brackloon Wood	*Mycena flavoalba*	*Mycena flavoalba*
Brackloon Wood	*Mycena galericulata*	*Mycena galericulata*
Brackloon Wood	*Mycena galopus*	*Mycena galopus* var. *galopus*
Brackloon Wood	*Mycena leucogala*	*Mycena galopus* var. *nigra*
Brackloon Wood	*Mycena polygramma*	*Mycena polygramma*
Brackloon Wood	*Mycena pura*	*Mycena pura*
Brackloon Wood	*Mycena setosa*	*Marasmius setosus*
Brackloon Wood	*Mycena tenella*	*Mycena vitrea* var. *tenella*
Brackloon Wood	*Mycena tenerrima*	*Mycena adscendens*
Brackloon Wood	*Mycena tenuis*	*Mycena tenuis*
Brackloon Wood	*Naucoria escharioides*	*Naucoria escharioides*
Brackloon Wood	*Naucoria melinoides*	*Galerina mniophila*
Brackloon Wood	*Nyctalis parasitica*	*Asterophora parasitica*
Brackloon Wood	*Omphalia umbellifera*	*Lichenomphalia umbellifera*
Brackloon Wood	*Orbilia leucostigma*	*Orbilia leucostigma*
Brackloon Wood	*Orbilia vinosa*	*Orbilia vinosa*
Brackloon Wood	*Otidea cochleata*	*Otidea cochleata*
Brackloon Wood	*Panaeolus campanulatus*	*Panaeolus papilionaceus* var. *papilionaceus*
Brackloon Wood	*Panaeolus sphinctrinus*	*Panaeolus papilionaceus* var. *papilionaceus*
Brackloon Wood	*Paxillus involutus*	*Paxillus involutus*
Brackloon Wood	*Phlebia vaga*	*Phlebiella sulphurea*
Brackloon Wood	*Pholiota marginata*	*Galerina marginata*
Brackloon Wood	*Pholiota mutabilis*	*Kuehneromyces mutabilis*
Brackloon Wood	*Phragmidium violaceum*	*Phragmidium violaceum*
Brackloon Wood	*Pilobolus crystallinus*	*Pilobolus crystallinus* var. *crystallinus*
Brackloon Wood	*Pistillaria puberula*	*Typhula quisquiliaris*
Brackloon Wood	*Pleospora herbarum*	*Pleospora herbarum*
Brackloon Wood	*Pleurotus acerosus*	*Arrhenia acerosa*
Brackloon Wood	*Pluteus cervinus*	*Pluteus cervinus*

(Continued)

Appendix 2 *(Continued)*

Locality	Original name	Current name
Brackloon Wood	*Polyporus betulinus*	*Piptoporus betulinus*
Brackloon Wood	*Polyporus caesius*	*Postia caesia*
Brackloon Wood	*Polyporus squamosus*	*Polyporus squamosus*
Brackloon Wood	*Polyporus varius*	*Polyporus leptocephalus*
Brackloon Wood	*Polystictus versicolor*	*Trametes versicolor*
Brackloon Wood	*Poria medulla-panis*	*Perenniporia medulla-panis*
Brackloon Wood	*Psalliota haemorrhoidaria*	*Agaricus silvaticus*
Brackloon Wood	*Psathyra bifrons*	*Psathyrella bifrons*
Brackloon Wood	*Psathyra corrugis*	*Psathyrella corrugis*
Brackloon Wood	*Psathyra fibrillosa*	*Psathyrella friesii*
Brackloon Wood	*Psathyra gossypina*	*Psathyrella gossypina*
Brackloon Wood	*Psathyrella atomata*	*Psathyrella atomata*
Brackloon Wood	*Psathyrella gracilis*	*Psathyrella corrugis*
Brackloon Wood	*Psilocybe semilanceata*	*Psilocybe semilanceata*
Brackloon Wood	*Puccinia menthae*	*Puccinia menthae*
Brackloon Wood	*Puccinia umbilici*	*Puccinia umbilici*
Brackloon Wood	*Rhopographus filicinus*	*Rhopographus filicinus*
Brackloon Wood	*Russula atropurpurea*	*Russula atropurpurea*
Brackloon Wood	*Russula cyanoxantha*	*Russula cyanoxantha*
Brackloon Wood	*Russula depallens*	*Russula exalbicans*
Brackloon Wood	*Russula drimeia*	*Russula sardonia*
Brackloon Wood	*Russula fallax*	*Russula violacea* var. *violacea*
Brackloon Wood	*Russula fellea*	*Russula fellea*
Brackloon Wood	*Russula fragilis*	*Russula silvestris*
Brackloon Wood	*Russula fragilis* var. *nivea*	*Russula raoultii*
Brackloon Wood	*Russula fragilis* var. *violascens*	*Russula fragilis* var. *fragilis*
Brackloon Wood	*Russula galochroa*	*Russula pseudoaeruginea*
Brackloon Wood	*Russula lepida*	*Russula rosea*
Brackloon Wood	*Russula lutea*	*Russula acetolens*
Brackloon Wood	*Russula nigricans*	*Russula nigricans*
Brackloon Wood	*Russula nitida*	*Russula cuprea*
Brackloon Wood	*Russula ochroleuca*	*Russula ochroleuca*
Brackloon Wood	*Russula puellaris*	*Russula puellaris*
Brackloon Wood	*Russula vesca*	*Russula vesca*
Brackloon Wood	*Russula xerampelina*	*Russula xerampelina*
Brackloon Wood	*Scleroderma verrucosum*	*Scleroderma verrucosum*
Brackloon Wood	*Scleroderma vulgare*	*Scleroderma citrinum*
Brackloon Wood	*Sepedonium chrysospermum*	*Hypomyces chrysospermus*
Brackloon Wood	*Sphaerobolus stellatus*	*Sphaerobolus stellatus*
Brackloon Wood	*Stereum rugosum*	*Stereum rugosum*
Brackloon Wood	*Stereum spadiceum*	*Stereum gausapatum*
Brackloon Wood	*Stropharia aeruginosa*	*Stropharia aeruginosa*
Brackloon Wood	*Stropharia semiglobata*	*Stropharia semiglobata*
Brackloon Wood	*Stropharia squamosa*	*Stropharia squamosa* var. *squamosa*
Brackloon Wood	*Stropharia stercoraria*	*Stropharia semiglobata*

(Continued)

Appendix 2 *(Continued)*

Locality	Original name	Current name
Brackloon Wood	*Stysanus stemonitis*	*Cephalotrichum stemonitis*
Brackloon Wood	*Torula herbarum*	*Torula herbarum*
Brackloon Wood	*Tremella mesenterica*	*Tremella mesenterica*
Brackloon Wood	*Trichoderma viride*	*Hypocrea schweinitzii*
Brackloon Wood	*Tricholoma album*	*Tricholoma stiparophyllum*
Brackloon Wood	*Tricholoma flavobrunneum*	*Tricholoma fulvum*
Brackloon Wood	*Tricholoma rutilans*	*Tricholomopsis rutilans*
Brackloon Wood	*Tricholoma saponaceum*	*Tricholoma saponaceum* var. *saponaceum*
Brackloon Wood	*Tricholoma virgatum*	*Tricholoma virgatum*
Brackloon Wood	*Trochila ilicina*	*Trochila ilicina*
Brackloon Wood	*Tubaria crobula*	*Psilocybe crobula*
Brackloon Wood	*Tubaria furfuracea*	*Tubaria furfuracea* var. *furfuracea*
Brackloon Wood	*Tubaria inquilina*	*Psilocybe inquilinus*
Brackloon Wood	*Ustilago hydropiperis*	*Sphacelotheca hydropiperis*
Brackloon Wood	*Ustilago violacea*	*Microbotryum violaceum*
Brackloon Wood	*Ustulina vulgaris*	*Kretzschmaria deusta*
Brackloon Wood	*Valsa lata*	*Eutypa lata*
Brackloon Wood	*Xylaria hypoxylon*	*Xylaria hypoxylon*
Brackloon Wood	*Zygodesmus fuscus*	*Zygodesmus fuscus*
Clare Island	*Acrothecium simplex*	*Pleurophragmium parvisporum*
Clare Island	*Androsaceus rotula*	*Marasmius rotula*
Clare Island	*Ascobolus argenteus*	*Coprotus argenteus*
Clare Island	*Aspergillus dubius*	*Aspergillus dubius*
Clare Island	*Belonidium deparculum*	*Phialina ulmariae*
Clare Island	*Bolbitius tener*	*Conocybe apala*
Clare Island	*Boletus luridus*	*Boletus luridus* var. *luridus*
Clare Island	*Boletus scaber*	*Leccinum scabrum*
Clare Island	*Botrytis cinerea*	*Botrytis cinerea*
Clare Island	*Bovista nigrescens*	*Bovista nigrescens*
Clare Island	*Brachysporium apicale*	*Brachysporium nigrum*
Clare Island	*Calloria fusarioides*	*Calloria neglecta*
Clare Island	*Candelospora ilicicola*	*Calonectria ilicicola*
Clare Island	*Cladosporium herbarum*	*Mycosphaerella tulasnei*
Clare Island	*Claudopus variabilis*	*Crepidotus variabilis*
Clare Island	*Clavaria acuta*	*Clavaria acuta*
Clare Island	*Clavaria amethystea*	*Clavulina amethystina*
Clare Island	*Clavaria cinerea*	*Clavulina cinerea f. cinerea*
Clare Island	*Clavaria cristata*	*Clavulina coralloides*
Clare Island	*Clavaria dissipabilis*	*Clavulinopsis helvola*
Clare Island	*Clavaria fumosa*	*Clavaria fumosa*
Clare Island	*Clavaria fusiformis*	*Clavulinopsis fusiformis*
Clare Island	*Clavaria kunzei*	*Ramariopsis kunzei*
Clare Island	*Clavaria luteoalba*	*Clavulinopsis luteoalba*
Clare Island	*Clavaria muscoides*	*Clavulinopsis corniculata*
Clare Island	*Clavaria persimilis*	*Clavulinopsis laeticolor*

(Continued)

Appendix 2 *(Continued)*

Locality	Original name	Current name
Clare Island	*Clavaria straminea*	*Clavaria straminea*
Clare Island	*Clavaria umbrinella*	*Clavulinopsis umbrinella*
Clare Island	*Clavaria vermicularis*	*Clavaria fragilis*
Clare Island	*Clypeosphaeria notarisii*	*Clypeosphaeria mamillana*
Clare Island	*Coleosporium euphrasiae*	*Coleosporium tussilaginis*
Clare Island	*Coleosporium senecionis*	*Coleosporium tussilaginis*
Clare Island	*Coleosporium sonchi*	*Coleosporium tussilaginis*
Clare Island	*Coprinus atramentarius*	*Coprinopsis atramentaria*
Clare Island	*Coprinus fimetarius*	*Coprinopsis cinerea*
Clare Island	*Coprinus friesii*	*Coprinopsis friesii*
Clare Island	*Coprinus hendersonii*	*Coprinus ephemeroides*
Clare Island	*Coprinus micaceus*	*Coprinellus truncorum*
Clare Island	*Coprinus plicatilis*	*Parasola plicatilis*
Clare Island	*Coprinus radiatus*	*Coprinopsis pseudoradiata*
Clare Island	*Cordyceps entomorrhiza*	*Cordyceps entomorrhiza*
Clare Island	*Cordyceps militaris*	*Cordyceps militaris*
Clare Island	*Corticium lividum*	*Phlebia livida*
Clare Island	*Corticium sambuci*	*Hyphodontia sambuci*
Clare Island	*Cortinarius anomalus*	*Cortinarius anomalus*
Clare Island	*Cortinarius biformis*	*Cortinarius tabacinus*
Clare Island	*Cortinarius iliopodius*	*Cortinarius parvannulatus*
Clare Island	*Cortinarius obtusus*	*Cortinarius obtusus*
Clare Island	*Cortinarius paleaceus*	*Cortinarius flexipes* var. *flabellus*
Clare Island	*Cortinarius torvus*	*Cortinarius torvus*
Clare Island	*Cortinarius triumphans*	*Cortinarius triumphans*
Clare Island	*Cortinarius uliginosus*	*Cortinarius uliginosus*
Clare Island	*Coryne sarcoides*	*Ascocoryne sarcoides*
Clare Island	*Crepidotus phillipsii*	*Melanotus phillipsii*
Clare Island	*Cryptospora corylina*	*Ophiovalsa corylina*
Clare Island	*Cyphella pimii*	*Calyptella capula*
Clare Island	*Cystopus candidus*	*Albugo candida*
Clare Island	*Cytospora salicis*	*Cytospora salicis*
Clare Island	*Dacrymyces deliquescens*	*Dacrymyces deliquescens*
Clare Island	*Dacrymyces stillatus*	*Dacrymyces stillatus*
Clare Island	*Diaporthe exasperans*	*Diaporthe eres*
Clare Island	*Diaporthe pulla*	*Diaporthe pulla*
Clare Island	*Diaporthe salicella*	*Cryptodiaporthe salicella*
Clare Island	*Diatrype stigma*	*Diatrype stigma*
Clare Island	*Diatrypella exigua*	*Diatrypella exigua*
Clare Island	*Didymosphaeria diplospora*	*Didymosphaeria oblitescens*
Clare Island	*Eccilia griseorubella*	*Entoloma griseorubellum*
Clare Island	*Entoloma costatum*	*Entoloma transvenosum*
Clare Island	*Entoloma jubatum*	*Entoloma porphyrophaeum*
Clare Island	*Entoloma prunuloides*	*Entoloma prunuloides*
Clare Island	*Entoloma sericeum*	*Entoloma sericeum* var. *sericeum*

(Continued)

Appendix 2 *(Continued)*

Locality	Original name	Current name
Clare Island	*Epicymatia balani*	Ascomycota
Clare Island	*Erinella apala*	*Lachnum apalum*
Clare Island	*Erinella nylanderi*	*Belonidium sulphureum*
Clare Island	*Erysiphe cichoracearum*	*Golovinomyces cichoracearum* var. *cichoracearum*
Clare Island	*Erysiphe galeopsidis*	*Neoerysiphe galeopsidis*
Clare Island	*Erysiphe pisi*	*Erysiphe pisi* var. *pisi*
Clare Island	*Galera hypnorum*	*Galerina hypnorum*
Clare Island	*Galera mycenoides*	*Galerina praticola*
Clare Island	*Galera tenera*	*Conocybe tenera*
Clare Island	*Geoglossum ophioglossoides*	*Geoglossum glabrum*
Clare Island	*Grandinia crustosa*	*Hyphodontia crustosa*
Clare Island	*Helotium citrinum*	*Bisporella citrina*
Clare Island	*Helotium rhodoleucum*	*Hymenoscyphus rhodoleucus*
Clare Island	*Helotium scutula*	*Hymenoscyphus scutula*
Clare Island	*Helotium terrigenum*	*Helotium terrigenum*
Clare Island	*Helvella lacunosa*	*Helvella lacunosa*
Clare Island	*Helvella pezizoides*	*Helvella pezizoides*
Clare Island	*Hendersonia arundinacea*	*Hendersonia arundinacea*
Clare Island	*Heterosphaeria patella*	*Heterosphaeria patella*
Clare Island	*Humaria granulata*	*Coprobia granulata*
Clare Island	*Humaria subhirsuta*	*Cheilymenia subhirsuta*
Clare Island	*Hydnum niveum*	*Trechispora nivea*
Clare Island	*Hygrophorus calyptriformis*	*Hygrocybe calyptriformis*
Clare Island	*Hygrophorus ceraceus*	*Hygrocybe ceracea*
Clare Island	*Hygrophorus chlorophanus*	*Hygrocybe chlorophana*
Clare Island	*Hygrophorus coccineus*	*Hygrocybe coccinea*
Clare Island	*Hygrophorus conicus*	*Hygrocybe conica*
Clare Island	*Hygrophorus fornicatus*	*Hygrocybe fornicata*
Clare Island	*Hygrophorus laetus*	*Hygrocybe laeta*
Clare Island	*Hygrophorus miniatus*	*Hygrocybe miniata*
Clare Island	*Hygrophorus ovinus*	*Hygrocybe ovina*
Clare Island	*Hygrophorus pratensis*	*Hygrocybe pratensis* var. *pratensis*
Clare Island	*Hygrophorus psittacinus*	*Hygrocybe psittacina* var. *psittacina*
Clare Island	*Hygrophorus puniceus*	*Hygrocybe punicea*
Clare Island	*Hygrophorus unguinosus*	*Hygrocybe irrigata*
Clare Island	*Hygrophorus virgineus*	*Hygrocybe virginea* var. *virginea*
Clare Island	*Hymenochaete fuliginosa*	*Hymenochaete fuliginosa*
Clare Island	*Hymenoscyphus cyathoideus*	*Crocicreas cyathoideum* var. *cyathoideum*
Clare Island	*Hymenoscyphus dumorum*	*Lachnum dumorum*
Clare Island	*Hypholoma dispersum*	*Hypholoma marginatum*
Clare Island	*Hypholoma hydrophila*	*Psathyrella piluliformis*
Clare Island	*Hypocrea rufa*	*Hypocrea rufa*
Clare Island	*Hypoderma virgultorum*	*Hypoderma rubi*
Clare Island	*Hypoxylon multiforme*	*Hypoxylon multiforme*

(Continued)

Appendix 2 *(Continued)*

Locality	Original name	Current name
Clare Island	*Inocybe eutheles*	*Inocybe sindonia*
Clare Island	*Inocybe rimosa*	*Inocybe rimosa*
Clare Island	*Isaria arachnophila*	*Isaria arachnophila*
Clare Island	*Laccaria laccata*	*Laccaria laccata*
Clare Island	*Lachnum virgineum*	*Lachnum virgineum*
Clare Island	*Lactarius pyrogalus*	*Lactarius pyrogalus*
Clare Island	*Leotia lubrica*	*Leotia lubrica*
Clare Island	*Lepiota amianthina*	*Cystoderma amianthinum*
Clare Island	*Lepiota granulosa*	*Cystoderma granulosum*
Clare Island	*Leptonia asprella*	*Entoloma asprellum*
Clare Island	*Leptonia lampropus*	*Entoloma sodale*
Clare Island	*Leptosphaeria acuta*	*Leptosphaeria acuta*
Clare Island	*Leptosphaeria chondri*	*Lautitia danica*
Clare Island	*Leptosphaeria culmifraga*	*Leptosphaeria culmifraga*
Clare Island	*Leptosphaeria derasa*	*Nodulosphaeria derasa*
Clare Island	*Leptosphaeria doliolum*	*Leptosphaeria doliolum*
Clare Island	*Leptosphaeria michotii*	*Paraphaeosphaeria michotii*
Clare Island	*Lophiostoma arundinis*	*Lophiostoma arundinis*
Clare Island	*Lophiostoma caulium*	*Lophiostoma caulium*
Clare Island	*Lycoperdon perlatum*	*Lycoperdon perlatum*
Clare Island	*Marasmius oreades*	*Marasmius oreades*
Clare Island	*Marasmius ramealis*	*Marasmiellus ramealis*
Clare Island	*Melampsora farinosa*	*Melampsora caprearum*
Clare Island	*Melampsora hypericorum*	*Melampsora hypericorum*
Clare Island	*Melampsora lini*	*Melampsora lini* var. *lini*
Clare Island	*Melampsora orchidis-repentis*	*Melampsora epitea* var. *epitea*
Clare Island	*Melampsora pustulata*	*Pucciniastrum epilobii*
Clare Island	*Melampsoridium betulinum*	*Melampsoridium betulinum*
Clare Island	*Melanconis stilbostoma*	*Melanconis stilbostoma*
Clare Island	*Melanomma pulvis-pyrius*	*Melanomma pulvis-pyrius*
Clare Island	*Menispora ciliata*	*Menispora ciliata*
Clare Island	*Merulius corium*	*Byssomerulius corium*
Clare Island	*Microglossum atropurpureum*	*Geoglossum atropurpureum*
Clare Island	*Mollisia atrata*	*Pyrenopeziza atrata*
Clare Island	*Mollisia cinerea*	*Mollisia cinerea*
Clare Island	*Mucor mucedo*	*Mucor mucedo*
Clare Island	*Mycena capillaris*	*Mycena capillaris*
Clare Island	*Mycena filopes*	*Mycena vitilis*
Clare Island	*Mycena flavoalba*	*Mycena flavoalba*
Clare Island	*Mycena luteoalba*	*Mycena luteoalba*
Clare Island	*Mycena olivaceomarginata*	*Mycena olivaceomarginata*
Clare Island	*Mycena peltata*	*Mycena peltata*
Clare Island	*Mycena pullata*	*Mycena pullata*
Clare Island	*Mycena rorida*	*Mycena rorida*
Clare Island	*Mycena rugosa*	*Mycena galericulata*

(Continued)

Appendix 2 *(Continued)*

Locality	Original name	Current name
Clare Island	*Mycena stylobates*	*Mycena stylobates*
Clare Island	*Mycena tenerrima*	*Mycena adscendens*
Clare Island	*Mycena vitilis*	*Mycena filopes*
Clare Island	*Mycosphaerella ascophylli*	*Mycosphaerella ascophylli*
Clare Island	*Mycosphaerella brassicicola*	*Mycosphaerella brassicicola*
Clare Island	*Mycosphaerella tassiana*	*Mycosphaerella tassiana*
Clare Island	*Naucoria myosotis*	*Hypholoma myosotis*
Clare Island	*Naucoria pediades*	*Agrocybe pediades*
Clare Island	*Naucoria semiorbicularis*	*Agrocybe pediades*
Clare Island	*Naucoria tabacina*	*Agrocybe tabacina*
Clare Island	*Nectria mammoidea*	*Nectria mammoidea* var. *mammoidea*
Clare Island	*Nectria sanguinea*	*Nectria episphaeria*
Clare Island	*Nolanea pascua*	*Entoloma pascuum*
Clare Island	*Omphalia fibula*	*Rickenella fibula*
Clare Island	*Omphalia integrella*	*Delicatula integrella*
Clare Island	*Omphalia oniscus*	*Arrhenia onisca*
Clare Island	*Omphalia sphagnicola*	*Arrhenia sphagnicola*
Clare Island	*Omphalia umbellifera*	*Lichenomphalia umbellifera*
Clare Island	*Ophiobolus acuminatus*	*Ophiobolus acuminatus*
Clare Island	*Panaeolus campanulatus*	*Panaeolus papilionaceus* var. *papilionaceus*
Clare Island	*Panaeolus papilionaceus*	*Panaeolus papilionaceus* var. *papilionaceus*
Clare Island	*Panaeolus phalaenarum*	*Panaeolus semiovatus* var. *phalaenarum*
Clare Island	*Panaeolus retirugis*	*Panaeolus papilionaceus* var. *papilionaceus*
Clare Island	*Patellaria atrata*	*Lecanidion atratum*
Clare Island	*Paxillus involutus*	*Paxillus involutus*
Clare Island	*Penicillium glaucum*	*Penicillium glaucum*
Clare Island	*Periconia pycnospora*	*Periconia byssoides*
Clare Island	*Peronospora calotheca*	*Peronospora calotheca*
Clare Island	*Peronospora effusa*	*Peronospora farinosa*
Clare Island	*Peronospora ficariae*	*Peronospora ficariae*
Clare Island	*Peronospora grisea*	*Peronospora grisea*
Clare Island	*Peziza badia*	*Peziza badia*
Clare Island	*Pholiota pumila*	*Galerina pumila*
Clare Island	*Phoma longissima*	*Phoma longissima*
Clare Island	*Phomatospora argentinae*	*Phomatospora argentinae*
Clare Island	*Phragmidium fragariastri*	*Phragmidium fragariae*
Clare Island	*Phragmidium violaceum*	*Phragmidium violaceum*
Clare Island	*Phyllachora junci*	*Phyllachora junci*
Clare Island	*Phytophthora infestans*	*Phytophthora infestans*
Clare Island	*Plasmopara densa*	*Plasmopara densa*
Clare Island	*Platystomum compressum*	*Lophiostoma compressum*
Clare Island	*Pleospora herbarum*	*Pleospora herbarum*
Clare Island	*Pleospora vulgaris*	*Lewia scrophulariae*
Clare Island	*Podosphaera oxyacanthae*	*Podosphaera clandestina* var. *clandestina*

(Continued)

Appendix 2 *(Continued)*

Locality	Original name	Current name
Clare Island	*Polyporus elegans*	*Polyporus leptocephalus*
Clare Island	*Poria vaporaria*	*Antrodia sinuosa*
Clare Island	*Poria vulgaris*	*Skeletocutis vulgaris*
Clare Island	*Protomyces macrosporus*	*Protomyces macrosporus*
Clare Island	*Psalliota campestris*	*Agaricus campestris* var. *campestris*
Clare Island	*Psathyra fibrillosa*	*Psathyrella friesii*
Clare Island	*Psathyra semivestita*	*Psathyrella microrhiza*
Clare Island	*Psathyrella atomata*	*Psathyrella atomata*
Clare Island	*Psathyrella gracilis*	*Psathyrella corrugis*
Clare Island	*Psilocybe bullacea*	*Psilocybe bullacea*
Clare Island	*Psilocybe ericaea*	*Hypholoma ericaeum*
Clare Island	*Psilocybe foenisecii*	*Panaeolina foenisecii*
Clare Island	*Psilocybe semilanceata*	*Psilocybe semilanceata*
Clare Island	*Psilocybe uda*	*Hypholoma udum*
Clare Island	*Puccinia baryi*	*Puccinia brachypodii* var. *brachypodii*
Clare Island	*Puccinia caricis*	*Puccinia caricina* var. *caricina*
Clare Island	*Puccinia centaureae*	*Puccinia calcitrapae*
Clare Island	*Puccinia chrysosplenii*	*Puccinia chrysosplenii*
Clare Island	*Puccinia graminis*	*Puccinia graminis* subsp. *graminis*
Clare Island	*Puccinia menthae*	*Puccinia menthae*
Clare Island	*Puccinia obscura*	*Puccinia obscura*
Clare Island	*Puccinia phragmitis*	*Puccinia phragmitis*
Clare Island	*Puccinia saxifragae*	*Puccinia saxifragae*
Clare Island	*Puccinia sonchi*	*Miyagia pseudosphaeria*
Clare Island	*Puccinia suaveolens*	*Puccinia punctiformis*
Clare Island	*Puccinia violae*	*Puccinia violae*
Clare Island	*Rhopographus pteridis*	*Rhopographus filicinus*
Clare Island	*Rhytisma salicinum*	*Rhytisma salicinum*
Clare Island	*Rosellinia anthostomoides*	*Rosellinia anthostomoides*
Clare Island	*Rosellinia clavariarum*	*Helminthosphaeria clavariarum*
Clare Island	*Russula adusta*	*Russula adusta*
Clare Island	*Russula nigricans*	*Russula nigricans*
Clare Island	*Scleroderma vulgare*	*Scleroderma citrinum*
Clare Island	*Sclerotinia sclerotiorum*	*Sclerotinia sclerotiorum*
Clare Island	*Sepedonium chrysospermum*	*Hypomyces chrysospermus*
Clare Island	*Septoria convolvuli*	*Septoria convolvuli*
Clare Island	*Septoria epilobii*	*Septoria epilobii*
Clare Island	*Septoria hederae*	*Mycosphaerella hedericola*
Clare Island	*Septoria scabiosicola*	*Septoria scabiosicola*
Clare Island	*Solenia anomala*	*Merismodes anomala*
Clare Island	*Sordaria decipiens*	*Podospora decipiens*
Clare Island	*Sporormia intermedia*	*Sporormiella intermedia*
Clare Island	*Stigmatea rumicis*	*Venturia rumicis*
Clare Island	*Stropharia merdaria*	*Psilocybe merdaria*
Clare Island	*Stropharia semiglobata*	*Stropharia semiglobata*

(Continued)

Appendix 2 *(Continued)*

Locality	Original name	Current name
Clare Island	*Stropharia stercoraria*	*Stropharia semiglobata*
Clare Island	*Thelephora anthocephala*	*Thelephora anthocephala*
Clare Island	*Tichothecium pygmaeum*	*Muellerella pygmaea* var. *pygmaea*
Clare Island	*Tomentella fusca*	*Tomentella fuscella*
Clare Island	*Torula herbarum*	*Torula herbarum*
Clare Island	*Torula ovalispora*	*Torula ovalispora*
Clare Island	*Trematosphaeria mastoidea*	*Melomastia mastoidea*
Clare Island	*Trematosphaeria pertusa*	*Trematosphaeria pertusa*
Clare Island	*Tricholoma panaeolum*	*Lepista panaeola*
Clare Island	*Triphragmium ulmariae*	*Triphragmium ulmariae*
Clare Island	*Trochila ilicina*	*Trochila ilicina*
Clare Island	*Tubaria stagnina*	*Phaeogalera stagnina*
Clare Island	*Typhula gyrans*	*Typhula setipes*
Clare Island	*Urceolella aspera*	*Unguicularia aspera*
Clare Island	*Uromyces poae*	*Uromyces dactylidis*
Clare Island	*Uromyces trifolii*	*Uromyces trifolii*
Clare Island	*Ustilago avenae*	*Ustilago avenae*
Clare Island	*Ustilago longissima*	*Ustilago filiformis*
Clare Island	*Ustilago nuda*	*Ustilago tritici*
Clare Island	*Ustilago utriculosa*	*Sphacelotheca hydropiperis*
Clare Island	*Ustilago violacea*	*Microbotryum violaceum*
Clare Island	*Valsa heteracantha*	*Eutypella scoparia*
Clare Island	*Valsa lata*	*Eutypa lata*
Clare Island	*Xylaria hypoxylon*	*Xylaria hypoxylon*
Cloonagh Wood	*Armillaria mellea*	*Armillaria mellea*
Cloonagh Wood	*Clitocybe fragrans*	*Clitocybe fragrans*
Cloonagh Wood	*Collybia velutipes*	*Flammulina velutipes* var. *velutipes*
Cloonagh Wood	*Cortinarius elatior*	*Cortinarius livido-ochraceus*
Cloonagh Wood	*Crepidotus calolepis*	*Crepidotus calolepis*
Cloonagh Wood	*Dacrymyces deliquescens*	*Dacrymyces stillatus*
Cloonagh Wood	*Grandinia granulosa*	*Dichostereum granulosum*
Cloonagh Wood	*Hygrophorus niveus*	*Hygrocybe virginea* var. *virginea*
Cloonagh Wood	*Hypholoma fasciculare*	*Hypholoma fasciculare* var. *fasciculare*
Cloonagh Wood	*Irpex obliquus*	*Schizopora paradoxa*
Cloonagh Wood	*Laccaria laccata*	*Laccaria laccata*
Cloonagh Wood	*Lactarius mitissimus*	*Lactarius aurantiacus*
Cloonagh Wood	*Lactarius pubescens*	*Lactarius pubescens*
Cloonagh Wood	*Mycena galericulata*	*Mycena galericulata*
Cloonagh Wood	*Mycena rorida*	*Mycena rorida*
Cloonagh Wood	*Mycena rugosa*	*Mycena galericulata*
Cloonagh Wood	*Nyctalis parasitica*	*Asterophora parasitica*
Cloonagh Wood	*Pholiota marginata*	*Galerina marginata*
Cloonagh Wood	*Polyporus adustus*	*Bjerkandera adusta*
Cloonagh Wood	*Polystictus versicolor*	*Trametes versicolor*

(Continued)

Appendix 2 *(Continued)*

Locality	Original name	Current name
Cloonagh Wood	*Psathyrella gracilis*	*Psathyrella corrugis*
Cloonagh Wood	*Rutstroemia firma*	*Rutstroemia firma*
Cloonagh Wood	*Stereum purpureum*	*Chondrostereum purpureum*
Cloonagh Wood	*Stropharia aeruginosa*	*Stropharia aeruginosa*
Cloonagh Wood	*Tricholoma personatum*	*Lepista saeva*
Cloonagh Wood	*Tricholoma saponaceum*	*Tricholoma saponaceum* var. *saponaceum*
Croagh Patrick	*Armillaria mellea*	*Armillaria mellea*
Croagh Patrick	*Coprinus radiatus*	*Coprinopsis pseudoradiata*
Croagh Patrick	*Galera hypnorum*	*Galerina hypnorum*
Croagh Patrick	*Hygrophorus coccineus*	*Hygrocybe coccinea*
Croagh Patrick	*Mycena ammoniaca*	*Mycena leptocephala*
Croagh Patrick	*Mycena flavoalba*	*Mycena flavoalba*
Croagh Patrick	*Mycena leucogala*	*Mycena galopus* var. *nigra*
Croagh Patrick	*Mycena rosella*	*Mycena rosella*
Croagh Patrick	*Nolanea pascua*	*Entoloma pascuum*
Croagh Patrick	*Omphalia umbellifera*	*Lichenomphalia umbellifera*
Croagh Patrick	*Psalliota arvensis*	*Agaricus arvensis*
Croagh Patrick	*Psathyra corrugis*	*Psathyrella corrugis*
Croagh Patrick	*Psathyrella gracilis*	*Psathyrella corrugis*
Croagh Patrick	*Psilocybe semilanceata*	*Psilocybe semilanceata*
Croagh Patrick	*Stropharia merdaria*	*Psilocybe merdaria*
Croagh Patrick	*Stropharia semiglobata*	*Stropharia semiglobata*
Croagh Patrick	*Tubaria furfuracea*	*Tubaria furfuracea* var. *furfuracea*
Croagh Patrick	*Tubaria paludosa*	*Galerina paludosa*
Derrygorman Wood	*Clavaria cristata*	*Clavulina coralloides*
Derrygorman Wood	*Clitocybe geotropa*	*Clitocybe geotropa*
Derrygorman Wood	*Collybia butyracea*	*Collybia butyracea* var. *butyracea*
Derrygorman Wood	*Cortinarius elatior*	*Cortinarius livido-ochraceus*
Derrygorman Wood	*Hydnum repandum*	*Hydnum repandum*
Derrygorman Wood	*Hypholoma fasciculare*	*Hypholoma fasciculare* var. *fasciculare*
Derrygorman Wood	*Lepiota carcharias*	*Cystoderma carcharias*
Derrygorman Wood	*Mycena galericulata*	*Mycena galericulata*
Derrygorman Wood	*Mycena hiemalis*	*Mycena olida*
Derrygorman Wood	*Mycena parabolica*	*Mycena parabolica*
Derrygorman Wood	*Mycena pura*	*Mycena pura*
Derrygorman Wood	*Mycena rugosa*	*Mycena galericulata*
Derrygorman Wood	*Polyporus adustus*	*Bjerkandera adusta*
Derrygorman Wood	*Psathyrella gracilis*	*Psathyrella corrugis*
Derrygorman Wood	*Russula fragilis*	*Russula silvestris*
Derrygorman Wood	*Stereum ochroleucum*	*Stereum ochroleucum*
Derrygorman Wood	*Stereum purpureum*	*Chondrostereum purpureum*
Derrygorman Wood	*Stropharia semiglobata*	*Stropharia semiglobata*
Kilboyne Wood	*Boletus variegatus*	*Fomes variegatus*
Kilboyne Wood	*Clitocybe ditopa*	*Clitocybe ditopa*
Kilboyne Wood	*Russula rubra*	*Russula atropurpurea*

(Continued)

Appendix 2 *(Continued)*

Locality	Original name	Current name
Knockranny Wood	*Amanita mappa*	*Amanita citrina* var. *citrina*
Knockranny Wood	*Amanita phalloides*	*Amanita phalloides*
Knockranny Wood	*Amanita rubescens*	*Amanita rubescens* var. *rubescens*
Knockranny Wood	*Amanita spissa*	*Amanita excelsa* var. *spissa*
Knockranny Wood	*Amanitopsis fulva*	*Amanita fulva*
Knockranny Wood	*Amanitopsis vaginata*	*Amanita vaginata* var. *vaginata*
Knockranny Wood	*Androsaceus epiphylloides*	*Marasmius epiphylloides*
Knockranny Wood	*Androsaceus graminum*	*Marasmius curreyi*
Knockranny Wood	*Androsaceus rotula*	*Marasmius rotula*
Knockranny Wood	*Anellaria fimiputris*	*Panaeolus semiovatus* var. *semiovatus*
Knockranny Wood	*Armillaria mellea*	*Armillaria mellea*
Knockranny Wood	*Ascobolus furfuraceus*	*Ascobolus stercorarius*
Knockranny Wood	*Bispora monilioides*	*Bispora antennata*
Knockranny Wood	*Boletus badius*	*Boletus badius*
Knockranny Wood	*Boletus bovinus*	*Suillus bovinus*
Knockranny Wood	*Boletus elegans*	*Suillus grevillei*
Knockranny Wood	*Boletus scaber*	*Leccinum scabrum*
Knockranny Wood	*Bombardia fasciculata*	*Bombardia bombarda*
Knockranny Wood	*Bovista nigrescens*	*Bovista nigrescens*
Knockranny Wood	*Bovista plumbea*	*Bovista plumbea*
Knockranny Wood	*Calocera cornea*	*Calocera cornea*
Knockranny Wood	*Cantharellus aurantiacus*	*Hygrophoropsis aurantiaca*
Knockranny Wood	*Cantharellus cibarius*	*Cantharellus cibarius*
Knockranny Wood	*Chlorosplenium aeruginosum*	*Chlorociboria aeruginascens*
Knockranny Wood	*Claudopus variabilis*	*Crepidotus variabilis*
Knockranny Wood	*Clavaria cristata*	*Clavulina coralloides*
Knockranny Wood	*Clitocybe clavipes*	*Ampulloclitocybe clavipes*
Knockranny Wood	*Clitocybe fragrans*	*Clitocybe fragrans*
Knockranny Wood	*Clitopilus prunulus*	*Clitopilus prunulus*
Knockranny Wood	*Coccomyces coronatus*	*Coccomyces coronatus*
Knockranny Wood	*Coleosporium petasitis*	*Coleosporium tussilaginis*
Knockranny Wood	*Collybia aquosa*	*Collybia aquosa*
Knockranny Wood	*Collybia butyracea*	*Collybia butyracea* var. *butyracea*
Knockranny Wood	*Collybia cirrhata*	*Collybia cookei*
Knockranny Wood	*Collybia conigena*	*Collybia conigena*
Knockranny Wood	*Collybia dryophila*	*Collybia dryophila*
Knockranny Wood	*Collybia fusipes*	*Collybia fusipes*
Knockranny Wood	*Collybia prolixa*	*Collybia prolixa*
Knockranny Wood	*Collybia radicata*	*Xerula radicata*
Knockranny Wood	*Collybia tuberosa*	*Collybia tuberosa*
Knockranny Wood	*Coniophora puteana*	*Coniophora puteana*
Knockranny Wood	*Coprinus atramentarius*	*Coprinopsis atramentaria*
Knockranny Wood	*Coprinus ephemerus*	*Coprinellus ephemerus*
Knockranny Wood	*Coprinus micaceus*	*Coprinellus truncorum*
Knockranny Wood	*Coprinus niveus*	*Coprinopsis nivea*

(Continued)

Appendix 2 *(Continued)*

Locality	Original name	Current name
Knockranny Wood	*Coprinus plicatilis*	*Parasola plicatilis*
Knockranny Wood	*Corticium arachnoideum*	*Athelia arachnoidea*
Knockranny Wood	*Corticium comedens*	*Vuilleminia comedens*
Knockranny Wood	*Corticium lacteum*	*Phanerochaete tuberculata*
Knockranny Wood	*Corticium laeve*	*Cylindrobasidium laeve*
Knockranny Wood	*Corticium porosum*	*Gloeocystidiellum porosum*
Knockranny Wood	*Corticium sambuci*	*Hyphodontia sambuci*
Knockranny Wood	*Cortinarius anomalus*	*Cortinarius anomalus*
Knockranny Wood	*Cortinarius castaneus*	*Cortinarius castaneus*
Knockranny Wood	*Cortinarius cinnamomeus*	*Cortinarius cinnamomeus*
Knockranny Wood	*Cortinarius decipiens*	*Cortinarius decipiens* var. *decipiens*
Knockranny Wood	*Cortinarius elatior*	*Cortinarius livido-ochraceus*
Knockranny Wood	*Cortinarius emollitus*	*Cortinarius emollitus*
Knockranny Wood	*Cortinarius hinnuleus*	*Cortinarius hinnuleus*
Knockranny Wood	*Cortinarius impennis*	*Cortinarius impennis*
Knockranny Wood	*Cortinarius infractus*	*Cortinarius infractus*
Knockranny Wood	*Cortinarius lepidopus*	*Cortinarius anomalus*
Knockranny Wood	*Cortinarius leucopus*	*Cortinarius leucopus*
Knockranny Wood	*Cortinarius paleaceus*	*Cortinarius flexipes* var. *flabellus*
Knockranny Wood	*Cortinarius pateriformis*	*Cortinarius pateriformis*
Knockranny Wood	*Cortinarius pluvius*	*Cortinarius pluvius*
Knockranny Wood	*Cortinarius psammocephalus*	*Cortinarius angelesianus*
Knockranny Wood	*Cortinarius rigidus*	*Cortinarius umbrinolens*
Knockranny Wood	*Cortinarius tabularis*	*Cortinarius tabularis*
Knockranny Wood	*Cortinarius torvus*	*Cortinarius torvus*
Knockranny Wood	*Coryne sarcoides*	*Ascocoryne sarcoides*
Knockranny Wood	*Coryne urnalis*	*Ascocoryne cylichnium*
Knockranny Wood	*Corynella glabrovirens*	*Claussenomyces prasinulus*
Knockranny Wood	*Crepidotus calolepis*	*Crepidotus calolepis*
Knockranny Wood	*Cyathus striatus*	*Cyathus striatus*
Knockranny Wood	*Cylindrium flavovirens*	*Cylindrium flavovirens*
Knockranny Wood	*Cyphella capula*	*Calyptella capula*
Knockranny Wood	*Dacrymyces deliquescens*	*Dacrymyces stillatus*
Knockranny Wood	*Dacrymyces stillatus*	*Dacrymyces stillatus*
Knockranny Wood	*Daedalea quercina*	*Daedalea quercina*
Knockranny Wood	*Diaporthe crustosa*	*Diaporthe crustosa*
Knockranny Wood	*Diatrype disciformis*	*Diatrype disciformis*
Knockranny Wood	*Diatrype stigma*	*Diatrype stigma*
Knockranny Wood	*Diatrypella quercina*	*Diatrypella quercina*
Knockranny Wood	*Dichaena quercina*	*Ascodichaena rugosa*
Knockranny Wood	*Entoloma ameides*	*Entoloma ameides*
Knockranny Wood	*Entoloma griseocyaneum*	*Entoloma griseocyaneum*
Knockranny Wood	*Entoloma jubatum*	*Entoloma porphyrophaeum*
Knockranny Wood	*Entoloma nidorosum*	*Entoloma politum*
Knockranny Wood	*Entoloma placenta*	*Entoloma placenta*

(Continued)

Appendix 2 *(Continued)*

Locality	Original name	Current name
Knockranny Wood	*Entoloma porphyrophaeum*	*Entoloma porphyrophaeum*
Knockranny Wood	*Entoloma prunuloides*	*Entoloma prunuloides*
Knockranny Wood	*Entoloma sericeum*	*Entoloma sericeum* var. *sericeum*
Knockranny Wood	*Entoloma speculum*	*Entoloma rhodopolium*
Knockranny Wood	*Erysiphe graminis*	*Blumeria graminis*
Knockranny Wood	*Eurotium herbariorum*	*Eurotium herbariorum*
Knockranny Wood	*Exidia albida*	*Exidia albida*
Knockranny Wood	*Fistulina hepatica*	*Fistulina hepatica*
Knockranny Wood	*Flammula sapinea*	*Gymnopilus sapineus*
Knockranny Wood	*Fomes annosus*	*Heterobasidion annosum*
Knockranny Wood	*Galera hypnorum*	*Galerina hypnorum*
Knockranny Wood	*Galera hypnorum* var. *sphagnorum*	*Galerina sphagnorum*
Knockranny Wood	*Galera rubiginosa*	*Galerina vittiformis*
Knockranny Wood	*Galera spartea*	*Conocybe rickeniana*
Knockranny Wood	*Galera tenera*	*Conocybe tenera*
Knockranny Wood	*Gomphidius roseus*	*Gomphidius roseus*
Knockranny Wood	*Grandinia granulosa*	*Dichostereum granulosum*
Knockranny Wood	*Grandinia mucida*	*Grandinia mucida*
Knockranny Wood	*Helotium claroflavum*	*Bisporella citrina*
Knockranny Wood	*Helotium phyllophilum*	*Hymenoscyphus phyllophilus*
Knockranny Wood	*Helotium virgultorum* var. *fructigenum*	*Hymenoscyphus fructigenus*
Knockranny Wood	*Humaria granulata*	*Coprobia granulata*
Knockranny Wood	*Hyaloscypha hyalina*	*Hyaloscypha hyalina*
Knockranny Wood	*Hydnum farinaceum*	*Trechispora farinacea*
Knockranny Wood	*Hydnum niveum*	*Trechispora nivea*
Knockranny Wood	*Hydnum ochraceum*	*Steccherinum ochraceum*
Knockranny Wood	*Hydnum repandum*	*Hydnum repandum*
Knockranny Wood	*Hydnum rufescens*	*Hydnum repandum*
Knockranny Wood	*Hygrophorus chlorophanus*	*Hygrocybe chlorophana*
Knockranny Wood	*Hygrophorus coccineus*	*Hygrocybe coccinea*
Knockranny Wood	*Hygrophorus fornicatus*	*Hygrocybe fornicata*
Knockranny Wood	*Hygrophorus miniatus*	*Hygrocybe miniata*
Knockranny Wood	*Hygrophorus nitratus*	*Hygrocybe nitrata*
Knockranny Wood	*Hygrophorus niveus*	*Hygrocybe virginea* var. *virginea*
Knockranny Wood	*Hygrophorus ovinus*	*Hygrocybe ovina*
Knockranny Wood	*Hygrophorus pratensis*	*Hygrocybe pratensis* var. *pratensis*
Knockranny Wood	*Hygrophorus psittacinus*	*Hygrocybe psittacina* var. *psittacina*
Knockranny Wood	*Hygrophorus reae*	*Hygrocybe mucronella*
Knockranny Wood	*Hygrophorus virgineus*	*Hygrocybe virginea* var. *virginea*
Knockranny Wood	*Hymenochaete rubiginosa*	*Hymenochaete rubiginosa*
Knockranny Wood	*Hypholoma appendiculatum*	*Psathyrella candolleana*
Knockranny Wood	*Hypholoma capnoides*	*Hypholoma capnoides*
Knockranny Wood	*Hypholoma epixanthum*	*Hypholoma epixanthum*
Knockranny Wood	*Hypholoma fasciculare*	*Hypholoma fasciculare* var. *fasciculare*
Knockranny Wood	*Hypholoma hydrophila*	*Psathyrella piluliformis*

(Continued)

Appendix 2 *(Continued)*

Locality	Original name	Current name
Knockranny Wood	*Hypholoma sublateritium*	*Hypholoma lateritium*
Knockranny Wood	*Hypoxylon multiforme*	*Hypoxylon multiforme*
Knockranny Wood	*Hysterium angustatum*	*Hysterium angustatum*
Knockranny Wood	*Inocybe geophylla*	*Inocybe geophylla* var. *geophylla*
Knockranny Wood	*Inocybe geophylla* var. *violacea*	*Inocybe geophylla* var. *lilacina*
Knockranny Wood	*Inocybe petiginosa*	*Inocybe petiginosa*
Knockranny Wood	*Inocybe rimosa*	*Inocybe rimosa*
Knockranny Wood	*Irpex fuscoviolaceus*	*Trichaptum fuscoviolaceum*
Knockranny Wood	*Irpex obliquus*	*Schizopora paradoxa*
Knockranny Wood	*Isaria farinosa*	*Paecilomyces farinosus*
Knockranny Wood	*Laccaria laccata*	*Laccaria laccata*
Knockranny Wood	*Laccaria laccata* var. *amethystea*	*Laccaria amethystina*
Knockranny Wood	*Laccaria laccata* var. *tortilis*	*Laccaria tortilis*
Knockranny Wood	*Lachnea scutellata*	*Scutellinia scutellata*
Knockranny Wood	*Lachnea stercorea*	*Cheilymenia stercorea*
Knockranny Wood	*Lachnum niveum*	*Lachnum niveum*
Knockranny Wood	*Lachnum virgineum*	*Lachnum virgineum*
Knockranny Wood	*Lactarius blennius*	*Lactarius blennius*
Knockranny Wood	*Lactarius camphoratus*	*Lactarius camphoratus*
Knockranny Wood	*Lactarius chrysorrheus*	*Lactarius chrysorrheus*
Knockranny Wood	*Lactarius cimicarius*	*Lactarius subumbonatus*
Knockranny Wood	*Lactarius deliciosus*	*Lactarius deliciosus*
Knockranny Wood	*Lactarius glyciosmus*	*Lactarius glyciosmus*
Knockranny Wood	*Lactarius mitissimus*	*Lactarius aurantiacus*
Knockranny Wood	*Lactarius pallidus*	*Lactarius pallidus*
Knockranny Wood	*Lactarius quietus*	*Lactarius quietus*
Knockranny Wood	*Lactarius rufus*	*Lactarius rufus*
Knockranny Wood	*Lactarius serifluus*	*Lactarius subumbonatus*
Knockranny Wood	*Lactarius subdulcis*	*Lactarius subdulcis*
Knockranny Wood	*Lactarius turpis*	*Lactarius turpis*
Knockranny Wood	*Lactarius vellereus*	*Lactarius vellereus*
Knockranny Wood	*Lasiosphaeria canescens*	*Lasiosphaeria canescens*
Knockranny Wood	*Lasiosphaeria ovina*	*Lasiosphaeria ovina*
Knockranny Wood	*Lenzites betulinus*	*Lenzites betulinus*
Knockranny Wood	*Leotia lubrica*	*Leotia lubrica*
Knockranny Wood	*Lepiota amianthina*	*Cystoderma amianthinum*
Knockranny Wood	*Lepiota carcharias*	*Cystoderma carcharias*
Knockranny Wood	*Leptonia chloropolia*	*Entoloma chloropolium*
Knockranny Wood	*Leptonia formosa*	*Entoloma formosum*
Knockranny Wood	*Leptonia lampropus*	*Entoloma sodale*
Knockranny Wood	*Leptonia sericella*	*Entoloma sericellum*
Knockranny Wood	*Lycoperdon depressum*	*Vascellum pratense*
Knockranny Wood	*Lycoperdon pyriforme*	*Lycoperdon pyriforme*
Knockranny Wood	*Marasmius erythropus*	*Collybia erythropus*
Knockranny Wood	*Marasmius ramealis*	*Marasmiellus ramealis*

(Continued)

Appendix 2 *(Continued)*

Locality	Original name	Current name
Knockranny Wood	*Melampsoridium betulinum*	*Melampsoridium betulinum*
Knockranny Wood	*Merulius tremellosus*	*Phlebia tremellosa*
Knockranny Wood	*Mollisia cinerea*	*Mollisia cinerea*
Knockranny Wood	*Mollisia melaleuca*	*Mollisia melaleuca*
Knockranny Wood	*Mucor mucedo*	*Mucor mucedo*
Knockranny Wood	*Mycena amicta*	*Mycena amicta*
Knockranny Wood	*Mycena corticola*	*Mycena hiemalis*
Knockranny Wood	*Mycena epipterygia*	*Mycena epipterygia*
Knockranny Wood	*Mycena filopes*	*Mycena vitilis*
Knockranny Wood	*Mycena galericulata*	*Mycena galericulata*
Knockranny Wood	*Mycena galopus*	*Mycena galopus* var. *galopus*
Knockranny Wood	*Mycena hiemalis*	*Mycena olida*
Knockranny Wood	*Mycena luteoalba*	*Mycena luteoalba*
Knockranny Wood	*Mycena metata*	*Mycena leptocephala*
Knockranny Wood	*Mycena polygramma*	*Mycena polygramma*
Knockranny Wood	*Mycena pura*	*Mycena pura*
Knockranny Wood	*Mycena rugosa*	*Mycena galericulata*
Knockranny Wood	*Mycena sanguinolenta*	*Mycena sanguinolenta*
Knockranny Wood	*Mycena tenella*	*Mycena vitrea* var. *tenella*
Knockranny Wood	*Mycena tenerrima*	*Mycena adscendens*
Knockranny Wood	*Mycena tenuis*	*Mycena tenuis*
Knockranny Wood	*Naucoria escharioides*	*Naucoria escharioides*
Knockranny Wood	*Naucoria melinoides*	*Galerina mniophila*
Knockranny Wood	*Naucoria myosotis*	*Hypholoma myosotis*
Knockranny Wood	*Nolanea pascua*	*Entoloma pascuum*
Knockranny Wood	*Nyctalis asterophora*	*Asterophora lycoperdoides*
Knockranny Wood	*Nyctalis parasitica*	*Asterophora parasitica*
Knockranny Wood	*Omphalia fibula*	*Rickenella fibula*
Knockranny Wood	*Omphalia grisea*	*Omphalia grisea*
Knockranny Wood	*Omphalia integrella*	*Delicatula integrella*
Knockranny Wood	*Omphalia striipilea*	*Tephrocybe striipilea*
Knockranny Wood	*Orbilia leucostigma*	*Orbilia leucostigma*
Knockranny Wood	*Orbilia xanthostigma*	*Orbilia xanthostigma*
Knockranny Wood	*Panaeolus campanulatus*	*Panaeolus papilionaceus* var. *papilionaceus*
Knockranny Wood	*Panaeolus sphinctrinus*	*Panaeolus papilionaceus* var. *papilionaceus*
Knockranny Wood	*Paxillus involutus*	*Paxillus involutus*
Knockranny Wood	*Penicillium crustaceum*	*Penicillium crustaceum*
Knockranny Wood	*Peniophora velutina*	*Phanerochaete velutina*
Knockranny Wood	*Peronospora calotheca*	*Peronospora calotheca*
Knockranny Wood	*Phacidium multivalve*	*Phacidium multivalve*
Knockranny Wood	*Phlebia merismoides*	*Phlebia radiata*
Knockranny Wood	*Pholiota marginata*	*Galerina marginata*
Knockranny Wood	*Pholiota mutabilis*	*Kuehneromyces mutabilis*
Knockranny Wood	*Pholiota squarrosa*	*Pholiota squarrosa*
Knockranny Wood	*Phragmidium violaceum*	*Phragmidium violaceum*

(Continued)

Appendix 2 *(Continued)*

Locality	Original name	Current name
Knockranny Wood	*Pilobolus crystallinus*	*Pilobolus crystallinus* var. *crystallinus*
Knockranny Wood	*Pleurotus chioneus*	*Pleurotellus chioneus*
Knockranny Wood	*Pleurotus ostreatus*	*Pleurotus ostreatus*
Knockranny Wood	*Polyporus betulinus*	*Piptoporus betulinus*
Knockranny Wood	*Polyporus caesius*	*Postia caesia*
Knockranny Wood	*Polyporus fragilis*	*Postia fragilis*
Knockranny Wood	*Polyporus squamosus*	*Polyporus squamosus*
Knockranny Wood	*Polyporus sulphureus*	*Laetiporus sulphureus*
Knockranny Wood	*Polystictus abietinus*	*Trichaptum abietinum*
Knockranny Wood	*Polystictus hirsutus*	*Trametes hirsuta*
Knockranny Wood	*Polystictus velutinus*	*Trametes pubescens*
Knockranny Wood	*Polystictus versicolor*	*Trametes versicolor*
Knockranny Wood	*Poria blepharistoma*	*Ceriporia viridans*
Knockranny Wood	*Poria mucida*	*Ceriporiopsis mucida*
Knockranny Wood	*Poria sanguinolenta*	*Physisporinus sanguinolentus*
Knockranny Wood	*Poria terrestris*	*Byssocorticium terrestre*
Knockranny Wood	*Poria vaporaria*	*Antrodia sinuosa*
Knockranny Wood	*Propolis faginea*	*Propolis farinosa*
Knockranny Wood	*Psathyra corrugis*	*Psathyrella corrugis*
Knockranny Wood	*Psathyra corrugis* var. *vinosa*	*Psathyrella candolleana*
Knockranny Wood	*Psathyra fibrillosa*	*Psathyrella friesii*
Knockranny Wood	*Psathyrella atomata*	*Psathyrella atomata*
Knockranny Wood	*Psathyrella gracilis*	*Psathyrella corrugis*
Knockranny Wood	*Psilocybe foenisecii*	*Panaeolina foenisecii*
Knockranny Wood	*Psilocybe semilanceata*	*Psilocybe semilanceata*
Knockranny Wood	*Puccinia bunii*	*Puccinia bulbocastani*
Knockranny Wood	*Puccinia graminis*	*Puccinia graminis* subsp. *graminis*
Knockranny Wood	*Puccinia menthae*	*Puccinia menthae*
Knockranny Wood	*Puccinia primulae*	*Puccinia primulae*
Knockranny Wood	*Puccinia violae*	*Puccinia violae*
Knockranny Wood	*Rhopographus filicinus*	*Rhopographus filicinus*
Knockranny Wood	*Rhytisma acerinum*	*Rhytisma acerinum*
Knockranny Wood	*Russula adusta*	*Russula adusta*
Knockranny Wood	*Russula caerulea*	*Russula caerulea*
Knockranny Wood	*Russula consobrina*	*Russula consobrina*
Knockranny Wood	*Russula consobrina* var. *sororia*	*Russula sororia*
Knockranny Wood	*Russula cyanoxantha*	*Russula cyanoxantha*
Knockranny Wood	*Russula depallens*	*Russula exalbicans*
Knockranny Wood	*Russula drimeia*	*Russula sardonia*
Knockranny Wood	*Russula emetica*	*Russula silvestris*
Knockranny Wood	*Russula fellea*	*Russula fellea*
Knockranny Wood	*Russula foetens*	*Russula foetens*
Knockranny Wood	*Russula fragilis*	*Russula silvestris*
Knockranny Wood	*Russula fragilis* var. *nivea*	*Russula raoultii*
Knockranny Wood	*Russula fragilis* var. *violascens*	*Russula fragilis* var. *fragilis*

(Continued)

Appendix 2 *(Continued)*

Locality	Original name	Current name
Knockranny Wood	*Russula lepida*	*Russula rosea*
Knockranny Wood	*Russula nigricans*	*Russula nigricans*
Knockranny Wood	*Russula nitida*	*Russula cuprea*
Knockranny Wood	*Russula ochroleuca*	*Russula ochroleuca*
Knockranny Wood	*Russula olivascens*	*Russula xerampelina* var. *olivascens*
Knockranny Wood	*Russula puellaris*	*Russula puellaris*
Knockranny Wood	*Russula roseipes*	*Russula nitida*
Knockranny Wood	*Russula sardonia*	*Russula luteotacta*
Knockranny Wood	*Russula vesca*	*Russula vesca*
Knockranny Wood	*Russula xerampelina*	*Russula xerampelina*
Knockranny Wood	*Sebacina incrustans*	*Sebacina incrustans*
Knockranny Wood	*Sepedonium chrysospermum*	*Hypomyces chrysospermus*
Knockranny Wood	*Septoria violae*	*Septoria violae-palustris*
Knockranny Wood	*Spinellus fusiger*	*Spinellus fusiger*
Knockranny Wood	*Sporodinia aspergillus*	*Syzygites megalocarpus*
Knockranny Wood	*Stereum ochroleucum*	*Stereum ochroleucum*
Knockranny Wood	*Stereum rugosum*	*Stereum rugosum*
Knockranny Wood	*Stigmatea ranunculi*	*Leptotrochila ranunculi*
Knockranny Wood	*Stilbella tomentosa*	*Byssostilbe stilbigera*
Knockranny Wood	*Stropharia aeruginosa*	*Stropharia aeruginosa*
Knockranny Wood	*Stropharia semiglobata*	*Stropharia semiglobata*
Knockranny Wood	*Stropharia stercoraria*	*Stropharia semiglobata*
Knockranny Wood	*Stysanus stemonitis*	*Cephalotrichum stemonitis*
Knockranny Wood	*Thelephora laciniata*	*Thelephora caryophyllea*
Knockranny Wood	*Tremella mesenterica*	*Tremella mesenterica*
Knockranny Wood	*Trichoderma viride*	*Hypocrea schweinitzii*
Knockranny Wood	*Tricholoma flavobrunneum*	*Tricholoma fulvum*
Knockranny Wood	*Tricholoma rutilans*	*Tricholomopsis rutilans*
Knockranny Wood	*Trochila ilicina*	*Trochila ilicina*
Knockranny Wood	*Tubaria furfuracea*	*Tubaria furfuracea* var. *furfuracea*
Knockranny Wood	*Tubaria paludosa*	*Galerina paludosa*
Knockranny Wood	*Typhula erythropus*	*Typhula erythropus*
Knockranny Wood	*Ustilago hydropiperis*	*Sphacelotheca hydropiperis*
Knockranny Wood	*Ustilago scabiosae*	*Bauhinus scabiosae*
Knockranny Wood	*Ustulina vulgaris*	*Kretzschmaria deusta*
Knockranny Wood	*Valsa lata*	*Eutypa lata*
Knockranny Wood	*Xylaria hypoxylon*	*Xylaria hypoxylon*
Knockranny Wood	*Zygodesmus fuscus*	*Zygodesmus fuscus*
Louisburgh	*Aegerita candida*	*Bulbillomyces farinosus*
Louisburgh	*Amanita muscaria*	*Amanita muscaria* var. *muscaria*
Louisburgh	*Anthostoma saprophilum*	*Anthostoma saprophilum*
Louisburgh	*Armillaria mellea*	*Armillaria mellea*
Louisburgh	*Armillaria mucida*	*Oudemansiella mucida*
Louisburgh	*Ascobolus furfuraceus*	*Ascobolus stercorarius*
Louisburgh	*Ascochyta pisi*	*Ascochyta pisi*

(Continued)

Appendix 2 *(Continued)*

Locality	Original name	Current name
Louisburgh	*Belonidium pruinosum*	*Belonidium pruinosum*
Louisburgh	*Boletus scaber*	*Leccinum scabrum*
Louisburgh	*Caldesiella ferruginosa*	*Tomentella crinalis*
Louisburgh	*Cantharellus aurantiacus*	*Hygrophoropsis aurantiaca*
Louisburgh	*Chlorosplenium aeruginosum*	*Chlorociboria aeruginascens*
Louisburgh	*Clavaria contorta*	*Macrotyphula fistulosa* var. *contorta*
Louisburgh	*Clavaria juncea*	*Macrotyphula juncea*
Louisburgh	*Claviceps nigricans*	*Claviceps nigricans*
Louisburgh	*Clitopilus undatus*	*Entoloma undatum*
Louisburgh	*Coleosporium senecionis*	*Coleosporium tussilaginis*
Louisburgh	*Coleosporium sonchi*	*Coleosporium tussilaginis*
Louisburgh	*Collybia confluens*	*Collybia confluens*
Louisburgh	*Coprinus lagopus*	*Coprinopsis radiata*
Louisburgh	*Coprinus plicatilis*	*Parasola plicatilis*
Louisburgh	*Coprinus radiatus*	*Coprinopsis pseudoradiata*
Louisburgh	*Cortinarius acutus*	*Cortinarius acutus*
Louisburgh	*Cortinarius iliopodius*	*Cortinarius parvannulatus*
Louisburgh	*Cortinarius torvus*	*Cortinarius torvus*
Louisburgh	*Cystopus lepigoni*	*Albugo lepigoni*
Louisburgh	*Diatrype disciformis*	*Diatrype disciformis*
Louisburgh	*Diatrypella favacea*	*Diatrypella favacea*
Louisburgh	*Dichaena quercina*	*Ascodichaena rugosa*
Louisburgh	*Entoloma nidorosum*	*Entoloma politum*
Louisburgh	*Exoascus turgidus*	*Taphrina betulina*
Louisburgh	*Flammula helomorpha*	*Ripartites tricholoma*
Louisburgh	*Fomes annosus*	*Heterobasidion annosum*
Louisburgh	*Geopora arenicola*	*Geopora arenicola*
Louisburgh	*Grandinia granulosa*	*Dichostereum granulosum*
Louisburgh	*Helminthosporium smithii*	*Corynespora smithii*
Louisburgh	*Humaria oocardii*	*Pachyella babingtonii*
Louisburgh	*Hydnum ochraceum*	*Steccherinum ochraceum*
Louisburgh	*Hygrophorus chlorophanus*	*Hygrocybe chlorophana*
Louisburgh	*Hygrophorus niveus*	*Hygrocybe virginea* var. *virginea*
Louisburgh	*Hygrophorus obrusseus*	*Hygrocybe obrussea*
Louisburgh	*Hygrophorus pratensis*	*Hygrocybe pratensis* var. *pratensis*
Louisburgh	*Hygrophorus puniceus*	*Hygrocybe punicea*
Louisburgh	*Hygrophorus virgineus*	*Hygrocybe virginea* var. *virginea*
Louisburgh	*Inocybe geophylla*	*Inocybe geophylla* var. *geophylla*
Louisburgh	*Lachnea scutellata*	*Scutellinia scutellata*
Louisburgh	*Lactarius blennius*	*Lactarius blennius*
Louisburgh	*Lactarius helvus*	*Lactarius helvus*
Louisburgh	*Lactarius piperatus*	*Lactarius piperatus*
Louisburgh	*Lactarius pyrogalus*	*Lactarius pyrogalus*
Louisburgh	*Lactarius torminosus*	*Lactarius torminosus*
Louisburgh	*Lepiota granulosa*	*Cystoderma granulosum*

(Continued)

Appendix 2 *(Continued)*

Locality	Original name	Current name
Louisburgh	*Leptonia chloropolia*	*Entoloma chloropolium*
Louisburgh	*Marasmius ramealis*	*Marasmiellus ramealis*
Louisburgh	*Mollisia cinerea*	*Mollisia cinerea*
Louisburgh	*Mycena ammoniaca*	*Mycena leptocephala*
Louisburgh	*Mycena polygramma*	*Mycena polygramma*
Louisburgh	*Mycena rugosa*	*Mycena galericulata*
Louisburgh	*Nyctalis asterophora*	*Asterophora lycoperdoides*
Louisburgh	*Omphalia camptophylla*	*Mycena speirea*
Louisburgh	*Orbilia inflatula*	*Orbilia auricolor*
Louisburgh	*Panaeolus papilionaceus*	*Panaeolus papilionaceus* var. *papilionaceus*
Louisburgh	*Paxillus involutus*	*Paxillus involutus*
Louisburgh	*Peronospora grisea*	*Peronospora grisea*
Louisburgh	*Pluteus cervinus*	*Pluteus cervinus*
Louisburgh	*Polystictus velutinus*	*Trametes pubescens*
Louisburgh	*Poria sanguinolenta*	*Physisporinus sanguinolentus*
Louisburgh	*Poria vaporaria*	*Antrodia sinuosa*
Louisburgh	*Psalliota campestris*	*Agaricus campestris* var. *campestris*
Louisburgh	*Psathyra fatua*	*Psathyrella fatua*
Louisburgh	*Psathyra spadiceogrisea*	*Psathyrella spadiceogrisea*
Louisburgh	*Psathyrella disseminata*	*Coprinellus disseminatus*
Louisburgh	*Psilocybe foenisecii*	*Panaeolina foenisecii*
Louisburgh	*Puccinia hypochaeridis*	*Puccinia hieracii* var. *hypochaeridis*
Louisburgh	*Puccinia porri*	*Puccinia porri*
Louisburgh	*Puccinia taraxaci*	*Puccinia hieracii* var. *hieracii*
Louisburgh	*Puccinia violae*	*Puccinia violae*
Louisburgh	*Rosellinia mammiformis*	*Rosellinia mammiformis*
Louisburgh	*Russula cyanoxantha*	*Russula cyanoxantha*
Louisburgh	*Russula drimeia*	*Russula sardonia*
Louisburgh	*Russula emetica*	*Russula silvestris*
Louisburgh	*Russula fellea*	*Russula fellea*
Louisburgh	*Russula lepida*	*Russula rosea*
Louisburgh	*Russula nigricans*	*Russula nigricans*
Louisburgh	*Sepedonium chrysospermum*	*Hypomyces chrysospermus*
Louisburgh	*Septoria scabiosicola*	*Septoria scabiosicola*
Louisburgh	*Sphaerospora trechispora*	*Scutellinia trechispora*
Louisburgh	*Sporidesmium anglicum*	*Sporidesmium anglicum*
Louisburgh	*Stereum rugosum*	*Stereum rugosum*
Louisburgh	*Tricholoma panaeolum*	*Lepista panaeola*
Louisburgh	*Tricholoma resplendens*	*Tricholoma sulphurescens*
Louisburgh	*Tricholoma sulphureum*	*Tricholoma sulphureum* var. *sulphureum*
Louisburgh	*Trochila ilicina*	*Trochila ilicina*
Louisburgh	*Uromyces trifolii*	*Uromyces trifolii*
Louisburgh	*Valsa lata*	*Eutypa lata*
Louisburgh	*Volvaria parvula*	*Volvariella pusilla]*
Louisburgh	*Volvaria speciosa*	*Volvariella gloiocephala*

(Continued)

Appendix 2 *(Continued)*

Locality	Original name	Current name
Mulranny	*Hypochnus solani*	*Thanatephorus cucumeris*
Mulranny	*Spongospora subterranea*	*Spongospora subterranea f. sp. subterranea*
Murrisk Abbey	*Russula lutea*	*Russula acetolens*
Near Leenane	*Cortinarius uliginosus*	*Cortinarius uliginosus*
Old Deer-Park Wood, Mount Browne	*Androsaceus epiphylloides*	*Marasmius epiphylloides*
Old Deer-Park Wood, Mount Browne	*Androsaceus rotula*	*Marasmius rotula*
Old Deer-Park Wood, Mount Browne	*Armillaria mellea*	*Armillaria mellea*
Old Deer-Park Wood, Mount Browne	*Armillaria mucida*	*Oudemansiella mucida*
Old Deer-Park Wood, Mount Browne	*Ascobolus furfuraceus*	*Ascobolus stercorarius*
Old Deer-Park Wood, Mount Browne	*Boletus chrysenteron*	*Boletus chrysenteron*
Old Deer-Park Wood, Mount Browne	*Boletus edulis*	*Boletus edulis*
Old Deer-Park Wood, Mount Browne	*Boletus elegans*	*Suillus grevillei*
Old Deer-Park Wood, Mount Browne	*Boletus luridus*	*Boletus luridus* var. *luridus*
Old Deer-Park Wood, Mount Browne	*Boletus piperatus*	*Chalciporus piperatus*
Old Deer-Park Wood, Mount Browne	*Boletus purpureus*	*Boletus rhodopurpureus*
Old Deer-Park Wood, Mount Browne	*Boletus subtomentosus*	*Boletus subtomentosus*
Old Deer-Park Wood, Mount Browne	*Bovista plumbea*	*Bovista plumbea*
Old Deer-Park Wood, Mount Browne	*Calocera cornea*	*Calocera cornea*
Old Deer-Park Wood, Mount Browne	*Calocera stricta*	*Calocera stricta*
Old Deer-Park Wood, Mount Browne	*Cantharellus aurantiacus*	*Hygrophoropsis aurantiaca*
Old Deer-Park Wood, Mount Browne	*Cantharellus cibarius*	*Cantharellus cibarius*
Old Deer-Park Wood, Mount Browne	*Chlorosplenium aeruginosum*	*Chlorociboria aeruginascens*
Old Deer-Park Wood, Mount Browne	*Claudopus variabilis*	*Crepidotus variabilis*
Old Deer-Park Wood, Mount Browne	*Clavaria cristata*	*Clavulina coralloides*
Old Deer-Park Wood, Mount Browne	*Clavaria muscoides*	*Clavulinopsis corniculata*
Old Deer-Park Wood, Mount Browne	*Clavaria rugosa*	*Clavulina rugosa*
Old Deer-Park Wood, Mount Browne	*Clitocybe geotropa*	*Clitocybe geotropa*
Old Deer-Park Wood, Mount Browne	*Clitopilus prunulus*	*Clitopilus prunulus*
Old Deer-Park Wood, Mount Browne	*Coniophora arida*	*Coniophora arida*
Old Deer-Park Wood, Mount Browne	*Coprinus atramentarius*	*Coprinopsis atramentaria*
Old Deer-Park Wood, Mount Browne	*Coprinus micaceus*	*Coprinellus truncorum*
Old Deer-Park Wood, Mount Browne	*Coprinus niveus*	*Coprinopsis nivea*
Old Deer-Park Wood, Mount Browne	*Coprinus plicatilis*	*Parasola plicatilis*
Old Deer-Park Wood, Mount Browne	*Coprinus radiatus*	*Coprinopsis pseudoradiata*
Old Deer-Park Wood, Mount Browne	*Corticium calceum*	*Exidiopsis calcea*
Old Deer-Park Wood, Mount Browne	*Corticium comedens*	*Vuilleminia comedens*
Old Deer-Park Wood, Mount Browne	*Cortinarius brunneus*	*Cortinarius disjungendus*
Old Deer-Park Wood, Mount Browne	*Cortinarius cinnamomeus*	*Cortinarius cinnamomeus*
Old Deer-Park Wood, Mount Browne	*Cortinarius decipiens*	*Cortinarius decipiens* var. *decipiens*
Old Deer-Park Wood, Mount Browne	*Cortinarius elatior*	*Cortinarius livido-ochraceus*
Old Deer-Park Wood, Mount Browne	*Cortinarius hinnuleus*	*Cortinarius hinnuleus*
Old Deer-Park Wood, Mount Browne	*Cortinarius paleaceus*	*Cortinarius flexipes* var. *flabellus*
Old Deer-Park Wood, Mount Browne	*Cortinarius torvus*	*Cortinarius torvus*
Old Deer-Park Wood, Mount Browne	*Coryne sarcoides*	*Ascocoryne sarcoides*
Old Deer-Park Wood, Mount Browne	*Crepidotus mollis*	*Crepidotus mollis*

(Continued)

Appendix 2 *(Continued)*

Locality	Original name	Current name
Old Deer-Park Wood, Mount Browne	*Dasyscyphus willkommii*	*Lachnellula willkommii*
Old Deer-Park Wood, Mount Browne	*Entoloma jubatum*	*Entoloma porphyrophaeum*
Old Deer-Park Wood, Mount Browne	*Entoloma porphyrophaeum*	*Entoloma porphyrophaeum*
Old Deer-Park Wood, Mount Browne	*Entoloma sericeum*	*Entoloma sericeum* var. *sericeum*
Old Deer-Park Wood, Mount Browne	*Exidia albida*	*Exidia albida*
Old Deer-Park Wood, Mount Browne	*Fistulina hepatica*	*Fistulina hepatica*
Old Deer-Park Wood, Mount Browne	*Fomes applanatus*	*Ganoderma applanatum*
Old Deer-Park Wood, Mount Browne	*Galera hypnorum*	*Galerina hypnorum*
Old Deer-Park Wood, Mount Browne	*Galera tenera*	*Conocybe tenera*
Old Deer-Park Wood, Mount Browne	*Grandinia mucida*	*Grandinia mucida*
Old Deer-Park Wood, Mount Browne	*Hebeloma crustuliniforme*	*Hebeloma crustuliniforme*
Old Deer-Park Wood, Mount Browne	*Hebeloma fastibile*	*Hebeloma mesophaeum* var. *crassipes*
Old Deer-Park Wood, Mount Browne	*Helotium claroflavum*	*Bisporella citrina*
Old Deer-Park Wood, Mount Browne	*Helotium herbarum*	*Calycina herbarum*
Old Deer-Park Wood, Mount Browne	*Helotium phyllophilum*	*Hymenoscyphus phyllophilus*
Old Deer-Park Wood, Mount Browne	*Helotium virgultorum*	*Hymenoscyphus calyculus*
Old Deer-Park Wood, Mount Browne	*Humaria granulata*	*Coprobia granulata*
Old Deer-Park Wood, Mount Browne	*Hyaloscypha leuconica*	*Hyaloscypha leuconica* var. *leuconica*
Old Deer-Park Wood, Mount Browne	*Hydnum alutaceum*	*Hyphodontia alutacea*
Old Deer-Park Wood, Mount Browne	*Hydnum repandum*	*Hydnum repandum*
Old Deer-Park Wood, Mount Browne	*Hygrophorus chlorophanus*	*Hygrocybe chlorophana*
Old Deer-Park Wood, Mount Browne	*Hygrophorus coccineus*	*Hygrocybe coccinea*
Old Deer-Park Wood, Mount Browne	*Hygrophorus conicus*	*Hygrocybe conica*
Old Deer-Park Wood, Mount Browne	*Hygrophorus miniatus*	*Hygrocybe miniata*
Old Deer-Park Wood, Mount Browne	*Hygrophorus niveus*	*Hygrocybe virginea* var. *virginea*
Old Deer-Park Wood, Mount Browne	*Hygrophorus obrusseus*	*Hygrocybe obrussea*
Old Deer-Park Wood, Mount Browne	*Hygrophorus psittacinus*	*Hygrocybe psittacina* var. *psittacina*
Old Deer-Park Wood, Mount Browne	*Hygrophorus puniceus*	*Hygrocybe punicea*
Old Deer-Park Wood, Mount Browne	*Hygrophorus squamulosus*	*Hygrophorus squamulosus*
Old Deer-Park Wood, Mount Browne	*Hygrophorus turundus*	*Hygrocybe turunda*
Old Deer-Park Wood, Mount Browne	*Hygrophorus unguinosus*	*Hygrocybe irrigata*
Old Deer-Park Wood, Mount Browne	*Hygrophorus virgineus*	*Hygrocybe virginea* var. *virginea*
Old Deer-Park Wood, Mount Browne	*Hypholoma epixanthum*	*Hypholoma epixanthum*
Old Deer-Park Wood, Mount Browne	*Hypholoma fasciculare*	*Hypholoma fasciculare* var. *fasciculare*
Old Deer-Park Wood, Mount Browne	*Hypholoma sublateritium*	*Hypholoma lateritium*
Old Deer-Park Wood, Mount Browne	*Hysterographium fraxini*	*Hysterographium fraxini*
Old Deer-Park Wood, Mount Browne	*Inocybe eutheles*	*Inocybe sindonia*
Old Deer-Park Wood, Mount Browne	*Inocybe geophylla*	*Inocybe geophylla* var. *geophylla*
Old Deer-Park Wood, Mount Browne	*Inocybe geophylla* var. *violacea*	*Inocybe geophylla* var. *lilacina*
Old Deer-Park Wood, Mount Browne	*Irpex obliquus*	*Schizopora paradoxa*
Old Deer-Park Wood, Mount Browne	*Laccaria laccata* var. *amethystea*	*Laccaria amethystina*
Old Deer-Park Wood, Mount Browne	*Lachnea scutellata*	*Scutellinia scutellata*
Old Deer-Park Wood, Mount Browne	*Lachnea stercorea*	*Cheilymenia stercorea*
Old Deer-Park Wood, Mount Browne	*Lachnum ciliare*	*Incrucipulum ciliare*
Old Deer-Park Wood, Mount Browne	*Lactarius blennius*	*Lactarius blennius*

(Continued)

Appendix 2 *(Continued)*

Locality	Original name	Current name
Old Deer-Park Wood, Mount Browne	*Lactarius chrysorrheus*	*Lactarius chrysorrheus*
Old Deer-Park Wood, Mount Browne	*Lactarius circellatus*	*Lactarius circellatus*
Old Deer-Park Wood, Mount Browne	*Lactarius glyciosmus*	*Lactarius glyciosmus*
Old Deer-Park Wood, Mount Browne	*Lactarius mitissimus*	*Lactarius aurantiacus*
Old Deer-Park Wood, Mount Browne	*Lactarius pallidus*	*Lactarius pallidus*
Old Deer-Park Wood, Mount Browne	*Lactarius serifluus*	*Lactarius subumbonatus*
Old Deer-Park Wood, Mount Browne	*Lactarius subdulcis*	*Lactarius subdulcis*
Old Deer-Park Wood, Mount Browne	*Lactarius vellereus*	*Lactarius vellereus*
Old Deer-Park Wood, Mount Browne	*Lasiosphaeria canescens*	*Lasiosphaeria canescens*
Old Deer-Park Wood, Mount Browne	*Lepiota amianthina*	*Cystoderma amianthinum*
Old Deer-Park Wood, Mount Browne	*Lepiota cristata*	*Lepiota cristata*
Old Deer-Park Wood, Mount Browne	*Leptonia formosa*	*Entoloma formosum*
Old Deer-Park Wood, Mount Browne	*Leptonia lampropus*	*Entoloma sodale*
Old Deer-Park Wood, Mount Browne	*Leptonia sericella*	*Entoloma sericellum*
Old Deer-Park Wood, Mount Browne	*Leptonia serrulata*	*Entoloma serrulatum*
Old Deer-Park Wood, Mount Browne	*Lycoperdon caelatum*	*Handkea utriformis*
Old Deer-Park Wood, Mount Browne	*Marasmius erythropus*	*Collybia erythropus*
Old Deer-Park Wood, Mount Browne	*Melampsora helioscopiae*	*Melampsora euphorbiae*
Old Deer-Park Wood, Mount Browne	*Melampsoridium betulinum*	*Melampsoridium betulinum*
Old Deer-Park Wood, Mount Browne	*Merulius tremellosus*	*Phlebia tremellosa*
Old Deer-Park Wood, Mount Browne	*Mollisia cinerea*	*Mollisia cinerea*
Old Deer-Park Wood, Mount Browne	*Mycena alcalina*	*Mycena maculata*
Old Deer-Park Wood, Mount Browne	*Mycena clavicularis*	*Mycena clavicularis*
Old Deer-Park Wood, Mount Browne	*Mycena filopes*	*Mycena vitilis*
Old Deer-Park Wood, Mount Browne	*Mycena galericulata*	*Mycena galericulata*
Old Deer-Park Wood, Mount Browne	*Mycena galopus*	*Mycena galopus* var. *galopus*
Old Deer-Park Wood, Mount Browne	*Mycena haematopus*	*Mycena haematopus*
Old Deer-Park Wood, Mount Browne	*Mycena hiemalis*	*Mycena olida*
Old Deer-Park Wood, Mount Browne	*Mycena polygramma*	*Mycena polygramma*
Old Deer-Park Wood, Mount Browne	*Mycena pterigena*	*Mycena pterigena*
Old Deer-Park Wood, Mount Browne	*Mycena pura*	*Mycena pura*
Old Deer-Park Wood, Mount Browne	*Mycena rubromarginata*	*Mycena capillaripes*
Old Deer-Park Wood, Mount Browne	*Mycena rugosa*	*Mycena galericulata*
Old Deer-Park Wood, Mount Browne	*Mycena tenella*	*Mycena vitrea* var. *tenella*
Old Deer-Park Wood, Mount Browne	*Mycena tenerrima*	*Mycena adscendens*
Old Deer-Park Wood, Mount Browne	*Naucoria escharioides*	*Naucoria escharioides*
Old Deer-Park Wood, Mount Browne	*Naucoria temulenta*	*Agrocybe pediades*
Old Deer-Park Wood, Mount Browne	*Nectria coccinea*	*Nectria coccinea*
Old Deer-Park Wood, Mount Browne	*Nectria episphaeria*	*Nectria episphaeria*
Old Deer-Park Wood, Mount Browne	*Nolanea pascua*	*Entoloma pascuum*
Old Deer-Park Wood, Mount Browne	*Nyctalis parasitica*	*Asterophora parasitica*
Old Deer-Park Wood, Mount Browne	*Omphalia integrella*	*Delicatula integrella*
Old Deer-Park Wood, Mount Browne	*Omphalia umbellifera*	*Lichenomphalia umbellifera*
Old Deer-Park Wood, Mount Browne	*Orbilia xanthostigma*	*Orbilia xanthostigma*
Old Deer-Park Wood, Mount Browne	*Panaeolus campanulatus*	*Panaeolus papilionaceus* var. *papilionaceus*

(Continued)

Appendix 2 *(Continued)*

Locality	Original name	Current name
Old Deer-Park Wood, Mount Browne	*Panaeolus papilionaceus*	*Panaeolus papilionaceus* var. *papilionaceus*
Old Deer-Park Wood, Mount Browne	*Peniophora cinerea*	*Peniophora cinerea*
Old Deer-Park Wood, Mount Browne	*Peziza cerea*	*Peziza cerea*
Old Deer-Park Wood, Mount Browne	*Phlebia merismoides*	*Phlebia radiata*
Old Deer-Park Wood, Mount Browne	*Pholiota marginata*	*Galerina marginata*
Old Deer-Park Wood, Mount Browne	*Phragmidium violaceum*	*Phragmidium violaceum*
Old Deer-Park Wood, Mount Browne	*Phyllachora graminis*	*Phyllachora graminis* var. *graminis*
Old Deer-Park Wood, Mount Browne	*Pilobolus crystallinus*	*Pilobolus crystallinus* var. *crystallinus*
Old Deer-Park Wood, Mount Browne	*Pleurotus acerosus*	*Arrhenia acerosa*
Old Deer-Park Wood, Mount Browne	*Pleurotus chioneus*	*Pleurotellus chioneus*
Old Deer-Park Wood, Mount Browne	*Polyporus adustus*	*Bjerkandera adusta*
Old Deer-Park Wood, Mount Browne	*Polyporus betulinus*	*Piptoporus betulinus*
Old Deer-Park Wood, Mount Browne	*Polyporus squamosus*	*Polyporus squamosus*
Old Deer-Park Wood, Mount Browne	*Polyporus sulphureus*	*Laetiporus sulphureus*
Old Deer-Park Wood, Mount Browne	*Polystictus abietinus*	*Trichaptum abietinum*
Old Deer-Park Wood, Mount Browne	*Polystictus velutinus*	*Trametes pubescens*
Old Deer-Park Wood, Mount Browne	*Polystictus versicolor*	*Trametes versicolor*
Old Deer-Park Wood, Mount Browne	*Propolis faginea*	*Propolis farinosa*
Old Deer-Park Wood, Mount Browne	*Psathyra bifrons* var. *semitincta*	*Psathyrella bifrons*
Old Deer-Park Wood, Mount Browne	*Psathyra corrugis*	*Psathyrella corrugis*
Old Deer-Park Wood, Mount Browne	*Psathyrella atomata*	*Psathyrella atomata*
Old Deer-Park Wood, Mount Browne	*Psathyrella disseminata*	*Coprinellus disseminatus*
Old Deer-Park Wood, Mount Browne	*Psathyrella gracilis*	*Psathyrella corrugis*
Old Deer-Park Wood, Mount Browne	*Psilocybe foenisecii*	*Panaeolina foenisecii*
Old Deer-Park Wood, Mount Browne	*Psilocybe semilanceata*	*Psilocybe semilanceata*
Old Deer-Park Wood, Mount Browne	*Radulum quercinum*	*Ascodichaena rugosa*
Old Deer-Park Wood, Mount Browne	*Rhytisma acerinum*	*Rhytisma acerinum*
Old Deer-Park Wood, Mount Browne	*Russula caerulea*	*Russula caerulea*
Old Deer-Park Wood, Mount Browne	*Russula cyanoxantha*	*Russula cyanoxantha*
Old Deer-Park Wood, Mount Browne	*Russula emetica*	*Russula silvestris*
Old Deer-Park Wood, Mount Browne	*Russula fellea*	*Russula fellea*
Old Deer-Park Wood, Mount Browne	*Russula foetens*	*Russula foetens*
Old Deer-Park Wood, Mount Browne	*Russula fragilis*	*Russula silvestris*
Old Deer-Park Wood, Mount Browne	*Russula fragilis* var. *violascens*	*Russula fragilis* var. *fragilis*
Old Deer-Park Wood, Mount Browne	*Russula nigricans*	*Russula nigricans*
Old Deer-Park Wood, Mount Browne	*Russula ochroleuca*	*Russula ochroleuca*
Old Deer-Park Wood, Mount Browne	*Russula rubra*	*Russula atropurpurea*
Old Deer-Park Wood, Mount Browne	*Russula sardonia*	*Russula luteotacta*
Old Deer-Park Wood, Mount Browne	*Russula vesca*	*Russula vesca*
Old Deer-Park Wood, Mount Browne	*Sepedonium chrysospermum*	*Hypomyces chrysospermus*
Old Deer-Park Wood, Mount Browne	*Stereum hirsutum*	*Stereum hirsutum*
Old Deer-Park Wood, Mount Browne	*Stereum purpureum*	*Chondrostereum purpureum*
Old Deer-Park Wood, Mount Browne	*Stereum spadiceum*	*Stereum gausapatum*
Old Deer-Park Wood, Mount Browne	*Stilbella tomentosa*	*Byssostilbe stilbigera*
Old Deer-Park Wood, Mount Browne	*Stropharia semiglobata*	*Stropharia semiglobata*

(Continued)

Appendix 2 *(Continued)*

Locality	Original name	Current name
Old Deer-Park Wood, Mount Browne	*Stropharia squamosa*	*Stropharia squamosa* var. *squamosa*
Old Deer-Park Wood, Mount Browne	*Tremella mesenterica*	*Tremella mesenterica*
Old Deer-Park Wood, Mount Browne	*Tricholoma flavobrunneum*	*Tricholoma fulvum*
Old Deer-Park Wood, Mount Browne	*Tricholoma rutilans*	*Tricholomopsis rutilans*
Old Deer-Park Wood, Mount Browne	*Trochila ilicina*	*Trochila ilicina*
Old Deer-Park Wood, Mount Browne	*Tubaria furfuracea*	*Tubaria furfuracea* var. *furfuracea*
Old Deer-Park Wood, Mount Browne	*Ustilago hydropiperis*	*Sphacelotheca hydropiperis*
Old Deer-Park Wood, Mount Browne	*Ustulina vulgaris*	*Kretzschmaria deusta*
Old Deer-Park Wood, Mount Browne	*Xylaria hypoxylon*	*Xylaria hypoxylon*
Old Deer-Park Wood, Mount Browne	*Zygodesmus fuscus*	*Zygodesmus fuscus*
Roonagh Quay	*Gibberella cyanogena*	*Gibberella cyanogena*
Westport	*Amanitopsis vaginata*	*Amanita vaginata* var. *vaginata*
Westport	*Boletus bovinus*	*Suillus bovinus*
Westport	*Coleosporium tussilaginis*	*Coleosporium tussilaginis*
Westport	*Coryne sarcoides*	*Ascocoryne sarcoides*
Westport	*Diatrypella verruciformis*	*Diatrypella favacea*
Westport	*Fomes applanatus*	*Ganoderma applanatum*
Westport	*Fusicladium depressum*	*Passalora depressa*
Westport	*Geopyxis cupularis*	*Tarzetta cupularis*
Westport	*Grandinia granulosa*	*Dichostereum granulosum*
Westport	*Helotium claroflavum*	*Bisporella citrina*
Westport	*Helotium herbarum*	*Calycina herbarum*
Westport	*Helotium scutula*	*Hymenoscyphus scutula*
Westport	*Helotium virgultorum* var. *fructigenum*	*Hymenoscyphus fructigenus*
Westport	*Hysterium angustatum*	*Hysterium angustatum*
Westport	*Inocybe asterospora*	*Inocybe asterospora*
Westport	*Inocybe pyriodora*	*Inocybe fraudans*
Westport	*Inocybe whitei*	*Inocybe whitei*
Westport	*Lactarius piperatus*	*Lactarius piperatus*
Westport	*Lactarius pubescens*	*Lactarius pubescens*
Westport	*Lasiosphaeria ovina*	*Lasiosphaeria ovina*
Westport	*Leotia lubrica*	*Leotia lubrica*
Westport	*Melampsora lini*	*Melampsora lini* var. *lini*
Westport	*Mycena capillaris*	*Mycena capillaris*
Westport	*Mycena gypsea*	*Hemimycena cucullata*
Westport	*Naucoria scolecina*	*Naucoria scolecina*
Westport	*Peziza succosa*	*Peziza succosa*
Westport	*Pholiota terrigena*	*Inocybe terrigena*
Westport	*Phragmidium violaceum*	*Phragmidium violaceum*
Westport	*Plasmopara nivea*	*Plasmopara crustosa*
Westport	*Poria vaporaria*	*Antrodia sinuosa*
Westport	*Psilocybe canobrunnea*	*Psilocybe canobrunnea*
Westport	*Puccinia centaureae*	*Puccinia calcitrapae*
Westport	*Puccinia hieracii*	*Puccinia hieracii* var. *hieracii*

(Continued)

Appendix 2 *(Continued)*

Locality	Original name	Current name
Westport	*Puccinia menthae*	*Puccinia menthae*
Westport	*Russula ochracea*	*Russula risigallina*
Westport	*Septoria scabiosicola*	*Septoria scabiosicola*
Westport	*Stigmatea rumicis*	*Venturia rumicis*
Westport	*Tichothecium pygmaeum*	*Muellerella pygmaea* var. *pygmaea*
Westport Park	*Amanitopsis vaginata*	*Amanita vaginata* var. *vaginata*
Westport Park	*Androsaceus graminum*	*Marasmius curreyi*
Westport Park	*Anellaria fimiputris*	*Panaeolus semiovatus* var. *semiovatus*
Westport Park	*Armillaria mellea*	*Armillaria mellea*
Westport Park	*Armillaria mucida*	*Oudemansiella mucida*
Westport Park	*Ascobolus furfuraceus*	*Ascobolus stercorarius*
Westport Park	*Boletus granulatus*	*Suillus granulatus*
Westport Park	*Boletus luridus*	*Boletus luridus* var. *luridus*
Westport Park	*Boletus scaber*	*Leccinum scabrum*
Westport Park	*Bovista plumbea*	*Bovista plumbea*
Westport Park	*Caldesiella ferruginosa*	*Tomentella crinalis*
Westport Park	*Cantharellus aurantiacus*	*Hygrophoropsis aurantiaca*
Westport Park	*Chlorosplenium aeruginosum*	*Chlorociboria aeruginascens*
Westport Park	*Claudopus variabilis*	*Crepidotus variabilis*
Westport Park	*Clavaria cinerea*	*Clavulina cinerea f. cinerea*
Westport Park	*Clavaria cristata*	*Clavulina coralloides*
Westport Park	*Clavaria dissipabilis*	*Clavulinopsis helvola*
Westport Park	*Clavaria muscoides*	*Clavulinopsis corniculata*
Westport Park	*Clavaria rugosa*	*Clavulina rugosa*
Westport Park	*Clavaria vermicularis*	*Clavaria fragilis*
Westport Park	*Clitocybe fragrans*	*Clitocybe fragrans*
Westport Park	*Clitocybe geotropa*	*Clitocybe geotropa*
Westport Park	*Clitocybe infundibuliformis*	*Lepista flaccida*
Westport Park	*Clitocybe metachroa*	*Clitocybe metachroa*
Westport Park	*Clitocybe nebularis*	*Clitocybe nebularis*
Westport Park	*Coccomyces coronatus*	*Coccomyces coronatus*
Westport Park	*Collybia butyracea*	*Collybia butyracea* var. *butyracea*
Westport Park	*Collybia conigena*	*Collybia conigena*
Westport Park	*Collybia dryophila*	*Collybia dryophila*
Westport Park	*Collybia tenacella*	*Strobilurus tenacellus*
Westport Park	*Coprinus atramentarius*	*Coprinopsis atramentaria*
Westport Park	*Coprinus comatus*	*Coprinus comatus*
Westport Park	*Coprinus lagopus*	*Coprinopsis radiata*
Westport Park	*Coprinus micaceus*	*Coprinellus truncorum*
Westport Park	*Coprinus niveus*	*Coprinopsis nivea*
Westport Park	*Coprinus plicatilis*	*Parasola plicatilis*
Westport Park	*Corticium lacteum*	*Phanerochaete tuberculata*
Westport Park	*Corticium laeve*	*Cylindrobasidium laeve*
Westport Park	*Corticium sambuci*	*Hyphodontia sambuci*

(Continued)

Appendix 2 *(Continued)*

Locality	Original name	Current name
Westport Park	*Cortinarius elatior*	*Cortinarius livido-ochraceus*
Westport Park	*Cortinarius obtusus*	*Cortinarius obtusus*
Westport Park	*Cortinarius psammocephalus*	*Cortinarius angelesianus*
Westport Park	*Cortinarius scandens*	*Cortinarius scandens*
Westport Park	*Coryne urnalis*	*Ascocoryne cylichnium*
Westport Park	*Crepidotus calolepis*	*Crepidotus calolepis*
Westport Park	*Cylindrium flavovirens*	*Cylindrium flavovirens*
Westport Park	*Dacrymyces deliquescens*	*Dacrymyces stillatus*
Westport Park	*Daedalea quercina*	*Daedalea quercina*
Westport Park	*Dasyscyphus willkommii*	*Lachnellula willkommii*
Westport Park	*Dichaena quercina*	*Ascodichaena rugosa*
Westport Park	*Entoloma griseocyaneum*	*Entoloma griseocyaneum*
Westport Park	*Entoloma jubatum*	*Entoloma porphyrophaeum*
Westport Park	*Entoloma prunuloides*	*Entoloma prunuloides*
Westport Park	*Eurotium herbariorum*	*Eurotium herbariorum*
Westport Park	*Exidia albida*	*Exidia albida*
Westport Park	*Flammula scamba*	*Pholiota scamba*
Westport Park	*Fomes annosus*	*Heterobasidion annosum*
Westport Park	*Fomes applanatus*	*Ganoderma applanatum*
Westport Park	*Fomes fomentarius*	*Fomes fomentarius*
Westport Park	*Galera hypnorum*	*Galerina hypnorum*
Westport Park	*Galera tenera*	*Conocybe tenera*
Westport Park	*Gomphidius viscidus*	*Gomphidius viscidus*
Westport Park	*Grandinia granulosa*	*Dichostereum granulosum*
Westport Park	*Grandinia mucida*	*Grandinia mucida*
Westport Park	*Hebeloma fastibile*	*Hebeloma mesophaeum* var. *crassipes*
Westport Park	*Humaria granulata*	*Coprobia granulata*
Westport Park	*Hydnum ochraceum*	*Steccherinum ochraceum*
Westport Park	*Hydnum udum*	*Mycoacia uda*
Westport Park	*Hygrophorus chlorophanus*	*Hygrocybe chlorophana*
Westport Park	*Hygrophorus conicus*	*Hygrocybe conica*
Westport Park	*Hygrophorus fornicatus*	*Hygrocybe fornicata*
Westport Park	*Hygrophorus miniatus*	*Hygrocybe miniata*
Westport Park	*Hygrophorus niveus*	*Hygrocybe virginea* var. *virginea*
Westport Park	*Hygrophorus pratensis*	*Hygrocybe pratensis* var. *pratensis*
Westport Park	*Hygrophorus psittacinus*	*Hygrocybe psittacina* var. *psittacina*
Westport Park	*Hygrophorus reae*	*Hygrocybe mucronella*
Westport Park	*Hygrophorus unguinosus*	*Hygrocybe irrigata*
Westport Park	*Hygrophorus virgineus*	*Hygrocybe virginea* var. *virginea*
Westport Park	*Hygrophorus virgineus* var. *roseipes*	*Hygrocybe virginea* var. *virginea*
Westport Park	*Hypholoma appendiculatum*	*Psathyrella candolleana*
Westport Park	*Hypholoma epixanthum*	*Hypholoma epixanthum*
Westport Park	*Hypholoma fasciculare*	*Hypholoma fasciculare* var. *fasciculare*
Westport Park	*Hypholoma hydrophila*	*Psathyrella piluliformis*
Westport Park	*Hypholoma lanaripes*	*Hypholoma lanaripes*

(Continued)

Appendix 2 *(Continued)*

Locality	Original name	Current name
Westport Park	*Hypholoma sublateritium*	*Hypholoma lateritium*
Westport Park	*Hypholoma velutinum*	*Lacrymaria lacrymabunda*
Westport Park	*Hypocrea rufa*	*Hypocrea rufa*
Westport Park	*Hypomyces aurantius*	*Hypomyces aurantius*
Westport Park	*Hypoxylon fuscum*	*Hypoxylon fuscum*
Westport Park	*Hypoxylon semi-immersum*	*Nemania confluens*
Westport Park	*Hysterographium fraxini*	*Hysterographium fraxini*
Westport Park	*Inocybe geophylla*	*Inocybe geophylla* var. *geophylla*
Westport Park	*Inocybe geophylla* var. *violacea*	*Inocybe geophylla* var. *lilacina*
Westport Park	*Irpex fuscoviolaceus*	*Trichaptum fuscoviolaceum*
Westport Park	*Irpex obliquus*	*Schizopora paradoxa*
Westport Park	*Laccaria laccata*	*Laccaria laccata*
Westport Park	*Lachnea stercorea*	*Cheilymenia stercorea*
Westport Park	*Lachnum ciliare*	*Incrucipulum ciliare*
Westport Park	*Lachnum sulphureum*	*Belonidium sulphureum*
Westport Park	*Lactarius blennius*	*Lactarius blennius*
Westport Park	*Lactarius camphoratus*	*Lactarius camphoratus*
Westport Park	*Lactarius chrysorrheus*	*Lactarius chrysorrheus*
Westport Park	*Lactarius deliciosus*	*Lactarius deliciosus*
Westport Park	*Lactarius glyciosmus*	*Lactarius glyciosmus*
Westport Park	*Lactarius mitissimus*	*Lactarius aurantiacus*
Westport Park	*Lactarius pallidus*	*Lactarius pallidus*
Westport Park	*Lactarius piperatus*	*Lactarius piperatus*
Westport Park	*Lactarius quietus*	*Lactarius quietus*
Westport Park	*Lactarius serifluus*	*Lactarius subumbonatus*
Westport Park	*Lactarius subdulcis*	*Lactarius subdulcis*
Westport Park	*Lactarius turpis*	*Lactarius turpis*
Westport Park	*Lactarius vellereus*	*Lactarius vellereus*
Westport Park	*Leptonia lampropus*	*Entoloma sodale*
Westport Park	*Leptonia sericella*	*Entoloma sericellum*
Westport Park	*Lophodermium pinastri*	*Lophodermium pinastri*
Westport Park	*Lycoperdon depressum*	*Vascellum pratense*
Westport Park	*Lycoperdon pyriforme*	*Lycoperdon pyriforme*
Westport Park	*Marasmius erythropus*	*Collybia erythropus*
Westport Park	*Marasmius ramealis*	*Marasmiellus ramealis*
Westport Park	*Melampsora laricis-populina*	*Melampsora laricis-populina*
Westport Park	*Mollisia cinerea*	*Mollisia cinerea*
Westport Park	*Mollisia melaleuca*	*Mollisia melaleuca*
Westport Park	*Mycena alcalina*	*Mycena maculata*
Westport Park	*Mycena ammoniaca*	*Mycena leptocephala*
Westport Park	*Mycena corticola*	*Mycena hiemalis*
Westport Park	*Mycena filopes*	*Mycena vitilis*
Westport Park	*Mycena flavoalba*	*Mycena flavoalba*
Westport Park	*Mycena galericulata*	*Mycena galericulata*
Westport Park	*Mycena galopus*	*Mycena galopus* var. *galopus*

(Continued)

Appendix 2 *(Continued)*

Locality	Original name	Current name
Westport Park	*Mycena lactea*	*Hemimycena lactea*
Westport Park	*Mycena leucogala*	*Mycena galopus* var. *nigra*
Westport Park	*Mycena luteoalba*	*Mycena luteoalba*
Westport Park	*Mycena polygramma*	*Mycena polygramma*
Westport Park	*Mycena pura*	*Mycena pura*
Westport Park	*Mycena rorida*	*Mycena rorida*
Westport Park	*Mycena rugosa*	*Mycena galericulata*
Westport Park	*Mycena sanguinolenta*	*Mycena sanguinolenta*
Westport Park	*Mycena setosa*	*Marasmius setosus*
Westport Park	*Mycena sudora*	*Mycena sudora*
Westport Park	*Mycena tenella*	*Mycena vitrea* var. *tenella*
Westport Park	*Mycena tenerrima*	*Mycena adscendens*
Westport Park	*Mycena tenuis*	*Mycena tenuis*
Westport Park	*Mycena virens*	*Mycena chlorantha*
Westport Park	*Naucoria badipes*	*Galerina badipes*
Westport Park	*Nectria aquifolii*	*Nectria aquifolii*
Westport Park	*Nectria cinnabarina*	*Nectria cinnabarina*
Westport Park	*Nolanea pascua*	*Entoloma pascuum*
Westport Park	*Nolanea pisciodora*	*Macrocystidia cucumis*
Westport Park	*Nyctalis parasitica*	*Asterophora parasitica*
Westport Park	*Omphalia gracillima*	*Hemimycena delectabilis*
Westport Park	*Omphalia stellata*	*Omphalia stellata*
Westport Park	*Orbilia xanthostigma*	*Orbilia xanthostigma*
Westport Park	*Panaeolus campanulatus*	*Panaeolus papilionaceus* var. *papilionaceus*
Westport Park	*Panaeolus papilionaceus*	*Panaeolus papilionaceus* var. *papilionaceus*
Westport Park	*Panaeolus sphinctrinus*	*Panaeolus papilionaceus* var. *papilionaceus*
Westport Park	*Paxillus involutus*	*Paxillus involutus*
Westport Park	*Phacidium multivalve*	*Phacidium multivalve*
Westport Park	*Pholiota marginata*	*Galerina marginata*
Westport Park	*Pholiota squarrosa*	*Pholiota squarrosa*
Westport Park	*Pholiota togularis*	*Conocybe arrhenii*
Westport Park	*Phragmidium violaceum*	*Phragmidium violaceum*
Westport Park	*Pistillaria pusilla*	*Ceratellopsis acuminata*
Westport Park	*Pleurotus applicatus*	*Resupinatus applicatus*
Westport Park	*Polyporus adustus*	*Bjerkandera adusta*
Westport Park	*Polyporus betulinus*	*Piptoporus betulinus*
Westport Park	*Polyporus caesius*	*Postia caesia*
Westport Park	*Polyporus chioneus*	*Tyromyces chioneus*
Westport Park	*Polyporus squamosus*	*Polyporus squamosus*
Westport Park	*Polystictus versicolor*	*Trametes versicolor*
Westport Park	*Poria mollusca*	*Trechispora mollusca*
Westport Park	*Poria terrestris*	*Byssocorticium terrestre*
Westport Park	*Psalliota arvensis*	*Agaricus arvensis*

(Continued)

Appendix 2 *(Continued)*

Locality	Original name	Current name
Westport Park	*Psalliota campestris*	*Agaricus campestris* var. *campestris*
Westport Park	*Psalliota haemorrhoidaria*	*Agaricus silvaticus*
Westport Park	*Psathyra bifrons*	*Psathyrella bifrons*
Westport Park	*Psathyra corrugis*	*Psathyrella corrugis*
Westport Park	*Psathyra fibrillosa*	*Psathyrella friesii*
Westport Park	*Psathyra gossypina*	*Psathyrella gossypina*
Westport Park	*Psathyrella atomata*	*Psathyrella atomata*
Westport Park	*Psathyrella disseminata*	*Coprinellus disseminatus*
Westport Park	*Psathyrella gracilis*	*Psathyrella corrugis*
Westport Park	*Psilocybe clivensis*	*Psathyrella clivensis*
Westport Park	*Psilocybe ericaea*	*Hypholoma ericaeum*
Westport Park	*Psilocybe semilanceata*	*Psilocybe semilanceata*
Westport Park	*Puccinia bunii*	*Puccinia bulbocastani*
Westport Park	*Rhinotrichum thwaitesii*	*Rhinotrichum thwaitesii*
Westport Park	*Rosellinia aquila*	*Rosellinia aquila*
Westport Park	*Russula caerulea*	*Russula caerulea*
Westport Park	*Russula cyanoxantha*	*Russula cyanoxantha*
Westport Park	*Russula depallens*	*Russula exalbicans*
Westport Park	*Russula drimeia*	*Russula sardonia*
Westport Park	*Russula emetica*	*Russula silvestris*
Westport Park	*Russula fellea*	*Russula fellea*
Westport Park	*Russula fragilis*	*Russula silvestris*
Westport Park	*Russula fragilis* var. *violascens*	*Russula fragilis* var. *fragilis*
Westport Park	*Russula nigricans*	*Russula nigricans*
Westport Park	*Russula ochroleuca*	*Russula ochroleuca*
Westport Park	*Russula puellaris*	*Russula puellaris*
Westport Park	*Russula xerampelina*	*Russula xerampelina*
Westport Park	*Rutstroemia firma*	*Rutstroemia firma*
Westport Park	*Scleroderma vulgare*	*Scleroderma citrinum*
Westport Park	*Spinellus fusiger*	*Spinellus fusiger*
Westport Park	*Stereum purpureum*	*Chondrostereum purpureum*
Westport Park	*Stropharia aeruginosa*	*Stropharia aeruginosa*
Westport Park	*Stropharia semiglobata*	*Stropharia semiglobata*
Westport Park	*Stropharia stercoraria*	*Stropharia semiglobata*
Westport Park	*Tremella frondosa*	*Tremella foliacea*
Westport Park	*Tricholoma imbricatum*	*Tricholoma imbricatum*
Westport Park	*Tricholoma melaleucum*	*Melanoleuca melaleuca*
Westport Park	*Tubaria furfuracea*	*Tubaria furfuracea* var. *furfuracea*
Westport Park	*Typhula erythropus*	*Typhula erythropus*
Westport Park	*Urceolella incarnatina*	*Unguicularia incarnatina*
Westport Park	*Ustilago hydropiperis*	*Sphacelotheca hydropiperis*
Westport Park	*Ustulina vulgaris*	*Kretzschmaria deusta*
Westport Park	*Zygodesmus fuscus*	*Zygodesmus fuscus*

BIBLIOGRAPHY

Adam, P. 1977 On the phytosociological status of *Juncus maritimus* on British saltmarshes. *Vegetatio* **35**, 81–94.

Adam, P. 1978 Geographical variation in British saltmarsh vegetation. *Journal of Ecology* **66**, 339–66.

Adamcík, S. and Kautmanová, I. 2005 *Hygrocybe* species as indicators of natural value of grasslands in Slovakia. *Catathelasma* **6**, 24–34.

Agri-environment Monitoring Unit 2004 Monitoring of the chough option in the Antrim Coast, Glens and Rathlin Environmentally Sensitive Area 1998–2002. Belfast. Queen's University Belfast.

Armitage, E. 1938 The British Bryological Society. *Journal of Botany* **76**, 171–4.

Arnolds, E. 1980 De oecologie en Sociologie van Wasplaten (*Hygrophorus* subgenus *Hygrocybe sensu lato*). *Natura* **77**, 17–44.

Arnolds, E. 1988 The changing macromycete flora in the Netherlands. *Transactions of the British Mycological Society* **90**, 391–406.

Arnolds, E. 1994 Paddestoelen en graslandbeheer. In T. Kuyper (ed.), *Paddestoelen en natuurbeheer: wat kan de beheerder?* 74–89. Hoogwoud, Holland. Wetenshappelijke Mededeling KNNV.

Arnolds, E. 1995 Conservation and management of natural populations of edible fungi. *Canadian Journal of Botany* **73**, S987–S998.

Ausden, M. and Treweek, J. 1995 Management of grasslands: techniques. In W.J. Sutherland and D.A. Hill (eds), *Managing habitats for conservation*, 205–17. Cambridge. Cambridge University Press.

Averis, A., Averis, B., Birk, J., Horsfield, D., Thompson, D. and Yeo, M. 2004 *An illustrated guide to British upland vegetation*. Peterborough. JNCC.

Bailey, J.S. 1994 Nutrient balance: the key to solving the phosphate problem. *Topics, Journal of the Milk Marketing Board for Northern Ireland* **16–17**.

Bailey, M.L. 1984 *Air quality in Ireland, the present position*. Dublin. An Foras Fórbatha.

Ball, D.F., Dale, J., Sheail, J. and Heal, O.W. 1982 *Vegetation change in upland landscapes*. Bangor. Institute of Terrestrial Ecology.

Bardgett, R.D. and McAlister, E. 1999 The measurement of soil fungal: bacterial biomass ratios as an indicator of ecosystem self-regulation in temperate meadow grassands. *Biology and Fertility of Soils* **29**, 282–90.

Barkmann, J.J., Moravec, J. and Rauschert, S. 1986 Code of phytosociological nomenclature. *Vegetatio* **67**, 145–58.

Beckers, A., Brock, T.H. and Klerkx, J. 1976 A vegetation study of some parts of Dooaghtry, Co. Mayo, Republic of Ireland. Unpublished MSc thesis, Laboratory for Geobotany, Catholic University, Nijmegen.

Birks, H.J.B. 1973 *Past and present vegetation of the Isle of Skye. A palaeoecological study.* Cambridge. Cambridge University Press.

Birks, H.J.B., Birks, H.H. and Ratcliffe, D.A. 1969 Mountain plants on Slieve League, Co. Donegal. *Irish Naturalists' Journal* **16**, 203.

Birse, E.L. 1980 *Plant communities of Scotland: a preliminary phytocoenonia.* Aberdeen. Macaulay Institute for Soil Research.

Birse, E.L. and Robertson, J.S. 1976 *Plant communities and soils of the Lowland and southern Uplands regions of Scotland.* Aberdeen. Macaulay Institute for Soil Research.

Bleasdale, A. 1998 Overgrazing in the west of Ireland—assessing solutions. In G. O'Leary and F. Gormley (eds), *Towards a conservation strategy for the bogs of Ireland*, 67–78. Dublin. Irish Peatland Conservation Council.

Bleasdale, A. and Conaghan, J. 1995 Flushes and springs in the Connemara hills and uplands, Co. Galway, Ireland. *Bulletin of the British Ecological Society* **26**, 28–35.

Bleasdale, A. and Sheehy-Skeffington, M. 1992 The influence of agricultural practices on plant communities in Connemara. In J. Feehan (ed.), *Environment and Development in Ireland*, 331–6. Dublin. University College Dublin Environment Institute.

Bleasdale, A. and Sheehy-Skeffington, M. 1995 The upland vegetation of north-east Connemara in relation to sheep grazing. In D.W. Jeffrey, M.B. Jones and J.H. McAdam (eds), *Irish grasslands—their biology and management*, 110–24. Dublin. Royal Irish Academy.

Blockeel, T.L. 1991 The *Racomitrium heterostichum* group in the British Isles. *Bulletin of the British Bryological Society* **58**, 29–35.

Blockeel, T.L. 1995 Summer field meeting, 1994, second week, Clifden. *Bulletin of the British Bryological Society* **65**, 12–18.

Bobbink, R., Hornung, M. and Roelofs, J.G.M. 1998 The effects of air-borne pollutants in natural and semi-natural European vegetation. *Journal of Ecology* **86**, 717–38.

Boertmann, D. 1995 *The genus Hygrocybe.* Copenhagen. The Danish Mycological society.

Boyle, G.M., Farrell, E.P. and Cummins, T. 1997 *Monitoring of forest ecosystems in Ireland, FOREM 3 Project Final Report.* Forest Ecosystem Research Group Number 21. Dublin. Department of Environmental resource management, UCD.

Braun-Blanquet, J. and Tüxen, T. 1952 Irische Plflanzengesellschaften. *Veroffentlichungen des Geobotanischen Institutes Rübel in Zurich* **25**, 224–415.

Brock, T., Frigge, P. and van der Ster, H. 1978 A vegetation study of the pools and surrounding wetlands in the Dooaghtry area, Co. Mayo, Republic of Ireland. Unpublished MSc thesis, Laboratory for Geobotany, Catholic University, Nijmegen.

Brodie, J. 1991 Some observations on the flora of Clare Island, western Ireland. *Irish Naturalists' Journal* **23** (9), 376–7.

Brodie, J. and Sheehy-Skeffington, M. 1990 Inishbofin: a resurvey of the flora. *Irish Naturalists' Journal* **23**, 293–8.

Browne, A. 1998 Vegetation–environment interactions in the vicinity of a pharmaceutical plant near Kinsale, Co. Cork. Unpublished PhD thesis, University College Dublin.

Browne, J.F. 1991 The glacial geomorphology of Clare Island, Co. Mayo. Unpublished BA thesis, University of Dublin, Trinity College.

Browne, P. 1986 Vegetational history of the Nephin Beg mountains, County Mayo. Unpublished PhD thesis, University of Dublin, Trinity College.

Bunce, R.G.H. 1989 *Heather in England and Wales*. Institute of Terrestrial Ecology, Research publication No. 3. London. Her Majesty's Stationery Office.

Carter, R.W.G. 1988 *Coastal environments: An introduction to the physical, ecological and cultural systems of coastlines*. London. Academic Press.

Caulfield, S. 1978 Neolithic fields: the Irish evidence. In H.C. Bowen and P.J. Fowler (eds), *Early land allotment in the British Isles*, 137–43. British Archeological Reports 48. Oxford.

Caulfield, S. 1983 The Neolithic settlement of N. Connaught. In T. Reeves-Smith and F. Hammond (eds), *Landscape archaeology in Ireland*, 195–215, British Archeological Reports 116. Oxford.

Clark, R.L. 1982 Point count estimation of charcoal in pollen preparations and thin sections of sediments. *Pollen et Spores* **24**, 523–35.

Cole, G.A.J., Kilroe, J.R., Hallissy, T. and Newell Arber, E.A. 1914 *The geology of Clare Island, Co. Mayo*. Memoirs of the Geological Survey of Ireland. Dublin. HMSO.

Colgan, N. 1901 Notes on Irish Topographical Botany, with some remarks on floral diversity. *Irish Naturalists' Journal* **10**, 232–40.

Colgan, N. and Scully, R.W. 1898 *Contributions towards a Cybele Hibernica*. 2nd edn. Dublin. Edward Ponsonby.

Collins, T. 1999 The Clare Island Survey of 1909–11: participants, papers and progress. *New Survey of Clare Island. Volume 1: history and cultural landscape*, 1–40. Dublin. Royal Irish Academy.

Cooper, E., Crawford, I., Malloch, A.J.C. and Rodwell, J. 1992 *Coastal vegetation survey of Northern Ireland*. Lancaster. Unit of Vegetation Science report to the Department of the Environment (Northern Ireland).

Coppins, B.J. 1983 A taxonomic study of the lichen genus *Micarea* in Europe. *Bulletin of the British Museum (Natural History)* **11**, 18–214.

Coppins, B.J. 2002 *Checklist of lichens of Great Britain and Ireland*. London British Lichen Society.

Corcoran, R. 2002 Palaeoenvironmental investigations of Clare Island, Co. Mayo. Unpublished MSc. thesis, Geography Department, Trinity College Dublin.

Cotton, J. 1975 The National Vegetation Survey of Ireland: Nardo-Callunetea. *Colloques Phytosociologiques* **2**, 237–44.

Coxon, P. (ed.) 1982 *A fieldguide to Clare Island, Co. Mayo*. Dublin. Irish Association for Quaternary Studies.

Coxon, P. 1987 A post-glacial pollen diagram from Clare Island, Co. Mayo. *Irish Naturalists' Journal* **22**, 219–23.

Coxon, P. 1994 The glacial geology of Clare Island. In P. Coxon and M. O'Connell (eds), *Clare Island and Inishbofin*. Field Guide No. 17. Dublin. Irish Association for Quaternary Studies.

Coxon, P. 2001 The Quaternary history of Clare Island. In John R. Graham (ed.), *New Survey of Clare Island. Volume 2: geology*, 87–112. Dublin. Royal Irish Academy.

Coxon, P. and McCarron, S.G. 2009 Cenozoic: Tertiary and Quaternary (until 11,700 years before 2000). In Charles H. Holland and Ian S. Sanders (eds), *The geology of Ireland*, 355–96. 2nd edn. Edinburgh. Dunedin Academic Press.

Coxon, P. and O'Connell, M. 1994 *Clare Island and Inishbofin*. Dublin. Irish Association for Quaternary Studies.

Cullen, C. and Gill, P. 1991 *Studying an island: Clare Island, Co. Mayo*. Clare Island Series 1. Clare Island. Centre for Island Studies.

Cullen, C. and Gill, P. 1992 *Recorded history: from Grace O'Malley to the present day on Clare Island, Co. Mayo*. Clare Island Series, No. 5. The Centre for Island Studies, Clare Island.

Curran, P.L., O'Toole, M.A. and Kelly, F.G. 1983 Vegetation of terraced hill grazings in north Galway and south Mayo. *Journal of Life Sciences, Royal Dublin Society* **4**, 195–202.

Curtis, T.G.F. and McGough, H.N. 1988 *The Irish Red Data Book 1. Vascular plants*. Dublin. Stationery Office.

Dahlberg, A. and Croneborg, H. 2003 *33 threatened fungi in Europe: complementary and revised information on candidates for listing in Appendix 1 of the Bern Conventtion, European Council for the Conservation of fungi*. Uppsala. Swedish Environmental Protection Agency and European Council for Conservation of Fungi.

Dargie, T.C.D. 1993 *Sand dune vegetation survey of Great Britain: a national inventory Part II: Scotland*. Peterborough. Joint nature Conservancy Committee.

David, J.C. and Hawksworth, D.L. 1995 Zevadia: a new lichenicolous hyphomycete from western Ireland. *Bibliotheca Lichenologica* **58**, 63–71.

Davies, B.E. 1974 Loss on ignition as an estimate of soil organic matter. *Soil Science Society of America Proceedings* **38**, 150–51.

Degelius, G. 1954 The lichen genus *Collema* in Europe. *Symbolae Botanicae Upsalienses* **13** (2), 1–499.

Dierßen, K. 1978 Die wichtigsten Pflanzengesellschaften der Moore NW-Europas. Doctoral thesis, Freiburg.

Dixon, H.N. 1897 *Handbook. Catalogue of British Mosses*. Eastbourne and London.

Doyle, G.J. 1982 Narrative of the excursion of the international society for vegetation science to Ireland, 21–31 July 1980. *Journal of Life Sciences, Royal Dublin Society* **3**, 43–64.

Doyle, G.J. 1982 The vegetation, ecology and productivity of Atlantic blanket bog in Mayo and Galway, western Ireland. *Journal of Life Sciences, Royal Dublin Society* **3**, 147–64.

Doyle, G.J. 1990 Phytosociology of Atlantic blanket bog complexes in north-west Mayo. In G.J. Doyle (ed.), *Ecology and conservation of Irish peatlands*, 75–90. Dublin. Royal Irish Academy.

Doyle, G.J. 1997 Blanket bogs: an interpretation based on Irish blanket bogs. In L. Parkyn, R.E. Stoneman and H.A.P. Ingram (eds), *Conserving peatlands*, 25–34. Oxford. CAB International.

Doyle, G.J. and Foss, P.J. 1986 A resurvey of the Clare Island flora. *Irish Naturalists' Journal* **22**, 85–9.

Doyle, G.J. and Foss, P.J. 1986 *Vaccinium oxycoccus* L. growing in the blanket bog area of West Mayo (H27). *Irish Naturalists' Journal* **22**, 101–4.

Doyle, G.J. and Moore, J.J. 1980 Western blanket bog (*Pleurozio purpureae-Ericetum tetralicis*) in Ireland and Great Britain. *Colloques Phytosociologiques* **7**, 217–23.

Doyle, G.J. and Whelan, S. 1991 Proposal for a new survey of Clare Island: 1991–1995. Unpublished report to Royal Irish Academy, Dublin.

Doyle, G.J., O'Connell, C.A. and Foss, P.J. 1987 The vegetation of peat islands in bog lakes in County Mayo, western Ireland. *Glasra* **10**, 23–5.

Duff, M. 1930 The ecology of the Moss Lane region, Lough Neagh. *Proceedings of the Royal Irish Academy* **39**B, 477–96.

Edwards, L.S. 1977 A modified pseudosection for resistivity and IP. *Geophysics* **42** (5), 1020–36.

Edwards, K.J. 1981 The separation of *Corylus* and *Myrica* pollen in modern and fossil samples. *Pollen Spores* **23**, 205–18.

Evans, E.E. 1957 *Irish folkways*. London. Rutledge and Kegan.

Evans, S. 2004 *Waxcap-grasslands—an assessment of English sites*. English Nature Research Reports. Report no 555. Peterborough. English Nature.

Faegri, K. and Iversen, J. 1989 *Textbook of pollen analysis*. 4th edn. Chichester. John Wiley and Sons.

Farrell, E.P., Cummins, T., Boyle, G.M., Smillie, G.W. and Collins, J.F. 1993 Intensive monitoring of forest ecosystems. *Irish Forestry* **50**, 70–83.

Feehan, J. and McHugh, R. 1992 The Curragh of Kildare as a *Hygrocybe* grassland. *Irish Naturalists' Journal* **24**, 13–17.

Forbes, A.C. 1914 Clare Island Survey, Part 9. Tree growth. *Proceedings of the Royal Irish Academy*, 31, section 1: part **9**, 32pp.

Foss, P.J. 1986 The distribution, phytosociology, autecology and post glacial history of *Erica erigena* R.Ross in Ireland. Unpublished PhD thesis, University College Dublin.

Foss, P.J. and Doyle, G.J. 1988 Why has *Erica erigena* (the Irish heather) such a markedly disjunct European distribution? *Plants Today* 161–8.

Foss, P.J., Doyle, G.J. and Nelson, E.C. 1987 The distribution of *Erica erigena* R. Ross in Ireland. *Watsonia* **16**, 311–27.

Fossitt, J.A. 2000 *A guide to habitats in Ireland*. Kilkenny. The Heritage Council.

Fox, H.F. 1996 Catalogue of Irish lichenicolous fungi. Unpublished MS thesis, University College Dublin.

Gaynor, K. 2008 The phytosociology and conservation value of Irish sand dunes. Unpublished PhD thesis, University College Dublin.

Géhu, J.-M. 1975 Essai pour un système de classification phytosociologiques des landes Atlantiques planitiares Françaises. *Colloques Phytosociologiques* **2**, 361–77.

Genney, D.R., Hale, A.D., Woods, R.G., and Wright, M.W. 2009 Chapter 20 Grassland fungi. In *Guidelines for selection of biological SSSIs Rationale Operational approach and criteria: detailed guidelines for habitats and species groups*. Peterborough. JNCC.

Gibson, P.J. and George, D.M. 2004 *Environmental applications of geophysical surveying techniques*. Hauppauge, NY. Nova Science Publishers Inc.

Gibson, P.J., Lyle, P. and George, D.M. 2004 Application of resistivity and magnetometry geophysical techniques for near-surface investigations in karstic terranes in Ireland. *Journal of Cave and Karst Studies* **66** (2), 35–8.

Gillham, M.E. 1953 An ecological account of the vegetation of Grassholm Island, Pembrokeshire. *Journal of Ecology* **41**, 84–99.

Gillham, M.E. 1955 Ecology of the Pembrokeshire Islands III. The effect of gazing on the vegetation. *Journal of Ecology* **43**, 172–206.

Gillham, M.E. 1956 Ecology of the Pembrokeshire Islands IV. Effects of treading and burrowing by birds and mammals. *Journal of Ecology* **44**, 51–82.

Gillham, M.E. 1956 Ecology of the Pembrokeshire Islands V. Manuring by colonial seabirds and mammals, with a note on seed distribution by gulls. *Journal of Ecology* **44**, 429–54.

Gimingham, C.H. 1964 Maritime and submaritime communities. In J.H. Burnett (ed.), *The vegetation of Scotland*, 67–141. Edinburgh. Oliver and Boyd.

Gimingham, C.H. 1989 Heather and heathlands. *Botanical Journal of the Linnean Society* **101**, 263–8.

Goodman, G.T. and Gillham, M.E. 1954 Ecology of the Pembrokeshire Islands II. Skolhom, environment and vegetation. *Journal of Ecology* **42**, 296–327.

Gosling, P. 2007 The human settlement history of Clare Island. In P. Gosling, C. Manning and J. Waddell (eds), *New Survey of Clare Island. Volume 5: archaeology*, 29–68. Dublin. Royal Irish Academy.

Gosling, P. 2007 Catalogue of archaeological and architectural monuments, sites and stray finds on Clare Island. In P. Gosling, C. Manning and J. Waddell (eds), *New Survey of Clare Island. Volume 5: archaeology*, 83–212. Dublin. Royal Irish Academy.

Gosling, P. 2007 A distributional and morphological analysis of fulachtaí fia on Clare Island. In P. Gosling, C. Manning and J. Waddell (eds), *New Survey of Clare Island. Volume 5: archaeology*, 69–80. Dublin. Royal Irish Academy.

Gosling, P. and Waddell, J. 2007 Appendix 3. Radiocarbon dates from archaeological sites on Clare Island. In P. Gosling, C. Manning and J. Waddell (eds), *New Survey of Clare Island. Volume 5: archaeology*, 331. Dublin. Royal Irish Academy.

Gosling, P., Manning, C. and Waddell, J. (eds) 2007 *New Survey of Clare Island. Volume 5: archaeology*. Dublin. Royal Irish Academy.

Graham, J.R. 1994 Pre-Pleistocene geology of Clare Island. In P. Coxon and M. O'Connell (eds), *Clare Island and Inishbofin*, 8–10. Dublin. Irish Association for Quaternary Studies.

Graham, J.R. (ed.) 2001 *New Survey of Clare Island. Volume 2: geology*. Dublin. Royal Irish Academy.

Grant, S.A., Lamb, W.I.C., Kerr, C.D. and Bolton, G.R. 1976 The utilisation of blanket bog vegetation by grazing sheep. *Journal of Applied Ecology* **13**, 857–69.

Grant, S.A., Barthram, G.T., Lamb, W.I.C. and Milne, J.A. 1978 Effect of season and level of grazing on the utilisation of heather by sheep. I. Responses of the sward. *Journal of the British Grassland Society* **33**, 289–300.

Grant, S.A., Suckling, D.E., Smith, H.K., Torvell, L., Forbes, T.D.A. and Hodgson, J. 1985 Comparative studies of diet selection by sheep and cattle: the hill grasslands. *Journal of Ecology* **73**, 987–1004.

Grant, S.A., Torvell, L., Smith, H.K., Suckling, D.E., Forbes, T.D.A. and Hodgson, J. 1987 Comparative studies of diet selection by sheep and cattle: blanket bog and heather moor. *Journal of Ecology* **75**, 947–60.

Grant, S.A., Torvell, L., Common, T.G., Sim, E.M. and Small, J.L. 1996 Controlled grazing studies on Molinia grassland: effects of different seasonal patterns and levels of defoliation on *Molinia* growth and responses of swards to controlled grazing by cattle. *Journal of Applied Ecology* **33**, 1267–80.

Grant, S.A., Torvell, L., Sim, E.M., Small, J.L. and Armstrong, R.H. 1996 Controlled grazing studies on *Nardus* grassland: effects of between-tussock sward height and species of grazer on *Nardus* utilization and floristic composition. *Journal of Applied Ecology* **33**, 1053–64.

Gray, J.M. and Coxon, P. 1991 The Loch Lomond Stadial Glaciation in Britain and Ireland. In J. Ehlers, P.L. Gibbard and J. Rose (eds), *Glacial deposits in Britain and Ireland*, 89–105. Rotterdam. Balkema.

Griffith, G.W. and Roderick, K. (eds) 2008 *Saprotrophic Basidiomycetes in grasslands: distribution and function*. London. Elsevier Ltd.

Griffith, G.W., Easton, G.L. and Jones, A.W. 2002 Ecology and diversity of waxcap (*Hygrocybe* spp.) fungi. *Botanical Journal of Scotland* **54**, 7–22.

Griffith, G.W., Bratton, J.L. and Easton, G.L. 2004 Charismatic megafungi: the conservation of waxcap grasslands. *British Wildlife* **15**, 31–43.

Griffith, G.W., Holden, L., Mitchel, D., Evans, D.E., Aron, C., Evans, S. and Graham, A. 2006 Mycological survey of selected semi-natural grasslands in Wales. Unpublished report for the Countryside Council for Wales.

Grime, J.P., Hodgson, J.G. and Hunt, R. 1988 *Comparative plant ecology*. London. Unwin Hyman.

Grimm, E.C. 1993 Tilia 2.0 and Tilia Graph 2. Illinois State Museum, Research Collections Centre, 1920 South 101/2 Street, IL 62703, USA.

Grolle, R. 1976 Verzlichnis der lebermoose Europas und benachbarter gebiete. *Feddes Repertorium* **87**, 171–279.

Hallissy, T. 1914 Clare Island Survey, part 7. Geology. *Proceedings of the Royal Irish Academy* **31**, 1–22.

Hammond, R.F. 1979 *The peatlands of Ireland*. Dublin An Foras Talúntais.

Hanrahan, J. 1997 The effects of grazing on vegetation and soils of an oak woodland in Glendalough, County Wicklow. Unpublished MSc thesis, University College Dublin.

Heil, G.W. and Bruggink, M. 1987 Competition for nutrients between *Calluna vulgaris* (L.) Hull and *Molinia caerulea* (L.) Moench. *Oecologia* **73**, 105–7.

Hill, M.O., Blackstock, T.H., Long, D.G. and Rothero, G.P. 2008 *A checklist and census catalogue of British and Irish bryophytes*. The British Bryological Society.

Holyoak, D.T. 2003 *The distribution of bryophytes in Ireland*. Dinas Powys. Broadleaf Books.

Huang, C.C. 2002 Holocene landscape development and human impact in the Connemara uplands, western Ireland. *Journal of Biogeography* **29**, 153–65.

Huntley, B. 1979 The past and present vegetation of the Caenlochan National Nature Reserve, Scotland. I. Present vegetation. *New Phytologist* **83**, 215–83.

Iremonger, S.F. 1986 An ecological account of Irish wetland woods; with particular reference to the principal tree species. Unpublished PhD thesis, University of Dublin, Trinity College.

Iremonger, S.F. 1990 Structural analysis of three Irish wooded wetlands. *Journal of Vegetation Science* **1**, 359–66.

Ivimey-Cook, R.B. and Proctor, M.C.F. 1966 The plant communities of the Burren, Co. Clare. *Proceedings of the Royal Irish Academy* **64**B, 211–301.

Jermy, A.C. and Crabbe, J.A. 1978 *The island of Mull: a survey of its flora and environment*. London. British Museum.

Jones, D.A. 1917 Muscineae of Achill Island. *Journal of Botany* **55**, 240–46.

Jordal, J.B. 1997 *Sopp i naturbeitemarker i Norge. En kunnskapsstatus over utbredelse, okologi, indikatorverdi og trusler i et europeisk perspektiv*. Direktoratet for naturforvaltning, Trondheim.

Jordan, C. 1997 Mapping of rainfall chemistry in Ireland 1972–4. *Biology and Environment: Proceedings of the Royal Irish Academy* **97**B, 53–73.

Keane, T. 1986 *Climate, weather and Irish agriculture*. Dublin. Agmet.

Kelly, D.L. 1975 Native woodland in western Ireland with especial reference to the region of Killarney. Unpublished PhD thesis: University of Dublin, Trinity College.

Kelly, D.L. and Kirby, E.N. 1982 Irish native woodlands over limestone. *Journal of Life Sciences, Royal Dublin Society* **3**, 181–98.

Kelly, D.L. and Moore, J.J. 1975 Preliminary sketch of the acidophilous oakwoods. In J.-M. Géhu (ed.), *La végétation des forêts caducifoliées acidophiles*, 375–87. Vaduz. Cramer.

Kent, M. and Coker, P. 1992 *Vegetation description and analysis—a practical approach*. Belhaven, London.

Kirby, E.N. and O'Connell, M. 1982 Shannawoneen wood, County Galway, Ireland: the woodland and saxicolous communities and the epiphytic flora. *Journal of Life Sciences, Royal Dublin Society* **4**, 73–96.

Klein, J. 1975 An Irish landscape. A study of natural and semi-natural vegetations in the Lough Ree area of the Shannon basin. PhD thesis, Rijksuniversiteit, Utrecht.

Klotzli, F. 1970 Eichen-, Edellaub- und Bruchwalder der Britischen Inseln. *Schweirerischen Zeitschrift fur Forstwesen* **121**, 329–66.

Knowles, M.C. 1929 The lichens of Ireland. *Proceedings of the Royal Irish Academy* **38**, 179–434.

Knudsen H. and Vesterholt, J. 2008 *Funga Nordica*. Copenhagen. Nordsvamp.

Lamb, H.H. 1977 *Climate—present, past and future*. Volume 1. London. Methuen.

Lee, J.A. and Caporn, S.J.M. 1998 Ecological effects of atmospheric reactive nitrogen deposition on semi-natural terrestrial ecosystems. *New Phytologist* **139**, 127–34.

Lee, J.A., Caporn, S.J.M. and Read, D.J. 1992 Effects of increasing nitrogen deposition and acidification on heathlands. In T. Schneider (ed.), *Acidification research, evaluation and policy applications*, 97–106. Amsterdam. Elsevier Science.

Legon, N.W. and Henrici, A. 2005 *Checklist of the British and Irish Basidiomycota*. London. Royal Botanic Gardens Kew.

Lett, H.W. 1912 Clare Island Survey. Parts 11–12. Musci and Hepaticae. *Proceedings of the Royal Irish Academy* **31**, (11–12), 1–18.

Lewis, S. 1837 *A topographical dictionary of Ireland*. Vol. I. London. Samuel Lewis and Co.

Little, D.J. and Collins, J.F. 1995 Anthropogenic influences on soil development at a site near Pontoon, Co. Mayo. *Irish Journal of Agriculture and Food Research* **34**, 151–63.

Lockhart, N. 1991 The phytosociology and ecology of blanket bog flushes in west Galway and north Mayo. Unpublished PhD thesis, University College Galway.

Loke, M. 2006 *RES2DINV version 3.55. Rapid 2-D Resistivity and IP inversion using the least-squares method. Instruction manual*. Geotomo Software.

Long, D.G. 1988 New vice-county records and amendments to the Census Catalogue. Hepaticae. *Bulletin of the British Bryological Society* **52**, 30–3.

Long, D.G. 1990 The bryophytes of Achill Island—Hepaticae. *Glasra* (new series) **1**, 47–54.

Lowday, J.E. 1984 The effects of cutting and Asulam on the frond and rhizome characteristics of bracken (*Pteridium aquilinum* (L.) Kuhn). *Aspects of Applied Biology* **5**, 275–82.

Lowday, J.E. and Marss, R.H. 1992 Control of bracken and restoration of heathland. I. Control of bracken. *Journal of Applied Ecology* **29**, 204–11.

MacArthur, R.H. and Wilson, E.O. 1967 *The theory of island biogeography*. New Jersey. Princeton University Press.

MacCárthaigh, C. 1999 Clare Island folklife. In C. MacCárthaigh and K. Whelan (eds), *New survey of Clare Island. Volume 1: history and cultural landscape*, 41–72. Dublin. Royal Irish Academy.

MacCárthaigh, C. and Whelan, K. (eds) 1999 *New survey of Clare Island. Volume 1: history and cultural landscape*. Dublin. Royal Irish Academy.

MacDonald, A. 1990 *Heather damage: a guide to the types of damage and their causes*. Peterborough. Nature Conservancy Council.

MacGowan, F. and Doyle, G.J. 1996 The effects of sheep grazing and trampling by tourists on lowland blanket bog in the west of Ireland. In P.S. Giller and A.A. Myers (eds), *Disturbance and recovery in ecological systems*, 20–32. Dublin. Royal Irish Academy.

MacGowan, F. and Doyle, G.J. 1997 Vegetation and soil characteristics of damaged Atlantic blanket bog in the west of Ireland. In J.H. Tallis, R. Meade and P.D. Hulme, *Blanket mire degradation: causes, consequences and challenges. Proceedings of the Mires research Group Conference, University of Manchester 9–11 April 1997*, 54–63. Aberdeen. British Ecological Society and Macaulay Land Use Research Institute.

Malloch, A.J.C. 1971 Vegetation of the maritime cliff-tops of the Lizard and Land's End Peninsulas, West Cornwall. *New Phytologist* **70**, 1155–97.

Malloch, A.J.C. 1972 Salt-spray deposition on the maritime cliffs of the Lizard peninsula. *Journal of Ecology* **60**, 103–12.

Marren, P. 1998 Fungal flowers: the waxcaps and their world. *British Wildlife* **9**, 164–72.

Marrs, R.H. 1987 Studies on the conservation of lowland *Calluna* heaths. I. Control of birch and bracken and its effects on heath vegetation. *Journal of Applied Ecology* **24**, 163–75.

Marrs, R.H. 1993 An assemblage of change in *Calluna* heathlands in Breckland, Eastern England between 1983 and 1991. *Biological Conservation* **65**, 133–9.

Marrs, R.H. and Hicks, M.J. 1986 Study of vegetation change at Lakenheath warren: a re-examination of A.S. Watt's theories of bracken dynamics in relation to succession and vegetation management. *Journal of Applied Ecology* **23**, 1029–46.

Marrs, R.H., Johnson, S.W. and Le Duc, M.G. 1998 Control of bracken and the restoration of heathlands. VI. The response of bracken to 18 years of continued control or 6 years of control followed by recovery. *Journal of Applied Ecology* **35**, 479–90.

Marrs, R.H., Johnson, S.W. and Le Duc, M.G. 1998 Control of bracken and the restoration of heathlands. VII. The response of bracken rhizomes to 18 years of continued control or 6 years of control followed by recovery. *Journal of Applied Ecology* **35**, 748–57.

Marrs, R.H., Johnson, S.W. and Le Duc, M.G. 1998 Control of bracken and the restoration of heathlands. VIII. The regeneration of the heathland community after 18 years of continued bracken control or 6 years of control followed by recovery. *Journal of Applied Ecology* **35**, 857–70.

McCarthy, P.M. 1988 The lichens of Inishbofin, Co. Galway. *Irish Naturalists' Journal* **22**, 403–7.

McClintock, D. 1969 Field meetings, 1968. Mayo and Galway, 14th–20th April. *Proceedings Botanical Society of the British Isles* **7**, 634.

McFerran, D.H., McAdam, J.H. and Montgomery, W.I. 1994 Seed bank associated with stands of heather moorland. *Irish Naturalists' Journal* **24**, 480–5.

McFerran, D.M., Montgomery, W.I. and McAdam, J.H. 1994 Effects of grazing intensity on heathland vegetation and ground beetle assemblages of the uplands of County Antrim, north-east Ireland. *Biology and Environment: Proceedings of the Royal Irish Academy* **94**B, 41–52.

McFerran, D.M., Montgomery, W.I. and McAdam, J.H. 1994 The impact of grazing on communities of ground dwelling spiders (Araneae) in upland vegetation types. *Biology and Environment: Proceedings of the Royal Irish Academy* **94**B: 119–26.

McHugh, R., Mitchel, D., Wright, M. and Anderson, R. 2001 The fungi of Irish grasslands and their value for nature conservation. *Biology and Environment: Proceedings of the Royal Irish Academy* **101**B, 225–42.

McKee, A.M., Bleasdale, A.J. and Sheehy-Skeffington, M. 1998 The effects of different grazing pressures on the above-ground biomass of vegetation in the Connemara uplands. In G. O'Leary and F. Gormley (eds), *Towards a conservation strategy for the bogs of Ireland*, 177–88. Dublin. Irish Peatland Conservancy Council.

McVean, D.N. and Ratcliffe, D.A. 1962 *Plant communities of the Scottish Highlands: a study of Scottish mountain, moorland and forest vegetation.* London. Her Majesty's Stationery Office.

Mhic Daeid, C. 1976 A phytosociological and ecological study of the vegetation of peatlands and heaths in the Killarney Valley. Unpublished PhD thesis, University of Dublin, Trinity College.

Miles, J. 1981 Problems in heathland and grasslands dynamics. *Vegetatio* **46**, 61–74.

Miles, J. 1985 The pedogenic effects of different species and vegetation types and the implications of succession. *Journal of Soil Science* **36**, 571–84.

Miles, J. 1986 What are the effects of trees on soils? In D. Jenkins (ed.), *Trees and wildlife in the Scottish uplands*. 55–62. Huntingdon. Institute of Terrestrial Ecology.

Miles, J. 1988 Vegetation and soil change in the uplands. In M.B. Usher and B.A. Thompson (eds), *Ecological change in the uplands*, 57–70. Oxford. Blackwell Scientific Publishers.

Minchin, D. and Minchin, C. 1996 The sea pea *Lathyrus japonicus* (Willd) in Ireland, and an addition to the flora of West Cork (H3) and Wexford (H12). *Irish Naturalists' Journal* **25**, 165–9.

Mitchel, D. 2006 Survey of the grassland fungi of County Clare. Unpublished report for the Heritage Council.

Mitchel, D. 2007 Survey of the grassland fungi of the vice county of west Cork. Unpublished report for the Heritage Council.

Mitchel, D. 2008 Survey of the grassland fungi of the vice county of west Mayo. Unpublished report for the Heritage Council.

Mitchell, F. 1986 *The Shell guide to reading the Irish landscape*. Dublin. Country House.

Mitchell, F.J.G. 2006 Where did Ireland's trees come from? *Biology and Environment: Proceedings of the Royal Irish Academy* **106**B, 251–9.

Mitchell, F.J.G. 2008 Tree migration into Ireland. *Irish Naturalists' Journal* (special supplement) **2008**, 73–5.

Mitchell, F.J.G. 2009 The Holocene. In Charles H. Holland and Ian S. Sanders (eds), *The geology of Ireland*. 2nd edn. 397–404. Edinburgh. Dunedin Academic Press.

Mitchell, M.E. 1993 *First records of Irish lichens 1696–1990*. Galway. Officina Typographica.

Mitchell, M.E. 1995 150 years of Irish lichenology: a concise survey. *Glasra* **2**, 139–55.

Molloy, K. 2007 Reconstruction of past vegetation history on Clare Island: palaeoenvironmental investigations undertaken in conjunction with the archaeological investigations. In P. Gosling, C. Manning and J. Waddell, (eds), *New survey of Clare Island. Volume 5: archaeology*, 297–310. Dublin. Royal Irish Academy.

Molloy, K. and O'Connell, M. 1987 The nature of the vegetational change at about 5000B.P. with particular reference to the Elm Decline: Fresh evidence from Connemara, western Ireland. *New Phytologist* **106**, 203–20.

Monthly weather report 1991 Monthly weather report. Dublin. Meteorological Service.

Moore, J.J. 1955 The distribution and ecology of *Scheuchzeria palustris* on a raised bog in Co. Offaly. *Irish Naturalists' Journal* **11**, 1–7.

Moore, J.J. 1960 A resurvey of the vegetation of the district lying south of Dublin (1905–1956). *Proceedings of the Royal Irish Academy* **61**B, 1–36.

Moore, J.J. 1968 A classification of the bogs and wet heaths of northern Europe (*Oxycocco–Sphagnetea* Br.-Bl. et Tx. 1943). In R. Tüxen (ed.), *Pflanzensoziologische systematik. Bericht über das internationale symposium in Stozenau/Weser 1964 der internationale vereinigung für vegetationskunde*, 306–20. Den Haag. Junk.

Moore, J.J., Fitzsimons, P., Lambe, E. and White, J. 1970 A comparison and evaluation of some phytosociological techniques. *Vegetatio* **20**, 1–20.

Moore, P.D., Webb, J.A. and Collinson, M.E. 1991 *Pollen analysis*. 2nd edn. Oxford. Blackwell Science.

More, D. and Moore, A.G. 1866 *Contributions towards a Cybele Hibernica*. Dublin. Curry.

Mueller-Dombois, D. and Ellenberg, H. 1974 *Aims and methods of vegetation ecology*. London. Wiley.

Newton, A.C., Davy, L.M., Holden, E., Silverside, A., Watling, R. and Ward, S.D. 2002 Status, distribution and definition of mycologically important grasslands in Scotland. *Biological Conservation* **111**, (1) 11–23.

Ní Ghráinne, E. 1993 Palaeoecological studies towards the reconstruction of vegetation and landuse history of Inishbofin, western Ireland. Unpublished PhD thesis, University College Galway.

Ní Lamhna, É. 1982 The vegetation of saltmarshes and sand-dunes at Malahide Island, County Dublin. *Journal of Life Sciences, Royal Dublin Society* **3**, 111–29.

Nitare, J. 1988 Jordtungor, en svampgrupp pa tillbakagang i naturliga fodermarker. *Svensk Botanisk Tidskrift* **82** (5), 485–89.

Ó Críodáin, C. 1988 Parvocaricetea in Ireland. Unpublished PhD thesis, University College Dublin.

Ó Críodáin, C. and Doyle, G.J. 1994 An overview of Irish small-sedge vegetation: syntaxonomy and a key to communities belonging to the *Scheuchzerio–Caricetea nigrae* (Nordh. 1936) Tx. 1937. *Biology and Environment: Proceedings of the Royal Irish Academy* **94**B, 127–44.

Ó Críodáin, C. and Doyle, G.J. 1997 *Schoenetum nigricantis*. The *Schoenus* fen and flush vegetation of Ireland. *Biology and Environment: Proceedings of the Royal Irish Academy* **97**B, 203–18.

O'Connell, M. 1977 A palaeoecological and phytosociological study of Scragh Bog, Co. Westmeath. Unpublished PhD thesis, National University of Ireland.

O'Connell, M. 1980 The developmental history of Scragh Bog, Co. Westmeath and the vegetational history of its hinterland. *New Phytologist* **85**, 301–19.

O'Connell, M. 1981 The phytosociology and ecology of Scragh Bog, Co. Westmeath. *New Phytologist* **87**, 139–87.

O'Connell, M. 1990 Origins of lowland blanket bog. In G.J. Doyle (ed.), *Ecology and conservation of Irish peatlands* 49–71. Dublin. Royal Irish Academy.

O'Connell, M. and Ni Ghráinne, E. 1994 Inishbofin palaeoecology. In P. Coxon and M. O'Connell (eds), *Clare Island and Inishbofin*, 61–101. Dublin. Irish Association for Quaternary Studies (IQUA).

O'Sullivan, A.M. 1965 A phytosociological survey of Irish lowland meadows and pastures. Unpublished PhD thesis, University College Dublin.

O'Sullivan, A.M. 1976 The phytosociology of the Irish wet grasslands belonging to the order Molinietalia. *Colloques Phytosociologiques* **5**, 259–67.

O'Sullivan, A.M. 1982 The lowland grasslands of Ireland. *Journal of Life Sciences, Royal Dublin Society* **3**, 131–42.

O'Toole, M.A. 1984 *Renovation of peat and hillland pastures*. Dublin. An Foras Talúntais.

Ordnance Survey 1920 1:10,560 or six-inches to one statute mile scale map of County Mayo. Surveyed 1838, levelled in 1895–6, revised 1915. Dublin. Ordnance Survey of Ireland. Sheets 75, 84, 84a and 85.

Öster, M. 2008 Low congruence between the diversity of waxcaps (*Hygrocybe* spp.) fungi and vascular plants in semi-natural grasslands. *Basic and Applied Ecology* **9**, 514–22.

Packham, J.R. and Willis, A.J. 1997 *Ecology of dunes, saltmarsh and shingle*. London. Chapman and Hall.

Pakeman, R.J. and Marrs, R.H. 1996 Modelling the effects of climate change on the growth of bracken (*Pteridium aquilinum*) in Britain. *Journal of Applied Ecology* **33**, 561–75.

Paton, J.A. 1999 *The liverwort flora of the British Isles*. Colchester. Harley Books.

Perring, F.H. and Walters, S.M. 1962 *Atlas of the distribution of British plants*. London and Edinburgh. Nelson.

Pethybridge, G.H. and Praeger, R.L. 1905 The vegetation of the district lying south of Dublin. *Proceedings of the Royal Irish Academy* 25**B**, 124–80.

Phillips, W.E.A. 1965 The geology of Clare Island, County Mayo. Unpublished PhD thesis, University of Dublin, Trinity College.

Phillips, W.E.A. 1973 The pre-Silurian rocks of Clare Island, Co. Mayo, Ireland and the age of metamorphism of the Dalradian in Ireland. *Journal of the Geological Society, London* **129**, 585–606.

Porter, L. 1948 The lichens of Ireland (supplement). *Proceedings of the Royal Irish Academy* **51**, 347–86.

Praeger, R.L. 1903 The flora of Clare Island. *The Irish Naturalist* **12**, 277–94.

Praeger, R.L. 1904 The flora of Achill Island. *The Irish Naturalist* **13**, 265–89.

Praeger, R.L. 1905 The flora of the Mullet and Inishkea. *The Irish Naturalist* **14**, 229–44.

Praeger, R.L. 1907 The flora of Inishturk. *The Irish Naturalist* **16**, 113–25.

Praeger, R.L. 1911 Clare Island Survey. Part 10 Phanerogamia and Pteridophyta. *Proceedings of the Royal Irish Academy* **31**, 1–112.

Praeger, R.L. 1911 Notes on the flora of Inishbofin. *The Irish Naturalist* **20**, 165–72.

Praeger, R.L. 1913 On the buoyancy of the seeds of some Britannic plants. *Scientific Proceedings of the Royal Dublin Society* **15**, 13–62.

Praeger, R.L. 1930 *Beyond soundings*. Dublin. Talbot Press Ltd.

Prentice, H.C. and Prentice, I.C. 1975 The hill vegetation of North Hoy, Orkney. *New Phytologist* **75**, 313–67.

Rald, E. 1985 Vokshatte som indikatorarter for mykologisk vaerdifulde overdrevslokaliteter. *Svampe* **11**, 1–9.

Randall, R.E. 1989 Shingle habitats in the British Isles. *Botanical Journal of the Linnean Society* **101**, 3–18.

Ratcliffe, D.A. 1962 The habitat of *Adelanthus unciformis* (Tayl. Mitt. and *Jamesoniella carringtonii* (Balf.) Spr. in Ireland. *Irish Naturalists' Journal* **14**, 38–40.

Ratcliffe, D.A. and Thompson, D.B.A. 1988 The British uplands: their ecological character and international significance. In M.B. Usher and D.B.A. Thompson, *Ecological change in the uplands*, 9–36. Oxford. Blackwell Scientific Publications.

Rawes, M. 1981 Further results of excluding sheep from high level grasslands in the North Pennines. *Journal of Ecology* **69**, 651–69.

Rea, C. and Hawley, H.C. 1912 Clare Island Survey. Part 13 Fungi. *Proceedings of the Royal Irish Academy* **31**, 1–13.

Reimer, P.J., Baillie, M.G.L., Bard, E., Bayliss, A., Beck, J.W., Blackwell, P.G., Bronk Ramsey, C., Buck, C.E., Burr, G.S., Edwards, R.L., Friedrich, M., Grootes, P.M., Guilderson, T.P., Hajdas, I., Heaton, T.J., Hogg, A.G., Hughen, K.A., Kaiser, K.F., Kromer, B., McCormac, F.G., Manning, S.W., Reimer, R.W., Richards, D.A., Southon, J.R., Talamo, S., Turney, C.S.M., van der Plicht, J., Weyhenmeyer, C.E. 2009 IntCal09 and Marine09 Radiocarbon Age Calibration Curves, 0–50,000 Years cal BP. *Radiocarbon* **51** (4), 1111–50.

Ridge, I. 1997 *Simplified key to Geoglossum.* North West Fungus Group.

Roberts, N. 1998 *The Holocene—an environmental history.* Oxford. Blackwell Publishers.

Rodwell, J.S. 1991–2000 *British Plant Communities.* Vols I–V. Cambridge. Cambridge University Press.

Rodwell, J.S., Dring, J.C., Averis, A.B.G., Proctor, M.C.F., Malloch, A.J.C., Schaminée, J.N.J. and Dargie, T.C.D. 2000 *Review of coverage of the National Vegetation Classification.* JNCC Report no. 302. Peterborough. JNCC.

Rohan, P.K. 1986 *The climate of Ireland.* 2nd edn. Dublin. Irish Stationery Office.

Rooney, D.C. and Clipson, N.J.W. 2009 Synthetic sheep urine alters fungal community structure in an upland grassland soil. *Fungal Ecology* **2**, 36–43.

Rothero, G.P. 1988 The summer meeting, 1987, Co. Mayo. Second week at Westport: 12–18 August. *Bulletin of the British Bryological Society* **51**, 12–15.

Rotheroe, M., Newton, A., Evans, S. and Feehan, J. 1996 Waxcap-grassland survey. *Mycologist* **10**, 23–5.

Rotheroe, M. 1999 *Mycological survey of selected semi-natural grasslands in Carmarthenshire.* Countryside Council for Wales.

Russell, P. 2005 Grassland fungi and the management history of St.Dunstan's Farm. *Field Mycology* **6**, 85–91.

Ruttledge-Fair, R. 1892 *Congested Districts Board: baseline reports of local Inspectors.* London. Her Majesties Stationery Office.

Ryle, T.J. 2000 Vegetation/environment interactions on Clare Island. Co. Mayo. Unpublished PhD thesis, National University of Ireland, Galway.

Ryle, T. and Doyle, G.J. 1998 The elusive sea-pea on Clare Island. *Newsletter of the New Survey of Clare Island* **4**.

Scannell, M.J.P. and Synnott, D.M. 1987 *Census catalogue of the flora of Ireland.* 2nd edn. Dublin. Irish Stationery Office.

Schoof van Pelt, M.M. 1973 Littorelletea, a study in the vegetation of some amphiphytic communities of western Europe. Dissertation, Catholic University, Nijmegen.

Schouten, M.G.C. and Nooren, M.J. 1977 Coastal vegetation types and soil features in south-east Ireland. *Acta Botanica Neerlandica* **26**, 357–8.

Seaward, M.R.D. 1994 Vice-county distribution of Irish lichens. *Proceedings of the Royal Irish Academy* **94**B, 177–94.

Seaward, M.R.D. 2010 *Census Catalogue of Irish Lichens.* 3rd edition. Holywood, Belfast. National Museums of Northern Ireland.

Seaward, M.R.D. and Richardson, D.H.S. 2000 Lichens of Lambay Island. *Glasra* **4**, 1–6.

Sheffield, E., Wolf, P.G. and Haufler, C.H. 1989 How big is a bracken plant? *Weed Research* **29**, 455–60.

Smith, A.J.E. 1990 The bryophytes of Achill Island—Musci. *Glasra* (new series) **1**, 27–46.

Smith, A.J.E. 1990 *The moss flora of Britain and Ireland.* 2nd edn. Cambridge. Cambridge University Press.

Smith, A.J.E. 2004 *The moss flora of Britain and Ireland.* Cambridge. Cambridge University Press.

Smith, A.L. 1911 Clare Island survey. Part 14 *Lichenes. Proceedings of the Royal Irish Academy* **31**, 1–14.

Smith, A.L. 1918 *A monograph of the British lichens.* Part 1. 2nd edn. London. British Museum (Natural History).

Smith, A.L. 1926 *A monograph of the British lichens.* Part 2. 2nd edn. London. British Museum (Natural History).

Spooner, B. 1998 Keys to the British Geoglossaceae (draft). Unpublished.

Stace, C. 1997 *New flora of the British Isles.* 2nd edn. Cambridge. Cambridge University Press.

Stevenson, A.C. and Birks, H.J.B. 1995 Heaths and moorland: long-term ecological changes and interactions with climate and people. In D.B.A. Thompson, A.J. Hester and M.B. Usher (eds), *Heaths and moorland: cultural landscapes*, 224–39. Edinburgh. Scottish Natural Heritage.

Stevenson, A.C. and Thompson, D.B.A. 1993 Long-term changes in the extent of heather moorland in upland Britain and Ireland: palaeoecological evidence for the importance of grazing. *The Holocene* **3**, 70–6.

Stockmarr, J. 1971 Tablets with spores used in absolute pollen analysis. *Pollen et Spores* **13**, 614–21.

Synge, F.M. 1968 The glaciation of west Mayo. *Irish Geography* **5**, 372–86.

Synnott, D.M. 1986 An outline of the flora of Mayo. *Glasra* **9**, 13–117.

Synnott, D.M. 1988 The summer meeting, 1987, County Mayo. First week in Achill Island: 5–12 August. *Bulletin of the British Bryological Society* **51**, 7–10.

Synnott, D.M. 1990 The bryophytes of Achill Island—a preliminary note. *Glasra* (new series) **1**, 21–6.

Synnott, D.S. 1990 The bryophytes of Lambay Island. *Glasra* (new series) **1**, 65–81.

Tallis, J.H. 1973 The terrestrialization of lake basins in north Chesire. *Journal of Ecology* **61**, 537–67.

Tansley, A.G. 1911 *Types of British vegetation.* Cambridge. Cambridge University Press.

Tansley, A.G. 1939 *The British Islands and their vegetation.* Cambridge. Cambridge University Press.

Taylor, J.A. 1986 The bracken problem: a local hazard and global issue. In R.T. Smith and J.A. Taylor (eds), *Bracken: ecology, landuse and control technology*, 21–42. Carnforth. Parthenon Press.

ten Cate, R.S. 1993 Clare Island Survey en paddestoelen in het westen van Ierland. *In-Nuachta* **IX**, 14–20.

Thompson, D.B.A. and Miles, J. 1995 Heaths and moorland: some conclusions and questions about environmental change. In D.B.A. Thompson, A.J. Hester and M.B. Usher (eds), *Heaths and moorland: cultural landscapes*, 362–85. Edinburgh. Scottish Natural Heritage.

Thompson, D.B.A., MacDonald, A.J. and Hudson, P.J. 1995 Upland moors and heaths. In W.J. Sutherland and D.A. Hill (eds), *Managing habitats for conservation*, 292–328. Cambridge. Cambridge University Press.

Thompson, D.B.A., Macdonald, A.J., Marsden, J.H. and Gailbraith, C.A. 1995 Upland heather moorland in Great Britain: a review of international importance, vegetation change and some objectives for nature conservation. *Biological Conservation* **71**, 163–78.

Tolonen, K. 1986 Charred particle analysis. In B.E. Berglund (ed.), *Handbook of Holocene palaeoecology and palaeohydrology*. Chichester. John Wiley and Sons.

Troels-Smith, J. 1955 Karakterising af løse jordater. *Danmarks Geologiske Undersøgelse* IV **3**, 1–73.

Tutin, T.G., Heywood, V.H., Burgess, N.A., Moore, D.M., Valentine, D.H., Walter, S.M. and Webb, D.A. 1964–1980 *Flora Europaea* Volumes 1–5. Cambridge. Cambridge University Press.

University of Glasgow 1992 1:7500 map of Clare Island. Unpublished map prepared for the Royal Irish Academy.

Usher, M.B. and Thompson, D.B.A. 1993 Variation in upland heathlands of Great Britain conservation importance. *Biological Conservation* **66**, 69–81.

Van Groenendael, J.M., Hochstenbach, S.M.H., Van Mansfeld, M.J.M. and Roozen, A.J.M. 1979 The influence of the sea and of parent material on wetlands and blanket bog in western Connemara, Ireland. Unpublished MSc thesis, Laboratory for Geobotany, Catholic University, Nijmegen.

Van Groenendael, J.M., Hochstenbach, S.M.H., Van Mansfeld, M.J.M. and Roozen, A.J.M. 1983 Soligenous influences on wetlands and blanket bog in western Connemara, Ireland. *Journal of Life Sciences, Royal Dublin Society* **4**, 103–28.

Van Groenendael, J.M., Hochstenbach, S.M.H., Van Mansfeld, M.J.M. and Roozen, A.J.M. 1983 Plant communities of lakes, wetlands and blanket bogs in western Connemara, Ireland. *Journal of Life Sciences, Royal Dublin Society* **4**, 129–37.

Vesterholt, J., Boertmann, D. and Tranberg, H. 1999 1998—et usaedvanlig godt ar for overdrevssvampe. Svampe **40**, 36–44.

Viney, M. 1990 Landscape history in a nutshell. *Irish Times*, Saturday June 30, 1990.

Vullings, L.A.E. 2000 Soil variability and pedogenetic trends, Clare Island, Co. Mayo. Unpublished report, Department of Crop Science, Horticulture and Forestry, University College Dublin.

Vullings, W., Collins, J.F. and Smillie, G. 2013 *Soils and soil associations on Clare Island*. New Survey of Clare Island. Dublin. Royal Irish Academy.

Waddell, C.H. 1897 *Moss Exchange Club Catalogue of British Hepaticae*. London.

Walker, M.J.C. 2005 *Quaternary dating methods*. Chichester: J. Wiley.

Warburg, E.F. 1963 Notes on the bryophytes of Achill Island. *Irish Naturalists' Journal* **14**, 139–45.

Warburg, E.F. 1966 New vice-county records and amendments—Musci. *Transactions of the British Bryological Society* **5**, 199.

Webb, D.A. 1980 The flora of the Aran Islands. *Journal of Life Sciences, Royal Dublin Society* **2**, 51–83.

Webb, D.A. 1982 The flora of Aran: additions and corrections. *Irish Naturalists' journal* **20**, 45.

Webb, D.A. 1983 The flora of Ireland in its European context. *Journal of Life Sciences, Royal Dublin Society* **4**, 143–60.

Webb, D.A. and Hodgson, J. 1968 The flora of Inishbofin and Inishshark. *Proceedings of the Botanical Society of the British Isles* **7** (3), 345–63.

Welch, D. 1984 Studies in the grazing of heather moorland in north-east Scotland. I. Site description and patterns of utilisation. *Journal of Applied Ecology* **21**, 179–95.

Welch, D. 1984 Studies in the grazing of heather moorland in north-east Scotland. II. Responses of heather. *Journal of Applied Ecology* **21**, 197–207.

Welch, D. 1986 Studies on the grazing of heather moorland in north-east Scotland. V. Trends in *Nardus stricta* and other unpalatable graminoids. *Journal of Applied Ecology* **23**, 1047–58.

Westhoff, V. and Den Held, A.J. 1969 *Plantengemeenschappen in Nederland*. Zutphen. Thieme.

Westhoff, V. and Van Der Maarel, E. 1973 The Braun-Blanquet approach. In R.H. Whittaker (ed.), *Handbook of vegetation science*, 617–726. The Hague. Junk.

Westhoff, V. and Van Der Maarel, E. 1978 The Braun-Blanquet approach. In R.H. Whittaker (ed.), *Classification of plant communities*, 2nd edn, 287–399. The Hague. Junk.

Wheeler, B.D. 1980 Plant communities of rich fen systems in England and Wales. II. Communities of calcareous mires. *Journal of Ecology* **64**, 405–20.

White, J.M. 1932 The fens of north Armagh. *Proceedings of the Royal Irish Academy* 40**B**: 233–83.

White, J. 1982 The *Plantago* sward, *Plantaginetum Coronopodo maritimi*. *Journal of Life Sciences, Royal Dublin Society* **3**, 105–10.

White, J. and Doyle, G.J. 1982 The vegetation of Ireland: a catalogue raisonné. *Journal of Life Sciences, Royal Dublin Society* **3**, 289–368.

Whitehead, S.J., Caporn, S.J.M. and Press, M.C. 1997 Effects of elevated CO_2, nitrogen and phosphorus on the growth and photosynthesis of two upland perennials: *Calluna vulgaris* and *Pteridium aquilinum*. *New Phytologist* **135**, 201–11.

Whittow, J.B. 1974 *Geology and scenery in Ireland*. Penguin. Middlesex.

Winder, F.G. and Moore, J.J. 1947 Some notes on the rarer plants of the Ben Bulben range. *Irish Naturalists' Journal* **9**, 68–71.

Wright, H.E., Livingstone, D.A. and Cushing, E.J. 1965 Coring devices for lake sediments. In B. Kummel and D.M. Raup (eds), *Handbook of palaeontological techniques*, 494–520. San Francisco. Freeman.

Zuidhoff, A.C., Rodwell, J.S. and Schaminée, J.H.J. 1995 The *Cynosurion cristati* Tx. 1947 of central, southern and western Europe: a tentative overview, based on the analysis of individual relevés. *Annali di Botanica* **53**, 25–47.

TAXONOMIC INDEX

Page references in bold indicate tables. Those in italics indicate illustrations.
Fungi that are not recorded for Clare Island in 'Original Clare Island Survey records of fungi' (Mitchel, Appendix 2, this volume, pp 279–319) do not appear in this index.

INDEX

Page references in bold indicate tables. Those in italics indicate illustrations. Latin family, genus or species names are included in parentheses for reference to the taxonomic index.